Springer
Tokyo
Berlin
Heidelberg
New York
Barcelona
Hong Kong
London
Milan
Paris
Singapore

T. Matsuzawa (Ed.)

Primate Origins of Human Cognition and Behavior

With 248 Figures, Including 4 in Color

 Springer

Tetsuro Matsuzawa
Professor
Primate Research Institute, Kyoto University
41 Kanrin, Inuyama, Aichi 484-8506
Japan

Cover: Wild chimpanzees at Bossou, Guinea, West Africa, use a pair of stones as hammer and anvil to crack open oil-palm nuts. Chimpanzees, like humans, have a cultural tradition unique to each community.

This book was financially supported by the Japan Society for the Promotion of Science (Grant-in-Aid for Publication of Scientific Research Result: Grant No.125346). The study was supported by grants (07102010 and 12002009) from the Ministry of Education, Culture, Sports, Science and Technology, Japan.

ISBN 4-431-70290-3 Springer-Verlag Tokyo Berlin Heidelberg New York

Library of Congress Cataloging-in-Publication Data

Matsuzawa, Tetsuro, 1950–
    Primate origins of human cognition and behavior / T. Matsuzawa (ed).
        p. cm.
Includes bibliographical references (p.   ).
    ISBN 4431702903
    1. Primates—Behavior. 2. Primates—Psychology. 3. Cognition. 4.
Psychology, Comparative. I. Title.
    QL737.P9 M345 2000
    599.8'15—dc21

                                                00-069258

Printed on acid-free paper

Printing and binding: Best-set Typesetter Ltd., Hong Kong
SPIN: 10854621

Sweet-potato washing. Japanese monkeys on Koshima Island take a potato to the edge of the water and wash the sand off it. This behavior was begun in September 1953 by a female named Imo (meaning "potato" in Japanese), 1½ years old at the time. The behavior gradually spread to the other monkeys. See Chapter 24. (Photo by Satoshi Hirata)

Nut-cracking with a pair of stones. Chimpanzees at Bossou, Guinea, West Africa, use a pair of stones as hammer and anvil to crack open oil-palm nuts so that they can remove the kernels and eat them. They show perfect laterality for hammering the stone. It takes 3 to 5 years to acquire this skill that is unique to their community. See Chapter 28. (Photo by Tetsuro Matsuzawa)

The chimpanzee Ai chooses an Arabic numeral that corresponds to the number of white dots appearing on the monitor. Human visitors watch through an observation window. See Chapter 10. (Photo by Tetsuro Matsuzawa)

Cognitive test of an infant chimpanzee held by his mother. The chimpanzee Ai gave birth to a son, Ayumu, on April 24, 2000. Here, Ayumu performs a unique series of cognitive tests, in which he opens his mouth while looking at the examiner's open mouth. (Photo by Nancy Enslin)

Foreword

The modern study of primate behavior—involving continuous, long-term observations of known individuals—began in 1948, when Kinji Imanishi and Jun'ichiro Itani began to observe free-ranging Japanese macaques at a number of sites in Japan. In the years that followed they were joined by several other investigators, and in 1956 they formed the Japan Monkey Centre.

Scientists in Europe and North America generally remained ignorant of these pioneering Japanese studies until 1957, when the Japan Monkey Centre began to publish the journal *Primates*. Although the first volume contained only articles written in Japanese, subsequent volumes were published in English, thanks to a grant from the Rockefeller Foundation. Contact between scientists in Japan and the West was further encouraged in 1965, when Stuart Altmann, then at the Yerkes Regional Primate Research Center, received a grant from the National Institute of Mental Health to translate and publish a number of seminal papers written by Japanese scientists between 1957 and 1961. These papers show clearly that Japanese primatologists were the first to document the existence of ranked, matrilineal kin groups in the societies of one Old World monkey, the Japanese macaque. Similar social organizations were soon documented among rhesus macaques, bonnet macaques, baboons, vervet monkeys, and many other species including elephants. Today, it is too often assumed that the significance of female-bonded primate groups was not recognized until the late 1960s and 1970s, when American and British women began to study primates. In fact, it was male Japanese scientists, themselves the products of a male-dominated society, who led the way in documenting the central leadership role played by females in our closest animal relatives.

There is no doubt, then, that Japanese biologists and anthropologists played a significant, if not crucial, role in the development of primatology as a scientific discipline. Publication of the present volume, *Primate Origins of Human Cognition and Behavior*, reaffirms the pervasive and creative role played by the intellectual descendants of Imanishi and Itani in the modern disciplines of behavioral ecology, psychology, and cognitive science.

We now know that genetic relatedness, through the maternal line, the paternal line, or both, plays a fundamental role in organizing behavior within primate groups. Because kinship affects social behavior throughout an individual's lifetime, and because monkeys and apes may live up to 40 years in the wild, research on nonhuman primates requires continuous, long-term observations, and these, in turn, require uninterrupted financial support, dedicated investigators, and a stable, cooperative local government. Perhaps because of the logistical difficulties involved, scientific research on monkeys and apes now relies for its data on only a handful of viable, long-term study sites throughout Asia, Africa, and Central and South America. Many of the most important and productive sites are represented in this volume. The most venerable and well-known, of course, is Koshima Island in Japan, home to the famous Imo (a Galileo among monkeys), and for years the longest-running primate field research site in the world. Also represented in this volume, however, are two of the five major research sites where chimpanzees are studied. All in all, the book contains data from field and laboratory studies of over 90 primate species.

Perhaps more important is the intellectual breadth represented in this volume, and for this we have its editor, Tetsuro Matsuzawa, to thank. Matsuzawa and his students attack the study of primate cognition and its evolutionary origins from a number of different directions. In their fieldwork, they begin with detailed observations, moving quickly to experiments whenever possible. If observations indicate that chimps use leaves to soak up water, sticks to "dip" for ants, and hammer stones to crack open nuts, Matsuzawa and colleagues conduct experiments by placing in the chimps' path water to be soaked, dead caterpillars that attract ants, and nuts to be cracked. In the laboratory, their work takes a more comparative approach, beginning with observations on young children, Old World monkeys, and chimpanzees. Once observations have identified an intriguing problem to be solved or a significant species difference, they devise experiments to probe the nature of each individual's knowledge. Their star subject, the chimpanzee Ai, has already been the source of many striking and important results in the study of primate cognition.

In most of its current guises, the modern discipline of cognitive science can hardly be called comparative. Funding of major research programs and articles in leading journals focus almost exclusively on humans, and even more narrowly on one human ability: linguistic syntax. Too often this has led to the view that language and cognition are interdependent, and that you can't have one without the other. *Primate Origins of Human Cognition and Behavior* begins with exactly the opposite premise. It makes no claims for the existence of language in nonhuman primates, but sets out instead to document the intelligence that nonhuman primates do have, and to offer hypotheses about the adaptive value of this cognition-without-language. The result is a major contribution to comparative cognitive science, and to current views on the origin of mind and behavior.

Robert M. Seyfarth   Dorothy L. Cheney
Department of Psychology   Department of Biology
University of Pennsylvania   University of Pennsylvania

Preface

The Japanese *kanji* character 人 , meaning "human," is derived from the shape of two persons leaning toward each other, thus lending support one to the other. Altruistic and mutual support may be fundamental elements of human nature. We cooperate with one another in a social group to survive, using the intelligence required for manipulating the physical environment, as in the manufacture and use of tools. Where did such behaviors arise in our evolutionary history?

Human cognition and behavior is as much a product of evolution as the human body is. Where did we come from? Why and how do we have the mind that we have at present? Through introspection, we recognize our anger, sorrow, fear, and joy, and can sense our own vivid emotions. We can even think about the minds of others, who may have different thoughts. We humans can have images in our mind from invisible genes and chromosomes, to the vastness of the cosmos. How could such a complex cognitive world emerge in the evolution of humans? This book aims to illuminate the evolution of human cognition and behavior from the primate background. We can trace the evolutionary process through which it has emerged by comparisons with other living primate species. Cross-species comparisons of this sort constitute a royal road in biology when studying the evolutionary process responsible for traits arising in morphology, physiology, behavior, and psychology. Comparing living primate species can be a fruitful approach to seeking to understand the nature of human cognition and behavior.

Humans are a species of the animal kingdom. This message is undoubtedly valid in a biological perspective, because humans obviously are not plants. However, a statement such as "A human is an animal" may still sound slightly unacceptable for many people. The dichotomy of "man and animal" is a sort of naïve belief that is built into our cognition. A cat is a cat, and differs from a dog. A monkey is a monkey; it is not human. Each species looks unique, and no intrinsic connections are perceived among creatures. Based on such a naïve belief, we had thought for a long time that these were God's creations. In the same way, such naïve belief also would tell us that the Earth is flat and the Sun revolves around the Earth.

Naïve belief is strongly supported by our experience in everyday life, but it is not true scientific understanding. We humans arrived at this scientific understanding about the Earth–Sun relationship in the sixteenth century. The Earth is not flat but round; it is the Earth that revolves around the Sun. This revolutionary understanding was brought about by astronomers like Copernicus, Galileo, Kepler, and others.

They made careful observations of the complex movements of the planets and created a splendid hypothesis explaining the details of the observed facts. This turning from a Ptolemaic system to the Copernican one provided a completely different understanding of our world. This shift in world-views is called the Copernican revolution. Now, some 400 years later, most people accept the scientific view although that understanding still goes against our innate naïve belief.

There was another paradigm shift in the nineteenth century, when we arrived at a true understanding of the evolution of living organisms. All creatures including humans are products of evolution. This breakthrough can be called the Darwinian revolution. It was about 150 years ago that this understanding was achieved. Therefore, many of us have already accepted evolution, although some misunderstanding remains, and there are even a few who still refuse to accept the truth of evolution.

Many people know that monkeys and humans share a common ancestor. Thanks to the scientific understanding of evolution, we know that although monkeys look different from us (supported by our naïve belief), we and monkeys were one in the past (supported by the scientific understanding). How about chimpanzees? Suppose that you are asked about the position of chimpanzees in relation to humans and monkeys. Many people still have no hesitation in lumping chimpanzees and monkeys together as opposed to humans. Even if you know that a chimpanzee is not a monkey but an ape, you tend to classify chimpanzees and monkeys together to make up one group, "nonhuman primates," pitted against another group, "humans." But this is completely wrong.

You can compare DNA sequences to calculate the genetic distances separating the three living species, humans (*Homo sapiens*), chimpanzees (*Pan troglodytes*), and monkeys (*Macaca fuscata,* for example). The genetic difference between humans and monkeys is estimated to be about 9% to 10%. However, humans and chimpanzees share about 98.3% of DNA sequences. That means that the difference between the two species is only about 1.7%, smaller than the distance between horses and zebras. If you think that zebras are "horses that have black and white stripes," chimpanzees must be seen as "humans that are fully covered by black hair."

The three species share a common ancestor that lived about 30 million years ago. Then there was a division into two lineages, one for the ancestor of living monkeys and the other for the ancestor of living humans and chimpanzees (they can be called "hominoids," which means "humans and apes"). Humans and chimpanzees shared a common ancestor about 5 million years ago. Suppose that it took about 3.6 billion years from the emergence of life on the Earth to the present. Humans and chimpanzees share 99.86% of that time, during which the two species had not yet diverged. The brain volume of *Australopithecus* was almost the same as that of present-day chimpanzees. From a biological point of view, there is no explicit reason for us to classify *Homo sapiens* as a unique species within a single genus of a single family. Chimpanzees are much closer to humans than to monkeys. This is true scientific understanding.

We have already reached an accurate understanding of the evolutionary history (that is, the phylogeny) of humans, chimpanzees, and monkeys. However, we got the definitive answer only about 10–20 years ago. Up to the 1980s, there was a prolonged dispute even among primatologists about which are closer to humans: chimpanzees, gorillas, or orangutans? At that time, there were few fossil records, no information from molecular biology, and a lot of controversy. Now everything has become clear. The answer is that chimpanzees are the closest relative of humans. However, scientific understanding is not strong enough to overcome our naïve belief. It now has been about 400 years since the Copernican revolution and about 150 years since the Darwinian revolution. Only one or two decades are not enough time for us to accept a truly scientific explanation that contradicts our naïve belief. I think it takes time for us to be convinced that we humans are not distinguished from other creatures, and that we are just one member of the living primates.

Japan is a special country in the world from the point of view of primatology. Why? Because Japan has an indigenous primate species (*Macaca fuscata*), known as snow monkeys, and also because there are many primatologists. The living nonhuman primates are distributed in Africa, Asia, and Central and South America although there are very few primatologists in those developing areas. In contrast, there are no indigenous monkeys in Europe and North America although there are a lot of primatologists. Japan is the exception in that it has both indigenous monkeys and many primatologists.

Monkeys do not appear in *Aesop's Fables, Grimm's Fairy Tales,* or *Mother Goose's Nursery Rhymes.* There are foxes, rabbits, bears, and geese playing important parts in these tales but no important roles for monkeys. In Japan, monkeys are one of the favorite characters of folklore and fairy tales. Almost all Japanese people have had a chance sometime in their life to see wild monkeys in nature: feeding on the ground, playing in the trees, or bathing in hot springs. Japanese people have long been fond of monkeys and, based on their own direct experience, have intuitively understood the close relationship between humans and nonhuman primates. This special affection of ordinary people toward monkeys in general seems to have supported the development of primatology in this country.

In Japan, the study of nonhuman primates began as the socioecological study of Japanese monkeys in their natural habitat. The late Dr. Kinji Imanishi (1902–92) of Kyoto University took the lead in promoting socioecological studies of wild Japanese monkeys. Imanishi and his students developed an original research method characterized by (1) long-term observation, (2) provisioning, i.e., feeding wild animals to habituate them to human observers for close observation, and (3) individual identification, i.e., identifying each individual by assigning names rather than alpha-numeric codes. His school was called the "Kyoto school" and it has produced many world-renowned primatologists such as Drs. Jun'ichiro Itani and Masao Kawai.

The survey of wild Japanese monkeys started in November 1948, when Imanishi and Itani visited the wild monkeys on Koshima Island. Itani and others succeeded in provisioning the wild monkeys on Koshima in 1952. Potato-washing behavior by the monkeys on Koshima is a well-known example of cultural (or precultural) behavior in nonhuman animals. There is a 50-year record of social behavior of wild monkeys on Koshima (see the chapters by Hirata et al. and by Watanabe in this volume). This must be one of the longest records of continuous observation of a population of wild animals. Itani and his colleagues also promoted a long-term project for studying wild Japanese monkeys from a socioecological perspective in other research sites such as Takasakiyama, Arashiyama, Shiga Heights, and Yakushima.

This socioecological research focused on social structure, leadership among the group's members, life history by gender, and so on. On the basis of their accumulated knowledge of wild Japanese monkeys, Imanishi and Itani first visited Africa in 1958 and started a socioecological study of African great apes, that is, gorillas and chimpanzees. Thereafter, Itani supported by Imanishi organized a research team for both the African great apes and living hunter–gatherers in Africa to understand the evolution of human society. In the case of chimpanzee research, they did extensive rather than intensive work in the first several years. Then in 1965 Dr. Kousei Izawa found the wild chimpanzees (*Pan troglodytes*) in the Mahale Mountains of Tanzania, East Africa, and Dr. Toshisada Nishida succeeded in provisioning them in 1967. Nishida and his colleagues have continued this study, now into its fourth decade. In West Africa, Dr. Yukimaru Sugiyama and his colleagues also are continuing a longitudinal study of chimpanzees, now into its third decade, at Bossou, Guinea. In their early stages, the studies on chimpanzees by Japanese researchers produced new findings such as the "unit group" characterized by "fission–fusion" of the subgroups, insect/meat eating and cannibalism, so-called infanticide, unique use of tools such as stone implements, and so on. Dr. Takayoshi Kano began his research on wild bonobos (pygmy chimpanzees, *Pan paniscus*) at Wamba, Zaire, in 1973. Kano and his colleagues were pioneers in the field study of bonobos and revealed much about the life of "the last ape."

In the shadow of socioecological studies, the study of cognition and behavior of nonhuman primates has not been widely recognized. However, there has been a stream of effort over the years. The experimental approach to cognition of nonhuman primates by Japanese psychologists can be traced back to the 1940s. The late Dr. Ben Yagi and his colleagues of the University of Tokyo might be considered the first researchers in Japan to test the cognitive ability of monkeys in the laboratory. They tested monkeys in a string-pattern test of pulling the correct string attached to a food reward at the opposite end. For example, two strings in front of the monkey were crossed and one of them was baited. In this case, the monkey had to be careful to choose the baited string because the strings were crossed. Since then, many Japanese psychologists have pursued various challenging issues related to the cognitive ability of nonhuman primates.

Based on the results of previous socioecological and psychological studies, an ape-language project was begun in 1978 under the leadership of Dr. Kiyoko Murofushi at the Primate Research Institute of Kyoto University, in collaboration with Toshio Asano and Tetsuro Matsuzawa. It later became known as the "Ai project," named for the female chimpanzee who was its main subject. The chimpanzee Ai learned to use Arabic numerals, Japanese *kanji* characters, geometric figures as visual symbols, and so on, to communicate with us about what she perceived and thought. The Ai project has continued for more than two decades. It has triggered multidisciplinary approaches toward direct understanding of the chimpanzee mind compared with the human mind.

In recent years, the studies of the Ai project have tried to clarify cognition and behavior of chimpanzees both in the laboratory and in the wild. We are focusing on the mechanisms of acquisition of knowledge and skills, and also on social transfer among individuals, particularly across generations. We have studied tool use, mostly concentrating on stone tool use in the wild. We also have explored the various kinds of cognitive skills exhibited in the laboratory. The following are some of the topics of our studies: comprehension of human speech and gestural signs, acquisition of visual symbols as a form of artificial language, comprehension and use of numbers, perception of biological motion, the mechanism of extracting features in a visual search paradigm, visual illusion and perceptual completion, auditory-visual cross-modal matching, memory span, establishment of natural concepts, comprehension and use of tokens, serial recognition of video images, and so forth. Cognitive behavior in social contexts is also studied: observation learning, imitation, deception, and so on. We believe that understanding chimpanzees will make it possible to create a bridge between humans and other nonhuman animals. Once we see the points shared by humans and chimpanzees, it will become much easier for us to understand those that also are shared by monkeys, and then by lemurs, and so on. All living organisms are interconnected through evolution.

Conducting work in the laboratory carries with it a necessary sensitivity to animal welfare. We have been successful in planting a large number of trees in our outdoor compound for a group of 14 chimpanzees, including three newborns, which also has a small stream and a 15-meter-high set of climbing frames. Captive chimpanzees need sun, soil, streams, trees, and grass, as well as conspecifics. An innovative structure in the compound that we refer to as an "outdoor booth," connected to the neighboring building through an underground tunnel, serves to keep human experimenters and apparatuses inside an enclosed space, while chimpanzees are free to roam outside. In such a seminatural situation, we now routinely conduct cognitive experiments where participation is based entirely on the free will of our chimpanzee subjects, whom we regard as our research partners.

For research in the wild, sensitivity to wildlife conservation is essential. There is only a small community of about 20 chimpanzees at Bossou, Guinea, living in isolation from adjacent communities. We have been conducting an extensive survey of chimpanzee habitats in the area and have embarked on a large-scale tree-planting project aimed at creating a green passage or natural corridor connecting adjacent communities that at present are separated by cultivated fields and savannah.

The core part of this book is derived from the long-term study of chimpanzee intelligence at the Primate Research Institute of Kyoto University, the Ai project. The project covers both laboratory and field studies of chimpanzees from a broad perspective. The book also contains studies of cognition and behavior of other primate species done by Japanese researchers and foreign collaborators, dealing with more than 90 species in total.

As I have explained earlier, this book aims to illuminate the evolution of human cognition and behavior from the primate background. There have been some books published with a similar purpose. However, this book may be unique in the following ways.

First, the book as a whole clearly aims to establish a new research field of comparative cognitive science. Human cognitive science has been trying to examine the human mind and the underlying brain mechanisms from various points of view. However, there remains a very interesting and important question: Why and how did our mind evolve? Minds and thoughts are not preserved in the fossil record. To explore the evolution of the human mind, we must compare cognitive functions in living representatives of related species. All the chapters of the book provide the original findings of this unique discipline. Comparative cognitive science covers various aspects of cognition from visual/auditory/taste sensation to memory span, concept formation, and self-awareness, to social cognition and cultural tradition. It is unique in covering both laboratory work and field work. The two different approaches—experimental vs. observational, psychological vs. ethological/ecological, laboratory vs. field—should be synthesized for understanding the evolutionary origins of the human mind.

Second, the book provides original findings characterized by an examination of a variety of living primate species, and by being based on long-term records. The book deals with the cognition and behavior of more than 90 living species, including apes such as chimpanzees, Old World monkeys, New World monkeys, and prosimians such as lemurs. The book not only covers a wide range of primates but also is deeply rooted in a long history. The studies of Japanese monkeys are based on data continuously collected for more than 50 years. The chimpanzee researches are also based on data accumulated for more than two decades both in the laboratory and in the wild.

Third, the book is the first to come out of Japan that talks about the evolution of the human mind. Japan has a long tradition of primate research. Japanese have published many papers in English academic journals such as *Nature, Animal Behaviour, Ethology, Journal of Comparative Psychology, Primates,* and *American Journal of Primatology*. However, there have been very few books written in English by Japanese researchers. This book is an attempt to look for primate origins of human cognition and behavior, and we believe it is the first one that shows the entire picture of Japanese efforts carried out with international collaboration. The modern Western discipline of writing academic papers in English has combined with Oriental wisdom and tradition to result in this book.

I deeply thank all the contributors for their efforts to make this book possible. Each of the authors may need a long list of acknowledgments because all research requires help from many people. Without the collaboration of Drs. Kazuo Fujita, Masaki Tomonaga, Masayuki Tanaka and other colleagues and students, I could not have continued my efforts for more than two decades to study chimpanzees, nor could I have dared to edit a book like this. Mr. Kiyoharu Nagumo has developed the computer programs for cognitive research in the laboratory. The directors of the Primate Research Institute, Drs. Shiro Kondo, Wataru Ohsawa, Masao Kawai, Kisou Kubota, Ken Nozawa, and Shozo Kojima, gave us continuous support. Dr.

Kiyoaki Matsubayashi and the staff of the veterinary and management section took care of more than 800 individuals of 25 species of primates in the institute, including a group of 14 chimpanzees. Special thanks are due to Kiyonori Kumazaki, the manager for environmental enrichment, and Norihiko Maeda, the head keeper of chimpanzees.

We want to thank the staff supporting the field work described in this book. Dr. Kunio Watanabe and the staff of the Field Research Center of the Primate Research Institute have taken the role of wildlife conservation of Japanese macaques. The field research of wild chimpanzees at Bossou, Guinea, also needed much support. We wish to thank Dr. Yukimaru Sugiyama and his colleagues for their efforts over a period of many years. Thanks are also due to the governments of the following three countries and the Japanese embassies in those countries: Guinea, Liberia, and Côte d'Ivoire. The people of Bossou, Nimba, and the surrounding areas also helped greatly. Each field study described in this book has its own unique background. When you read the text, please imagine the long-term efforts made by many people to see that the field work continues.

In closing, I would like to mention three names. Prof. Kiyoko Murofushi was responsible for initiating the study of chimpanzee intelligence in Japan in the mid-1970s and thus provided the basis for the long-term research reported here. Her help and dedication drove the project forward year after year. Dr. David Premack has always been a stimulating force behind my experimental ideas ever since I spent a sabbatical leave at his lab in Pennsylvania in 1985–87. I learned a great deal from conversations with him, particularly concerning the importance of social aspects of intelligence. Finally, Dr. Jane Goodall's words have stayed with me ever since our first meeting at a symposium in Chicago in 1986 entitled "Understanding chimpanzees": to study "for the chimpanzees." Without love and respect for chimpanzees, it does not make sense to study them.

In that respect, I truly appreciate all the chimpanzees, the Japanese monkeys, and other primates that I have met in my life. I have learned a lot of things from them living in the wild as well as in the laboratory. I am convinced that the same feeling is shared by all of the authors of this book. Please enjoy all the chapters that follow in this book—inspired by long-term study, they touch upon the hearts of our evolutionary relatives. Thank you.

Tetsuro Matsuzawa

Contents

Part 4 Learning and Memory

Part 5 Recognition of Self, Others, and Species

Part 6 Society and Social Interaction

Part 7 Culture

List of Authors

ANDERSON, JAMES R. (CHAPTER 16)
Department of Psychology, University of Stirling, Stirling FK9 4LA, Scotland, UK

BIRO, DORA (CHAPTERS 10, 28)
Animal Behaviour Research Group, Department of Zoology, University of Oxford, Oxford OX1 3PS, UK

DELIUS, JUAN D. (CHAPTER 13)
Allgemeine Psychologie, Universität Konstanz, D-78434 Konstanz, Germany

DERUELLE, CHRISTINE (CHAPTER 4)
CNRS, 31 ch. Joseph Aiguier 13402, Marseille cedex 20, France

FAGOT, JOËL (CHAPTER 4)
CNRS, 31 ch. Joseph Aiguier 13402, Marseille cedex 20, France

FUJITA, KAZUO (CHAPTERS 2, 18)
Department of Psychology, Graduate School of Letters, Kyoto University, Yoshida Honmachi, Sakyo-ku, Kyoto 606-8501, Japan

HAGIWARA, TOSHIO (CHAPTER 25)
Jigokudani Monkey Park, Yamanouchi-machi, Nagano 381-0401, Japan

HASHIYA, KAZUHIDE (CHAPTER 8)
Department of Cognitive Psychology, Graduate School of Education, Kyoto University, Yoshida Honmachi, Sakyo-ku, Kyoto 606-8501, Japan

HIRATA, SATOSHI (CHAPTER 24)
Primate Research Institute, Kyoto University, 41 Kanrin, Inuyama, Aichi 484-8506, Japan

VAN HOOFF, JAN A.R.A.M. (CHAPTER 26)
Ethology and Socio-Ecology Group, Utrecht University, Centrumgebouw Noord, Padualaan 14, Pb. 80.086, 3508TB, Utrecht, The Netherlands

HUMLE, TATYANA (CHAPTER 28)
Department of Psychology, University of Stirling, Stirling FK9 4LA, Scotland, UK

INOUE-NAKAMURA, NORIKO (CHAPTERS 14, 28)
Primate Research Institute, Kyoto University, 41 Kanrin, Inuyama, Aichi 484-8506, Japan

ITAKURA, SHOJI (CHAPTER 15)
Department of Psychology, Faculty of Letters, Kyoto University, Yoshida Honmachi, Sakyo-ku, Kyoto 606-8501, Japan

IVERSEN, IVER H. (CHAPTER 12)
Department of Psychology, University of North Florida, Jacksonville, FL 32224, USA

JITSUMORI, MASAKO (CHAPTER 13)
Department of Cognitive and Information Sciences, Chiba University, 1-33 Yayoi-cho, Inage-ku, Chiba 263-8522, Japan

KAWAI, MASAO (CHAPTER 24)
Museum of Nature and Human Activities, 6 Yayoigaoka, Sanda, Hyogo 669-1546, Japan

KAWAI, NOBUYUKI (CHAPTER 11)
Primate Research Institute, Kyoto University, 41 Kanrin, Inuyama, Aichi 484-8506, Japan

KOBAYASHI, HIROMI (CHAPTER 19)
Basic Biology, Faculty of Bioscience and Biotechnology, Tokyo Institute of Technology, 2-12-1 Ookayama, Meguro-ku, Tokyo 152-8551, Japan

KOHSHIMA, SHIRO (CHAPTER 19)
Basic Biology, Faculty of Bioscience and Biotechnology, Tokyo Institute of Technology, 2-12-1 Ookayama, Meguro-ku, Tokyo 152-8551, Japan

KOJIMA, SHOZO (CHAPTERS 8, 9)
Primate Research Institute, Kyoto University, 41 Kanrin, Inuyama, Aichi 484-8506, Japan

MATSUMURA, SHUICHI (CHAPTER 22)
Primate Research Institute, Kyoto University, 41 Kanrin, Inuyama, Aichi 484-8506, Japan

MATSUZAWA, TETSURO (CHAPTERS 1, 10, 11, 12, 28)
Primate Research Institute, Kyoto University, 41 Kanrin, Inuyama, Aichi 484-8506, Japan

MUROYAMA, YASUYUKI (CHAPTER 23)
Field Research Center, Primate Research Institute, Kyoto University, 41 Kanrin, Inuyama, Aichi 484-8506, Japan

MYOWA-YAMAKOSHI, MASAKO (CHAPTER 17)
Primate Research Institute, Kyoto University, 41 Kanrin, Inuyama, Aichi 484-8506, Japan

NAKAMICHI, MASAYUKI (CHAPTER 21)
Laboratory of Ethological Studies, Faculty of Human Sciences, Osaka University, 1-2 Yamada-oka, Suita, Osaka 565-0871, Japan

ODA, RYO (CHAPTER 6)
Department of Humanities and Social Sciences, Nagoya Institute of Technology, Gokiso-cho, Showa-ku, Nagoya 466-8555, Japan

SUGIURA, HIDEKI (CHAPTER 7)
Primate Research Institute, Kyoto University, 41 Kanrin, Inuyama, Aichi 484-8506, Japan

TAKEFUSHI, HARUO (CHAPTER 25)
Jigokudani Monkey Park, Yamanouchi-machi, Nagano 381-0401, Japan

TAKESHITA, HIDEKO (CHAPTER 26)
School of Human Cultures, The University of Shiga Prefecture, 2500 Hassakacho, Hikone, Shiga 522-8533, Japan
Ethology and Socio-Ecology Group, Utrecht University, Centrumgebouw Noord, Padualaan 14, Pb. 80.086, 3508TB, Utrecht, The Netherlands

TANAKA, ICHIROU (CHAPTER 25)
Section of Language and Intelligence, Section of Ecology, Primate Research Institute, Kyoto University, 41 Kanrin, Inuyama, Aichi 484-8506, Japan

TOKIDA, EISHI (CHAPTER 25)
Jigokudani Monkey Park, Yamanouchi-machi, Nagano 381-0401, Japan

TOMONAGA, MASAKI (CHAPTERS 3, 4)
Primate Research Institute, Kyoto University, 41 Kanrin, Inuyama, Aichi 484-8506, Japan

TONOOKA, RIKAKO (CHAPTER 28)
Primate Research Institute, Kyoto University, 41 Kanrin, Inuyama, Aichi 484-8506, Japan

UENO, YOSHIKAZU (CHAPTER 5)
Primate Research Institute, Kyoto University, 41 Kanrin, Inuyama, Aichi 484-8506, Japan

YAMAKOSHI, GEN (CHAPTERS 27, 28)
Center for African Area Studies, Kyoto University, 46 Yoshida Shimoadachi-cho, Sakyo-ku, Kyoto 606-8501, Japan

WATANABE, KUNIO (CHAPTERS 20, 24)
Field Research Center, Primate Research Institute, Kyoto University, 41 Kanrin, Inuyama, Aichi 484-8506, Japan

Part 1
Introduction to Comparative Cognitive Science

1
Primate Foundations of Human Intelligence: A View of Tool Use in Nonhuman Primates and Fossil Hominids

Tetsuro Matsuzawa

1 Introduction

My research has focused primarily on the cognition and behavior of chimpanzees both in the laboratory and in the wild (Matsuzawa 1985a, b, 1990, 1994, 1999; Biro and Matsuzawa 1999, Kawai and Matsuzawa 2000). This chapter deals with the evolution of human intelligence as viewed from the perspective of "comparative cognitive science" (Matsuzawa 1998)—a discipline that compares cognitive functions in living species. Some human traits seem to be shared by chimpanzees, the closest living relatives of humans, and to some extent by other nonhuman primates. To give the study of human intelligence an evolutionary perspective, the chapter will present a framework for examining intelligence with reference to tool use as its main focus. Tool use provides a unitary measure for comparing the intelligence of living primates with that of extinct hominids. Based on its primate foundations, human intelligence has acquired unique characteristics such as a "self-embedding hierarchical structure in cognition," supported by an increase in the number of levels and relationships that can be comprehended and handled simultaneously by the brain.

2 Phylogeny of Primates and Fossil Hominids

Human evolution can be studied by two different approaches. One is an examination of the fossil record, and the other is a comparative study of living species. The former is in search of "missing links," while the latter seeks "living links." Comparative cognitive science is based on the latter approach, but an effort to draw together evidence from the two different approaches is certain to be of great benefit when attempting to shed light on the evolution of human intelligence.

First, knowledge of primate phylogeny—including humans—is of fundamental importance. Recent advances in paleontology and molecular biology provide data in the form of the fossil record and DNA analysis in support of the outline of primate phylogeny shown in Fig. 1. Living primates are usually divided into four

Primate Research Institute, Kyoto University, 41 Kanrin, Inuyama, Aichi 484-8506, Japan

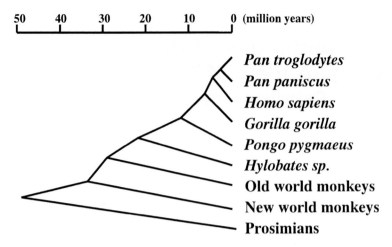

Fig. 1. Phylogeny of living primates based on data from paleontology and molecular biology

groups: prosimians (PRO), New World monkeys (NWM, platyrrhina), Old World monkeys (OWM), and hominoids, including apes and humans. Our hominoid-like ancestors are differentiated into various hominoid species over evolutionary time, with humans and chimpanzees having shared their most recent common ancestor an estimated 5 million years ago (White et al. 1994).

The split in the ancestral chimpanzee lineage is thought to have occurred approximately 2.5 million years ago, giving rise to the two species alive today: *Pan troglodytes* and *Pan paniscus* (called "bonobo"). At present, we recognize four subspecies among wild populations of *Pan troglodytes* (Morin et al. 1994; Gonder et al. 1997). Figure 2 draws together data from the analysis of human and chimpanzee DNA and the estimated time range of the hominid fossil record. This figure thus provides the rationale for comparing human cognition and behavior with those of living *Pan* species and extinct fossil hominids.

Manifestations of human intelligence include their linguistic capabilities and an extremely varied tool-making ability (Berthelet and Chavaillon 1993; Byrne 1995; Gibson and Ingold 1993; Langer and Killen 1998; Parker and Gibson 1990; Tomasello and Call 1997). The question that arises is how and why human evolutionary history has given rise to creatures who can make and use the most complex tools and communicate in the most complex ways among all the species of the animal kingdom. Recent progress in the study of chimpanzees has succeeded in demonstrating examples of both continuity and discontinuity between human and nonhuman animals. Chimpanzees are immature at birth, are weaned at the age of 3 years (i.e., the inter-birth interval is about 5 years), and establish strong bonds, especially between mothers and offspring (Goodall 1986). Through the long process of socialization, wild chimpanzees learn how to make and use various kinds of tools which are unique to each community. Flexibility of intelligence allows captive chimpanzees to acquire two-way communication with human companions, mainly through arbitrary visual and gestural signs (Matsuzawa 1996).

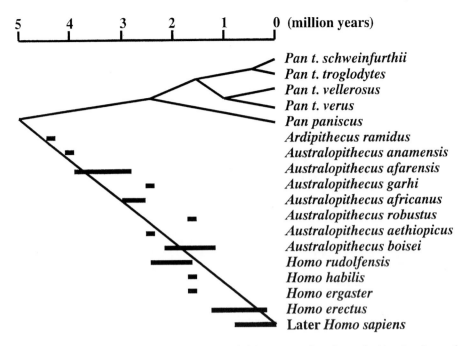

5   4   3   2   1   0 (million years)

Pan t. schweinfurthii
Pan t. troglodytes
Pan t. vellerosus
Pan t. verus
Pan paniscus
Ardipithecus ramidus
Australopithecus anamensis
Australopithecus afarensis
Australopithecus garhi
Australopithecus africanus
Australopithecus robustus
Australopithecus aethiopicus
Australopithecus boisei
Homo rudolfensis
Homo habilis
Homo ergaster
Homo erectus
Later *Homo sapiens*

Fig. 2. Data from DNA analysis of humans and chimpanzees (Morin et al. 1994; Gonder et al. 1997) juxtaposed on the estimated time range of the hominid fossil record (Asfaw et al. 1999; Aiello et al. 1999)

3 Triadic Relationship: A Framework for Analyzing Intelligence

First, I would like to give a brief outline of a framework for discussing various intellectual behaviors, including human language and technology. From a behavioral and biological viewpoint, intelligence can be defined as a way of modulating behavior to adapt to an ever-changing ecological environment. In a naïve categorization of the world from an egocentric perspective, the perceptual world may be divided into three components: self, conspecifics (other individuals of the same species), and objects (either animate or inanimate; either detached object or substrate). These three components make up a triadic relationship (Fig. 3). Following this scheme, intelligence possessed by "self" can be broken down into three constituent aspects: social intelligence, corresponding to self–conspecific relationships; material intelligence, corresponding to self-object relationships; and intelligence concerning the intelligence of other individuals. These then combine to build self–conspecifics–objects triadic relationships.

The first of these three aspects, social intelligence, is the intelligence underlying the modulation of social relationships with other members of a group. In many primate societies, the repertoire of behaviors in such social contexts includes fighting, alliance formation, reassurance of others (postconflict behavior), and so forth (Byrne and Whiten 1988; Cheney and Seyfarth 1990; de Waal 1989). The second

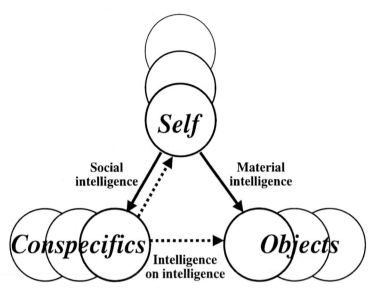

Fig. 3. Framework for analyzing intelligence. The perceptual world from an egocentric perspective is represented by the triadic relationship of self–conspecifics–objects

aspect of intelligence, material intelligence, is the intelligence implicated in utilizing objects in the environment. An advanced form of material intelligence underlies manipulative skills in which different objects are handled in relation to each other, such as in tool manufacture and tool use. The third aspect of intelligence completes the triadic relationship and implies intelligence concerning other conspecifics' intelligence. As described later, the recursive or self-embedding structure of these different aspects of intelligence is what renders human intelligence unique.

Human language and advanced technology are shaped through the social transmission of information from one generation to the next. In this chapter, tool use will be defined as a set of behaviors utilizing a detached object to obtain a goal that is adaptive in the biological sense. The following sections will provide detailed explanations of intelligence, focusing on tool use rather than symbol use or communication. Why tool use? Why not symbol use? The reasons are three-fold. First, chimpanzee communication, especially symbolic communication, is still open to question and requires more research (Boesch 1991; Mitani et al. 1999). Second, the recent discovery of "mirror neurons" suggests that rather than vocal communication having evolved directly, the evolution of human speech may have been derived from a communication system of manual gestures. The neurons located in the monkey premotor cortex (area F5) discharge both when the monkey grasps or manipulates objects and when the monkey observes the experimenter performing a similar action (Kurata and Tanji 1986; Rizzolatti et al. 1988). Transcranial magnetic stimulation and positron emission tomography (PET) experiments suggest that a mirror system for gesture recognition also exists in humans in Broca's area, which is responsible for speech (Rizzolatti and Arbib 1998). Broca's area and its counterpart in the monkey brain both possess neural structures for controlling orolaringeal, facial, and hand–arm movements, and both are provided with mechanisms linking action perception and action production. Third, recent advances in the study of chimpanzee technology allow us to compare material intelligence from a phylogenetic perspective, supported by archeological evidence of behavior. Such evidence cannot be obtained about communication.

What are the advantages of focusing on tool use? There are three main ones. First, tool use is based on manipulation by hands, which is a unique characteristic setting primates apart from other mammals. Second, tool use can reveal critical differences in intelligence among species of living primates. Third, tools can be recovered from the fossil record to provide us with objective evidence for estimating the intelligence of fossil hominids. Taken together, tool use can thus be a unitary measure for comparing the intelligence of living and fossil, and human and nonhuman primates.

4 Postural Development and Object Manipulation in Primates

Members of the order Primates possess two major characteristics in cognition and behavior that distinguish them from other mammals: manipulation by hand and a highly developed visual system. Primates are dexterous "quadramana" who have four hands for manipulation. Although descended from a common ancestor with a primarily nocturnal lifestyle, primates are able to perceive color (Matsuzawa 1985b) and binocular depth (Fuji and Kojima 1981; Sarmiento 1975). These are clearly adaptations to the ecological environment associated with arboreal life and to the shift from nocturnal to diurnal life, which is common to the primates.

4.1 Comparison of Postural Development in Primates

Primates in general have to climb up and down vertical tree trunks. Sitting in primates is characterized by an upright posture, which is a preadaptation for human upright bipedal locomotion. Although adult-type positional behaviors appear at first sight to differ among primates, there are common characteristics to be found in their respective developmental courses. A test of postural development was devised, and applied to primate infants (Takeshita et al. 1989). The test consisted of 11 types of treatment, each of which induced certain postural reactions (Fig. 4). Although spontaneous postures and patterns of locomotion differ among primate species, a common developmental process was observed in the postural reactions induced. Such reactions developed in at least three stages (left to right across columns in Fig. 4) as follows. In one of the inducing techniques, referred to as "Collis-horizontal," the tester lifts up the infant lying on the floor by holding the limbs on one side of the body. The first stage is characterized by the flexion of forelimbs and hind-limbs. No reaction of the limbs to support the body are induced. The second stage is characterized by the extension of the forelimbs while

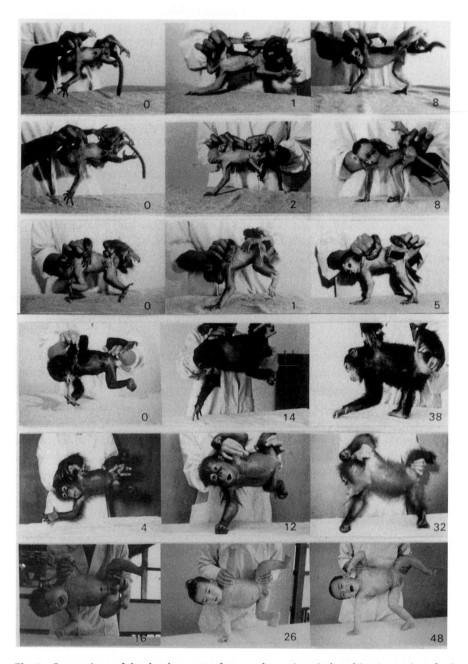

Fig. 4. Comparison of the development of postural reactions induced in six species of primates: from the top, three species of Old World monkeys (*Macaca fascicularis, Macaca radiata, Macaca mulatta*), two species of great apes (*Pan troglodytes, Pongo pygameus*), and a human (*Homo sapiens*). The treatment used to induce the postural reactions illustrated here is called Collis-horizontal, one of the 11 techniques utilized in the study by Takeshita et al. (1989). Common developmental processes across primate species were noted in the postural reactions induced. Such reactions developed in at least three stages, as shown from left to right across columns. The age in weeks of subjects is represented by the number at the right-bottom corner of each picture

the hind-limbs are flexed. The elbow extends with the pronation of the forearm. The third stage is characterized by the extension of both fore- and hind-limbs. The knee extends with the pronation of the foot. In sum, there is a progression from flexion to extension of limbs to support the body. The maturation of motor functions proceeds from the forelimbs toward the hind-limbs. Such successive developmental stages in postural reactions are not only common to various species, but are also irreversible.

4.2 Comparison of Object Manipulation in Primates

The development of object manipulation has also been compared among primate species within the frame of reference provided by the distinct developmental stages of postural reactions. Approach behavior toward objects appears in the second stage, while a variety of manipulations and increased dexterity are marked in the third stage in all species. However, object manipulation by human infants always exceeds that of nonhuman primates in a number of ways. Human infants begin manipulating objects by hand in the supine posture.

Object manipulation has been examined in nonhuman primates under comparable conditions (Parker 1974; Takeshita and Walraven 1996; Torigoe 1985). Torigoe (1985) reported comparisons among 74 species of nonhuman primates. They were presented with a nylon rope and a wooden cube, and their subsequent manipulations were recorded in detail. In total, 506 manipulation patterns were distinguished on the basis of the actions performed ($n = 21$), body-parts used, and relation to other objects. In sum, a 74 (species) × 21 (actions) 1–0 matrix was subjected to multidimensional scaling analysis (Fig. 5). This allowed the 11 species groups to be further classified into three groups. The first consists of lemurs, NWM except *Cebus*, and leaf-eating OWM: this group is characterized by the smallest repertoire of manipulations. The second is OWM except leaf-eaters: this group has more varied modes of manipulation and is characterized by actions against substrates and the use of fingers. The third is *Cebus* and the apes: this group has the most varied modes of manipulation and is characterized, except for the lesser ape, by actions in which one object is related to others.

Takeshita and Walraven (1996) compared object manipulation between 27 chimpanzees at Arnhem and 8 bonobos (*Pan paniscus*) at Planckendael. They identified 582 unique manipulation patterns which were distinguished by the actions performed, the types of objects or body-parts used, the number of objects manipulated, and the types of orienting manipulation. They identified three major differences between the species. First, chimpanzees preferred only one hand during manipulation, whereas bonobos often used both hands. Second, chimpanzees performed more orienting manipulation (relating one thing to another) than bonobos. Third, manipulation by chimpanzees was more substrate-oriented than was that of bonobos. These characteristics are reflected in differences in posture and tool use between the two species.

Not only humans, but also some nonhuman primates such as apes, OWM except leaf-eaters, and *Cebus*, have the dexterity to manipulate small objects. Macaques

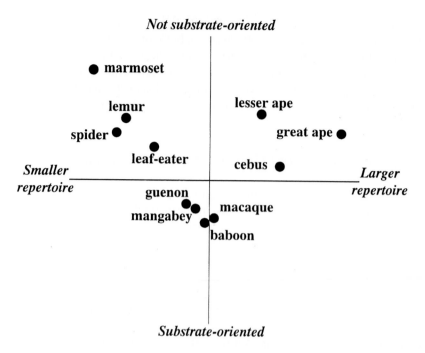

Fig. 5. Object manipulation compared among 74 species of nonhuman primates. The species are further classified into 11 species groups. Actions toward a nylon rope and a wooden cube were compared. Multidimensional scaling was applied to evaluate the similarities. Drawn from the original data provided by Torigoe (1985)

can pick up tiny objects such as wheat grains or sesame seeds with the thumb and index finger (Fig. 6). In a way, chimpanzees and other great apes appear to be less dexterous than macaques. Chimpanzees prefer to handle very small objects with their lips rather than their fingers, and the most advanced stage of precision grip they display is the pincer grip that utilizes the index and middle fingers (Tonooka and Matsuzawa 1995). In sum, dexterity in manual skills does not provide an accurate reflection of material intelligence. However, skills of relating objects and the manipulative repertoire that includes combinatorial or relational acts (Fragaszy and Adams-Curtis 1991) are sufficient predictors of species differences in tool use, and may reflect a form of material intelligence. Prosimians seldom relate one thing to another; they simply hold an object or put it in their mouth. On the other hand, simians sometimes relate one object to another. Such skills in object manipulation provide the basis for the development of tool use in hominoids.

Fig. 6. A young Japanese monkey (*Macaca fuscata*) picking up a grain of wheat. Manual dexterity is not a clear predictor of material intelligence

5 Tool Use in Nonhuman Primates

The taxonomic distribution of tool use is limited to primates. According to the definition described above, there are no reports of tool use in prosimians, almost none in New World monkeys except *Cebus*, and very few in Old World monkeys. Although a few exceptions can be found, these tend to entail incidents in captive or provisioned groups (stick use by crab-eating monkeys, Zuberbuhler et al. 1996; stick or stone use by Japanese monkeys, Tokida et al. 1994). Tool-using behavior is exhibited by humans and the great apes. In particular, chimpanzees are known to use a variety of different tools both in the wild (McGrew 1992) and in captivity (Takeshita and van Hooff 1996).

Goodall (1968) was the first to discover tool use in wild chimpanzees. In 1960, at Gombe in Tanzania, East Africa, she observed an adult male inserting a twig into a termite mound, which was subsequently bitten by termites (*Macrotermes bellicosus*) attempting to defend their nests. Retracting the twig now had the effect of essentially "fishing" out such defenders. Goodall's discovery has since been followed by reports of numerous kinds of tool use from different research sites (Alp 1997; Boesch and Boesch 1983; Hirata et al. 1998; Matsuzawa and Yamakoshi 1996; McGrew 1992; Nishida 1990; Sugiyama and Koman 1979; Whiten et al. 1999).

There are five sites in Africa at which long-term research is carried out: Gombe (Goodall 1986), Mahale (Nishida 1990), and Kibale (Wrangham et al. 1994, 1996) in East Africa, and Tai (Boesch and Boesch-Achermann 2000) and Bossou in West Africa. Data accumulated at these sites clearly show that there are unique cultural traditions of tool use in each chimpanzee community. The term "culture" is defined as a set of behaviors that are shared by members of a community, and are transmitted from one generation to the next through nongenetic channels (Matsuzawa 1999). For example, chimps at Mahale fish not for termites, but for ants in tree trunks (Fig. 7). Chimpanzees at Bossou in Guinea, West Africa, use a stick to scoop up algae floating on a pond—a behavior which has never been reported in other chimpanzee communities (Matsuzawa et al. 1996). They eat termites, but they seldom use twigs in the fishing manner seen at Gombe (Humle 1999). They

Fig. 7. A chimpanzee at Mahale, Tanzania, fishing for ants in a tree trunk. This is an example of level-1-type tool use. (Photograph provided by the ANC Corporation)

simply wait for the termites to leave their mound, and pick them up individually with the aid of their fingers.

A behavior referred to as "ant dipping" is found at both locations. However, details of the behavior are different. Chimpanzees at Gombe hold a relatively long (66 cm on average) stick in one hand, which they dip among safari ants of the species *Dorylus molestus* (McGrew 1974). In a highly coordinated, bimanual pattern of behavior, the other hand is used to sweep ants that climb or bite the stick into the chimpanzee's mouth. This contrasts with the Bossou chimpanzees' use of a shorter stick (47 cm on average) manufactured by a unique technique in which a twig is broken off, bitten in half, and its outer layer removed by mouth to produce a fishing rod. Bossou chimpanzees use this tool in a single-handed maneuver, holding a stick between the index and middle fingers to catch safari ants.

In addition to algae scooping, ant dipping (Sugiyama 1995a), pestle pounding (Sugiyama 1994; Yamakoshi and Sugiyama 1995), and the use of leaves for drinking water from a tree hollow (Sugiyama 1995b; Tonooka et al. 1997), the Bossou chimpanzees' unique repertoire of tool-using behavior also includes nut cracking with stone tools (Matsuzawa 1994; Inoue-Nakamura and Matsuzawa 1997). Using a pair of stones as a hammer and anvil, they crack open oil-palm nuts, thus gaining access to their nutritious kernel (Fig. 8). At Bossou, chimpanzees always use one hand for hammering, demonstrating perfect laterality, which is in contrast to Tai chimpanzees who sometimes use both hands for hammering. The target nuts for cracking are also different in the two communities. Tai chimpanzees never crack open the oil-palm nuts favored at Bossou, even though oil-palm nuts are available at Tai. Moreover, such stone tool use or hammering techniques cannot be found in East African communities. In Gombe, chimpanzees do not crack oil-palm nuts with stones, but only eat the outer thin layer of the fruit, which is soft and contains oil; they discard the edible part covered by a hard shell. In Mahale, chimpanzees do not utilize oil palm at all even though the trees are available.

Fig. 8. Nut-cracking by wild chimpanzees at Bossou, Guinea. This is an example of level-2-type tool use

No explicit reason has been found as to why Gombe or Mahale chimpanzees should not use stone tools to crack the hard shell of oil-palm nuts just as Bossou chimpanzees do. Nuts as well as suitable stones are readily available in Gombe and Mahale. One possible explanation is provided by the chimpanzees' propensity to learn from others what to eat and what kind of objects to use as tools. Community-specific cultural traditions are maintained through such learning, and serve as the basis of the intercommunity differences in tool-using behavior described above (Matsuzawa 1999; Chapter 28 by Matsuzawa, this volume).

6 Tree-Structure Analysis of Tool Use

The material intelligence of wild chimpanzees is manifest in their use of various kinds of tools. Furthermore, cultural variation serves to illustrate the point that every occurrence of tool use has its own unique characteristics when the behavior is examined in detail. Labels such as "ant-dipping" should therefore be applied with caution, as cultural differences among communities allow for only superficial similarities between behaviors. Target foods, tool materials, tool sizes, methods of tool manufacture, tool-using techniques, and so forth, may differ from community to community. Thus, the flexible nature of chimpanzee intelligence combines with ecological and cultural constraints to produce the different types of tool use found at different sites.

6.1 A Novel Way of Analyzing Material Intelligence

This section will introduce a novel way of analyzing material intelligence in chimpanzees. Although much variation exists among different types of tool use in the wild, it is possible to extract a common theme from these cases if we bear in mind the following statement. The essence of tool use is to relate one thing (the tool) to another (the target). Consider, for instance, the case of the stick and the ant. There is no intrinsic relationship between these two objects, and only when the chimpanzee picks up the stick and puts one end into a stream of safari ants is the stick transformed into a tool. This is a universal aspect of tool use. A twig is used to fish for termites; a stick is used to scoop algae; and leaves are used to get water.

Tool use is an adaptive behavior in which one object is related to another (or self or conspecifics in social contexts) to obtain a goal. Here, "adaptive" refers to the idea that by performing the behavior, the individual's fitness, or that of its kin, is increased, for example by obtaining food or a mating partner. In this scheme, tool use in self–object relationships at their simplest level involves two objects. For example, in using a twig to fish for termites, chimpanzees relate one object to another in an adaptive fashion. I call this type of behavior "level 1" tool use, because only a single relationship between objects exists: the association relating one of the two objects to the other. Despite the relatively large variety of different types of tool use that have been observed, level 1 tool use appears to be predominant, and describes almost all known examples of tool use by wild chimpanzees.

Nut cracking by chimpanzees is a rare example of a more complicated form of tool use. Chimpanzees use a hammer stone to crack open a nut that is placed on an anvil stone. I call this type of tool use "level 2" tool use, as there are two kinds of relationships among the objects involved: the nut is related to the anvil stone in the positioning phase, and the hammer stone is related to the nut in the cracking phase. Moreover, the two relationships are not equivalent, but can be viewed as having a hierarchical arrangement in a specific temporal order: the nut should be placed on the anvil stone prior to applying force with the hammer stone. By virtue of this pattern, the relationship between two of the three objects—the nut and the anvil stone in this case—should form a small cluster. In turn, the clustered set forms the second relationship, this time with the third object, the hammer stone.

The most complex form of tool use ever encountered among nonhuman primates is the use of metatools by chimpanzees. On some occasions, Bossou chimpanzees employ a third stone in the nut-cracking process (Matsuzawa 1991, 1994). Suppose that an anvil stone has a slanting upper surface. A nut placed on such an anvil will not remain balanced, but will roll off the stone before it can be hit by the hammer. Chimpanzees aged 6.5 years and above have been observed to insert a third stone underneath the anvil to serve as a wedge, thereby keeping it stable and flat—a strategy identical to that employed by humans (Fig. 9). This third stone is thus a part of the anvil apparatus; it is an anvil for another anvil. The name "metatool" has been coined to describe a tool with such a role. In instances of metatool use, three relationships between objects can be discerned. (1) A chimpanzee uses a hammer stone to hit a nut, where (2) the nut is placed on an anvil stone, and (3) the

Fig. 9. Metatool use by a boy from the Manon tribe at Bossou, Guinea, who was 6 years 9 months old. A slanting anvil stone was provided. The boy spontaneously applied a third stone as a wedge to keep the surface of the anvil stone flat and stable, in a manner similar to the solution used by chimpanzees

anvil stone itself is supported by a wedge stone. Thus, this complex type of tool use is referred to as "level 3."

Step by step, therefore, just as a linguist applies the tree-structure analysis of generative grammar to dissect the structure of sentences, it is possible to analyze tool-using behaviors in an analogous fashion. The important points are summarized below. In the analysis, we pay attention only to the relationships among the objects involved: we neglect who performed the behavior, how and where it was performed, and so forth. In other words, we omit the actor, the action, the location, and instead focus on the objects to determine the relationships among them, along with the appropriate temporal order. Such relationships are depicted by a node connecting a pair of objects located on the terminal branches. Figure 10 illustrates the representation of the three levels of tool use elucidated above in terms of hierarchical tree-structure analysis.

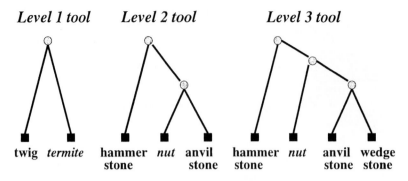

Level 1 tool   *Level 2 tool*   *Level 3 tool*

twig *termite*   hammer *nut* anvil   hammer *nut*   anvil wedge
                  stone   stone   stone   stone stone

Fig. 10. In tree-structure analysis, actions involved in tool use were analyzed by focusing on the objects alone (see text for details). In the notation system of tree-structure analysis, hierarchical levels are shown by the depth of nodes, i.e., the number of nested clusters in the tree structure. A node is represented by a solid circle and an object is represented by a solid square

6.2 Validity of the Tree-Structure Analysis of Tool Use

To further validate the depiction of different types of tool use in terms of hierarchical levels, it may be useful to consider the development of these behaviors in young chimpanzees. In Bossou, initial, level-0 object manipulation (where individual objects are handled singly) is followed by the appearance of level-1-type tool use relating two objects to each other, such as using leaves to drink water or a stick to scoop algae from a pond surface. The minimum age for the execution of such level-1 tasks appears to be approximately 2 years. In the case of level-2 tools, such as stone tool use to crack open nuts, the handling of individual stones or nuts precedes actions on multiple objects (again, first seen at around 2 years), which in turn gradually develop into coordinated nut cracking behavior (Inoue-Nakamura and Matsuzawa 1997). The youngest subject demonstrating such level-2-type tool use was 3.5 years old at the time of observation. Furthermore, the use of metatools (level-3-type tool) has never been seen in a chimpanzee younger than 6.5 years. There is no evidence in the wild that chimpanzees can use tools at the level-4 stage or above. The increase in complexity and cognitive difficulty involved in higher hierarchical levels of tool use is therefore clearly reflected in the minimum age at which the corresponding behaviors are acquired.

6.3 Comparison of Material Intelligence of Living Primates and Fossil Hominids

Table 1 summarizes the cognitive performance of living primates and fossil hominids as revealed by the manufacture and use of tools. The comparison shows evidence of a clear gap between chimpanzees and Old World monkeys such as macaques. In turn, differences between humans and chimpanzees are highlighted by the depths of nodes in relating objects to one another within hierarchical structures.

Table 1. Comparisons among living primates and fossil hominids in cognitive performance as shown by aspects of tool manufacture and use

	PRO	NWM	OWM	*Pt*	*Aa*	*He*	*Hs*
Relate one thing to the other	N	Y	Y	Y	Y*	Y	Y
Use of level-1 tool	N	N*	y	Y	Y*	Y	Y
Use of level-2 tool	N	N*	N	Y	Y*	Y	Y
Tool use for making tool	N	N	N	N*	-	Y	Y
Use of level-3 tool	N	N	N	y	-	Y	Y
Use of level-4 tool	N	N	N	N	-	Y	Y
Throw an object toward an enemy	N	N	N	Y	-	-	Y
Touch oneself with an object	N	N	N	y	-	-	Y
Use of social tool	N	N	N	Y	-	-	Y
Use of multilevel social tool	N	N	N	N	-	-	Y

N, No evidence; N*, no evidence with exceptionally rare cases reported; y, possible but difficult to ascertain; Y*, no clear evidence but inferred; Y, clear evidence; –, unknown; PRO, prosimians; NWM, New World monkeys; OWM, Old World monkeys; *Pt*, chimpanzees; *Aa*, *Australopithecus afarensis*; *He*, *Homo erectus*; *Hs*, *Homo sapiens*.

Chimpanzees also make use of objects in social contexts, such as aimed throwing, leaf clipping for courtship display, and charging displays with branches or rocks. Such manipulation, where objects are directed toward other individuals, represents "social tool" use. Examples of social tool use are found less commonly than examples of "material tools," and are limited to the level-1 type (relating one object to a conspecific) in wild chimpanzees.

7 Tool Use in Fossil Hominids

A stone-tool-using hominid was recovered from the Orduvai Gorge in Tanzania (from site Bed I, age an estimated 1.8 million years) and became known as *Homo habilis* (Leakey et al. 1964). The stone tools of the Ordowan were the products of lithic technology, prepared using stone-flaking techniques. Using small stones as raw material, this hominid was able to manufacture tools that served as choppers and scrapers with the aid of another stone. Such metatools—used in the production of other tools—have never been found in wild chimpanzees. In captivity, a male bonobo named Kanzi acquired the basic skills required to produce usable flakes and fragments by hard-hammer percussion as well as throwing (Toth et al. 1993). In the case of wild chimpanzees, episodes have been noted where individuals broke an anvil stone by a bout of forceful hammering during nut cracking, and then went on to use the broken piece as a superior hammer (Matsuzawa 1994). The brush-sticks used for catching termites in southwest Cameroon might have been produced by chimpanzees hitting the brush end with a stone (Sugiyama 1985). However, neither the manufacture nor the use of this brush-stick tool has ever been observed directly. In sum, chimpanzees in the wild have not evolved a flaked-stone technology characteristic of the Ordowan industry.

The African hominid record is extensive for 3–4 million years prior to the present. Numerous fossils of *Australopithecus afarensis* (also known as *Praeanthropus africanus*; see Wood and Collard 1999)—widely thought to be ancestral to *Homo*—have been recovered from various sites around Africa. However, the East African record is relatively sparse for the period covering the past 2–3 million years, which coincides with the emergence of lithic technology and *Homo*. Recently, Asfaw et al. (1999) recovered a hominid fossil from Ethiopia's Middle Awash, dated 2.5 million years old. This species, named *Australopithecus garhi*, is a descendent of *A. afarensis* and is a candidate for the ancestor of early *Homo*. In a recent report, evidence is presented demonstrating that *A. garhi* used stone tools to butcher large mammal carcasses for meat and marrow in an open lake margin habitat (de Heinzelin et al. 1999). However, it is still controversial which of these extinct species was responsible, and when, for the emergence of the lithic technology exemplified by the flake production seen in the Ordowan—a tool-using technique that is clearly more advanced than that of wild chimpanzees (Roche et al. 1999).

Two different phases of hominid technology are recognized, corresponding to the periods before and after the critical phase approximately 2.5 million years ago during which both lithic technology and *Homo* first appeared. Subsequently, *H. erectus* adopted the use of a variety of highly complex sets of tools. In contrast, before this critical point, *A. afarensis* demonstrates very little evidence of tool use. Table 1 summarizes cognitive performance as revealed in the tool manufacturing and using skills of living primates and fossil hominids. Tool use in chimpanzees implies some form of tool manufacture in most cases, such as stripping side leaves to make ant-catching wands. Although in the case of stone-tool use the active production of tools is more difficult to imagine, these are, in fact, easily distinguishable from natural stones, even down to clear distinctions between hammers and anvils (Sakura and Matsuzawa 1991). Sets of such stones are usually found scattered over a limited area, often under palm trees, and a shallow depression is formed on the striking surface of the tools habitually used by the chimpanzees. It seems reasonable to assume that *A. afarensis* also possessed a rudimentary form of lithic technology which was similar to the stone-tool use shown by Bossou chimpanzees.

There is little doubt that the common ancestor of chimpanzees and humans, 5 million years ago, already possessed stone tools similar to those of extant chimpanzees. Provided that *A. afarensis* is at least equal to living chimpanzees in terms of the capacity for hierarchical organization of material tools, they most likely utilized comparable stone tools. Besides stone as raw material, they also used various kinds of tools made from more perishable substances such as bones, teeth, horns, shells, skins, trees, bark, leaves, thorns, grass, mud, and the like. In addition, they are likely to have been users of social tools with as much propensity as chimpanzees, although no comparable record of social tools is available. Evidence so far gathered for chimpanzees and bonobos in the wild suggests that the fossil hominids were able to perform aimed throwing, threatening of an enemy by holding up a stick or a rock, pointing out directions with the help of an object, and leaving marks on the ground (Ingmanson 1996).

Table 2. Comparisons among living primates (including humans) in cognitive performance related to social intelligence

	PRO	NWM	OWM	Pt	Hs
Social tool: using an object in a social context	N	N	N	Y	Y
Looking into eyes in an affectionate context	N	N	N	Y	Y
Cross-modal matching to sample	N	N	N	Y	Y
Mirror self-recognition	N	N*	N*	Y	Y
Deception	N	N	N	Y	Y
Allowing others to take one's food	N	N	N*	Y	Y
Actively sharing one's food with others	N	N	N	y	Y
Imitation: immediately imitate an action	N	N	N	y	Y
Cooperation: coordinated action to get a goal	N	N	N*	y	Y
Spontaneous turn-taking	N	N	N	N*	Y
Pretense or pretend play	N	N	N	N*	Y
Teaching (especially active teaching)	N	N	N	N*	Y
"Theory of mind" in a strict sense	N	N	N	N	Y

N, No evidence; N*, no evidence with exceptionally rare cases reported; y, possible but difficult to ascertain; Y, clear evidence .
Other abbreviations are the same as those in Table 1.

8 Social Intelligence

8.1 Comparison of Social Intelligence in Primates

This chapter has so far focused on material aspects of intelligence based on the self–object relationship of the triadic scheme. This section briefly summarizes social aspects of intelligence based on the self–conspecific relationship. In the case of humans, objects are often involved in our interactions with other individuals. For instance, we actively share food with conspecifics. Even infants less than 1 year old often exchange objects with their mothers in a face-to-face situation. However, nonhuman primates very rarely engage in such interactions. They may manipulate objects, but these are neither used nor exchanged within a social context. In other words, social tools are highly uncommon. With the exception of hominoids, primates do not look into the eyes of one another in an affectionate context. Instead, such gazes express a mild threat in primates in general. Table 2 summarizes the evidence for comparisons of cognitive performance related to social intelligence among the living primates. Once again, a sizeable gap between chimpanzees and Old World monkeys is evident. This discontinuity has stood up to rigorous and repeated testing in a variety of cognitive tasks.

8.2 Recognition of Self and Others

One of the clearest examples supporting the existence of this gap may be provided by the mirror self-recognition (MSR) task, i.e., the recognition of self in a reflected image in the mirror (Gallup 1970). MSR seems to be limited to humans and the great apes, and even studies involving gorillas remain controversial (Parker et al. 1994; Inoue-Nakamura 1997). The behavioral measure of this particular ability is the presence of self-directed behavior when facing a reflected image, as opposed to responding directly toward the image. For example, a subject noticing an unusual feature on the reflection would, in the case of true mirror self-recognition, reach out towards the appropriate spot on him or her self—invisible though it may be to that individual—rather than towards the image. In a sense, this represents the use of an object (a mirror) to guide behavior toward oneself. The lack of MSR in monkeys is in concordance with the fact that they seldom succeed in cross-modal matching-to-sample tasks, in which visual, auditory, or tactile cues from an identical source are to be matched together.

The discrimination of "self" and "conspecifics" leads to the next step in cognitive performance: imitation. Monkeys fail to show MSR and cannot imitate the behavior of other individuals immediately. Such immediate imitation is difficult even for chimpanzees (Myowa-Yamakoshi and Matsuzawa 1999; Tomasello et al. 1993), despite them having shown evidence of MSR. Imitation refers to the mimetic process in which the behavior of individual A is immediately copied in the behavior of individual B. Even when chimpanzees demonstrate imitative abilities, careful observation by individual B is essential, while Individual A seldom shows active involvement such as "teaching," "molding," or "guidance." The lack of the latter process in social interaction is supported by the observation that cooperation and turn-taking are also tasks that chimpanzees find difficult.

Thus, while social intelligence is a characteristic common to many species of primate, cognitive performance evidenced in social tool use or self–conspecific discrimination are highly undeveloped in nonhuman primates. Based on the comparative data presented in Table 2, it is reasonable to assume that fossil hominids were not only superior to extant chimpanzees in terms of material intelligence, but also possessed superior social intelligence despite the lack of fossil evidence illuminating social behavior. Our current knowledge of chimpanzees and bonobos in the wild suggests that fossil hominids were capable of self-recognition, imitation, pretense, and deception. They shared and exchanged food, cooperated, and spontaneously traded social roles. They began using various kinds of objects in the context of social exchange, including communication.

9 Uniquely Human Intelligence

Humans possess by far the largest brain within the order Primates. Their average brain volume (approximately 1450 cc) is almost three times that of the chimpanzees (450 cc). While there are no large differences in sensory and perceptual functions such as vision and hearing among primates, the neocortex, and especially the associative cortex, has undergone remarkable development in *Homo*. The massive enlargement in size of the human associative cortex has led to an increase in (1) the number of levels and (2) the number of relationships that can be handled simultaneously. This resulted in uniquely human characteristics such as (1) "self-embedding hierarchical structure in cognition" and (2) "cognition in terms of the self–conspecifics–objects triadic relationship."

9.1 Self-Embedding Hierarchical Structure in Cognition

Chimpanzees, as well as other great apes, are able to use tools. However, the vast majority of the examples observed belong to the level-1 type, while level 2, and especially level 3, cases of tool use are extremely rare occurrences. Constraints on complexity are to be expected, as represented by a limit on the depth and number of nodes in the tree-structure analysis. In contrast, tool use in humans is well developed, and is characterized by a large number of hierarchical levels. Accordingly, the number of nodes in the tree-structure analysis far exceeds the level-3 maximum so far observed in chimpanzees. Relating one thing to another, and then such clusters to yet other clusters, leads to further clusters. In particular, complex cognitive processes such as those required to use an object as a tool for another tool have a self-embedding and recursive structure that is a prominent feature of human intelligence. As an extreme case, consider that we can construct an infinitely complex machine, consisting of an infinite number of tools, simply by applying the self-embedding rule.

With the aid of tree-structure analysis, it is possible to pinpoint the main difference between human and chimpanzee intelligence as lying merely in the depth of hierarchical levels. Thus, this difference is not domain-specific, which would suggest a complete lack of particular abilities. Instead, it should be seen as being of a domain-free nature. The apparent gap between human intelligence and chimpanzee intelligence can thus be accounted for in terms of numbers of levels, as described above in relation to the tree-structure analysis of material intelligence.

9.2 Cognition in Terms of the Self–Conspecifics–Objects Triadic Relationship

Human neonates innately possess unique reflex behaviors that are lacking in non-human primates (except chimpanzees): neonatal facial/gestural imitation (Meltzoff and Moore 1977; Myowa 1996), neonatal smiling reflex, and entrainment (the synchrony of behavior in mother–infant interaction). These reflexes provide the basis of self–conspecific interactions such as face-to-face communication and imitation. Human infants, from a very early age, start to manipulate objects in social contexts, and interact with conspecifics such as the mother. In other words, human infants are able to handle two relationships in parallel: self–object and self–conspecific combine to produce a triadic relationship.

In the built-in triadic relationship consisting of self, conspecifics, and objects, humans combine two aspects of intelligence, material intelligence and social intelligence, to produce a third: intelligence of intelligence, or the ability to perceive and understand the social and material intelligence of other individuals. This type of intelligence was first formulated by David Premack and Guy Woodruff (Premack and Woodruff 1978), who coined the term "theory of mind" to describe the skill. In essence, humans over the age of 4 or 5 years can distinguish the beliefs of other individuals from those of their own, while children younger than 4 years easily confuse the two. So far there have been no convincing reports that the great apes demonstrate this type of intelligence. Possession of a theory of mind may stem from abilities such as imitation, turn-taking, and teaching, the existence of which among the great apes are likewise controversial.

Humans perceive the color and depth of three-dimensional space in a way which is comparable to nonhuman primates. However, there seem to be uniquely human perspectives when humans look at the outer world: the perception of relationships. First, consider stationary objects. Humans tend to perceive relationships between the latter: large–small, up–down, left–right, front–rear, same–different, inclusion, contact–separate, and so forth. Second, regarding moving objects, humans spontaneously perceive social relationships even in the physical world of interacting objects (Michotte 1963; Premack and Premack 1997). For example, imagine a red ball approaching a blue ball. Next, upon impact, the red ball stops and in turn the blue ball starts to move. When perceiving such an event, humans develop a vivid impression of the red ball having "pushed" the blue ball forward. Events in the outer world are perceived not only in terms of their physical features, but also in an intentional/social context, as human cognition locates the built-in triadic relationship of self–conspecifics–objects.

10 Conclusion

In conclusion, in examining the nature of intelligence by focusing mainly on tool use, this chapter has shown the uniquely human characteristics such as (1) self-embedding hierarchical structures in cognition, and (2) cognition in terms of self–conspecifics–objects triadic relationships. Hence, the main difference between human and chimpanzee intelligence does not lie in a complete lack of specific cognitive modules such as the linguistic ability of speech, but in the depth of hierarchical levels in cognition as suggested by tree-structure analysis. The nature of the difference is not domain-specific but domain-free. Humans and chimpanzees cannot be differentiated on the basis of particular abilities such as language and tool use. The apparent gap between human intelligence and chimpanzee intelligence can be accounted for in terms of the number of levels and the number of relationships that can be handled simultaneously. This difference appears in all domains of cognition, and is physically based on the enlargement of the associative cortex over the course of human evolution. The built-in mechanism for perceiving the world in terms of the triadic relationship of self–conspecifics–objects provides the basis for both advanced technology and spoken language in humans. Overall, this provides a novel way of perceiving the distinctness of humans within the animal kingdom.

Acknowledgments

This study was financed by grants from the Ministry of Education, Science, and Culture in Japan to the author (Nos. 07102010, 08044006, and 12002009). I sincerely thank Dr. April Nowell for encouragement and careful reading of the text. Thanks are also due to Dora Biro, Gen Yamakoshi, Mitsuru Aimi, Gen Suwa, and Yukimaru Sugiyama for preparing the manuscript.

References

Aiello LC, Wood B, Key C, and Lewis M (1999) Morphological and taxonomic affinities of the Olduvai Ulna (OH36) Am J Phys Anthropol 109:89–110

Alp R (1997) "Stepping-sticks" and "seat-sticks": new types of tools used by wild chimpanzees (*Pan troglodytes*) in Sierra Leone. Am J Primatol 41:45–52

Asfaw B, White T, Lovejoy O, Latimer B, Simpson S, Suwa G (1999) *Australopithecus garhi*: a new species of early hominid from Ethiopia. Science 284:629–635

Berthelet A, Chavaillon J (1993) The use of tools by human and non-human primates. Clarendon Press, Oxford

Biro D, Matsuzawa T (1999) Numerical ordering in a chimpanzee (*Pan troglodytes*): planning, executing, and monitoring. J Comp Psychol 113:178–185

Boesch C (1991) Teaching among wild chimpanzees. Anim Behav 41:530–532

Boesch C, Boesch H (1983) Optimization of nut-cracking with natural hammers by wild chimpanzees. Behaviour 83:265–286

Boesch C, Boesch-Achermann H (2000) The chimpanzees of the Tai forest. Oxford University Press, Oxford

Byrne R (1995) The thinking ape. Oxford University Press, Oxford

Byrne R, Whiten A (1988) Machiavellian intelligence: social expertise and the evolution of intellect in monkeys, apes, and humans. Oxford University Press, Oxford

Cheney DL, Seyfarth RM (1990) How monkeys see the world. University of Chicago Press, Chicago

De Heinzelin J, Clark D, White T, Hart W, Renne P, WoldeGabriel G, Beyenne Y, Vrba E (1999) Environment and behavior of 2.5-million-year-old Bouri hominidsScience 284:625–629

De Waal F (1989) Peace making among primates. Harvard University Press, Cambridge

Fragaszy DM, Adams-Curtis LE (1991) Generative aspects of manipulation in tufted capuchin monkeys (*Cebus apella*). J Comp Psychol 105:387–397

Fuji K, Kojima S (1981) Acquisition of depth discrimination in a Japanese macaque: a preliminary study. Percept Motor Skills 52:827–830

Gallup GG Jr (1970) Chimpanzees: self recognition. Science 167:86–87

Gibson K, Ingold T (1993) Tools, language and cognition in human evolution. Cambridge University Press, Cambridge

Gonder MK, Oates JF, Disotell TR, Forstner MRJ, Morales JC, Melnick DJ (1997) A new West African chimpanzee subspecies? Nature 388:337

Goodall J (1968) The behaviour of free-living chimpanzees in the Gombe Stream Reserve. Anim Behav Monogr 1:161–311

Goodall J (1986) The chimpanzees of Gombe: patterns of behavior. Harvard University Press, Cambridge

Hirata S, Myowa M, Matsuzawa T (1998) Use of leaves as cushions to sit on wet ground by wild chimpanzees. Am J Primatol 44:215–220

Humle T (1999) New record of fishing for termites (*Macrotemes*) by the chimpanzees of Bossou (*Pan troglodytes verus*), Guinea Pan Africa News 6:3–4

Ingmanson EJ (1996) Tool-using behavior in wild *Pan paniscus*: social and ecological consideration. In: Russon A, Bard K, Palker ST (eds) Reaching into thought. Cambridge University Press, New York, pp 190–210

Inoue-Nakamura N (1997) Mirror self-recognition in nonhuman primates: a phylogenetic approach. Jpn Psychol Res 39:266–275

Inoue-Nakamura N, Matsuzawa T (1997) Development of stone tool use by wild chimpanzees (*Pan troglodytes*). J Comp Psychol 111:159–173

Kawai N, Matsuzawa T (2000) Numerical memory span in a chimpanzee. Nature 403:39–40

Kurata K, Tanji J (1986) Premotor cortex neurons in macaques: activity before distal and proximal forelimb movements. J Neurosci 6:403–411

Langer J, Killen M (1998) Piaget, evolution, and development. Lawrence Erlbaum, Hillsdale

Leakey L, Tobias P, Napier J (1964) A new species of the genus *Homo* from Orduvai Gorge. Nature 202:7–9

Matsuzawa T (1985a) Use of numbers by a chimpanzee. Nature 315:57–59

Matsuzawa T (1985b) Color naming and classification in a chimpanzee (*Pan troglodytes*). J Hum Evol 14:283–291

Matsuzawa T (1990) Form perception and visual acuity in a chimpanzee. Folia Primatol 55:24–32

Matsuzawa T (1991) Nesting cups and metatools in chimpanzees. Behav Brain Sci 14:570–571

Matsuzawa T (1994) Field experiments on the use of stone tools by chimpanzees in the wild. In: Wrangham R, de Waal F, Heltne P (eds) Chimpanzee cultures. Cambridge University Press, Cambridge, pp 196–209

Matsuzawa T (1996) Chimpanzee intelligence in nature and in captivity: isomorphism of symbol use and tool use. In: McGrew W, Marchant L, Nishida T (eds) Great ape societies. Cambridge University Press, Cambridge

Matsuzawa T (1998) Chimpanzee behavior: a comparative cognitive perspective. In: Greenberg G, Haraway MM (eds) Comparative psychology: a handbook. Garland, New York, pp 360–375

Matsuzawa T (1999) Communication and tool use in chimpanzees: cultural and social contexts. In: Hauser M, Konishi M (eds) Neural mechanisms of communication. MIT Press, Cambridge, pp 645–671

Matsuzawa T, Yamakoshi G (1996) Comparison of chimpanzee material culture between Bossou and Nimba, West Africa. In: Russon AE, Bard KA, Parker ST (eds) Reaching into thought: the minds of the great apes. Cambridge University Press, Cambridge, pp 211–232

Matsuzawa T, Yamakoshi G, Humle T (1996) A newly found tool-use by wild chimpanzees: algae scooping. Primate Res 12:283

McGrew WC (1974) Tool use by wild chimpanzees in feeding upon driver ants. J Hum Evol 3:501–508

McGrew WC (1992) Chimpanzee material culture. Cambridge University Press, Cambridge

Meltzoff AN Moore MK (1977) Imitation of facial and manual gestures by human neonates. Science 198:775–778

Michotte A (1963) The perception of causality. Methuen, Andover

Mitani J, Humley K, Murdock M (1999) Geographic variation in the calls of wild chimpanzees: a reassessment. Am J Primatol 47:133–151

Morin PA, Moore JJ, Chakraborty R, Jin L, Goodall J, Woodruff DS (1994) Kin selection, social structure, gene flow, and the evolution of chimpanzees. Science 265:1193–1201

Myowa M (1996) Imitation of facial gestures by an infant chimpanzee. Primates 37:207–213

Myowa-Yamakoshi M, Matsuzawa T (1999) Factors influencing imitation in chimpanzees. J Comp Psychol 113:128–136

Nishida T (1990) The chimpanzees of the Mahale mountains: sexual and life history strategies. University of Tokyo Press, Tokyo

Parker C (1974) The antecedents of man the manipulator. J Hum Evol 3:493–500

Parker S, Gibson K (1990) "Language" and intelligence in monkeys and apes: comparative developmental perspectives. Cambridge University Press, Cambridge

Parker S, Mitchell R, Boccia M (eds) (1994) Self-awareness in animals and humans: developmental perspectives. Cambridge University Press, Cambridge

Portmann A (1951) Biologische Fragmente zu einer Lehre vom Menschen. Benno Schwabe, Basel

Premack D, Premack AJ (1997) Infants attribute value+/- to the goal-directed actions of self-propelled objects. J Cog Neurosci 9:848–856

Premack D, Woodruff G (1978) Does the chimpanzee have a theory of mind? Behav Brain Sci 4:515–526.

Rizzolati G, Arbib M (1998) Language within our grasp. Trends Neurosci 21:188–194

Rizzolatti G, Camarda R, Foggasi L, Gentilucci M, Luppino G, Matelli M (1988)Functional organization of inferior area 6 in the macaque monkey. II: Area F5 and the control of distal movements. Exp Brain Res 71:491–507

Roche H, Delagnes A, Brugal JP, Feibel C, Kibunjia M, Mourre V, Texier PJ (1999) Early hominid stone tool production and technical skill 2.34 Myr ago in West Turkana, Kenya. Nature 399:57–60

Sakura O, Matsuzawa T (1991) Flexibility of wild chimpanzee nut-cracking behavior using stone hammers and anvils: an experimental analysis. Ethology 87:237–248

Sarmiento R (1975) The stereoacuity of macaque monkeys. Vision Res 15:493–498

Sugiyama Y (1985) The brush-sticks of chimpanzees found in southwest Cameroon and their cultural characteristics. Primates 26:361–374

Sugiyama Y (1994) Tool use by wild chimpanzees. Nature 367:327

Sugiyama Y (1995a) Tool-use for catching ants by chimpanzees at Bossou and Monts Nimba, West Africa. Primates, 36:193–205

Sugiyama Y (1995b) Drinking tools of wild chimpanzees at Bossou. Am J Primatol 37:263–269

Sugiyama Y, Koman J (1979) Tool-using and making behavior in wild chimpanzees at Bossou, Guinea. Primates 20:513–524

Takeshita H, van Hooff J (1996) Tool use by chimpanzees (*Pan troglodytes*) of the Arnhem Zoo community. Jpn Psychol Res 38:163–173

Takeshita H, Walraven V (1996) A comparative study of the variety and complexity of object manipulation in captive chimpanzees (*Pan troglodytes*) and bonobos (*Pan paniscus*). Primates 37:423–441

Takeshita H, Tanaka M, Matsuzawa T (1989) Development of postural reactions and object manipulation in primate infants (in Japanese with English summary). Primate Res 5:111–120

Tokida E, Tanaka I, Takefushi H, Hagiwara T (1994) Tool-using in Japanese macaques: use of stones to obtain fruit from a pipe. Anim Behav 47:1023–1030

Tomasello M, Call J (1997) Primate cognition. Oxford University Press, Oxford

Tomasello M, Savage-Rumbaugh S, Kruger AC (1993) Imitative learning of actions on objects by children, chimpanzees, and enculturated chimpanzees. Child Dev 64:1688–1705

Tonooka R, Matsuzawa T (1995) Hand preferences of captive chimpanzees in simple reaching for food. Int J Primatol 16:17–35

Tonooka R, Tomonaga M, Matsuzawa T (1997) Acquisition and transmission of tool making and use for drinking juice in a group of captive chimpanzees (*Pan troglodytes*). Jpn Psychol Res 39:253–265

Torigoe T (1985) Comparison of object manipulation among 74 species of nonhuman primates. Primates 26:182–194

Toth N, Schick K, Savage-Rumbaugh S, Sevcik R, Rumbaugh D (1993) Pan the tool-maker: investigations into the stone tool-making and tool-using capabilities of a bonobo (*Pan paniscus*). J Archaeol Sci 20:81–91

White TD, Suwa G, Asfaw B (1994) *Australopithecus ramidus*, a new species of early hominid from Aramis, Ethiopia. Nature 371:306–312

Whiten A, Goodall J, McGrew W, Nishida T, Reynolds V, Sugiyama Y, Tutin C,

Wrangham R, Boesch C (1999) Cultures in chimpanzees. Nature 399:682–685

Wood B, Collard M (1999) The human genus. Science 284:65–71

Wrangham R, McGrew W, de Waal F, Heltne P (1994) Chimpanzee cultures. Harvard University Press, Cambridge

Wrangham RW, Chapman CA, Clark-Arcadi AP, Isabirye-Basuta G (1996) Social ecology of Kanyawara chimpanzees: implications for understanding the costs of great ape groups. In: McGrew WC, Marchant LF, Nishida T (eds) Great ape societies. Cambridge University Press, Cambridge

Yamakoshi G, Sugiyama Y (1995) Pestle-pounding behavior of wild chimpanzees at Bossou, Guinea: a newly observed tool-using behavior. Primates 36:489–500

Zuberbuhler K, Gygax L, Harley N, Kummer H (1996) Stimulus enhancement and spread of a spontaneous tool use in a colony of long-tailed macaques. Primates 37:1–12

Part 2
Phylogeny of Perception and Cognition

2
What You See is Different from What I See: Species Differences in Visual Perception

KAZUO FUJITA

Human perceptions of environmental events often differ from what they physically are. Suppose a man who has been talking to you at a distance of 1 m leaves you and walks away for a distance of 10 m. He should shrink to 1/10 of his real size because the retinal size of his image becomes 1/10 of what it was before. In fact, this idea never occurs to us; he looks almost as tall as he did at 1 m. This well-known phenomenon is called size constancy, and gives us an impressive example that what we see is different from what the real world is. In a sense, all perception is an illusion like this.

Is this a characteristic of human perception? What about other animal species? Do they experience a similar illusory world? If they do, in what manner? These questions are not only interesting in their own right, but also seem to be essential in our attempts to understand the characteristics and evolution of human perception. We humans tend to regard our own perception as global, natural, general, and, in extreme cases, the single best way to recognize the environment. This is not necessarily true for two major reasons. One is that human perception must be constrained by the history of evolution. We have to use physical devices inherited from ancestors which are common to ourselves and our close animal relatives (historical constraints). The other reason is that human perception has been adapted to the particular way people live in this world (adaptational constraints). The reasons why humans have the type of perception we do will never be discovered unless we understand the other types of perception observed in nonhuman animals who may or may not share these constraints with humans. Historical constraints should probably be addressed by testing our phylogenetic neighbors such as nonhuman primates. However, adaptational constraints can be studied by testing a wider variety of animals who might have evolved in parallel with humans.

This chapter describes studies of two different aspects of visual perception. One is geometric illusions and the other is perceptual completion. In both studies, primates and pigeons were tested to try to assess the influence of the two types of constraints described above.

Department of Psychology, Graduate School of Letters, Kyoto University, Yoshida Honmachi, Sakyo-ku, Kyoto 606-8501, Japan

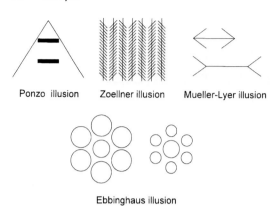

Ponzo illusion　　Zoellner illusion　　Mueller-Lyer illusion

Ebbinghaus illusion

Fig. 1. Some examples of visual illusions

1 Geometric Illusions

When we observe Fig. 1, we perceive the length (Mueller–Lyer and Ponzo illusions), the orientation (Zoellner illusion), and the size (Ebbinghaus illusion) of certain portions of the figure as greatly distorted from reality. Such geometric illusions have received much attention from students of human perception (e.g., Coren and Girgus 1978; Frisby 1979; Imai 1984), and a variety of illusory figures have been invented. These illusions are regarded as being cases in which the general characteristics of human visual perception reveal themselves in extreme ways. The overwhelming effects of geometric illusions can provide a useful tool to compare the perceptual processes of human and nonhuman animals. However, the question of how nonhuman observers perceive such illusory figures has not been asked as frequently as how human observers do it.

There have been several reports claiming that some nonhuman animals perceive several illusions. For example, Dominguez (1954) demonstrated that monkeys experienced a horizontal–vertical illusion in which vertical lines are perceived as longer than horizontal lines. Dominguez (1954) and Harris (1968) showed that monkeys perceived a breadth-of-rectangles illusion in which rectangles are judged to be taller than squares. Benhar and Samuel (1982) reported that olive baboons perceived a Zoellner illusion (see Fig. 1). Bayne and Davis (1983) showed that rhesus monkeys perceived a version of the Ponzo illusion (Fig. 1). Malott et al. (1967) and Malott and Malott (1970) reported some evidence, although it was not convincing, that pigeons perceived a Mueller–Lyer illusion (Fig. 1). However, in all of these studies, we have not learned much about the relationship between the illusory perception of these animals and that of humans. Is this relationship homology or analogy?

To answer this question, we ought to compare the effects of a variety of figural parameters among different species. We would expect that the effects of many parameters would be similar in two species if the illusory perception of one species is homologous to that of the other. Otherwise, such effects may change according to the phylogenetic relationship among the species being compared. However, if the perceptions are analogous among the species compared, the effects are not necessarily similar and will not necessarily change according to the phylogeny.

1.1 Pigeons See the Ponzo Illusion

We first investigated whether pigeons could see a common version of the Ponzo illusion (Fujita et al. 1991). Vertical black bars, of different lengths, were displayed in pairs on a touch-sensitive computer monitor. We used three bars, referred to as S, M, and L. The pigeons were trained to peck at the longer one of the two; i.e., to peck at M in the presence of S and M and to peck at L in the presence of M and L (Fig. 2a). They learned this distinction easily. The pigeons were still able to discriminate when two parallel lines or two converging lines were added outside the bars (Fig. 2b). However, when the difference in bar lengths was reduced to half that of the original bars, the pigeons had serious difficulty in recognizing the distinction when bar M was placed close to the apex of the converging lines and paired with L (Fig. 2c, bottom-left). This was not because the pigeons failed to recognize the lengths of the bars, because they did well when the bars were between horizontal lines. A reasonable explanation of this deterioration of performance is that pigeons perceived the bar close to the apex of the converging lines as being longer than the bar further away.

The procedure above seems good enough to demonstrate the illusion, but it may not be suitable to test the effects of parametric change for two reasons. One is that where there are two bars in the figure at the same time, animals always have to compare the two. A complete relational judgment is required when we change the length of the bars. Such judgment is often difficult for nonhuman animals (Carter and Werner 1978; Premack 1978). The other reason is that where there are two bars, the illusion is a result of the combined effect of the bar closer to the apex and the one further away. It may be better not to have to compare the bars in order to

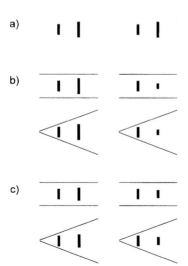

Fig. 2. Examples of stimuli used in the first experiment with pigeons on perception of the Ponzo illusion

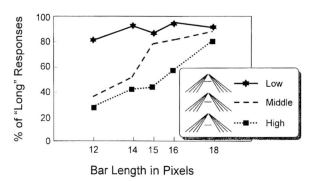

Fig. 3. Results from the second experiment with pigeons using an absolute judgment of the length of a single bar. (From Fujita et al. 1991, with permission)

address the homology–analogy issue. Therefore, in the second experiment, we devised a different procedure in which animals are required to judge the absolute length of a single bar.

The trials started with a horizontal target bar at the center of a touch-sensitive monitor. Two choice locations appeared at the bottom corners of the monitor after the subject had pecked at the target several times. A peck at one location was reinforced if the target bar was longer than a predetermined length, and a peck at the other location was reinforced if the bar was shorter than the predetermined value. There were six different lengths; three of them were "short" and the other three were "long." At first the pigeons were trained on the two extreme values. Next they were trained on all lengths with four vertical parallel lines on each side as context stimuli. The vertical location of the target bar changed from trial to trial among three possible locations, which are referred to as high-context, middle-context, and low-context conditions. The subjects were then tested in probe trials. The target bars were placed in one of the same three possible locations, but the context lines converged upward instead of being parallel. All the choice responses, long or short, were nondifferentially reinforced in these test trials.

Figure 3 shows the results of one such test. The data are the average of three birds. The proportion of responses on the "long" choice location (vertical axis) for these test stimuli was consistently highest in the low-context condition, in which the target bar appeared close to the apex of the converging lines, and lowest in the high-context condition, in which the target appeared far from the apex of the converging lines. This suggests that the pigeons perceived the target bar close to the apex of the converging lines as being longer than the others.

These results strongly suggest that pigeons perceive the Ponzo illusion as humans do. This is interesting because pigeons have poorly developed neocortices and have tectum-based vision instead (e.g., Donovan 1978). Von der Heydt et al. (1984) have suggested that V2 of the primate brain contributes to at least some of the visual illusions. Clearly V2 is not needed to induce the Ponzo illusion, although the illusion demonstrated here may simply be an analogy through parallel evolution. However, it is not immediately clear what kinds of selection pressure have favored this type of visual illusion. One possibility is that some sort of computational algorithm may be more successful if it incorporates the logic for this illusion when the visual system is required to assess the length or size of stimuli.

1.2 Inclination of the Lines and the Magnitude of the Illusion in Pigeons

In humans, the magnitude of the Ponzo illusion changes with the inclination of the context lines, reaching a maximum at about 30°–60° (Pressey et al. 1971; Pressey 1974). It also changes with the orientation of the whole figure and is slightly larger for upward-converging contexts (Leibowitz et al.1969; Brislin 1974). Is a pigeon's perception of this illusion similarly affected by these factors?

The context lines were pairs of lines converging either upward or downward. There were seven inclinations from 54.6° to 125.4°, as shown in Fig. 4. The pigeons were trained on the same absolute discrimination task as described above. Training was always done with the middle-context figures of one inclination first, followed by test sessions in which high- or low-context figures were presented as probe trials. Seven different inclinations were tested one at a time, in different orders, for all birds.

Figure 5 shows the result of tests for five pigeons with each inclination. As is clear, the proportion of responses on the "long" choice location (vertical axis) was consistently higher in the low-context condition than in the other conditions for the upward converging inclinations less than 90°. This tendency became stronger as the inclination decreased. There is no such difference for the parallel context lines (90°). On the other hand, the results for inclinations larger than 90° mirrored those

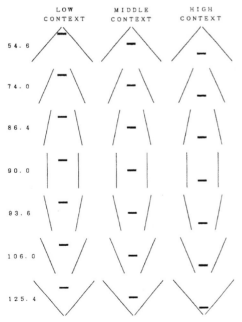

Fig. 4. Examples of stimuli used to study the effects of the inclination and orientation of context stimuli on the Ponzo illusion in pigeons. (From Fujita et al. 1993, with permission)

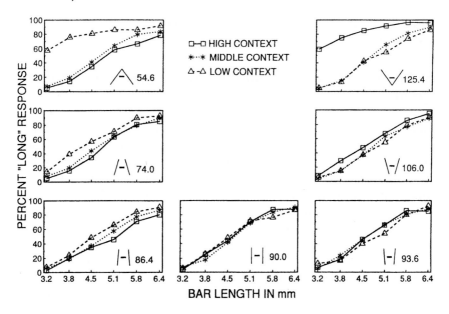

Fig. 5. Effects of the inclination and orientation of context stimuli on the magnitude of the Ponzo illusion in pigeons. The digits in each graph denote the inclination of the context. Data are the average for five birds. Others details as in Fig. 3. (From Fujita et al. 1993, with permission)

for inclinations smaller than 90°. This shows that (a) the pigeons always overestimated the length of the bars closer to the apex of the converging lines regardless of the orientation of the V-shaped context lines, (b) the magnitude of this illusion changed as the inclination of the lines changed, and (c) there is no difference between upward-converging contexts and downward-converging contexts. When we calculated the magnitude of the illusion, we found that it increased linearly with the ratio of the bar length and the gap between the bar and the context lines. These findings suggest that, at least in pigeons, this illusion is determined by simple figural parameters, and that the contribution of linear perspective implied in the illusory figure suggested by Gregory (1963) is not large.

1.3 Perception of the Ponzo Illusion in Primates

Two species of primates, rhesus monkeys and chimpanzees, were tested for their perception of the Ponzo illusion with the same task (Fujita 1997). Like the pigeons of the preceding section, they were first trained to classify the absolute length of the bars presented alone into "long" or "short" by touching either of the two choice locations. They were then transferred to the task in which upward-converging context stimuli were placed around the target bar. The training was done with the middle-context condition of the inverted-V figures shown in Fig. 6, and the probe testing was done with the high- and low-context conditions.

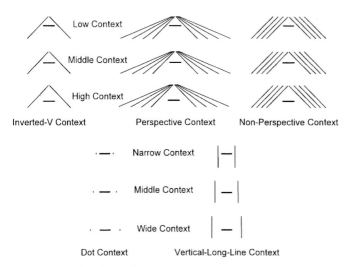

Fig. 6. Examples of stimuli used in the series of comparative studies on perception of the Ponzo illusion. (From Fujita 1997, with permission)

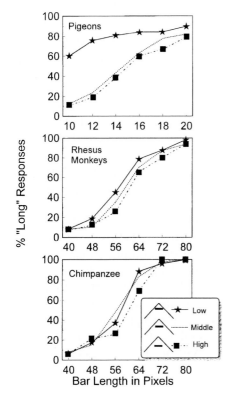

Fig. 7. A direct comparison of the Ponzo illusion between pigeons, rhesus monkeys, and chimpanzees. Other details as in Fig. 3. (From Fujita 1997, and Fujita et al. 1993, with permission)

Figure 7 shows the results of the tests. The data for the pigeons from the previous experiment using the same context are also shown at the top of the figure. The data for rhesus monkeys (middle) are an average for three individuals, and those for the chimpanzees (bottom) are for one individual. As is clear from Fig. 7, both species tended to report "long" more often for the low-context condition than for the high-context condition. This suggests that these two species of primate perceive the Ponzo illusion in the same way as pigeons and humans do. The results for these two species are qualitatively similar to the pigeon data from the experiment using the same context stimuli, although there is a substantial difference in the magnitude of the illusion across the three species. That is, the pigeons' illusion is far greater than that of the nonhuman primates. This is probably due to considerable differences in the visual systems of birds and primates. The reason for this big quantitative difference should be addressed in the future from both behavioral and physiological points of view.

1.4 Similarities and Differences in Perception of the Ponzo Illusion Across Species: Homology or Analogy?

1.4.1 Contribution of Linear Perspective

As noted earlier, it is suggested that the perspective cues implied in the illusory figures induce several types of visual illusion (e.g., Gregory 1963; Gillam 1971). Although this theory is not particularly successful in accounting for human perception of the Ponzo illusion (Humphrey and Morgan 1965; Fineman and Carlson 1973; Georgeson and Blakemore 1973; Newman and Newman 1974), it nevertheless provides a good tool to investigate the homology and analogy of the illusory perception across species.

Three primate species (humans, chimpanzees, and rhesus monkeys) and pigeons served as subjects (Fujita et al.1991; Fujita 1997). The stimuli had two types of contexts: the perspective context and the nonperspective context (see Fig.6). Both types of stimuli had four lines on each side of the target bar. These lines converged to one point in the perspective context; in the nonperspective context they were parallel. The former gives a stronger impression of depth to human eyes, and hence is expected to be stronger at inducing the illusion if the perspective implied in the stimulus figure is important in producing the illusion.

The two types of stimulus appeared randomly in the session. For both types of stimulus, the nonhuman subjects were trained with the middle-context condition. They were then tested with the high- and low-context conditions. Human subjects followed a 2-up/2-down titration schedule of bar length in which the bar was extended after two consecutive choice responses of "short," and shortened after two consecutive choices of "long." Six independent titration schedules were run for each type (perspective or nonperspective) and location (high, middle, and low) of the context. In fact, this titration session was conducted with all the five types of context shown in Fig. 6. Therefore, in humans 15 independent titration schedules were run at the same time.

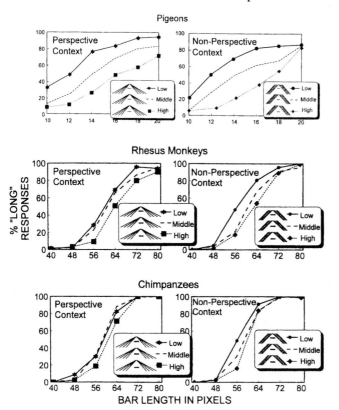

Fig. 8. The effects of the strength of the linear perspective implied in an illusory figure on the magnitude of the Ponzo illusion. Details as in Fig. 3. (From Fujita 1997, and Fujita et al. 1991, with permission)

Figure 8 shows the results of the tests for nonhuman subjects. As is clear from the figures, all the species tended to report "long" more often for the low-context condition than for the high-context condition. This was true for both perspective and nonperspective contexts. This shows that the three species perceived the illusion irrespective of the type of context. However, there seems to be no difference in the magnitude of the illusion between the two types of context for any species. Apparently, the stronger linear perspective in the perspective context had no effect of enhancing the illusion in these species.

Figure 9 shows the results for the human subjects. In this case, the magnitude of the illusion is shown as a function of the types of context tested at the same time. The magnitude of the illusion was calculated based on the bar length averaged across the last ten titration trials for each type and condition of the stimuli. The lines denote individual subjects, and the histograms denote the average for the six humans.

As shown in the figure, there is very little difference between the perspective and nonperspective contexts. This was consistent with the data from nonhuman animals shown in Fig. 8. In fact, the magnitude of the illusion is comparable to that of

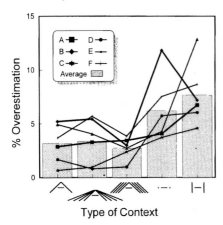

Fig. 9. Magnitude of the Ponzo illusion for human observers for the stimuli shown in Fig. 6. The type of stimulus is shown on the horizontal axis and the magnitude of the illusion is shown on the vertical axis. The line graphs are for individual subjects and the histogram shows the average for each context. (From Fujita 1997, with permission)

the inverted-V context with a single line on each side. That is to say, the linear perspective implied in the stimulus figure had no effect of enhancing this illusion in any species tested here. Although this does not necessarily negate any effect of linear perspective for any types of Ponzo illusion, the important finding here is that there is no species differences in the effect of this parameter as systematically modified in this experiment. This result may favor the homology of the illusory perception among the species tested, but we have to be cautious about reaching this conclusion because, as before, the magnitude of the illusion is very different between pigeons and the three primate species.

1.4.2 Contribution of the Size of the Gap Between Target Bar and the Context

Another parameter that has been suggested to induce the Ponzo illusion is the proximity of the contours (Fisher 1969, 1973). In literature relating to humans, the strength of the illusion is little affected by replacing the converging lines of the standard Ponzo figure with four separate dots placed on both sides of the two target bars (Yamagami 1978).

Three primate species, humans, chimpanzees, and rhesus monkeys, served as subjects in this experiment (Fujita 1997). A new type of context, labeled the dot context, was introduced (see Fig. 6). This context had two very short vertical lines on each side of the target bar. The size of the gap exactly matched those of the inverted-V context and the wide-, middle-, and narrow-contexts corresponded to the high-, middle-, and low-contexts of the inverted-V context.

Nonhuman subjects were trained on the middle-context conditions of both the inverted-V and the dot contexts. They were then tested with the high and low contexts of the inverted-V and the wide and narrow contexts of the dot context. Human subjects were tested as described earlier, including a fifth type of context labeled the vertical-long-line context (see Fig. 6). This type of context has two long lines instead of the short lines of the dot context.

Figure 10 shows the results for nonhumans, i.e., rhesus monkeys and two indi-

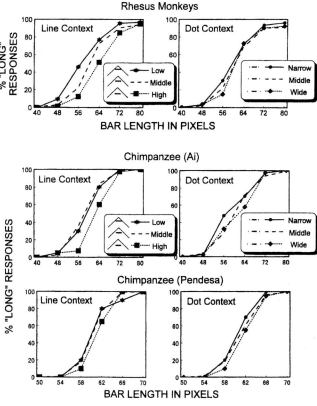

Fig. 10. The effects of a dot context on the magnitude of the Ponzo illusion. The left-hand column is for an inverted-V context and the right-hand panel is for a dot context. Other details as in Fig. 3. (From Fujita 1997, with permission)

vidual chimpanzees. Because the two chimpanzee subjects differed slightly in their choice of bar length because of an extremely accurate performance by one of them (Pendesa), the data for the individual subjects are shown. As before, the data for both species suggested a clear perception of the illusion in the inverted-V context. However, for the dot context, there was very little evidence that the illusion was observed by rhesus monkeys. Only chimpanzees showed that they had noticed an illusion in the dot context. The magnitudes of the illusions in the dot and the inverted-V contexts were comparable for the chimpanzees. In humans (see Fig. 8), the magnitude of the illusion in the dot context was much larger than that in the inverted-V context. It is possible that the rhesus monkey subjects did not perceive the illusion in the dot context because the two short lines were too weak. However, this does not seem likely because in humans, at least, the effect of the dot context was comparable to that of the vertical-long-line context. There was an indication that the size of the gap between the target bar and the lines or dots used to induce the illusion is related to the phylogeny of the three primate species tested. This seems to support the view that perception of the Ponzo illusion is homological

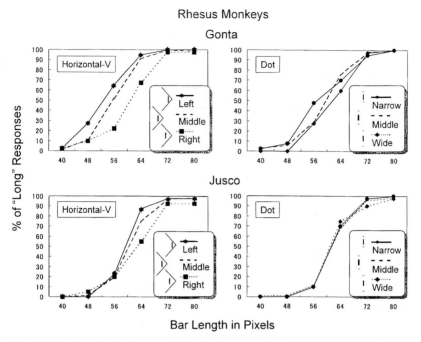

Fig. 11. Illusions in the inverted-V and dot contexts rotated 90° clockwise for individual rhesus monkeys. The left-hand column is for the rotated inverted-V (labeled horizontal-V) context, and the right-hand column is for the rotated dot context. Other details as in Fig. 3

among these primates, and that the effects of some of the parameters contributing to this illusion change according to the phylogeny of the viewer.

A clue as to the source of this species difference might be the spatial anisotropy of the interaction among the figural elements. We perceive the illusion because there is an interaction between the target bar and the context in which it appears. This interaction may not be homogeneous for all orientations. For example, vertical Ponzo figures are more powerful than horizontal figures (Brislin 1974). Such spatial anisotropy may be different across species. If rhesus monkeys are less susceptible to the interaction between the elements located near each other horizontally, they may not perceive the illusion in the dot context. To address this hypothesis, the subjects were tested in the dot and inverted-V contexts rotated through 90° clockwise.

Figure 11 shows the result of this test for each individual rhesus monkey. One of the monkeys (Gonta) showed that it perceived an illusion in the dot context, but the second monkey (Jusco) did not. This result only partly supports the hypothesis outlined above.

What is still puzzling is that the results for the two individual chimpanzees are completely different. As shown in Fig. 12, one of the chimpanzees (Ai) now saw no illusion in the inverted-V context, but did perceive a substantial illusion in the dot context. On the other hand, the second chimpanzee (Pendesa) still saw the illusion

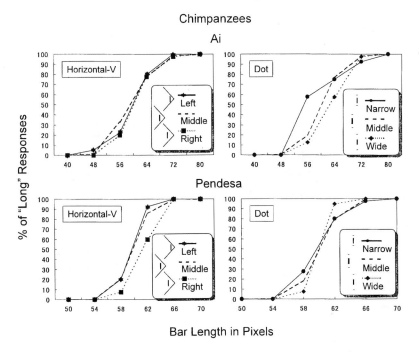

Fig. 12. Illusions in the inverted-V and dot contexts rotated 90° clockwise for individual chimpanzees. Details as in Fig. 11

in the inverted-V context and no illusion in the dot context. There seem to be large individual differences in the effects of the orientation of the figures. We are still unable to provide a satisfactory answer to the question of the source of the species differences in the effects of the gap.

1.4.3 The Contribution of Natural Perspective

The Ponzo illusion is stronger when the two bars are located on a photograph of a broad highway, since this provides a stronger impression of perspective than when they are located between simple converging lines. The effect of such natural perspective was compared between rhesus monkeys and humans in the next experiment.

A photograph of a natural highway was prepared. In one type of stimulus (labeled "upright picture + lines"), this photograph was superimposed on the inverted-V Ponzo figure. In the second type ("upside-down picture + lines"), the upside-down image of the photograph was superimposed on the Ponzo figure. The photograph and the lines conform to one unitary context in these two cases. In the third case ("upright picture only"), an isolated target bar was placed in the photograph. In the fourth case ("upside-down picture only"), an isolated target bar was placed in the upside-down version of the photograph.

Rhesus monkeys were trained on the middle-context condition and then tested with the others. Humans received one session of the same 2-up/2-down titration

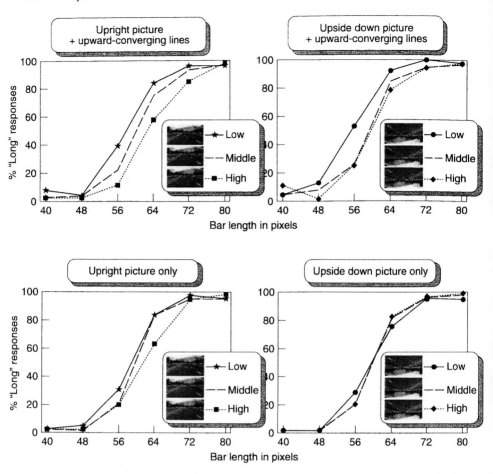

Fig. 13. Effects of perspective suggested by a natural photograph of a scene for rhesus monkeys. Details as in Fig. 3. (From Fujita 1996, with permission)

during which the four types of stimulus described above and the inverted-V context appeared.

Figure 13 shows the results for rhesus monkeys. The subjects perceived an illusion with the two types of stimulus with lines (upright picture + lines and upside-down picture + lines), irrespective of the orientation of the perspective. This seems to suggest that the rhesus monkeys were not susceptible to the natural perspective in the photograph. However, the same subjects also perceived an illusion, although somewhat weaker, for the stimulus of an upright picture with no lines (upright picture only). No illusion was perceived with the stimulus of the upside-down picture with no lines (upside-down picture only). This means that the natural perspective provided by a photograph has the effect of inducing an illusion, but that this effect is overshadowed by two converging lines as the context.

Humans, on the other hand, perceived a strong illusion with the upright picture

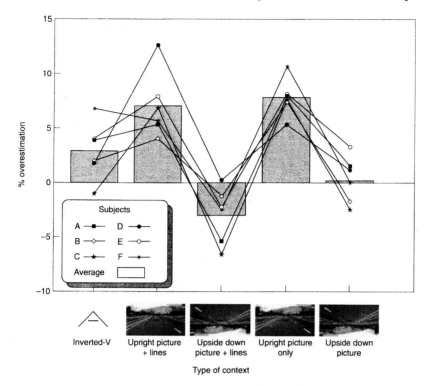

Fig. 14. Effects of perspective suggested by a natural photograph of a scene for humans. Details as in Fig. 9. (From Fujita 1996, with permission)

(Fig. 14) irrespective of the presence or absence of the lines (upright picture + lines and upright picture only), and a moderate illusion in the inverted-V context.

In combination, these results suggest that both the line context and the natural perspective have the effect of inducing the illusion in both species, but that the relative strength of these effects differs between the two species. In humans, the effect of natural perspective is stronger than the line context, while the opposite is true in rhesus monkeys. This again suggests that the illusion addressed here is homologous between the two species, but that there is a quantitative difference in the effect of the figural/photographic parameters contributing to the illusion.

1.5 Summary of the Illusory Perception in Humans and Nonhumans

Four species of animals, three primates and one bird, were tested for their perception of the Ponzo illusion. Figure 15 summarizes the results of the experiments in which direct comparisons were made between species. Both similarities and differences were found.

One thing the species had in common was that they all clearly perceived the

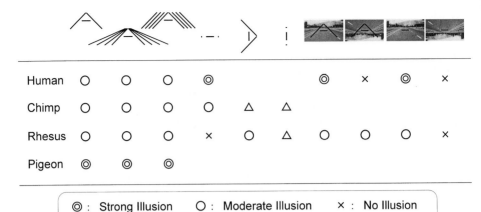

Human	○	○	○	◎			◎	×	◎	×
Chimp	○	○	○	○	△	△				
Rhesus	○	○	○	×	○	△	○	○	○	×
Pigeon	◎	◎	◎							

◎ : Strong Illusion   ○ : Moderate Illusion   × : No Illusion
△ : Not Clear / Individual Difference

Fig. 15. Summary of the series of comparative studies on perception of the Ponzo illusion. The top row shows the figures used, and the lower four rows show the strength of the illusion

Ponzo illusion. Another was that in no species did the stronger perspective provided by the converging lines enhance the illusion.

One difference across species was that in pigeons the magnitude of the illusion was far greater than in primates. Another difference is that the effect of the gap between the target bar and the context changes according to phylogeny. A third point is that the relative contribution of a line context and a photographic perspective is opposite in rhesus monkeys and in humans.

Although we have not filled all the cells in Fig. 15, this series of experiments has given us some insights into how this illusory perception has evolved, and how the illusion starts in the brain. First, it is not difficult to see the continuity of this perception among primates. It is probably homologous among the primate species tested, with slight quantitative modifications of the effects of some parameters contributing to the illusion. These differences may have resulted from adaptations of the species to their niche, although we cannot yet identify what that might have been. Second, it is interesting that pigeons, which have a completely different visual system, perceive a similar illusion. Clearly well-developed neocortices are not needed to perceive this illusion. However, this does not necessarily mean that primate perception of this illusion occurs in a neural area other than the neocortices, because it may be a consequence of the same fruitful computational algorithm that can be realized in other parts of the brain. Pigeons have been shown to have a variety of other cognitive skills which are comparable to those shown by humans (e.g., Herrnstein et al. 1976; Straub et al. 1979; Maki and Hegvik 1980; Wright et al. 1985; Blough and Blough 1997; Wright 1997). Perhaps the pigeons' algorithm allowing them to perceive this illusion also means that they adapt to their environment better.

2 Perceptual Completion

When a pacman figure is placed adjacent to a triangle, humans perceive a triangle on a complete circle. We spontaneously complete the portion "occluded" by the triangle. Some forms of the ability to recognize the occluded part of an object develop very early in human infants. For example, Kellman and Spelke (1983) showed that 4-month-old infants recognized two rods moving in concert behind the occluder as a unitary rod. This is called perception of *object unity*.

This ability to complete invisible portions of an object seems essential for any animals because in the natural environment objects are rarely visible in full. However, our knowledge of this ability in nonhumans is very limited. We still do not know how widespread the ability is in the animal kingdom or how it has evolved. Several previous reports have claimed to have established this ability in several species of nonhumans; Kanizsa et al. (1993) reported that mice showed choice responses consistent with perceptual completion after being trained to discriminate complete and incomplete figures. Lea et al. (1996) and Regolin and Vallortigara (1995) used imprinting procedures to show that domestic chicks perceive object unity.

On the other hand, pigeons have repeatedly been reported to have failed to complete occluded portions. Cerella (1980) trained pigeons to discriminate a triangle from other figures. When the pigeons were tested with an incomplete triangle touching a rectangle, they consistently reported the figure as not-triangle. Using a similar procedure, Sekuler et al. (1996) showed that pigeons failed to complete a circle and a rectangular figure.

The following section describes two types of study on perceptual completion ability in pigeons and primates.

2.1 Completion of the Still Figures

In the pigeon studies described above, the subject had to complete a figure to form a "good" shape, as humans do. However, it is possible that, for example, a pacman figure may be as "good" as a complete circle for pigeons. That is to say, procedures that require subjects to complete the occluded portions as humans do could be unfairly testing perceptual completion ability in nonhuman animals.

As long as we ask animals to report what the figure is, we are not liberated from the "good shape" problem mentioned above. I used a procedure in which animals judge the length or the size of the figure instead. The idea is based on one of the illusory figures of Kanizsa (1979). When we observe two small rectangles of the same size, one of which touches a large rectangle, we overestimate the size of the one touching the large rectangle. This is because, as Kanizsa explains, humans perceive a figure which touches another as continuing behind the larger occluder; hence they complete the "occluded" portion. Note that this is a judgment of the size, not the shape.

Rhesus monkeys and pigeons were trained on the same bar length discrimination task that was used in the Ponzo study (Fujita 2001). Briefly, their task was to

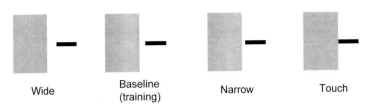

Fig. 16. Examples of stimuli used in the study of perceptual completion of still figures. (From Fujita in press, with permission)

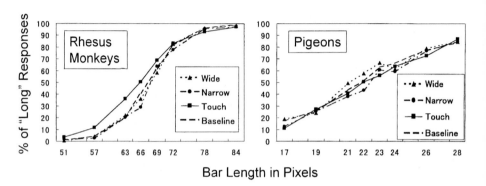

Fig. 17. A direct comparison between rhesus monkeys and pigeons of perceptual completion of a still figure. (From Fujita in press, with permission)

classify black horizontal bars into "long" or "short," based on the absolute length of the bars presented at the center of the display, by touching one of two choice locations at the bottom. Sixteen different bar lengths were used.

After this discrimination training was complete, a large gray rectangle was added either to the right or the left of the target bar. The size of the gap between the target bar and the rectangle was fixed during this stage ("baseline" condition in Fig. 16). Then the subjects were tested in probe trials with stimuli which had different-sized gaps from the baseline condition. With one stimulus, called the "wide" condition, the gap was wider than the baseline. With the second stimulus, called the "narrow" condition, it was narrower. In the last stimulus, called the "touch" condition, there was no gap, i.e., the target bar touched a rectangle in this condition. Both choice responses, i.e., "long" or "short," were nondifferentially reinforced in the test trials. If the subjects perceive the same illusion, supposedly as a consequence of completing the occluded portions, the subjects will choose "long" more often in the touch condition than in the other conditions.

Figure 17 shows the results of this test. The left panel is an average for two rhesus monkeys, and the right panel is an average for three pigeons. The horizontal axis is the bar length and the vertical axis is the proportion of "long" choices. As is clear from the figure, rhesus monkeys showed a consistent bias for more "long" choices only in the touch condition. On the other hand, pigeons showed no such bias. That is, the rhesus monkeys overestimated the length of the bar touching a rectangle but the pigeons did not. This suggests that rhesus monkeys completed the "occluded" portion of the target bar behind the rectangle but pigeons did not. As this procedure does not require the same Gestalt principle of "good shape," the failure by pigeons to complete the target bar in this situation may suggest that they do not even recognize continuation of the target bar behind the rectangle. There may be a substantial difference between primates and pigeons in the ability to complete static figures.

2.2 Completion of Moving Figures: Recognition of Object Unity in Nonhumans

As noted above, human infants start to perceive a unity in two rods moving in concert at 4 months old. In adults and older infants, this perception occurs without common motion. That is, they recognize a unity in two static rods aligned in good continuity. If the same sequence of behavioral development repeats in the phylogeny, pigeons may show some ability to unite figures having common motion.

The subjects were one chimpanzee (Sato et al. 1997) and three pigeons (Yamanaka and Fujita 1999). First, they were trained to match a complete rod to a complete rod and two broken rods to two broken rods in a two-choice matching-to-sample procedure. Next, they were trained with the sample stimuli moving from left to right at a constant speed. After this, a narrow horizontal belt was superimposed on the screen. During this training, the belt never occluded any portions of the sample stimuli.

After the subjects had shown a consistent matching performance, they were

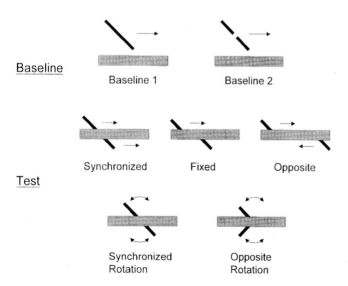

Fig. 18. Examples of stimuli used to test perceptual completion of moving figures, or perception of object unity. The *top line* shows the training stimuli and the *lower two lines* show the test stimuli

tested with the three types of test stimuli shown in the middle panel of Fig. 18, which had the central portion of the rods occluded by a horizontal belt. The top portion of the rod moved from left to right at the same constant speed in all three test stimuli. In the "synchronized" condition, the bottom portion moved in concert with the top portion. In the "fixed" condition, the bottom portion stayed still at the center. In the "opposite" condition, the bottom portion moved from right to left at the same speed as the top portion. All the matching responses (unitary rod or broken rods) are nondifferentially reinforced in these test trials.

The chimpanzee was given an additional test with the two types of stimulus shown in the bottom panel of Fig. 18. One was "synchronized rotation," in which top and bottom portions swung repeatedly left and right in concert; the other was "opposite rotation", in which the two portions swung in opposite directions to each other.

The proportion of "unity" choices in each of the test trials is shown in Fig. 19. The chimpanzee overwhelmingly chose "unity" in the synchronized condition and the synchronized rotation condition, and chose "nonunity" in the other conditions. On the other hand, all the pigeons chose "nonunity" in all three conditions

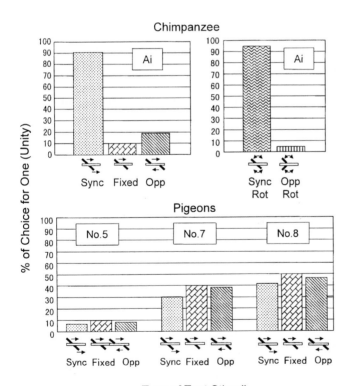

Fig. 19. Proportion of choice responses for "unity" for each type of test stimulus in a chimpanzee (*top*) and individual pigeons (*bottom*). The type of test stimulus is shown on the horizontal axis. (From Sato et al. 1997, and Yamanaka and Fujita 1999, with permission)

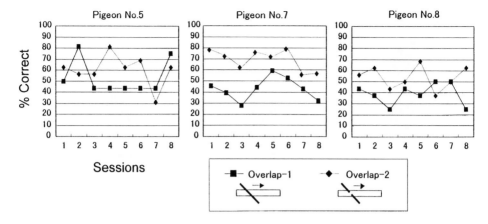

Fig. 20. Proportion of correct responses during the first eight sessions of training to match overlapping stimuli, in which bars moved this side of the horizontal belt, for individual pigeons. (From Yamanaka and Fujita 1999, with permission)

tested. Interestingly, the order of the strength of this choice was exactly the same among the three pigeons, and the choice of "nonunity" was strongest in the synchronized condition. Apparently the pigeons recognized the synchronized condition as being most similar to the broken rods. This is, in a sense, reasonable if the pigeons perceive the top and bottom portions as separate. In any case, the pigeons did not show any evidence of seeing object unity, while the chimpanzee clearly did.

The failure of the pigeons might be because they recognized the test situation in which the horizontal belt came on at the center of the sample rods as completely novel. This novelty might have prevented the pigeons from showing any generalization from the original matching to sample. After the test described above, we trained the pigeons with stimuli in which the unitary rod or the broken rods moved on the subjects side of the screen. These conditions are called "overlap-1" and "overlap-2," respectively.

Figure 20 shows the results of the first eight sessions of this training. All the pigeons performed moderately well for the overlap-2 condition, but they did not match the overlap-1 condition to the unitary rod. Their performances with the overlap-1 condition gradually improved, but they never reached the same level as the baseline condition even after several dozen sessions. We finally gave up and tested them with the synchronized condition.

Figure 21 shows the results of this final test. In the synchronized condition, the pigeons' choices of "unity" were still well below the chance level. This result seems strange, but may be understandable if the pigeons perceived the portions above, on, and below the horizontal belt as separate figures. It is likely that the pigeons may have divided the figure at the crossings. Again, no evidence for completion was obtained with pigeons.

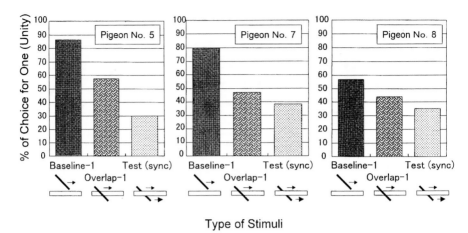

Fig. 21. Proportion of choice responses for "unity" for each type of training and test stimulus in pigeons after training with the overlapping stimuli. The type of stimulus is shown on the horizontal axis. (From Sato et al. 1997, and Yamanaka and Fujita 1999, with permission)

2.3 Summary of Perceptual Completion in Nonhumans

In the first experiment in which completion of static figures was tested, rhesus monkeys, like human observers, showed clear evidence for completion, but pigeons failed to do so. They do not even seem to recognize that one figure continues behind another. In the second experiment, in which completion of moving figures was tested, the chimpanzee united the two moving rods which had consistent motion, but the pigeons again showed no evidence for completion. These results are consistent with those of previous studies (Cerella 1980; Sekuler et al.1996). A plausible explanation of the repeated failure of the pigeons seems to be that they may perceive the figures as being divided at the crossings.

As noted earlier, domestic chicks have been shown to perceive object unity (Regolin and Vallortigara 1995; Lea et al. 1996). This again suggests that well-developed neocortices are not required to complete occluded portions. Interestingly, some previous reports have demonstrated that recognition of object permanence may be poorly developed in pigeons. Plowright et al. (1998) showed that pigeons would not follow a truck with food on it when it went into a short tunnel and was temporarily invisible, whereas mynahs did. Ring doves, a species closely related to pigeons, have also been shown not to look for food if the food is occluded in front of their eyes before they started to move toward it (Dumas and Wilkie 1995). Several psittacine species, including parrots, macaws, cockatiels, and parakeets, show better performances in these object-permanence tasks (Pepperberg and Funk 1990; Funk 1996). These differences in ability to find invisible objects may be partly understood by considering the feeding habits of the birds. Pigeons and doves feed mostly on grains and seeds on the ground. On the other hand, domestic fowls eat worms as well as grains, and mynahs eat insects and berries. Psittacine birds feed on seeds, but larger species crack open hard nuts with their beak. This predatory and extractive feeding requires better recognition of invisible food. Systematic testing of a variety of species with different feeding habits might give some support to this theory.

Although there seems to be a huge difference in the capacity for perceptual completion between primates and pigeons, it is still premature to conclude that pigeons never complete. This is because pigeons have been shown to have difficulty in recognizing depth from two-dimensional stimuli (Blough 1984; Cerella 1990) unless they are specially trained to recognize rotating two-dimensional representations of objects (Cook and Katz 1999). We may need to test pigeons with real three-dimensional objects.

3 Summary and Conclusion

We have investigated two aspects of visual perception, illusion and perceptual completion. We have found that the Ponzo illusion is commonly observed in primates and pigeons. As we compared the effects of several figural parameters that may contribute to the illusion, however, we found points which were common to all the species tested and points which differed across species. That is, the stronger perspective implied in the figural context did not enhance the illusion in any species. On the other hand, the magnitude of the illusion is overwhelmingly larger in pigeons than in primates. The effects of the size of the gap between the target and the context changed with the phylogeny in primates. Photographic perspective induced the illusion in both monkeys and humans, but the relative contribution of this and the line contexts differed between the two species. These findings suggest that observation of the Ponzo illusion is probably homological, at least among primates, but that at the same time the relative contribution of parameters inducing the illusion may change with the phylogeny and, probably, with adaptation to their niche. The results of the same tests with pigeons suggests that well-developed neocortices are not needed in order to perceive this illusion. However, this does not mean that the primates' illusion occurs in an area other than the neocortices. Rather, it implies that a similar computational algorithm allowing this illusion may be fruitful for visual information processing and easily realizable both in brains with well-developed neocortices and in those without them.

We next found that the processes of perceptual completion may differ greatly between primates and pigeons. In the first experiment with still figures, rhesus monkeys readily completed portions of one figure occluded by another; pigeons failed to do this. They do not even seem to recognize that one figure continues behind another. In the second experiment, in which completion of figures having common motion was tested, a chimpanzee perceived object unity as humans do, but again the pigeons failed to do so. They even chose nonunity for the unitary rod moving "this side" of the horizontal belt (overlap-1 condition). The pigeons seemed to perceive the figures as divided at the crossings. Given that at least one species of bird, domestic chicks, have been shown to perceive object unity, it is clear that perceptual completion is not beyond the capacity of the brain of birds. This ability may be related to the feeding habits of the species. This suggestion, if correct, means that the fundamental way in which different species of birds recognize environments may vary according to the adaptation of the species to their niche.

In summary, the two aspects of visual information processing addressed in this paper show how species have evolved their own way of perceiving their environment under historical and adaptational constraints. Our knowledge is still limited to a few species and perceptual aspects, but as it accumulates we will probably be able to understand why our own perception has developed in the way it has.

References

Bayne KAL, Davis RT (1983) Susceptibility of rhesus monkeys (*Macaca mulatta*) to the Ponzo illusion. Bull Psychonomic Soc 21:476–478

Benhar E, Samuel D (1982) Visual illusions in the baboon (*Papio anubis*). Anim Learn Behav 10:115–118

Blough DS (1984) Form recognition in pigeons. In: Roitblat HL, Bever TG, Terrace HS (eds) Animal Cognition, Lawrence Erlbaum, Hillsdale, pp 277–289

Blough DS, Blough PM (1997) Form perception and attention in pigeons. Anim Learn Behav 25:1–20

Brislin RW (1974) The Ponzo illusion: additional cues, age, orientation, and culture. J Cross Cult Psychol 5:139–61

Carter DE, Werner TJ (1978) Complex learning and information processing by pigeons: a critical analysis. J Exp Anal Behav 29:565-01

Cerella J (1980) The pigeon's analysis of picture. Pattern Recognition 12:1–6

Cerella J (1990) Shape constancy in the pigeon: the perspective transformations decomposed. In: Commons ML, Herrnstein RJ, Kosslyn SM, Mumford DB (eds) Quantitative analyses of behavior, vol 8. Lawrence Erlbaum, Hillsdale, pp 145–163

Cook RG, Katz JS (1999) Dynamic object perception by pigeons. J Exp Psychol: Anim Behav Process 25:194–210

Coren S, Girgus JS (1978) Seeing is deceiving: the psychology of visual illusions. Lawrence Erlbaum, Hillsdale

Dominguez KE (1954) A study of visual illusions in the monkey. J Genet Psychol 85:105–127

Donovan WJ (1978) Structure and function of the pigeon visual system. Physiol Psychol 6:403–437

Dumas C, Wilkie DM (1995) Object permanence in ring doves (*Streptopelia risoria*). J Comp Psychol 109:142–150

Fineman MB, Carlson J (1973) A comparison of the Ponzo illusion with a textural analogue. Percept Psychophys 14:31–33

Fisher GH (1969) Towards a new explanation for the geometrical illusions. I. The properties of contours which induce illusory distortion. Br J Psychol 60:179–185

Fisher GH (1973) Towards a new explanation for the geometrical illusions. II. Apparent depth or contour proximity? Br J Psychol 64:607–621

Frisby JP (1979) Seeing: illusion, brain, and mind. Oxford University Press, Oxford

Fujita K (1996) Linear perspective and the Ponzo illusion: a comparison between rhesus monkeys and humans. Jpn Psychol Res 38:136–145

Fujita K (1997) Perception of the Ponzo illusion by rhesus monkeys, chimpanzees, and humans: similarity and difference in the three primate species. Percept Psychophys 59:284–292

Fujita K (2001) Perceptual completion in rhesus monkeys (*Macaca mulatta*) and pigeons (*Columba livia*). Percept Psychophys 63:115–125

Fujita K, Blough DS, Blough PM (1991) Pigeons see the Ponzo illusion. Anim Learn Behav 19:283–293

Fujita K, Blough DS, Blough PM (1993) Effects of the inclination of context lines on perception of the Ponzo illusion by pigeons. Anim Learn Behav 21:29–34

Funk MS (1996) Development of object permanence in the New Zealand parakeet (*Cyanoramphus auriceps*). Anim Learn Behav 24:375-383

Georgeson MA, Blakemore C (1973) Apparent depth and the Mueller-Lyer illusion. Perception 2:225-234

Gillam B (1971) A depth processing theory of the Poggendorff illusion. Percept Psychophys 10(4A):211-216

Gregory RL (1963) Distortion of visual space as inappropriate constancy scaling. Nature 199:678-680

Harris AV (1968) Perception of the horizontal-vertical illusion by stump-tailed monkeys. Radford Rev 22:61-72

Herrnstein RJ, Loveland DH, Cable C (1976) Natural concepts in pigeons. J Exp Psychol: Anim Behav Process 2:285-302

Humphrey NK, Morgan MJ (1965) Constancy and the geometric illusions. Nature 206:744-745

Imai S (1984) Sakushi zukei: miekata no shinrigaku (Figures of optical illusions) (in Japanese). Science-sha, Tokyo

Kanizsa G (1979) Organization in vision: essays on Gestalt perception. Praeger, New York

Kanizsa G, Renzi P, Conte S, Compostela C, Guerani L (1993) Amodal completion in mouse vision. Perception 22:713-721

Kellman PJ, Spelke ES (1983) Perception of partly occluded objects in infancy. Cogn Psychol 15:483-524

Lea SEG, Slater AM, Ryan CME (1996) Perception of object unity in chicks: a comparison with the human infant. Infant Behav Dev 19:501-504

Leibowitz H, Brislin R, Perlmutter L, Hennessy R (1969) Ponzo perspective illusions as a manifestation of space perception. Science 166:1174-1176

Maki WS, Hegvik DK (1980) Directed forgetting in pigeons. Anim Learn Behav 8:567-574

Malott RW, Malott MK (1970) Perception and stimulus generalization. In: Stebbins WC (ed) Animal psychophysics. Plenum, New York, pp 363-400

Malott RW, Malott MK, Pokrzywinski J (1967) The effects of outward-pointing arrowheads on the Mueller-Lyer illusion in pigeons. Psychonomic Sci 9:55-56

Newman CV, Newman BM (1974) The Ponzo illusion in pictures with and without suggested depth. Am J Psychol 87:511-516

Pepperberg IM, Funk MS (1990) Object permanence in four species of psittacine birds: an African gray parrot (*Psittacus erithacus*), an Illiger mini macaw (*Ara maracana*), a parakeet (*Melopsittacus undulatus*), and a cockatiel (*Nymphicus hollandicus*). Anim Learn Behav 18:97-108

Plowright CMS, Reid S, Kilian T (1998) Finding hidden food: behavior on visible displacement tasks by mynahs (*Gracula religiosa*) and pigeons (*Columba livia*). J Comp Psychol 112:13-25

Premack D (1978) On the abstractness of human concepts: why it would be difficult to talk to a pigeon. In: Hulse SH, Fowler H, Honig WK (eds) Cognitive processes in animal behavior. Erlbaum, Hillsdale, pp 423-451

Pressey A (1974) Age changes in the Ponzo and filled-space illusions. Percept Psychophys 15:315-319

Pressey A, Butchard N, Scrivner L (1971) Measuring the Ponzo illusion with the method of production. Behav Res Methods Instrum 6:424-426

Regolin L, Vallortigara G (1995) Perception of partly occluded objects by young chicks. Percept Psychophys 57:971-976

Sato A, Kanazawa S, Fujita K (1997) Perception of object unity in a chimpanzee (*Pan troglodytes*). Jpn Psychol Res 39:191-199

Sekuler AB, Lee JAJ, Shettleworth SJ (1996) Pigeons do not complete partly occluded figures. Perception 25:1109-1120

Straub RO, Seidenberg MS, Bever TG, Terrace HS (1979) Serial learning in the pigeon. J Exp Anal Behav 32:137-148

von der Heydt R, Peterhans E, Baumgartner G (1984) Illusory contours and cortical neuron responses. Science 224:1260-1262

Wright AA (1997) Concept learning and learning strategies. Psychol Sci 8:119-123

Wright AA, Santiago HC, Sands SF, Kendrick DF, Cook RG (1985) Memory processing of serial lists by pigeons, monkeys, and people. Science 229:287–289

Yamagami A (1978) Two kinds of apparent size distortion in the Ponzo illusion (in Japanese with English summary). Jpn J Psychol 49:273–279

Yamanaka R, Fujita K (1999) Perception of object unity by pigeons. Paper Presented at the 59th Annual Meeting of the Japan Society for Animal Psychology, May, Kanazawa, Japan

3
Investigating Visual Perception and Cognition in Chimpanzees (*Pan troglodytes*) Through Visual Search and Related Tasks: From Basic to Complex Processes

Masaki Tomonaga

1 Introduction

Our visual environment is filled with an abundance of objects. When we see these objects, our visual system initially processes their features in a parallel manner and then integrates these features into objects using selective attention. It has been about 20 years since Treisman first proposed this "feature integration theory" (Treisman and Gelade 1980). Visual search and texture segregation tasks have revealed many interesting phenomena directly related to the feature integration theory: "pop-out", search asymmetry, conjunction search, etc. Based on these findings, this theory was modified (Treisman and Sato 1990), and alternative models have also been proposed (e.g., guided search models, Wolfe 1994a; Wolfe et al. 1989).

These models are also linked with findings from neurophysiological studies using nonhuman animals (especially nonhuman primates; e.g., Treisman et al. 1990). There is, however, relatively less behavioral evidence to assess the validity of these models in nonhuman animals. If these models are valid for human visual systems, do they also have "phylogenetic" validity? Are these models also valid for nonhuman primates, the closest relatives of humans? Furthermore, were we to find evidence for these models in animals phylogenetically far removed from humans (such as birds), how would we interpret these results?

In general, these questions have been the main topics of comparative cognitive science. In fact, visual search or texture segregation tasks functionally identical to those for humans, or modified versions of these tasks for animals, have been employed to obtain data that are directly comparable to human data. Pigeons (Allan and Blough 1989; Blough 1977, 1979, 1989, 1992; Blough and Franklin 1985; Cook 1992a, b; Cook et al. 1996), cats (Wilkinson 1986, 1990), macaques (Dursteler and von der Heydt 1992; Fujita and Kanazawa 1994), and baboons (Deruelle and Fagot 1998) have shown remarkable differences from, as well as similarities to, humans.

I have also been conducting research on chimpanzees using visual search tasks. In this chapter, I present a summary of these results, and discuss the properties of

Primate Research Institute, Kyoto University, 41 Kanrin, Inuyama, Aichi 484-8506, Japan

visual perception and cognition in chimpanzees from the perspective of comparative cognitive science.

2 Initial Training of Visual Search Tasks: From Matching-to-Sample to Odd-Item Search

As noted in the introduction, researchers in comparative studies have modified visual search tasks from the standard procedures employed in human studies. In a standard task, human subjects are presented with a search display in which several items are presented. When this display contains the *target* (as defined by the experimenter) among distractors, the subjects make a "yes" response by pressing one key or reporting verbally (Treisman and Gormican 1988). When the display contains only *distractors*, they make a "no" response by pressing the other key or responding verbally. This type of procedure is identical to the "go left/go right" discrimination, or conditional position discrimination (Mackay 1991). To date, there have been no visual-search studies with nonhuman animals in which this type of procedure has been employed. This may be due to the relative difficulty of training animals in this type of conditional discrimination.

Instead, animal researchers have used relatively "easy" visual search tasks. For example, Blough (1977, 1979) trained pigeons to *peck* the target among distractors on the search display. He did not, however, show the "target-absent" display. This task can easily be modified from the simple two-item simultaneous discrimination. In addition, he also trained pigeons in the "odd-item search," a variant of oddity discrimination (Blough 1989). Deruelle and Fagot (1998, see Chapter 4 in this volume) trained baboons to respond when the search display presented a target (GO), but not to respond when the display did not present a target (NO-GO). This task is a variant of the GO/NO-GO discrimination that is frequently used in neurophysiological studies.

In the initial phase of training chimpanzees in visual search tasks, I designed a shift from the traditional two-alternative identity matching-to-sample (MTS) to a multiple-alternative version, in which the sample stimulus was followed by the one correct and several uniform incorrect stimuli (Fig. 1). If chimpanzees are presented with multiple-choice stimuli, they should search these items to detect the correct stimulus (i.e., the target). The subject who had been extensively trained on standard MTS tasks successfully acquired multiple-alternative MTS skills (Tomonaga 1993a). This chimpanzee showed increasing response times as a function of display size (the total number of stimuli in the search display) when the target and the distractors (incorrect choice stimuli) were perceptually similar to each other.

In the standard two-alternative MTS, the correct stimulus is defined by the identity relationship between the sample stimulus and the correct choice stimulus (Fig. 1). In contrast, two different controlling relations are potentially operating in the multiple-alternative MTS. One is the identity relation between the sample and the correct choice stimulus, and the other is the "oddity" relation between the correct

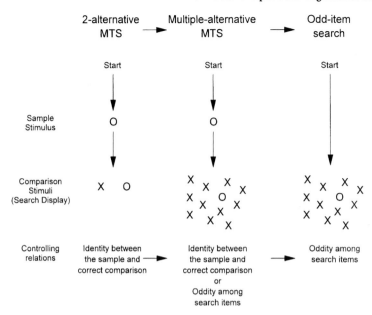

Fig. 1. Schematic representations of standard matching to sample (MTS), multiple-alternative MTS, and odd-item search procedures. *Vertical arrows* indicate the sequence of events in a trial. *Horizontal arrows* indicate the training order for the chimpanzees. (Based on Tomonaga 1993a, 1995a, b)

and incorrect choice stimuli. Before progressing to the next step, I attempted to assess the controlling relations in the chimpanzees' performance.

As shown in Fig. 2, the three chimpanzees I tested were given three types of test trials in addition to the baseline multiple-alternative MTS trials (Tomonaga 1995a): *odd-item search trials*, in which the sample stimulus was not presented and the search display contained one target and several uniform distractors, *MTS trials with nonuniform distractors*, in which the sample stimulus was followed by a display containing one target and various nonuniform distractors, and *no-sample trials with nonuniform distractors*, in which the sample was not presented and the display contained one target and various nonuniform distractors. To detect the target, subjects must use oddity relations among stimuli in the search display on odd-item search trials, but identity relations between the sample and correct choice stimulus on the MTS trials with nonuniform distractors. The no-sample trials with nonuniform distractors were used as the control condition. Figure 2 presents the percentage of correct responses for each type of trial as a function of display size averaged across the subjects. The chimpanzees showed a high degree of accuracy in both the odd-item search trials and MTS trials with nonuniform distractors. As a result of the multiple-alternative MTS training, the chimpanzees acquired the control of (generalized) oddity relations between the target and distractors that is necessary for the odd-item search in which trial–unique target–distractor pairs appeared (Blough 1989; Tomonaga 1995b). In addition, the chimpanzees were

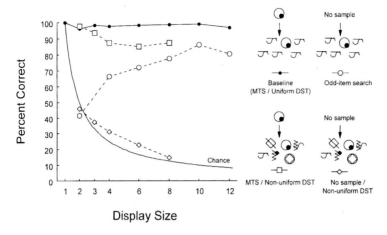

Fig. 2. Results of probe test trials for assessing the controlling relations in multiple-alternative MTS (adapted from Tomonaga 1995a). Examples of the four types of trials are shown at the right. *DST,* Distractors

able to use both the identity and oddity relations flexibly depending on the task requirements.

3 Basic Processes in the Visual Search: Tests for the Search Asymmetries

After testing the controlling relations of visual search performance in the chimpanzees, I went on to test the validity of Treisman's feature integration theory. As described briefly in the introduction, the feature integration theory proposes a two-stage model of visual information processing; parallel (or preattentive) processing for coding "features" and then mapping them onto feature maps, and serial (or attentional) processing for integrating these features into objects. Behaviorally, parallel processing is identified with a flat response-time function irrespective of display size, whereas serial processing is identified with linearly increasing response times as a function of display size.

What then are the features involved in visual processing? To identify these features, phenomena known as search asymmetries are often used as a diagnostic tool (Treisman and Gormican 1988; Treisman and Souther 1985). When the target contains features and the distractors do not, search response times are fast and constant, irrespective of the display size. On the other hand, when the target does not contain the feature but the distractors do, search response times increase as a function of display size. Many researchers have identified various features that cause search asymmetries. Treisman and Gormican (1988) classified these features into four types (Table 1). Although these features are varied in form, the feature integration theory provided a coherent explanation to account for the search asymmetries (Treisman and Gormican 1988). One way to test the phylogenetic validity

Table 1. Low-level features causing search asymmetries (Treisman and Gormican 1988) and the summary of results from humans, chimpanzees, macaques, and pigeons

Type of features (targets popping out)	Examples	Humans[a]	Chimpanzees[b]	Macaques	Pigeons
Targets defined by an added component	Q vs. O $ vs. S	O	O	?	×[e,f]
Targets with a categorical feature that can be only present or absent	C vs. O	O	O	?	△[e], O[g]
Targets with more of a quantitative property	Long vs. short Large vs. small Bright vs. dark	O	O	O[c]	?
Targets that deviate from a "standard" value	Tilt vs. vertical Ellipse vs. circle Irregular vs. regular	O	O	O[d]	△[f]

O, At least one feature caused the search asymmetry; △, features causing the search asymmetry in this species but not or reversed in humans; ×, no features caused search asymmetries; ?, not conducted.

[a] Treisman and Gormican (1988).

[b] The present chapter (Tomonaga 1993b, 1994b, 1999a).

[c] Hasegawa et al. (1997).

[d] Fujita and Kanazawa (1994).

[e] Allan and Blough (1989).

[f] Kuroshima (1999).

[g] Blough and Franklin (1985) under the texture segregation experiments.

of this theory is by comparing these features which cause similar search asymmetries in humans and other animals (Blough 1992). From the list of features shown for humans in Table 1, I selected a set of features for chimpanzees and conducted the visual search experiments. To replicate the results across different tasks, I also used the texture segregation procedures that are frequently used in experiments with humans.

3.1 Targets Defined by an Added Component

The first of the four types of features is the target defined by an added component. For example, humans show faster and more accurate responses in detecting a Q among Os than in detecting an O among Qs. An added component (a short line crossing O, in this case) yields an emergent feature that pops out. In pigeons, however, these types of features did not cause search asymmetries (Allan and Blough 1989; Kuroshima 1999). I used these types of features and investigated search asymmetry in two chimpanzees (Tomonaga 1993b) utilizing the two types of visual search procedures: multiple-alternative MTS and odd-item search. The stimuli with features were a diamond with a horizontal line through it and a wavy line with a bisecting diagonal line, and the stimuli without features were the diamond and the wavy line (Fig. 3). In the feature target trials, the targets were stimuli with features and the distractors were stimuli without features. The situation was reversed in no-feature target trials. The number of display items (display size) varied between 7 and 12. Figure 3 shows the correct response times for each type of trial as a function of display size averaged across the subjects. Both subjects displayed clear search asymmetries consistent with human responses. They showed fast and relatively flat response time functions for the feature target trials, but increased response times as a function of display size for the no-feature target trials. There were no strong differences between the MTS and odd-item search trials.

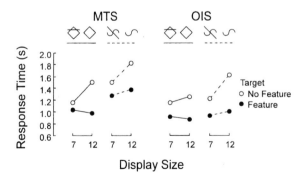

Fig. 3. Search asymmetries observed in the chimpanzees. Mean correct response times for each condition are shown. The stimulus pairs used in this experiment are shown at the top of the panels. The *left* stimulus of each pair is the stimulus with a feature, and the *right* is the stimulus without. *MTS*, Data from multiple-alternative MTS trials; *OIS*, data from odd-item search trials. (Adapted from Tomonaga 1993b)

Human subjects also participated in the same experiments and showed similar patterns to the chimpanzees.

3.2 Targets with a Categorical Feature that Can Be only Present or Absent

The second of Treisman and Gormican's (1988) four types of feature is the target with a categorical feature that can only be present or absent. A good example is discrimination between C and O. In this case, the gap in the C is the categorical feature. Detecting a C among Os is easier than detecting an O among Cs for humans. For the chimpanzee, I conducted the texture segregation experiment to observe any asymmetry of performance. It is well known that humans also show asymmetries in texture segregation, and the patterns are quite similar to those observed in visual search experiments (Gurnsey and Browse 1989). Pigeons also showed asymmetry in texture segregation between C and O (Blough and Franklin 1985). I trained one chimpanzee to segregate the target texture area from the surrounding background texture area by using various types of stimuli to replicate the results of visual search experiments (Fig. 4; Tomonaga 1999a). During this experiment, I also presented a texture display consisting of Cs and Os. The subject showed clear asymmetry in segregating the target area with respect to both accuracy and response times (Fig. 4). Interestingly, pigeons showed the reverse pattern

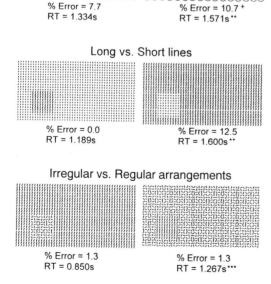

C vs. O

% Error = 7.7
RT = 1.334s

% Error = 10.7 +
RT = 1.571s **

Long vs. Short lines

% Error = 0.0
RT = 1.189s

% Error = 12.5
RT = 1.600s **

Irregular vs. Regular arrangements

% Error = 1.3
RT = 0.850s

% Error = 1.3
RT = 1.267s ***

Fig. 4. Asymmetries of performance observed in texture segregation experiments by the chimpanzee. Each block shows an example of each texture condition. The target area appears in the bottom left-hand area of each block. Mean percentage error and correct response times (*RT*) are shown at the bottom of each block. Significance levels of paired *t*-tests between data from left and right-hand blocks are as follows: $^{+}P < 0.07$; $^{**}P < 0.01$; $^{***}P < 0.001$. (Adapted from Tomonaga 1999a)

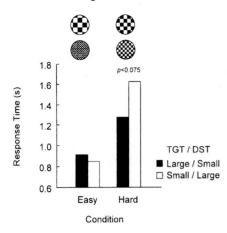

Fig. 5. Search asymmetry in the size of checker patterns for the chimpanzee. *Upper* stimuli are the large sized stimuli, and *lower* ones are the small sized stimuli

of visual search asymmetry when the stimuli were triangles with or without a gap (Allan and Blough 1989). Thus, pigeons may have visual features which are different from those of humans and chimpanzees.

3.3 Targets with More of a Quantitative Property

Search asymmetry is also observed when the target and distractor exhibit different quantitative properties in a continuum. For example, a brighter stimulus is easier to detect than a darker one, a longer line than a shorter one, a larger stimulus than a smaller one, and so on. Mikami and I also conducted a visual search experiment with four checkered pattern stimuli (Mikami 1999). Each pattern had checks of a different size. When the difference was large enough to differentiate, the chimpanzees showed no signs of search asymmetry, whereas they showed clear asymmetry in visual search performance when the target and distractor had similar check sizes (Fig. 5). In other words, they detected the larger-check target better than the smaller-check target. The same pattern of results was obtained in rhesus macaques (Hasegawa et al. 1997; Mikami 1999).

In the texture segregation experiment, I found asymmetry in segregation performance when using other types of stimuli such as long and short lines (Tomonaga 1999a). The chimpanzee showed better performance when the target area consisted of longer lines rather than shorter lines (Fig. 4).

3.4 Targets that Deviate from a "Standard" Value

The last category of features causing search asymmetry classified by Treisman and Gormican (1988) is the target that deviates from a *standard* value. The problem lies in defining what the standard values in a visual system are. In humans, for example, a tilted line pops out among vertical or horizontal lines. However, it is plausible that different species have different standard values, so it is worth conducting comparative studies. By using line orientations as stimuli, I conducted

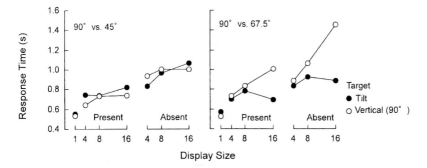

Fig. 6. Search asymmetries for line orientations in the chimpanzee. (Adapted from Tomonaga 1994a, b)

such search experiments with a single chimpanzee (M. Tomonaga, 1994, unpublished doctoral thesis; Tomonaga 1994b).

In this experiment, I prepared two conditions to see the effects of ease of differentiation as in the previous checker-pattern experiment. The first condition was discrimination between the vertical (90°) and lines tilted to 45°, and the second was between the vertical and lines tilted to 67.5°. Furthermore, I made procedural modifications. In addition to target-present trials, I also gave target-absent trials to the subject. When humans search a display serially, they use the self-terminating search strategy to find the target, whereas the exhaustive search strategy is used during target-absent trials. This difference in search strategies can be identified by differences in the slopes of response time functions (that is, the search rate). If the subjects search in an exhaustive manner, the search rate is twice as steep as that of the self-terminating search (Egeth et al. 1984; Treisman and Gelade 1980). In the target-absent trials of this experiment, the subject was allowed to touch any distractor, and the response to any stimulus was nondifferentially reinforced.

The results are shown in Fig. 6. When using easily distinguishable pairs (left panel), the subject showed no search asymmetry. When using less easily distinguishable pairs (right panel), however, the subject showed very clear search asymmetries in both target-present and target-absent trials. For less distinguishable pairs, when the target was the tilted line, the search rate was almost flat, indicating parallel processing. When the target was the vertical line, however, the search rate was steeper than that of the tilted line target. Furthermore, the ratio of search rates between target-present and target-absent trials was 1:2 (23.0 ms per item and 47.4 ms per item for target-present and target-absent trials, respectively). These results suggest that the chimpanzee searched the display in a self-terminating manner during the target-present trials, but in an exhaustive manner during target-absent trials when serial processing was required to detect the target.

To replicate the present results across individuals, I conducted further experiments. The other chimpanzee subject showed the same patterns of search asymmetry during the visual search for line orientations when vertical and/or horizontal lines were used as standard values (Tomonaga 1997b). Fujita and Kanazawa (1994) also found search asymmetry for line orientations in Japanese macaques.

The other feature defined by deviation from the standard, regularity of line arrangements, caused asymmetry in texture segregation performance (Fig. 4). One chimpanzee showed better performance detecting a target area of irregularly arranged lines from the background area of regularly arranged lines than vice versa (Tomonaga 1999a). This was probably because regular arrangements of lines played the role of the standard value (cf. Treisman and Gormican 1988).

Recently, Kuroshima (1999) found that pigeons displayed search asymmetry when the stimuli were zigzag and straight lines. Pigeons showed better performance when the target was the zigzag line. Interestingly, human subjects showed no such search asymmetry in a similar experimental setting. Zigzag lines deviate from standard straight lines to cause search asymmetry in pigeons, but the perceptual difference between these stimuli may be much larger for humans. Zigzag lines produce some other candidates for emergent features such as the number of line segments (the more–less feature) and the presence of angles (categorical features). It is necessary to identify which type(s) of properties actually caused the search asymmetry in pigeons.

3.5 Summary of Search Asymmetry Experiments

This section summarizes the search asymmetry experiments and related studies using chimpanzees as subjects. Table 1 presents a summary of results from various species. Chimpanzees showed clear search asymmetries for the same types of features as humans. In contrast, pigeons showed completely different patterns of results, with no signs or reversed patterns of search asymmetries when using stimuli that caused asymmetries in humans. Moreover, they showed search asymmetry with some stimuli where humans did not. Some possible explanations for this species difference include the ease of differentiation of stimuli (Pashler 1987), practice or the perceptual learning effect (Ahissar and Hochstein 1996; Sireteanu and Rettenbach 1995), and the difference in the repertoire of features between primates and pigeons (Blough 1992). More importantly, however, is the fact that pigeons actually showed search asymmetry for some types of stimuli, which indicates that they might also have a visual information processing system that can be explained by the feature integration theory or related theories such as guided search models. Cook et al. (1996) found that pigeons showed similar results to humans in the discrimination of textures defined by the conjunctions of two or three features. These results can be explained by the guided search model (Wolfe 1994a; Wolfe et al. 1989). Pigeons have a brain with a different architecture from that of the primate brain. The same (or similar) functional "modules" in different hardware may lead us to speculate on the possibility that the same environmental selective pressures had affected both aves and primates after the phylogenetic differentiation of birds and mammals.

The feature integration theory and related models also explain some other phenomena such as conjunction search (Egeth et al. 1984; Treisman and Gelade 1980;

Treisman and Sato 1990), illusory conjunctions (Treisman and Schmidt 1982), and guided search (Wolfe et al. 1989). Unfortunately there is very little evidence of these phenomena occurring in other animals (Bolster and Pribram 1993; Cook et al. 1996; Dursteler and von der Heydt 1992; Fujita and Kanazawa 1994). To test the phylogenetic validity of these influential models, we need further comparative studies using those animals close to and/or far from humans.

4 Attention and Visual Search

The feature integration theory concerns visual attention. Attentional processes are involved in visual search tasks. The term "attention" has various meanings in the field of psychological research. Generally, attention refers to those processes concerning the selection of information of sensory inputs. In this section, I summarize the results from research pertaining to attentional processes during the chimpanzees' visual search performances.

4.1 Action-Centered Attention

As mentioned before, pointing responses were required in our visual-search procedures instead of the yes–no reporting responses frequently used in human experiments. We often observed manual response biases among the target locations, and considered these as biases which should be counterbalanced (e.g., Tomonaga 1993b, 1997a). However, Tipper et al. (1992) found that these manual response biases were the result of interactions between attention and action-centered internal representations. Manual responses, such as reaching directly to the target, are affected by the target–distractor spatial configurations (distractor effects). Tipper et al. (1992) found two major distractor effects. First, the reaching response to the target in the presence of the distractor (selective reaching) was slower when the distractor was located on the trajectory of manual movement to the target. Second, manual responses were faster when the hand moved to its ipsilateral side (i.e., the right side when the right hand was used) than to the contralateral side (Berlucchi et al. 1977; Stins and Michaels 1997). Furthermore, when the distractor was located ipsilateral to the target, the reaching response was slower than for the contralateral side.

In the pointing-type visual search tasks, response biases among target locations might be interpreted as reflecting these distractor effects. I reanalyzed data from the search-asymmetry experiments in chimpanzees (Tomonaga 1993b), picking trials in which one target and three distractors appeared (i.e., display size 4), and calculated mean response times on the basis of spatial configurations. These trials had been omitted from the previous data analyses because of high variability (Tomonaga 1993b). As a result of reanalysis, however, this variability in response times was clearly due to target–distractor configurations. Distractor effects were identified by the difference in response times subtracted from those response times for target-only trials in which only the target appeared in a search display. Two

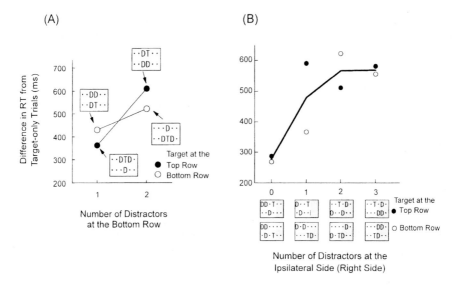

Fig. 7. Distractor effects on the visual search in chimpanzees. Vertical axes indicate the difference in response times (*RT*) between trials with and without distractors (target-only trials). **A** Distractor effect as a function of the number of distractors between the start point (warning signal) and the target. **B** Distractor effect as a function of the number of distractors on the ipsilateral side of the hand that subjects used. Both chimpanzee subjects used their right hand during the experiment

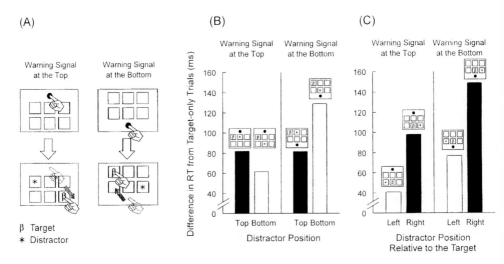

Fig. 8. Distractor effects on simple discrimination in the chimpanzee. **A** Schematic diagrams of the trials. **B** Distractor effect as a function of the position of the distractor relative to the warning signal and the target. **C** Distractor effect as a function of the horizontal position of the distractor relative to the target. Vertical axes of each panel indicate the difference in response times for target-only trials. (Adapted from Tomonaga 1998b)

chimpanzees who used their right hand during the experiment displayed two types of distractor effects. Distractor effects increased as a function of the number of distractors between the start point (a warning signal located at the bottom of the display) and the target (Fig. 7A), and increased as a function of number of distractors on the ipsilateral side of the target (Fig. 7B).

To investigate these distractor effects further, I conducted a simple discrimination experiment by using manual pointing as a response with a chimpanzee (Tomonaga 1998b). In this experiment, only the target ("β") and a single distractor ("*") appeared on the CRT display. The start point of the manual response (warning signal) appeared randomly at the top or bottom of the display from trial to trial (Fig. 8A). As in the visual search task, one "right-handed" chimpanzee showed similar patterns of distractor effects. Distractor effects were larger when the distractor was located between the start point and the target. This effect was consistent irrespective of the location of the start point (Fig. 8B). Furthermore, the distractor located on the ipsilateral side yielded a larger distractor effect (Fig. 8C).

In summary, it is highly likely that the chimpanzees' discriminative manual responses were affected by the spatial configurations of the target and distractors. Pointing-type visual search tasks would therefore be one of the best procedures for investigating the action-centered attention processes in both humans and chimpanzees.

4.2 Automatic and Controlled Processing of Precues

As mentioned already, in the feature integration theory, visual processing is separated into two stages: parallel, preattentive processing, and serial, attentional processing. Such a dichotomy is frequently observed in human attention processes. For example, priming effects also involve "automatic" and "controlled" processes (Neely 1977; Posner and Snyder 1975; cf., Schneider and Shiffrin 1977; Shiffrin and Schneider 1977). In contrast to the abundance of human literature, there have been very few comparative studies on automatic and controlled processing of primes or precues (Blough 1989; Blough and Blough 1997; Hopkins et al. 1991; Pineda and Nava 1993). I conducted an experiment to investigate the precuing effect on visual search performance in the chimpanzee (Tomonaga 1997a). In this experiment, I used a multiple-alternative MTS version of the visual search task. When the sample was presented, the subject was required to touch it five times (observing responses). During these observing responses, a precue (gray square) briefly appeared at one of the predefined locations (peripheral precuing, Posner 1980). In some trials, subsequent search display contained the target at the precued location (valid trials), while it was at a different location in other trials (invalid trials). Under some conditions, the validity of the precue (predictability of the following target location) was very low (20%), whereas under other conditions it was high (80%). Facilitation of visual search performance by the valid precue, and inhibition by the invalid precue, were identified by the difference in response times from the control trials in which the precue was not presented.

Figure 9 shows the results of the experiment. When validity was high, the precue

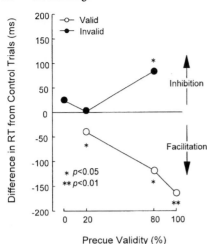

Fig. 9. Effects of the precue on the chimpanzee's visual search performance as a function of precue validity. Vertical axis indicates the difference in response times between cued and noncued (control) trials. The *upper panel* highlights inhibition in the cued trials and the *lower panel* highlights facilitation. *Asterisks* near the symbols indicate significant differences in response times between cued and control trials as determined by analyses of variances. (Adapted from Tomonaga 1997a)

caused both facilitatory and inhibitory effects on visual search performance. However, when validity was low, only the facilitatory effect was observed. This differential effect of precue validity on facilitation and inhibition was also observed in human experiments (Posner and Snyder 1975). When the validity is low, subjects ignore the precue, but its automatic processing resulted in facilitation without inhibition. On the other hand, when subjects "attend" to the precue, that precued location is positively activated (resulting in facilitated performance), while the other locations are negatively activated (resulting in inhibited performance). The results of Tomonaga's (1997a) experiment clearly demonstrated that the chimpanzee has the same attention processes as humans.

4.3 Negative Priming

Selective attention facilitates the processing of attended locations and items. So, how are the ignored locations and/or items processed? The results of a study by Tomonaga (1997a) suggested that the subsequent processing of these stimulus properties were inhibited. One of the direct ways to see how the ignored properties are processed is using the *negative priming* paradigm (Tipper 1985). In a traditional priming experiment, a prime was followed by the probe stimulus. If there is some (perceptual or semantic) relationship between the prime and probe stimuli, processing of this probe stimulus is facilitated. If there is no relationship, processing of the probe stimulus is "inhibited." Tipper (1985) presented two line drawings as stimuli. These stimuli were colored differently (e.g., green and red), and superimposed. The human subjects were instructed to name one of the line drawings ("attend to the red one and name it"). Tipper manipulated the relationship between two successive trials. Under the control condition, the attended to and ignored items in the preceding trial never appeared in the subsequent trial. Under the ignored repetition condition, the ignored item from the preceding trial appeared as the to-be-attended-to item in the subsequent trial. Tipper found an in-

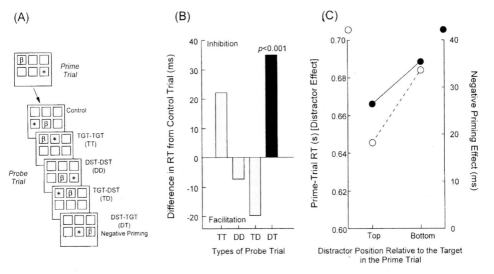

Fig. 10. Spatial negative priming during simple discrimination by the chimpanzee. **A** Schematic diagrams of prime and probe trials. **B** Difference in response times between probe trials and target-only trials. The *upper panel* indicates inhibition and the *lower panel* indicates facilitation. **C** Relationship between distractor effects in the prime trials and negative priming effects in the probe trials under the distractor–target (DST–TGT) condition. *Left* vertical axis indicates the response times for the prime trials (distractor effect, *open circles*) and the *right* axis indicates the difference in response times between the probe and target-only trials (negative priming effect, *filled circles*). (Adapted from Tomonaga 1998b)

hibitory effect of the ignored item on the subsequent processing of that item. This effect is called negative priming.

To investigate the inhibitory effect of the ignored stimulus in chimpanzees, I conducted two negative priming experiments. In the first experiment, the spatial location of the target stimulus was primed under the simple discrimination procedure described above. In this experiment, two trials were successively presented (cf. Maylor and Hockey 1985; Tipper et al. 1992; Fig. 10A); the first was the prime trial, and the second was the probe trial. Based on the relationship of target and distractor locations between the prime and target trials, five conditions were prepared: (1) *control condition*, in which there were no overlaps of stimulus locations between the prime and probe trials; (2) *target–target (TGT–TGT) condition* (TT), in which the target location was repeated between the prime and probe trials; (3) *distractor–distractor (DST–DST) condition* (DD), in which the distractor location was repeated; (4) *TGT–DST condition* (TD), in which the probe distractor appeared in the location where the prime target had appeared; (5) *DST–TGT condition* (DT), in which probe target appeared in the location where the prime distractor had appeared. The last condition (DT) tests for the negative priming effect. As shown in Fig. 10B, the subject displayed a statistically significant inhibitory response (identified by the difference in response time from the control condition)

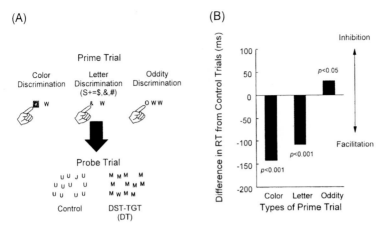

Fig. 11. Priming of stimulus identity in visual search by the chimpanzee. **A** Schematic representations of the three types of prime trials and the two types of probe visual-search trials. **B.** Mean difference in probe response times between DST–TGT and control conditions. (Adapted from Tomonaga 1998b)

only under the DT condition. This inhibitory effect is considered as evidence for negative priming (Tomonaga 1998b).

As noted in the previous section, the chimpanzee's manual response was modulated by distractor effects. Differential distractor effects imply a differential attentional load. Thus, it might be plausible that a stronger distractor effect would predict for a stronger negative priming effect. I therefore calculated the response times in the prime trials (distractor effect) and for the negative priming effect as a function of the distractor position relative to the target (Fig. 10C). When the prime distractor was located between the start point and the target the distractor effect was larger, as was the negative priming effect. Hence, the distractor effect and negative priming effect were positively correlated.

The second experiment examined the negative priming effect of stimulus identity (Tomonaga 1998b). Figure 11A illustrates the experimental procedure. The probe trial was a visual search task (odd-item search for letters), while three different tasks were used as prime trials. The first was color discrimination, in which the target was a white letter on a blue background. The second task was letter discrimination, in which the target was either "$," "&," or "#," and the distractor was a letter of the alphabet. The third prime task was oddity discrimination, in which the target was defined by the odd item in three stimuli. In each prime-trial condition, two types of probe condition were prepared: control trials and DT trials.

The results are shown in Fig. 11B. The priming effect was based on the difference in probe response times from those under DT and control conditions. Surprisingly, the chimpanzee showed a positive priming effect under the first two prime conditions. A negative priming effect was observed only when the prime trial was that of oddity discrimination. In the color and letter discrimination tasks, discrimination between the target and distractor was quite easy and might require no selective attention. Therefore, in these tasks the distractor was automatically processed as well as the target, such that positive priming effects occurred. On the other hand, oddity discrimination required selective attention, and resulted in both the attentional (facilitatory) processing of the target and the inhibitory processing of the distractor.

4.4 Summary

Although generally increasing in number, comparative studies of attention are still very rare and provide limited information about the attention processes of animals. As shown in this section, visual search and related tasks are quite useful for investigating the attention processes of nonhuman animals. These attention studies may provide us with a new way of studying comparative cognition. For example, if we can modify the priming tasks (e.g., semantic priming tasks) so that they are suitable for animals, we can examine different aspects of natural categorizations in animals instead of traditional synthetic studies which use concept formation procedures.

5 Visual Search Tasks for More Complex Processes

Human visual search experiments usually employ simple visual stimuli consisting of only a few features (e.g., Treisman and Gormican 1988). In our everyday life, however, we search much more complicated visual environments. Recently, human visual search experiments have been shifting focus to higher-order vision, such as object recognition (e.g., Wolfe 1994b; Suzuki and Cavanagh 1995).

Are visual search tasks still useful for studies of higher-order vision? Visual search tasks have some unique properties that other simple discrimination tasks do not. For example, visual search tasks can distinguish between parallel and serial processing of visual stimuli by examining the slope of response times as a function of the number of items. Furthermore, we often observe asymmetry of performance in visual search and related tasks when target–distractor mapping is reversed. By using such behavioral measures, we can examine the processing of complex visual stimuli, such as more naturalistic ones, under the visual search tasks. In this section, I describe the results of a series of visual search experiments in chimpanzees using more complex stimuli.

5.1 Frame of Reference and Search Asymmetry for Line Orientations

Treisman (1985) reported that search asymmetries for line orientations were modified by the addition of a local frame of reference. She presented a search display containing line stimuli to human subjects. When the display was surrounded by an untilted, upright frame, the subjects showed clear search asymmetry: detection

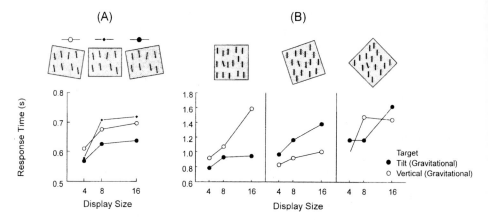

Fig. 12A,B. Effects of the frame of reference on search asymmetries for line orientations in the chimpanzees. Examples of the search display for each condition are shown at the top of the panels

of the tilted target among the vertical distractors was faster than vice versa. When a tilted frame surrounded the search display, however, search asymmetry for line orientation was reversed; i.e., detection of the gravitationally tilted target among the gravitationally vertical distractors was slower than vice versa (cf. Fig. 12). Such an effect of the frame of reference is usually observed in the rod-and-frame illusion (e.g., Spinelli et al. 1991; Wenderoth 1973). The results of Treisman's experiment suggest that the "standard" value for visual information processing is not an absolute one; nor is it based on retinal coordinates. Instead, the standard is relative.

I also conducted experiments on the frame of reference in chimpanzees. Under the first condition, the search display contained two types of symmetrically tilted lines (Fig. 12A). This search display was surrounded by either upright or tilted frames. As shown in Fig. 12A, the chimpanzee subjects showed search asymmetry on the basis of line orientations relative to the frame. When the target and local frame were tilted at the same angle (9°), response times were slower than under the condition using the other tilted frame (Tomonaga 1997b).

In the first condition, gravitationally vertical lines never appeared as stimuli. The lack of an "absolute" standard might therefore facilitate the effect of a frame of reference. Thus, I replicated Treisman's conditions with a chimpanzee in the second experiment and found results consistent with Treisman's (1985) study (Fig. 12B). When the upright frame was used, the chimpanzee showed the usual patterns of search asymmetry. However, when the tilted frame was used, search asymmetry was based on the coordinates of the local frame. Furthermore, when the local frame was rotated to 45° (and therefore all stimuli were "tilted" on the basis of this frame), search asymmetry disappeared.

These two experiments neatly replicated Treisman's (1985) study with humans as subjects. For both species then, the standard line orientation is neither gravitational nor retinal, but is based on the local frame of reference.

5.2 Perception of Shape from Shading

When humans see stimuli with shading, we perceive depth from that shading under certain conditions. For example, if a shading pattern is graded from bright at the top to dark at the bottom, we see a convex shape, whereas we see a concave shape from the reversed pattern of shading (Fig. 13). Surprisingly, if these patterns of shading are rotated by 90°, we no longer perceive the depth. Ramachandran (1988) and Kleffner and Ramachandran (1992) conducted a series of visual search and texture segregation experiments using these shading patterns, and found that the shape or depth from shading was processed in a parallel manner. Humans showed faster detection of the target with vertical shading (top bright/bottom dark or vice versa) than with horizontal shading (left bright/right dark or vice versa). Ramachandran suggested that the human visual system has two constraints for the processing of shape from shading. First, a single source of light illuminates the whole scene. Second, light must be shining from "above" in relation to retinal coordinates, not gravitational ones. These two constraints might have been built into the human visual system through evolutionary processes when, during the course of adaptation to an open environment, the human visual system was modified. If this is the case, animals that are adapted to a completely different environment from humans would be expected to show different patterns of shape perception from shading.

Humans and chimpanzees have adapted to rather different environments: tropical forest for chimpanzees and open land for humans. Common ancestors of humans and great apes might have been living in tropical forests just as chimpanzees do. Structural differences in environments have led to differences in anatomical structures as a result of different locomotor patterns. Visual perception might also be influenced by adaptations to different environments. Investigating the perception of shape from shading is quite a good method for comparative studies of visual perception. Thus, I conducted a series of experiments in chimpanzees on the perception of shape from shading (Tomonaga 1998a) using visual search tasks.

In the first experiment, I examined the effects of shading directions. As shown in Fig. 13A, two chimpanzees were given three types of visual search tasks for shading directions. Under each condition, the target and distractors were shaded with same type of shading, vertical, diagonal, or horizontal, but the polarity was reversed. For example, the target was bright at the top and the distractors were bright at the bottom under the vertical shading condition. Five human subjects also participated in the same experiment. Figure 13A shows the mean correct response times for each condition as a function of display size. In humans, detecting the vertically shaded target was easier than with the horizontally shaded target. This was consistent with Ramachandran's experiment. On the other hand, both chimpanzees showed completely opposite patterns from humans, with a better performance for horizontal than for vertical shading.

An additional experiment confirmed that this species difference could not be due to viewing positions or head rotations (Tomonaga 1998a; cf. Kleffner & Ramachandran 1992). Furthermore, when the shading type was changed from

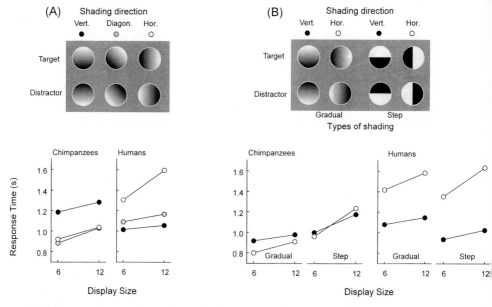

Fig. 13. Perception of shape from shading in chimpanzees and humans tested with visual search tasks. *Upper panels* show examples of stimuli. *Lower panels* show mean correct response times as a function of display size for each species. **A** Effect of shading directions. **B** Effect of shading types. *Vert.*, Vertical shading; *Diagon.*, diagonal shading; *Hor.*, horizontal shading. (Adapted from Tomonaga 1998a)

gradual to stepwise, the effects of shading direction disappeared in chimpanzees but remained in humans (Fig, 13B). The type of contour (circle or rectangle) also affected the effects of shading direction, but only in chimpanzees (Tomonaga 1998a). These results suggest that chimpanzees process shading information differently from humans (cf. Aks and Enns 1992). It is still unclear, however, whether chimpanzees perceive *depth* from the horizontal shading patterns but not from the vertical shading patterns. Are the present results due to lack of perception of depth from shading, or to a shift of constraints from vertical to horizontal shading in chimpanzees? The next step would be to investigate the relationship between recognition (or identification) of three-dimensional shapes from two-dimensional shading, and the processing of monocular shading information from a comparative perspective (e.g., Miura and Kawabata 1999).

5.3 Discrimination of the Orientation of Faces

Face perception in nonhuman primates is of great interest to researchers in comparative cognition, ethology, and neuroscience. Faces are quite important as visual features of individuals and as media for visual communications. Faces contain information such as species, identity, sex, age, emotion, and so on. Regarding face perception, one of the most striking phenomena in humans is the "inversion effect," which refers to the hampered recognition of upside-down faces in comparison with upright faces (e.g., Bruce 1988; Bruce and Young 1986; Diamond and Carey 1986; Ellis 1986; Yin 1969). The results of many comparative studies on the inversion effect in nonhuman primates have been confusing, with some researchers obtaining evidence against an inversion effect (Bruce 1982; Rosenfeld and Van Hoesen 1979; Tomonaga et al. 1993), but others finding evidence for it (Overman and Doty 1982; Wright and Roberts 1996). However, recent progress clearly indicates that nonhuman primates readily show the face-inversion effect under the appropriate experimental setting (Parr et al. 1998; Tomonaga 1994a, 1999b).

In everyday life, we often search for a face in the crowd (Hansen and Hansen 1988). The use of faces in visual search studies will provide more information both about the visual search in naturalistic situations and face perception per se. Unfortunately, visual search for faces has not been so intensively examined, even in humans (Kuehn and Jolicoeur 1994; Nothdurft 1993; Suzuki and Cavanagh 1995; Tong and Nakayama 1999). Here, I describe a part of a series of experiments on visual search for faces by chimpanzees (Tomonaga 1999c).

The present topic is related to the inversion effect: the visual search for the orientation of faces. For chimpanzees, is it easier to find an upright face among disoriented faces, or vice versa (analogy to search asymmetry)? If so, is this specific to faces? By using visual search tasks, we may be able to argue differences in the processing of upright and disoriented faces in chimpanzees.

In the first experiment, I tested one experienced chimpanzee on the visual search for orientations of human faces. Six conditions were given to the subject by pairing two of the three types of orientations (upright, horizontal, and inverted, see upper panels in Fig. 14). The subject was required to touch the face with a different orientation from the others. I used a large number of photographs of faces of Japanese male individuals who were not known to the subject in order to prevent her from searching for the target on the basis of specific facial features. Figure 14 shows the mean correct response times under each condition. I derived two main findings from this experiment. First, the chimpanzee showed better performance when upright faces were included in the search display (such as upright and horizontal faces) than when only horizontal and inverted faces were used. Second, when the upright face was the target, the subject showed faster response times than when upright faces were the distractors (Tomonaga 1999c). This latter effect was seemingly analogous to search asymmetry (see Sect. 3).

To verify the between-individuals validity, I also trained another chimpanzee to distinguish facial orientations under the visual search tasks. One target and four distractors appeared in each trial, and six conditions were trained concurrently. Figure 15 shows the acquisition processes. Clearly, the subject acquired discrimination more quickly when the upright face was the target than when the upright faces were the distractors. There was no difference in the speed of acquisition between conditions in which the horizontal and inverted faces were used.

These effects are not specific to human faces. When using chimpanzee faces, the subject again showed better performance when upright faces were used, and showed faster response times when the upright face was the target rather than when up-

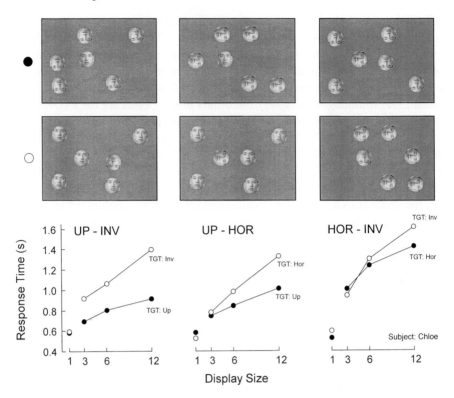

Fig. 14. Visual search for the orientation of faces by the chimpanzee. *Upper panels* show examples of the search display. *Lower panels* show the mean correct response times for each condition as a function of display size. *Up*, Upright; *Hor*, horizontal; *Inv*, inverted. (Adapted from Tomonaga 1999c)

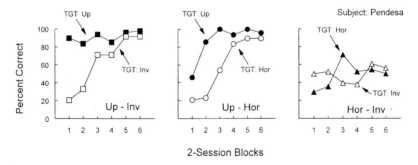

Fig. 15. Acquisition of the visual search for orientation of faces by the chimpanzee. Data were obtained from a different chimpanzee to the one used in Fig. 14. Vertical axes indicate the percentage of correct trials averaged across two successive sessions. *TGT*, Target stimuli; *Up*, upright; *Hor*, horizontal; *Inv*, inverted

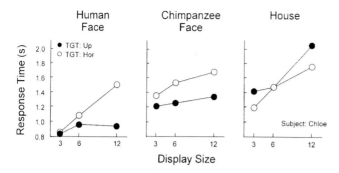

Fig. 16. Visual search in the chimpanzee for the orientation of human faces, chimpanzee faces, and photographs of houses. Mean correct response times for discrimination between upright and horizontal orientations are shown. (Adapted from Tomonaga 1999c)

right faces were the distractors (especially when paring upright and horizontal faces). Interestingly, the subject showed no such effects when photographs of houses that were as complex as faces were used (Tomonaga 1999c; Fig. 16).

For chimpanzees, it is easy to find an upright face among disoriented faces. Upright faces might pop out from the disoriented faces. It is unlikely, however, that this search asymmetry-like phenomenon in the visual search for the orientation of faces is governed by processes identical to search asymmetries for low-level features. Identification or recognition of faces occurs at a later stage, after the integration of low-level features (Bruce and Young 1986; Young 1995). However, recent studies report that more complex properties cause search asymmetry. For example, humans show faster detection of an unfamiliar letter (mirror-reversed Z) among familiar letters (Z) than vice versa (Wang et al. 1994). For humans, familiarity plays the role of "standard." For face perception, Tong and Nakayama (1999) also found that the "self face" played the role of the standard that caused search asymmetry (easier detection of the stranger's face among self faces). Upright faces are much more familiar than disoriented faces in everyday life. Humans (and also nonhuman primates) are expert at processing upright faces. This great familiarity, however, did not play the role of the standard for the chimpanzee. Some other factors, such as meaningfulness or biological significance, might have caused an efficient search for upright faces. For example, Hansen and Hansen (1988) found that humans quickly detected an angry face in visual search tasks. Needless to say, many questions remain to be addressed. Further investigations from the comparative perspective are necessary.

5.4 Visual Search for Biological Motion Patterns

In the visual search for orientations of faces, a familiar orientation caused search asymmetry or an efficient search. Is it plausible, then, that some other naturalistic properties of stimuli familiar to chimpanzees cause an efficient search? In other experiments, I examined biological motion patterns (Tomonaga 1997c, 1998c).

Johansson (1973) clearly demonstrated that humans perceived human locomotion when seeing the movements of point lights on the joints of a human body. This ability is already established in 4-month-old human infants (Fox and McDaniel 1982): they preferentially looked at normal biological motion patterns rather than mechanical patterns. Recently, there has been increasing evidence that nonhuman animals can perceive biological motions (Blake 1993; Dittrich et al. 1998; Fujita et al. 1999; Herman et al. 1990; Omori 1997; Oram & Perrett 1994; Yamaguchi & Fujita 1999).

In the present experiments, I used point light displays or stick picture patterns of a chimpanzee's quadrupedal walking cycle (Fig. 17). In a search display, several (3, 5, and 8) biological motion patterns were presented. These patterns all moved in the same place (i.e., they did not move from left to right). As a mechanical pattern, I prepared a random movement pattern in which each point of the original pattern was randomly shifted from frame to frame. In one condition, the original pattern (biological motion pattern) was the target with the random patterns as distractors, and vice versa in the other condition. The target–distractor mapping was changed alternately from session to session. The point light displays and stick picture patterns appeared equally but randomly in a session.

One chimpanzee subject was able to distinguish biological motion patterns after approximately 20 sessions of training. Furthermore, the subject showed (1) better

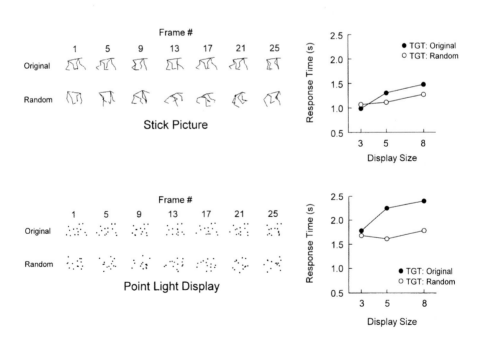

Fig. 17. Visual search for biological motion patterns by the chimpanzee. To the *left* are examples of biological motion stimuli. *Right-hand panels* show the mean correct response times for each condition as a function of display size. *Upper panels* show the results from the stick picture patterns and the *lower panels* show those from the point light display patterns. (Adapted from Tomonaga 1997c)

performance for the stick picture patterns than the point light display patterns, and (2) better performance for the random target than for the biological motion target (Fig. 17; Tomonaga 1997c). In this experiment, all stimuli started at the initial frame. These synchronized movements among the distractor items might have resulted in easier detection of the deviated movements. However, even when each stimulus (including the target) was started at different frames, the patterns of results were unchanged (Tomonaga 1997c).

It seems that deviations from biological motion patterns popped out among random motion patterns in the chimpanzee. Thus, the complex but familiar patterns of movements of dots might play the role of the standard. It is interesting that this tentative conclusion is opposite to that of the previous study on the visual search for the orientation of faces.

However, it is still premature to conclude from these experiments that the biological motion pattern acted as a "standard" in the chimpanzees' visual system. The random movement pattern used was quite different in various aspects from the original biological motion pattern. For example, the overall speed of each point was 1.5 times faster, on average, in the random pattern than in the biological mo-

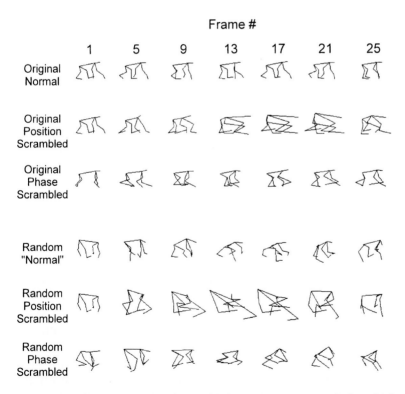

Fig. 18. Examples of normal and scrambled motion patterns made from biological motion and random movement patterns. Only stick picture patterns are shown. (Adapted from Tomonaga 1998c)

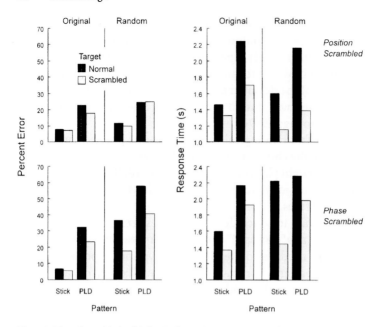

Fig. 19. Visual search for biological motion patterns in the chimpanzee. Mean percentage error (*left*) and mean correct response times (*right*) for each condition averaged across display size are shown. The *left-hand side* of each panel shows data from the original–scramble discriminations, and the *right-hand side* shows those from random–scramble discriminations. *Upper panels* show data from position-scrambled conditions, and *lower panels* show those from phase-scramble conditions. *Stick,* Stick picture patterns; *PLD,* point light display patterns. (Adapted from Tomonaga 1998c)

tion pattern. Furthermore, some points often seemed to "jump out" in the random pattern, but not in the biological motion pattern. These global and local features of the random pattern that are not related to the properties of biological motion itself caused search asymmetry in the early vision.

Traditionally, control patterns for biological motion have been prepared on the basis of methods different to those of the present experiments (Blake 1993; Fox and McDaniel 1982; Fujita et al. 1999). To keep the trajectory and speed of each point, their spatial locations are either scrambled (position scrambled), or the start frames of these points are scrambled (phase scrambled). In the next experiment, these transformations were applied both to the original biological motion pattern and to the random movement pattern, as shown in Fig. 18. If biological motion was the standard, scrambled patterns from the original should pop out. On the other hand, since the random pattern already deviates from the biological motion patterns, scrambled random patterns should cause no clear effects on search asymmetries.

The results of the experiments are summarized in Fig. 19 (Tomonaga 1998c). Each bar represents the mean percentage of errors or mean correct response times averaged across the three display size conditions (3, 5, and 8). Overall, the subject showed better performance for original–scrambled discrimination than for random–scrambled discrimination. When comparing the two scramble conditions, the position scrambled condition was distinguished better than the phase scrambled condition. This might reflect the greater distortions of overall shape in the position scramble compared with the phase scramble (Fig. 18). Concerning search asymmetries assessed with response times, scramble target detection was faster both in the original–scrambled and random–scrambled differentiations. These results were inconsistent with the hypothesis that only the scrambled pattern of biological motion would cause search asymmetry. At present, it is unclear why the scrambled patterns of the random pattern also caused search asymmetries. One plausible reason is the effect of familiarization (Johnston and Schwarting 1997; Wang et al. 1994). As discussed in Sect. 5.3, deviation from familiar patterns causes search asymmetry in humans (Wang et al. 1994). The subject of the experiment was exposed to the random pattern in more than 12 000 trials. Such a prolonged experience would form an extensive familiarity with this pattern (cf. Tong and Nakayama 1999). Speculation along these lines might be of interest in considering the role of familiarity in visual perception and cognition, although we need further investigation to clarify which factors influenced these chimpanzee experiments.

6 Conclusion

This chapter has described a series of experiments on chimpanzee visual perception and cognition using visual search and related tasks. Although these results strongly highlight the usefulness of these tasks in comparative studies of visual perception and cognition, there are still some problems to be considered. In the visual search task, subjects are only required to "discriminate" between the target and distractors when detecting the target. This property of visual search tasks raises questions about the relationship between "discrimination" between objects and "recognition" of them. Do chimpanzees recognize the objects presented on the search display while performing the visual search tasks? More specifically, do chimpanzees, for example, actually perceive depth from shading, do they recognize (upright) faces in a search display, or do they understand the complex movement patterns of dots as the type of locomotion of their own species? To address these questions more directly, we should use tasks other than the visual search, such as transfer of discrimination, category discrimination, and so on. However, it is plausible that the representations of objects affect perceptual processes. For example, familiarity is the "nonvisual" property of the stimuli because familiarity is not a characteristic of the visual pattern itself. As discussed above, however, familiarity causes search asymmetry in humans (Wang et al. 1994). Familiarity is in some part a characteristic of memory. In Treisman's feature integration theory, access to memory (object representations) occurred in the last stage of visual recognition. However, as Wang et al. suggested, memory access may occur at an early step in visual processing. Object representations, or top-down "knowledge," might affect simple visual perception. We should further examine the relationship between discrimination and recognition. To this end, comparative study would be fruitful for at least two reasons. First, comparative study will facilitate the connections between human studies and neurophysiological evidence from the standpoint of comparative cognitive science. Second, comparative study will tell us how nonhuman animals perceive objects.

Visual search for complex visual patterns by the chimpanzee provided a set of evidence for search asymmetry-like effects in these patterns. Previously, search asymmetry, pop-out, and parallel search were considered as topics relevant for only the early stages of visual processing and have been related to low-level features. However, recent progress is revealing that humans show search asymmetries, efficient search, and parallel processing when complex visual patterns are used, e.g., shape from shading (Kleffner and Ramachandran 1992), surfaces (He and Nakayama 1992), three-dimensional structures (Enns and Rensink 1990), faces (Tong and Nakayama 1999), familiarity (Wang et al. 1994), and symmetry (Olivers and van der Helm 1998). Can models of visual processing successfully and coherently explain these phenomena from simple to complex stimuli? I strongly believe that animal studies from the standpoint of comparative cognitive science will be one of the more powerful tools to test and extend models of visual information processing.

Acknowledgments

I thank the staff of the Department of Behavioral and Brain Sciences, Primate Research Institute, Kyoto University. In particular, I thank Sumiharu Nagumo for his technical assistance during experiments. I also thank Dr. Vanessa Hayes for her critical reading of an earlier version of the English manuscript. The "*Guide for the Care and Use of Laboratory Primates*" of the Primate Research Institute, Kyoto University (1986 version), was adhered to throughout all experimental procedures.

References

Ahissar M, Hochstein S (1996) Learning pop-out detection: specificities to stimulus characteristics. Vision Res 36:3487–3500

Aks DJ, Enns JT (1992) Visual search for direction of shade is influenced by apparent depth. Percept Psychophys 52:63–74

Allan SE, Blough DS (1989) Feature-based search asymmetries in pigeons and humans. Percept Psychophys 46:456–464

Berlucchi G, Crea F, di-Stefano M, Tassinari G (1977) Influence of spatial stimulus–response compatibility on reaction time of ipsilateral and contralateral hand to lateralized light stimuli. J Exp Psychol Hum Percept Perform 3:505–517

Blake R (1993) Cats perceive biological motion. Psychol Sci 4:54–57

Blough DS (1977) Visual search in the pigeon: hunt and peck method. Science 196:1013–1014

Blough DS (1979) Effects of the number and form of stimuli on visual search in the pigeon. J Exp Psychol Anim Behav Process 5:211–223

Blough DS (1989) Odd-item search in pigeons: display size and transfer effects. J Exp Psychol Anim Behav Process 15:14–22

Blough DS (1992) Features of forms in pigeon perception. In: Honig WK, Fetterman JG (eds) Cognitive aspects of stimulus control. Erlbaum, Hillsdale, pp 263–277

Blough DS, Blough PM (1997) Form perception and attention in pigeons. Anim Learn Behav 25:1–20

Blough DS, Franklin JJ (1985) Pigeon discrimination of letters and other forms in texture displays. Percept Psychophys 38:523–532

Blough PM (1989) Attentional priming and visual search in pigeons. J Exp Psychol Anim Behav Process 15:358–365

Bolster RB, Pribram KH (1993) Cortical involvement in visual scan in the monkey. Percept Psychophys 53:505–518

Bruce C (1982) Face recognition by monkeys: absence of inversion effect. Neuropsychologia 20:515–521

Bruce V (1988) Recognising faces. Erlbaum, Hillsdale

Bruce V, Young A (1986) Understanding face recognition. Br J Psychol 77:305–327

Cook RG (1992a) Acquisition and transfer of visual texture discriminations by pigeons. J Exp Psychol Anim Behav Process 18:341–353

Cook RG (1992b) Dimensional organization and texture discrimination in pigeons. J Exp Psychol Anim Behav Process 18:354–363

Cook RG, Cavoto KK, Cavoto BR (1996) Mechanisms of multidimensional grouping, fusion, and search in avian texture discrimination. Anim Learn Behav 24:150–167

Deruelle C, Fagot J (1998) Visual search for global/local stimulus features in humans and baboons. Psychon Bull Rev 5:476–481

Diamond R, Carey S (1986) Why faces are and are not special: an effect of expertise. J Exp Psychol: Gen 115:107–117

Dittrich WH, Lea SEG, Barrett J, Gurr PR (1998) Categorization of natural movements by pigeons: visual concept discrimination and biological motion. J Exp Anal Behav 70:281–299

Dursteler MR, von der Heydt R (1992, November) Visual search strategies of monkey and man. Paper presented at the meeting of the Society of Neuroscience, Anaheim, CA

Egeth HE, Virzi RA, Garbart H (1984) Searching for conjunctively defined targets. J Exp Psychol Hum Percept Perform 10:32–39

Ellis HD (1986) Processes underlying face recognition. In: Bruyer R (ed) The neuropsychology of face perception and facial expression. Erlbaum, Hillsdale, pp 1–27

Enns JT, Rensink RA (1990) Sensitivity to three-dimensional orientation in visual search. Psychol Sci 1:323–326

Fox R, McDaniel C (1982) The perception of biological motion by human infants. Science 218:486–487

Fujita K, Kanazawa S (1994) Visual search in Japanese macaques. In: Research report of the 1992–1993 Grant-in-Aid for Scientific Research from the Monbusho (in Japanese). Inuyama, Japan, pp 12–20

Fujita K, Ishikawa S, Tomonaga M, Matsuzawa T (1999) Development of initial knowledge in primate infants. In: Research report of the Grant-in-Aid for Scientific Research from the Monbusho (in Japanese). Tokyo, pp 59–66

Gurnsey R, Browse RA (1989) Asymmetries in visual texture discrimination. Spat Vision 4:31–44

Hansen CH, Hansen RD (1988) Finding the face in the crowd: an anger superiority effect. J Personality Soc Psychol 54:917–924

Hasegawa R, Kato M, Mikami A (1997) Delayed visual search on a rhesus monkey (abstract only in Japanese). Primate Res 13:284

He ZJ, Nakayama K (1992) Surfaces versus features in visual search. Nature 359:231–233

Herman LM, Morrel-Samuels P, Pack AA (1990) Bottlenosed dolphin and human recognition of veridical and degraded video displays of an artificial gestural language. J Exp Psychol Gen 119:215–230

Hopkins WD, Morris RD, Savage-Rumbaugh ES (1991) Evidence for asymmetrical hemispheric priming using known and unknown warning stimuli in two language-trained chimpanzees (Pan troglodytes). J Exp Psychol: Gen 120:46–56

Johansson G (1973) Visual perception of biological motion and a model for its analysis. Percept Psychophys 14:201–211

Johnston WA, Schwarting IS (1997) Novel popout: an enigma for conventional theories of attention. J Exp Psychol Hum Percept Perform 23:622–631

Kleffner DA, Ramachandran VS (1992) On the perception of shape from shading. Percept Psychophys 52:18–36

Kuehn SM, Jolicoeur P (1994) Impact of quality of the image, orientation, and similarity of the stimuli on visual search for faces. Perception 23:95–122

Kuroshima H (1999, September) Comparison of search asymmetries between humans and pigeons. Paper presented at the 63rd annual meeting of the Japanese Psychological Association, Nagoya, Japan

Mackay HA (1991) Conditional stimulus control. In: Iversen IH, Lattal KA (eds) Experimental analysis of behavior, Part 1. Elsevier, Amsterdam, pp 301–350

Maylor EA, Hockey R (1985) Inhibitory component of externally controlled covert orienting in visual space. J Exp Psychol Hum Percept Perform 11:777–787

Mikami A (1999) Analyses of brain mechanisms underlying the visual search behavior of primates. In: The Mitsubishi Foundation annual report 1998 (in Japanese). Tokyo, pp 257–259

Miura K, Kawabata H (1999) Effects of relative orientation differences on the detection of shading information: tests with visual search tasks. Paper presented at the 63rd annual meeting of the Japanese Psychological Association, Nagoya, Japan

Neely JH (1977) Semantic priming and retrieval from lexical memory: roles of inhibitionless spreading activation and limited-capacity attention. J Exp Psychol Gen 106:226–254

Nothdurft HC (1993) Faces and facial expressions do not pop out. Perception 22:1287–1298

Olivers CNL, van der Helm PA (1998) Symmetry and selective attention: a dissociation between effortless perception and serial search. Percept Psychophys 60:1101–1116

Omori E (1997) Comparative study of visual perception using Johansson's stimuli. In: Watanabe S, Chase S (eds) Pattern recognition in humans and animals. Keio University, Tokyo, pp 27–30

Oram MW, Perrett DI (1994) Responses of anterior superior temporal polysensory (STPa) neurons to "biological motion" stimuli. J Cogn Neurosci 6:99–116

Overman WA, Doty RW (1982) Hemispheric specialization displayed by man but not macaques for analysis of faces. Neuropsychologia 20:113–128

Parr LA, Dove T, Hopkins WD (1998) Why faces may be special: evidence of the inversion effect in chimpanzees. J Cogn Neurosci 10:615–622

Pashler H (1987) Target–distractor discriminability in visual search. Percept Psychophys 41:285–302

Pineda JA, Nava C (1993) Event-related potentials in macaque monkey during passive and attentional processing of faces in a priming paradigm. Behav Brain Res 53:177–187

Posner MI (1980) Orienting of attention. Q J Exp Psychol 32:3–25

Posner MI, Snyder CCR (1975) Facilitation and inhibition in the processing of signals. In: Rabbitt PMA, Dornic S (eds) Attention and performance, V. Academic Press, New York, pp 669–682

Ramachandran VS (1988) Perception of shape from shading. Nature 331:163–166

Rosenfeld SA, Van Hoesen GW (1979) Face recognition in the rhesus monkey. Neuropsychologia 17:503–509

Schneider W, Shiffrin RM (1977) Controlled and automatic human information processing. I. Detection, search, and attention. Psychol Rev 84:1–66

Shiffrin RM, Schneider W (1977) Controlled and automatic human information processing. II. Perceptual learning, automatic attending and a general theory. Psychol Rev 84:127–190

Sireteanu R, Rettenbach R (1995) Perceptual learning in visual search: fast, enduring, but, non-specific. Vision Res 35:2037–2043

Spinelli D, Antonucci G, Goodenough DR, Pizzamiglio L, Zoccolotti P (1991) Psychological mechanisms underlying the rod-and-frame illusion. In: Wapner S, Demick J (eds) Field dependence–independence: cognitive style across the life span. Erlbaum, Hillsdale, pp 37–60

Stins JF, Michaels CF (1997) Stimulus–target compatibility for reaching movements. J Exp Psychol Hum Percept Perform 23:756–767

Suzuki, S, Cavanagh P (1995) Facial organization blocks access to low-level features: an object inferiority effect. J Exp Psychol Hum Percept Perform 21:901–913

Tipper SP (1985) The negative priming effect: inhibitory priming by ignored objects. Q J Exp Psychol 37A:571–590

Tipper SP, Lortie C, Baylis GC (1992) Selective reaching: evidence for action-centered attention. J Exp Psychol Hum Percept Perform 18:891–905

Tomonaga M (1993a) Use of multiple-alternative matching-to-sample in the study of visual search in a chimpanzee (Pan troglodytes). J Comp Psychol 107:75–83

Tomonaga M (1993b) A search for search asymmetry in chimpanzees (Pan troglodytes). Percept Motor Skills 76:1287–1295

Tomonaga M (1994a) How laboratory-raised Japanese monkeys (Macaca fuscata) perceive rotated photographs of monkeys: evidence for an inversion effect in face perception. Primates 35:155–165

Tomonaga M (1994b, October) Search asymmetry in the chimpanzee. II. Does a tilted line pop out among the vertical lines? Paper presented at the 58th annual meeting of the Japanese Psychological Association, Tokyo, Japan

Tomonaga M (1995a) Visual search by chimpanzees (Pan): assessment of controlling relations. J Exp Anal Behav 63:175–186

Tomonaga M (1995b) Transfer of odd-item search performance in a chimpanzee (Pan troglodytes). Percept Motor Skills 80:35–42

Tomonaga M (1997a) Precuing the target location in visual searching by a chimpanzee (Pan troglodytes): effects of precue validity. Jpn Psychol Res 39:200–211

Tomonaga M (1997b) Search asymmetry in the chimpanzee. III (abstract only in Japanese). Jpn J Anim Psychol 47:200

Tomonaga M (1997c, September) Visual search for biological motion patterns in the chimpanzee. Paper presented at the 61st annual meeting of the Japanese Psychological Association, Nishinomiya, Japan

Tomonaga M (1998a) Perception of shape from shading in chimpanzees (Pan troglodytes) and humans (Homo sapiens). Anim Cogn 1:25–35

Tomonaga M (1998b, October) Priming effects on the discrimination performance in the chimpanzees. Paper presented at the 62nd annual meeting of the Japanese Psychological Association, Koganei, Japan

Tomonaga M (1998c) Visual search for biological motion patterns in the chimpanzee. II (Japanese abstract). Jpn J Anim Psychol 48:106

Tomonaga M (1999a) Visual texture segregation by the chimpanzee (Pan troglodytes). Behav Brain Res 99:209–218

Tomonaga M (1999b) Inversion effect in perception of human faces in a chimpanzee (Pan troglodytes). Primates 40:417–438

Tomonaga M (1999c) Visual search for orientation of faces by a chimpanzee (Pan troglodytes) (in Japanese with English summary). Primate Res 15:215–229

Tomonaga M, Itakura S, Matsuzawa T (1993) Superiority of conspecific faces and reduced inversion effect in face perception by a chimpanzee (Pan troglodytes). Folia Primatol 61:110–114

Tong F, Nakayama K (1999) Robust representations for faces: evidence from visual search. J Exp Psychol Hum Percept Perform 25:1016–1035

Treisman A (1985) Preattentive processing in vision. Comput Vision, Graphics, Image Process 31:156–177

Treisman A, Gelade G (1980) A feature-integration theory of attention. Cogn Psychol 12:97–136

Treisman A, Gormican S (1988) Feature analysis in early vision: evidence from search asymmetries. Psychol Rev 95:15–48

Treisman A, Sato S (1990) Conjunction search revisited. J Exp Psychol Hum Percept Perform 16:459–478

Treisman A, Schmidt H (1982) Illusory conjunctions in the perception of objects. Cogn Psychol 14:107–141

Treisman A, Souther J (1985) Search asymmetry: a diagnostic for preattentive processing of separable features. J Exp Psychol Gen 114:285–310

Treisman A, Cavanagh P, Fischer B, Ramachandran VS von der Heydt R (1990) Form perception and attention: striate cortex and beyond. In: Spillmann L, Werner JS (eds) Visual perception: the neurophysiological foundations. Academic Press, San Diego, pp 273–316

Wang Q, Cavanagh P, Green M (1994) Familiarity and pop-out in visual search. Percept Psychophys 56:495–500

Wenderoth PM (1973) The effects of tilted outline frames and intersecting line patterns on judgments of vertical. Percept Psychophys 14:242–248

Wilkinson F (1986) Visual texture segmentation in cats. Behav Brain Res 19:71–82

Wilkinson F (1990) Texture segmentation. In: Stebbins WC, Berkley MA (eds) Comparative perception, vol 2. Wiley, New York, pp 125–156

Wolfe JM (1994a) Guided search 2.0: a revised model of visual search. Psychon Bull Rev 1:202–238

Wolfe JM (1994b) Visual search in continuous, naturalistic stimuli. Vision Res 34:1187–1195

Wolfe JM, Cave KR, Franzel SL (1989) Guided search: an alternative to the feature integration model for visual search. J Exp Psychol Hum Percept Perform 15:419–433

Wright AA, Roberts WA (1996) Monkey and human face perception: inversion effects for human faces but not for monkey faces or scenes. J Cogn Neurosci 8:278–290

Yamaguchi MK, Fujita K (1999) Perception of biological motion by newly hatched chicks and quail. Perception 28(supplement):23–24

Yin RK (1969) Looking at upside-down faces. J Exp Psychol 81:141–145

Young MP (1995) Open questions about the neural mechanisms of visual pattern recognition. In: Gazzaniga MS (ed) The cognitive neurosciences. MIT Press, Cambridge, pp 463–474

4
Processing of the Global and Local Dimensions of Visual Hierarchical Stimuli by Humans (*Homo sapiens*), Chimpanzees (*Pan troglodytes*), and Baboons (*Papio papio*)

JOËL FAGOT[1], MASAKI TOMONAGA[2], and CHRISTINE DERUELLE[1]

1 Global-to-Local Precedence Effects

A well-known phenomenon in the literature on human perception is the global precedence effect, which was initially reported by Navon in 1977. Navon presented human subjects with hierarchical stimuli such as those shown in Fig. 1. These were large letters (global level) made up of smaller letters (local level) which had to be identified as quickly as possible. Use of these forms showed that response latencies were shorter on average for the global trials, involving identification of the global letter, than for the local trials, involving identification of the local letter. It was also shown that response times (RTs) in global trials remained unchanged whatever the identity of the letter shown at the local level, whereas RTs in local trials were higher when the global and local levels showed different letters than when they showed the same letter. On the basis of these results, Navon (1977) suggested that the processing of visual stimuli by humans proceeds from an analysis of the global structure of the visual form prior to the analysis of its more local details. According to Navon (1977), this global-to-local order of visual processing is a general trait of visual perception, and is independent of the use of hierarchical letter stimuli, as revealed by the title of his article "Forest before the tree: the precedence of global features in visual perception."

Navon's (1977) influential work has triggered a large body of literature on the reliability of global precedence, its perceptual or cognitive origins, and the experimental conditions under which this effect appears in humans. Briefly stated, global precedence was found to be a reliable effect (e.g., Boer and Keuss 1982; Kimchi 1988), although some authors found that it may turn into a local advantage under certain circumstances, for instance when the display is larger than 6–9° of visual angle (Kinchla and Wolfe 1979). This relative ability to process the local and global dimensions of hierarchical stimuli was also found to have diagnostic value for

---

[1] CNRS, 31 ch. Joseph Aiguier 13402, Marseille cedex 20, France
[2] Primate Research Institute, Kyoto University, 41 Kanrin, Inuyama, Aichi 484-8506, Japan

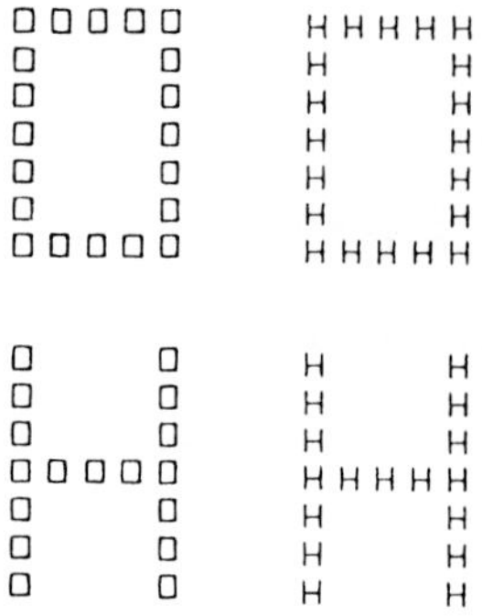

Fig. 1. Some of the stimuli used in Navon's (1977) original study. (Redrawn from Navon 1977)

some diseases such as Alzheimer's disease (Filoteo et al. 1992) or depression (Basso et al. 1996). Still uncertain, however, is the relative importance of purely sensory-perceptual or post-perceptual mechanisms in global precedence.

According to Navon (1991), it is likely that precedence effects have some sort of adaptive advantages. The prioritization of the global aspect of the shape, Navon argued, may allow reliable identification of the perceived object, because the global shape is often more unique than the isolated features, thus allowing the viewer quickly to narrow the range of possible identities of the stimulus. If this is true, we would expect global precedence to exist in animals as well, because there is a priori no reason to believe that global shape is a more discriminant dimension for humans than for animals. So what does the animal literature tells us about this possibly adaptive effect?

Unfortunately, very few studies have investigated precedence effects in animals in a manner comparable to what Navon did with humans. In birds, Cook et al. (1996) demonstrated that pigeons may learn to attend selectively to the global aspects of stimuli differing in their texture and color and ignore changes in local features, which suggests a precedence of the global dimension over the local one. Cerella (1982), by contrast, observed that pigeons strongly focus on local features to distinguish complex forms. Fulbright-Cavato (unpublished thesis, cited in Fremouw et al. 1998) showed that pigeons learned the local aspects of hierarchical stimuli more rapidly than the global ones. Fremouw et al. (1998) showed that responses to global and local targets in hierarchical stimuli are sensitive to a base-rate type of priming. In such cases, there were shorter RTs to targets (global or local) at the primed level.

In nonhuman primates, chimpanzees constructed figures composed of two elements by first selecting the outer contour of the figures, and then by selecting the local element (Fujita and Matsuzawa 1990), which demonstrates a global-to-local order of processing. Horel (1994) demonstrated that cooling the dorsal inferotemporal cortex of fascicularis macaques inhibited their ability to process the global level of these stimuli but not the local level, thus suggesting that these two stimulus levels are processed by distinct neural networks. Using a video formatted task in which hierarchical forms of the Navon (1977) type were to be matched by considering either their global or their local shape, Hopkins (1997) found a left hemisphere advantage in chimpanzees for processing local targets, and no clear-cut lateralization for the processing of global cues. In a similar task with baboons, Fagot and Deruelle (1997) reported a significant right hemisphere advantage for global matching, and an insignificant left hemispheric advantage for local matching. The findings by Hopkins (1977) and Fagot and Deruelle (1997) are consistent with previous findings on lateralization in humans (e.g., Delis et al. 1986). Fagot and Deruelle (1997) also reported a score and speed advantage in baboons to process the local level of the stimuli compared with the global level. Interestingly, humans tested in the same conditions as the baboons showed better scores and shorter RTs for global than for local trials, suggesting that the processing of these forms may differ between species.

In brief, two statements emerge from the animal literature on precedence effects. First, although the real world is not strictly dichotomized into global and local levels, the global/local distinction appears to be heuristically valid for an understanding of animal perception and cognition, as demonstrated in both the animal cognition (e.g., Fremouw et al. 1998) and neuroscientific (e.g., Hopkins 1997) literature. Second, there is very weak evidence for global precedence effects in animals, and available data show that animals may selectively attend to either the global or local stimulus level (e.g., Fremouw et al. 1998), or have a strong propensity to focus on the local stimulus level (e.g., Fagot and Deruelle 1997). The elusiveness of global precedence effects in animals, compared with the apparent pervasiveness of this effect in humans, suggests that this phenomenon has a recent phylogenetic origin. Unfortunately, comparisons between species are often difficult because of variations in the testing procedures and stimuli (but see Fagot and Deruelle (1977) for the use of similar procedures with humans and baboons). Examination of the conditions under which a global or local precedence effect may emerge in animals, and a direct comparison with humans, seems critical to a better understanding of visual perception in both humans and animals, and its evolution.

Bearing the preceding remarks in mind, the aim of this chapter is to present a synthesis of some of our collaborative work on global and local processing in humans and nonhuman primates (see Deruelle and Fagot (1998) and Fagot and Tomonaga (1999) for a more detailed presentation of these studies). Two lines of research will be presented. The first will focus on the comparison between humans and baboons. The second will focus on the comparison between humans and chimpanzees, a species which is more closely related to humans than are baboons. The use of a common visual search procedure with these different groups of primates allows direct comparisons between species, in order to pinpoint similarities and differences in their processing of visual stimuli. In addition, these two lines of research shed some light on the nature of the processing applied to the stimuli by each species, and allow identification of some of the factors that may contribute to the prioritization of one stimulus level (global or local) over the other.

2 Principle of the Visual Search Task

The experiments reported in this chapter are all based on the visual search procedure which is commonly employed to assess the relative involvement of attentional processes in visual discrimination tasks in humans (e.g., Treisman and Gelade 1980). In these visual search tasks, subjects report whether or not a target stimulus is present in a display containing a variable number of distractors. Of particular interest are variations in RTs with the number of distractors in the display, the rationale being that baseline RTs reflect preattentive stages of processing, which are the product of reflexive, data-driven mechanisms. Complementary, more inferential processes that might be associated with selective attention are revealed by a linear increment of RTs with display size.

In the context of our experiments, subjects (a human or nonhuman primate) were to detect a hierarchical stimulus of a Navon type among distractors of the same type. Differences between targets and distractors differed from trial to trial. In some trials, hereafter referred to as the "global trials," the target and distractors had the same local elements but differed in their global structure, implying an analysis of the global dimension of the shape for discrimination. In other trials, the "local trials," the target differed from the distractors at its local level only. Such a procedure is particularly well suited for studying global/local processing because it permits the detection of precedence effects by comparing mean RTs in these two conditions. Moreover, variations of response times across display sizes indicate whether these two stimulus levels are coded by early visual mechanisms, without selective attention, or are processed at a higher, more deliberate, level, involving an attentional consideration of form variations.

3 Comparative Assessment of Global and Local Processing in Humans and Baboons

This first line of research consists of two experiments, the second experiment serving as a control for the first one. The first experiment was conducted on eight adult humans (four males and four females) and eight adult baboons (*Papio papio*; five males and three females). The subjects were tested with the same apparatus and experimental design for the purpose of species comparison. In all these experiments, the care and use of animals adhered to the "Guide for the Care and Use of Laboratory Primates" (1986), of the Primate Research Institute, Kyoto University. Figure 2 illustrates the experimental set-up. The experiment used a go/no go procedure, and involved an apparatus comprising a joystick and a monitor screen.

The subject initiated the trials by manipulating the joystick in order to move a cursor appearing in the center of the screen on a small fixation point, appearing either above or below the cursor. Once done, a go or a no go display, comprising several forms, appeared on the screen. In the go trials, the forms were all large squares made up of smaller squares except for the target, which differed from the distractors in its global (global trial) or local (local trial) structure (Fig. 3). In the no

Fig. 2. Photographic view of the experimental setup. The picture shows a baboon manipulating a joystick in response to visual displays presented on the monitor screen

go trials, all the forms of the display were large squares made up of smaller squares. There were also three possible display sizes, corresponding to 4, 8, or 12 hierarchical stimuli on the screen. The task was to move the joystick in the go trial, when the target was present, and to refrain from moving it in the no go trial, when the target was absent. A go response was considered as correct when the subject moved the joystick within 3 s after form presentation. A no go response was considered correct when the subject refrained from moving the joystick within 3 s. Correct go or no go responses gave rise to the delivery of a tone for the two species, and of an additional food reward for the baboons.

Prior to the test, the baboons received training sessions with the same stimuli as those shown in Fig. 3, but with only 1, 3, or 5 stimuli in each display. The training consisted of sessions of 120 trials, each comprising 30 global-go and 30 local-go trials intermixed with 60 no go trials. There was the same number of trials with 1, 3 or 5 stimuli per display in each condition. Training sessions were repeated until the subject performed 80% correct or more over two consecutive sessions. After this performance was achieved, the subjects from the two species received four test sessions of 120 trials each, which were similar to those of the training sessions except that the displays contained 4, 8, or 12 stimuli.

Inspection of Fig. 3 shows several differences between the global and local levels of our test stimuli. First, these two levels differ in visual size. Second, the local level is made of continuous forms, whereas the global shape is discontinuous. Because the main experiment does not disentangle the possible effects of these two factors (size and discontinuity), we proposed a control experiment which used the stimuli shown at the top of Fig. 3. Control stimuli were of the same visual size as the global structure of the hierarchical stimuli, but consisted of continuous circles or squares.

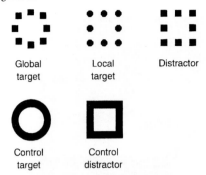

Global   Local   Distractor
target   target

Control   Control
target   distractor

Fig. 3. (*Top*) Stimuli used in the visual search task with humans and baboons, and (*bottom*) an illustration of a test display as perceived by the subject. The experimental stimuli were 3° × 3° of visual angle yellow forms displayed on a black background. The local elements subtended 0.6° of visual angle

During the control test, continuous stimuli were presented to the baboons within the context of the same visual search task as before, and in the same number of trials. The control experiment was run immediately after the main experiment.

Performance levels were very high in these experiments, with very few individual variations in scores (humans, 99% correct on average, range 98–100%; baboons, 90.1%, range 81–99%). Given such a high level of performance, data analyses focused on variations in RTs across species and test condition.

To analyze the speed of the responses, median RTs for successful go trials were subjected to an analysis of variance (ANOVA) in which the species (human, baboon), the type of trial (global, local), and the display size (4, 8, 12) were considered as factors. This analysis revealed a significant interaction between the species and the type of trial ($F(1,14) = 21$, $P < 0.001$). Post hoc tests based on this significant interaction (Tukey HSD, $P < 0.05$) indicated that humans exhibited a significant

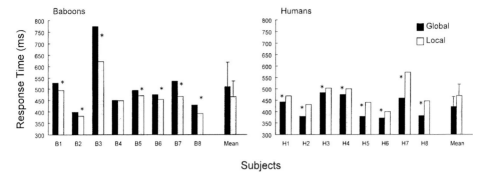

Fig. 4. Individual median response time (RT) for the global and local trials presented to baboons (*left*) and humans (*right*). *Stars* indicate the results of two-tailed *t*-tests ($P < 0.05$) comparing individual mean RTs for local trials with mean response times for global trials (regardless of display size). Note that seven baboons exhibited shorter RTs for local than for global trials, and that all the eight humans showed an opposite speed advantage for the global trials

advantage in the global trials compared with the local ones, whereas baboons showed a strong and reliable advantage in the local trials. This effect is illustrated in Fig. 4, which reports individual RTs for both human and baboon subjects. Inspection of this figure reveals that seven of the eight baboons demonstrated significantly shorter RTs for local compared with global trials. Individual data from humans show a significant global advantage for all of them. Clearly, *humans and baboons did not process our stimuli in the same way*.

Another interesting result from the ANOVA was a significant species-by-trial type (global, local) by display size (4, 8, 12) interaction ($F(2,28) = 3.3$, $P = 0.05$). That interaction is shown in Fig. 5, along with the results from the use of continuous stimuli with baboons (control experiment).

Inspection of Fig. 5 first confirms the reliability of the global advantage for humans and of the local advantage for baboons, as these two effects appear regardless of the display size. Figure 5 also reveals that the search slope for the global trials of baboons is much steeper than the search slope for local trials made by the same animals. In order to verify whether RTs increased linearly with display size, we computed trend analyses independently for each condition. In the case of global trials, linearity accounted for 99% of the observed variance ($P < 0.05$), showing the linear increment of response times with display size. By contrast, the search slope is not significant in the local condition. This result demonstrates qualitative differences in the processing of global and local cues by baboons. It appears that the detection of the global target demands great attention, whereas the detection of the local cues involves little or no attention, these cues being automatically perceived by the animals.

Of particular importance for our purpose is the comparison between the findings from baboons in the global and local trials and those obtained with the continuous stimuli (control condition). Thus, a first display size (4, 8, 12) by experi-

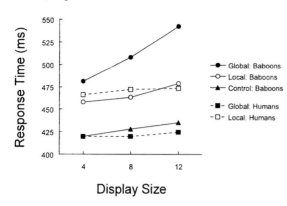

Fig. 5. Median RT for each species for each display size (4, 8, 12) and trial type (global or local) condition

ment (main experiment, control) ANOVA was conducted to compare median RTs in the local condition with median RTs in the control condition. That analysis revealed that the effect of the experiment was significant ($F(1,7) = 7.04, P < 0.05$), showing faster response times on average in the control condition than in the main experiment (Fig. 5). Note that the display size by experiment interaction was not significant, which suggests that these two conditions involved a similar type of presumably nonattentional processing.

A similar ANOVA comparing the global condition involving discontinuous stimuli with the control condition showed that the effect of display size was larger for the global trials than for the control trials ($F(2,14) = 3.71, P < 0.01$), suggesting that, although the forms were of the same size in the two experimental conditions, the processing of the global discontinuous stimuli differed from that of the continuous forms. Taken together, these results demonstrate that the attentional nature of the global processing is unrelated to the size of the stimuli per se. Instead, results indicate that the search slope observed for global trials reflects the attentional nature of the grouping operations which are necessary to perceive the global shape of the discontinuous stimuli.

It is of interest to compare the results from baboons with those of humans who were tested under the same conditions as the baboons. A visual inspection of data shown in Fig. 5 reveals flat search slopes for humans, whatever the testing condition. This lack of effect of display size was confirmed by trend analyses ($P < 0.1$). It appears, therefore, that the global shape of our hierarchical stimuli could be detected in parallel by humans, without focal attention.

This first series of experiments pinpoints some important human–baboon species differences in the processing of complex hierarchical stimuli. On the one hand, the baboons have a greater facility for processing the local stimulus level compared with the global one, whereas humans show the opposite pattern (i.e., global precedence). On the other hand, the processing of the global stimulus level demands great attention from baboons, whereas humans detect the global targets without focal attention. At this point, experimental manipulations reveal that large stimuli may pop out to both humans and baboons, suggesting that species differences cannot be explained by variations in the size of their attentional windows. Instead, the findings suggest that these species differ in their facility for perceptual grouping. The significance of these results will be discussed later, after presentation of the second line of research which provides complementary information on a different primate species.

4 Comparative Assessment of Global and Local Processing in Humans and Chimpanzees

This line of research was intended to serve as a bridge between our previous studies with humans and baboons. It involved two adult chimpanzees (*Pan troglodytes*), Chloe (a 16-year-old female) and Akira (a 21-year-old male), both from the Primate Research Institute of Kyoto University. These experiments had two main objectives. First, to discover whether chimpanzees would behave like baboons or humans when they were tested with the same hierarchical stimuli as those employed in the previous experiments. Second, to provide additional information on the possible relation between the need for perceptual grouping and precedence effects. A small group of two adult human subjects was also tested in this experiment. Inclusion of these subjects served as a test to verify whether our procedure was instrumental in revealing precedence effects.

In their general principle, the experiments with chimpanzees were identical to those with baboons. Thus, hierarchical stimuli, which were the same as those at the top of Fig. 3, were presented in the context of a visual search task. Animals had to detect the target forms, depending on the trials, considering either their global or their local structure. However, there was one main difference in the procedure employed with baboons and chimpanzees; chimpanzees had to respond to the target by pointing at it on a touch screen, rather than by manipulating a joystick (Fig. 6).

Fig. 6. Photographic view of the experimental set-up used for the chimpanzees. The animals were seated in front of a tactile screen and had to give their response by touching the target stimulus

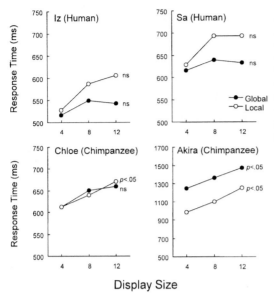

Fig. 7. Individual RTs for humans (*top*) and chimpanzees (*bottom*) as a function of display size (4, 8, 12) and trial type (global, local). Significant linear relations between RTs and display size, as inferred from a trend analysis ($P < 0.05$), are indicated on the right of each curve. *ns*, Nonsignificant linear relation between RTs and display size

In practice, a trial started by the presentation of a warning stimulus which appeared at the bottom of the monitor screen. Immediately after the subject had touched this, the warning signal disappeared and the search display appeared. The task was then to detect the target and to touch it on the screen. Correct responses gave rise to the delivery of a small food item. An incorrect response induced a correction trial in which the same target–distractor was displayed again. If another incorrect response was given to this display, only the target was shown in a second correction trial.

After an initial training phase in which the animals had to learn the identity of the target stimuli, chimpanzees and humans were tested in three consecutive experimental sessions of 168 trials each. These sessions used the stimuli shown at the top of Fig. 3. Within a test session, half of the trials were global and the other half were local. For each trial type there were 12 one-display-size trials, and 24 trials with 4, 8, or 12 displays. Trial order was randomly determined prior to each session, and thus varied from one session to the other.

Because of the small number of subjects, only individual data will be presented in this chapter. Individual performance was very high on average (more than 98% correct for each individual), which prevents score analyses owing to a likely ceiling effect. Individual RTs as a function of display size are shown in Fig. 7 for each individual subject (human and chimpanzee) and display size.

The results for the two humans are shown at the top of Fig. 7. They demonstrated (1) a significant advantage in processing the global shape compared with the local one, and (2) no significant linear relation between display size and response times. These two results confirm our previous findings with human subjects (Sect. 3). They both demonstrate the reliability of these two effects as well as the adequacy of our testing procedure to reveal them.

The results from the two chimpanzees differ from those of the humans in two respects (Fig. 7). Unlike humans, there was no evidence of any advantage for processing the global stimulus level in the two chimpanzees. Indeed, Chloe showed no significant difference in response speed between the global and local trials ($P >$ 0.1). Akira showed a significant advantage for processing the local stimulus level ($P < 0.05$), as did the baboons in their first experiment. Also noticeable is the significant linear relation between RTs and display size in the local condition for both the chimpanzees. Although similar findings have occasionally been reported for humans (e.g., Saarinen 1994; Found and Müller 1997), this is the first time we have observed it in the context of our comparative experiments.

In brief there were both differences and similarities between our findings for chimpanzees and those obtained with baboons. On the one hand, not unlike baboons, there is no evidence in chimpanzees for a global precedence effect, as Akira showed a local advantage and Chloe showed no significant advantage for processing either local or global cues. As with baboons, there is also some evidence for attentional processing at the global stimulus level, as revealed by the steepness of the search curves in that condition. On the other hand, processing at the local stimulus level demanded sustained attention from chimpanzees, as revealed by the linear search slopes, whereas the stimulus popped out for baboons, thus suggesting that differences exist between nonhuman primate species in the processing of this type of visual cue.

One limitation of the above study is that it does not indicate why global precedence effects never emerged in chimpanzees. In line with the findings from baboons, it can be hypothesized that the chimpanzees showed some difficulty in processing the global shapes because they were made of discontinuous local elements which had to be grouped into a unitary percept. Under this hypothesis, the additional operation of grouping is presumed to be very time consuming for chimpanzees, which would slow down the processing of the global dimension of the forms, and thus provide a speed advantage to the local trials compared with the global ones.

In order to test this "grouping hypothesis" directly, the chimpanzees were tested with stimuli similar to those of the previous experiment. On some trials, however, dotted lines were added in between adjacent local elements in order to connect them (Fig. 8). According to the gestalt theory (e.g., Koffka 1935), such lines make the forms close and continuous, which should facilitate the perception of their global shape. Because Chloe showed no significant advantage in the previous experiment, this test used stimuli which were larger than the previous ones in order to make it easier to demonstrate shifts in precedence effects, depending on the stimulus type. Accordingly, the experimental stimuli had $4 \times 4°$ of visual angle, and contained eight local elements.

The two chimpanzees received three sessions of 336 trials each. Within a session, there were an equal number of trials per stimulus set (connected, disconnected), by stimulus level (global, local), by display size (4, 8, 12) condition, trial order being randomly determined prior to each experimental session.

Individual response times are shown graphically in Fig. 9. Inspection of this

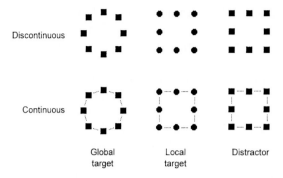

Discontinuous

Continuous

Global   Local   Distractor
target   target

Fig. 8. Stimuli used in the control experiment conducted with chimpanzees. The discontinuous stimuli (*top*) were identical to those of the first experiment with chimpanzees. The continuous stimuli (*bottom*) were similar to the discontinuous ones, except that line segments were added between the local elements. Experimental stimuli were 4 × 4° of visual angle yellow forms displayed on a black background. The local elements subtended 0.6° of visual angle

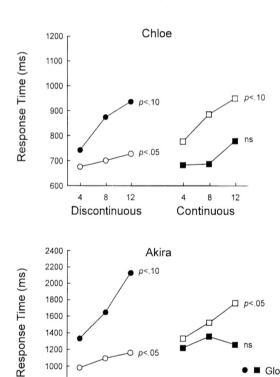

Fig. 9. Mean RTs for Chloe and Akira as a function of display size (4, 8, 12), type of stimulus (discontinuous, continuous), and type of trial (global, local). Significant linear relations between RTs and display size, as inferred from a trend analysis, are shown on the right of each curve. *ns,* Nonsignificant linear relation between RTs and display size

figure reveals several interesting effects. First consider the results from the discontinuous condition, which are shown on the left-hand side of each graph. In line with our previous findings, the two chimpanzees showed (statically significant, $P < 0.05$) advantages for processing local cues compared with global ones. Because Chloe did not show that advantage in the previous experiment, this result supports the conclusion that increasing the interelement distance favors local precedence. Figure 9 also shows the linearity of the search slope in the local condition involving discontinuous stimuli, as well as the steepness of the search slopes in the global condition, suggesting that these two stimulus levels are processed under focal attention. Again, the last two findings confirm the results of the previous experiment.

Interestingly, a very different pattern of results emerges from the use of the continuous stimulus set; for each chimpanzee, use of these stimuli revealed an advantage for processing the global stimulus level. Therefore, *the local advantage observed with the discontinuous forms turned into a global one when line segments were added to the stimulus.* Also important is the observation that the search slope disappeared, in the global condition, when segments were added to the stimuli. This suggests that the continuous stimuli were perceived in a pop-out manner. The results for chimpanzees will be discussed below, along with those previously obtained with baboons and humans. In that discussion, we will first emphasize some of the mechanisms accounting for precedence effects in humans, chimpanzees, and baboons. We will then focus our attention on the observed species differences in precedence effects, and their theoretical and practical significance for our understanding of primate visual perception and cognition.

5 What Mechanisms Are the Basis of Precedence Effects?

Several aspects of our data suggest that precedence effects in chimpanzees and baboons are affected by changes in stimulus input. The strongest evidence for such an effect derives from the test with chimpanzees, in which lines were added in between the local elements. In that experiment, the use of lines induced an advantage for processing the global shape, whereas a local advantage emerged when the line segments were removed from the stimulus. Although deriving from a different experimental approach, comparable effects can be inferred from the experiments conducted with baboons. Indeed, these animals showed faster processing of the local cues compared with the global ones when discontinuous stimuli were used (main experiment). By contrast, the RTs of baboons became minimal, or even significantly faster than for the local trials, when continuous large stimuli were used as targets and distractors (control condition).

Why did line segments make the local advantage turn into a global one? It might be that the main effect of line segments is to enhance the continuity of the global shape, thus favoring its perception as an integrated stimulus dimension. The use of lines might also have the additional effect of increasing the colinearity of the local elements, which might again favor perception of the global stimulus level as a "good form," according to the gestalt rule of stimulus perception (e.g., Koffka 1935). Although the current experiments did not attempt to disentangle the respective influences of the factors of stimulus continuity and local element linearity, results overall reveal that the cohesiveness of the global stimulus dimension is a determinant factor affecting precedence effects in chimpanzees and baboons. In line with this remark, several other experiments carried out by our research group revealed that the local advantage for both baboons (Deruelle and Fagot 1998, Experiment 4) and chimpanzees (Fagot and Tomonaga 1999, Experiment 2) may vanish when the distance between the local elements is reduced. Chloe's data showed the same effect, because Chloe had no significant local or global advantage with stimuli of 3° of visual angle (main experiment), but she had a local advantage with larger stimuli of 4°, for which the interelement distance was increased (discontinuous condition).

Interestingly, stimulus cohesiveness also controls precedence effects in humans, as reported by Martin (1979) who showed a shift from a global to a local advantage depending on the distance separating the local elements. These findings support the contention that both the global and local information of disconnected patterns are processed in parallel, the relative availability of the information at each level depending on the ease with which the global (and presumably local) information is processed. They also demonstrate that precedence effects are, at least in part, under the control of input-driven mechanisms.

Although the current experiments strongly emphasize the influence of stimulus factors on precedence effects, they do not rule out the possibility that these effects are complementarily determined by more inferential top-down processes at a higher level. Remember that Fremouw et al. (1998) demonstrated that pigeons may shift their attention from one stimulus level to the other depending on the relative probability of the global and local trials. In one of our experiments on categorization processes (Fagot et al. 1998), baboons selectively processed different subsets of stimulus features depending on the constraints of the task, thus showing comparable abilities for attentional shift. It is worth noting that some aspects of the human literature also point out the possibility that precedence may have an attentional source. Thus, Ward (1982) reported that identification of a form at a given level (local or global) is faster when the previous trials involved processing of the same level (see also Kinchla et al. (1983) for a base-rate type of global/local priming). Generally, results from both the animal and human literature support the idea that precedence effects are codetermined by bottom-up (perceptual) and top-down (attentional) processes. We believe that this dual influence of bottom-up and top-down processes is important in order to understand species differences in precedence effects.

6 Species Differences in Precedence Effects

In order to understand species differences in global/local processing, it is important to ensure that our three species under study have equally sensitive visual systems. Unfortunately, there are very few psychophysical data on baboon visual perception. The available information on primate perceptual abilities derives largely from studies of macaques species, mainly rhesus macaques (*Macaca mulatta*; e.g., De Valois and De Valois 1990) which are Old-World species closely related to baboons. Macaque studies have demonstrated that the visual system of these species share some important properties with that of humans. The visual field of macaques is the same size as the human one (e.g., Wilson et al. 1989). At a photopic luminance level, their visual acuity is slightly lower than that of humans, but these two species can still perceive details as small as 0.65 min of visual angle (Fobes and King 1982). The contrast sensitivity functions of macaques and humans are very similar, and are both characterized by a high sensitivity for stimuli of 2–6 cycles per degree (De Valois and De Valois 1990), and their temporal acuity critical fusion frequencies are also similar (Fobes and King 1982). The functional characteristics of the visual perception of chimpanzees are still largely unknown, although one study has demonstrated that these animals have an acuity comparable to that of humans (Matsuzawa 1990). Overall, the evolution of visual systems appears to have reached a plateau in these primate species, since they are all of roughly similarly sensitivity. In the context of our study, although it is impossible to completely rule out the possibility that subtle interspecies variations in perceptual abilities alter the results, a reasonable and safe position is to interpret interspecies differences in global/local processing as determined by postperceptual (cognitive) factors, rather than by early sensory factors.

The current experiments have demonstrated important differences in the way humans and nonhuman primates (i.e., baboons and chimpanzees) process hierarchical stimuli: only in humans did the global precedence effect emerge whenever we used discontinuous stimuli. In the other two primate species, the use of discontinuous stimuli showed either no reliable difference between the two stimulus levels, or a local advantage. Also striking is the observation that the perception of the global structure of our stimuli popped out for humans, but involved a serial search strategy in baboons and in one chimpanzee (i.e., Akira), as demonstrated by significant search slopes. As already stated above, our control experiments suggest that significant search slopes in global trials reveal the animals' difficulty in grouping the local elements into a unitary percept.

To account for the differences between humans and nonhuman primates, it might be suggested that these species differ in their conceptual knowledge of the test stimuli. The human and animal cognitive literature has demonstrated that conceptualization and categorization may affect perception. The largest bulk of evidence of this sort is derived from the fields of perceptual learning (e.g., Gibson 1969) and categorical perception (e.g., Harnad 1987), which have demonstrated that people increase their perceptual sensitivity to stimuli by categorizing or identifying them. In the context of our experiments, it might be suggested that human subjects were able to detect changes in global shapes easily because they were familiar with the shape (square or circle) depicted at the global level. Baboons or chimpanzees would not have such a facility because of their lack of a conceptual representation of a square or a circle, or their lack of symbolic tags to refer to them. Although our baboons and chimpanzees were sophisticated animals, in the sense that they were trained and had previously been involved in a variety of cognitive discrimination problems, they had never been asked to sort squares or circles into distinct categories, or to differentially associate symbols to circles or squares. However, it is interesting to note that humans and baboons had very similar (flat) visual search curves in local trials, although the local elements were squares or circles. These findings suggest that categorical perception effects, if any, helped the grouping process of the local elements into coherent circles or squares, but had minimal effects on the perception of physically well-defined squares or circles.

It is also interesting to compare the results from monkeys (the baboons) and apes (the chimpanzees). Two aspects of the data suggest that these two species were not similarly sensitive to precedence effects. First, seven of the eight baboons showed a significant advantage for processing the local stimulus level of discontinuous stimuli, whereas only one of the two chimpanzees (i.e., Akira) showed this local effect. Second, the two chimpanzees showed a statistically significant linear relation between RTs and display size in their local trials, whereas baboons did not. Although humans showed flat search slopes in the local trials, visual search studies have repeatedly revealed attentional processing of the local aspects of hierarchical stimuli (e.g., Saarinen 1994; Found and Müller 1997). The results from chimpanzees are therefore in line with the main findings of the human literature. Further comparative experiments will be needed to evaluate the reliability of these findings, and to hypothesize a possible evolutionary path, within the primate order, of precedence effects and their underlying mechanisms.

Acknowledgments

The authors wish to thank the staff of both the CRNC and the Primate Research Institute for their care of the animals. Special thanks are also due to M. Chiambretto (CRNC) and S. Nagumo (PRI) for their technical advice, and T. Matsuzawa and M. Huffman for their helpful comments on an earlier version of this paper.

References

Basso MR, Schefft BK, Ris MD, Dember WN (1996) Mood and global–local visual processing. J Int Neuropsychol Soc 3:249–255
Boer LC, Keuss PJG (1982) Global precedence as a postperceptual effect: an analysis of speed–accuracy tradeoff functions. Percept Psychophys 31:352–366
Cerella J (1982) Mechanisms of concept formation in pigeon. In: Ingle DJ, Goodale MA, Mansfield RJ (eds) Analysis of visual behaviour. MIT Press, Cambridge
Cook RG, Cavato KK, Cavato, BR (1996) Mechanisms of multidimensional grouping, fusion, and search in avian texture discrimination. J Exp Psychol, Anim Behav Process 24:150–167
Delis DC, Roberston LC, Efron R (1986) Hemispheric specialization of memory for visual hierarchical stimuli. Neuropsychologia 24:205–214

Deruelle C, Fagot J (1998) Visual search for global/local stimulus features in humans and baboons. Psychonom Bull Rev 5:476–481

De Valois RL, De Valois KK (1990) Spatial vision. Oxford University Press, New York

Fagot J, Deruelle C (1997) Processing of global and local visual information and hemispheric specialization in humans (*Homo sapiens*) and baboons (*Papio papio*). J Exp Psychol: Hum Percept Performance 23:429–442

Fagot J, Tomonaga M (1999) Comparative assessment of global–local processing in humans (*Homo sapiens*) and chimpanzees (*Pan troglodytes*): use of a visual search task with compound stimuli. J Comp Physiol 113:3–12

Fagot J, Kruschke JK, Dépy D, Vauclair J (1998) Associative learning in humans (*Homo sapiens*) and baboons (*Papio papio*): species differences in learned attention to features. Anim Cogn 1:123–133

Filoteo JV, Delis DC, Massman PJ, Demadura T (1992). Directed and divided attention in Alzheimer's disease: impairment in shifting attention to global and local stimuli. J Clin Exp Neuropsychol 14:871–883

Fobes JL, King JE (1982) Vision: the dominant primate modality. In: Fobes JL, King JE (eds) Primate behavior. Academic Press, New York, pp 219–243

Found A, Müller HJ (1997) Local and global orientation in visual search. Percept Psychophys 59:941–963

Fremouw T, Walter TH, Shimp CP (1998) Priming of attention to local and global levels of visual analysis. J Exp Psychol: Anim Behav Process 24:278–290

Fujita K, Matsuzawa T (1990) Delayed figure reconstruction by a chimpanzee (*Pan troglodytes*) and humans (*Homo sapiens*). J Comp Psychol 104:345–351

Gibson EJ (1969) Principle of perceptual learning and development. MIT Press, Cambridge

Harnad S (1987) Categorical perception. Oxford University Press, Oxford

Hopkins WD (1997) Hemispheric specialization for local and global processing of hierarchical visual stimuli in chimpanzees (*Pan troglodytes*). Neuropsychologia 35:343–348

Horel JA (1994) Local and global perception examined by reversible suppression of temporal cortex with cold. Behav Brain Res 65:157–164

Kimchi R (1988) Selective attention to global and local levels in the comparison of hierarchical patterns. Percept Psychophys 43:189–198

Kinchla RA, Wolfe JM (1979) The order of visual processing: "top down," "bottom up," or "middle-out." Percept Psychophys 25:225–231

Kinchla RA, Solis-Macias V, Hoffman J (1983) Attending to different levels of structure in the visual image. Percept Psychophys 33:1–10

Koffka KA (1935) Principles of gestalt psychology. Harcourt, Brace & World, New York

Martin M (1979) Local and global processing: the role of sparsity. Mem Cogn 7:476–484

Matsuzawa T (1990) Form perception and visual acuity in a chimpanzee. Folia Primatol 55:24–32

Navon D (1977) Forest before the tree: the precedence of global feature in visual perception. Cogn Psychol 9:353–383

Navon D (1991) Testing a queue hypothesis for the processing of global and local information. J Exp Psychol: Gen 120:173–189

Saarinen J (1994) Visual search for global and local stimulus features. Perception 23:237–243

Treisman A, Gelade G (1980) A feature integration theory of attention. Cogn Psychol 12:97–136

Ward LM (1982) Determinants of attention to local and global features of visual forms. J Exp Psychol: Hum Percept Performance 8:562–581

Wilson JR, Lavallee KA, Joosse MV, Hendrickson AE, Boothe RG, Harwerth RS (1989) Visual field of monocularly deprived macaque monkeys. Behav Brain Res 33:13–22

5
How Do We Eat? Hypothesis of Foraging Strategy from the Viewpoint of Gustation in Primates

Yoshikazu Ueno

1 Relation Between Tastes and Foods

Eating is one of the fundamental activities of animals, including humans. We humans usually have to eat several times in a day. Humans can consume quite a wide range of plants and animal foods and so are called omnivorous apes. Also, humans cannot tolerate continuous ingestion of monotonous foods irrespective of nutritional sufficiency. Although this behavior is conspicuous in humans, many primates other than humans also consume a variety of foods, including plants and animals such as insects. Chimpanzees and baboons, for example, even hunt small mammals. As a result of this feeding strategy, the primates are generally labeled as omnivores.

However, it is not always biologically beneficial to be able to consume a variety of foods because foods do not always include only nourishment but also toxins. The greater variety in the diet, the more unpredictable the risk of ingesting toxic substances, that is, the "packaging problem" (Altman 1998). Actually, unripe fruits and leaves, for example, include plant secondary compounds such as alkaloids and tannins, which cause poisoning or dyspepsia in mammals. Animals select foods by the senses of vision, smell, and taste, which decreases the risk of ingesting too many toxic compounds. In particular, the sense of taste does not start working until foods directly contact the tongue and thus functions as an important checkpoint for the decision of intake.

The substances humans perceive to be sweet include sugars, amino acids, and proteins. These foods do not include toxic substances for humans but are utilizable as energy resources. Therefore, these substances are palatable for many animals as well as humans, which facilitates their ingestion as foods. However, carnivores such as felines, which depend on amino acids rather than carbohydrates as their energy resource, do not show a preference for sugars such as glucose. Animals prefer the tastes that are related to their resource and thus those tastes enable them to accomplish efficient foraging or energy intake.

On the other hand, plant secondary compounds such as alkaloids, terpenes, glycosides, and peptides are generally perceived as bitter or astringent in taste by

Primate Research Institute, Kyoto University, 41 Kanrin, Inuyama, Aichi 484-8506, Japan

humans. These substances often have physiological effects, and they can cause ill effects or even lethal effects if they are ingested excessively. That is, animals must avoid excessive intake of such substances. The tastes of these substances must function as a cue to inhibit ingestion. Actually, many animals reject them as unpalatable as to taste and vomit reflexively if they swallow them.

We can say that fundamentally sweet and bitter or astringent tastes function as cues for "facilitation" and "inhibition," respectively, of ingestion. However it cannot be always determined unconditionally how animals choose foods or perceive the taste of foods. To elucidate these problems, it is necessary to investigate the evolutional background of foraging. Therefore, this study aimed to investigate the evolution of feeding in primates, including humans, considering data concerning the sense of taste and the ecology of feeding from previous studies.

2 Response to Sweet Taste in Primates

Most species of primates eat fruits as a major food item and show a high preference for sweet taste in general. Some species, for example, squirrel monkeys, prefer such extremely high concentrations of sweet solution that it seems as if they do not have an upper limit to elicit rejection of sweetness (Laska 1996). Although sweet taste functions as a cue to facilitate feeding behavior, as already described, there have been fewer studies that compare the response to sweetness among various species of primates. In 1996, Hladik and Simmen reported how 25 species of primates from prosimian to simian responded to different concentrations of sucrose solution by a two-bottle test: comparison of the intake of sucrose solution with that of water. The lowest preferred concentrations varied among species: for example, preference thresholds were 6 mM in rhesus macaques and 330 mM in slow lorises. However, these thresholds are significantly correlated with species body weight. That is, as body size increased, the taste acuity of primates became better so that they could perceive a wide range of concentration as palatable (Fig. 1). Primates also responded to fructose with a similar tendency. Simmen and Hladik interpreted this relationship as reflecting the importance of taste acuity in improving foraging efficiency.

Energy requirements vary with the species body size, so that two strategies to meet this requirement can be hypothesized as follows. First: animals prefer only high sugar concentration and selectively use sweeter foods or a narrow range of foodstuffs. Second: animals perceive even a low sugar concentration as palatable and use a wide range of foodstuffs regardless of the intensity of sweetness. The former strategy allows animals to forage efficiently when foods with a high sugar concentration, for example, ripe fruits, are always available. In general, however, it is not actually possible for animals with large body size to acquire consistently sufficient foods with a high sugar concentration. In contrast, the latter strategy allows animals to perform rather adaptive foraging because they can use sufficient amounts of available foods, including some with low sugar. Consequently, we can agree with the interpretation by Simmen and Hladik that primates with large body

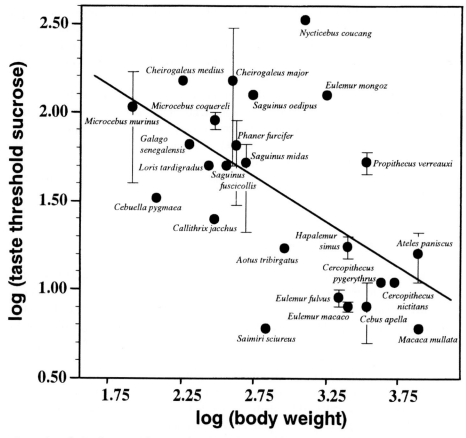

Fig. 1. Correlation between the taste threshold for sucrose and the body weight for 25 species of primates (all logarithms to base 10). Correlation are significant: $r = -0.58$, 24 df, $F_{1,23} = 11.7$, $P < 0.003$. (Redrawn from Simmen and Hladik 1998)

size are enabled to choose more varied foods by this foraging strategy regardless of the intensity of sweet taste.

3 Response to Bitter/Astringent Taste in Primates

Anecdotally, it had been generally assumed that bitter or astringent taste functions as an important clue to avoid poisons present in foods (Bate-Smith 1972). Most animals reject bitter or astringent taste. However, Glendinning (1994) showed that bitter/astringent tastes do not necessarily elicit the rejection response (e.g., withdrawal reflex, negative affective response) equally in any mammalian species by their review of the response to bitter/astringent taste in various mammalian species. In addition, Glendinning suggested that mammalian species have basically the following two strategies for coping with the plant secondary compounds according to the trophic feeding strategy: (1) carnivores and omnivores, whose intrinsic food items do not include the secondary compounds, are sensitive to bitter/astringent tastes and select foods by clues of taste; and (2) grazers and browsers, whose intrinsic food items tend to include the secondary compounds, are very tolerant of bitter/astringent taste, and they can digest or detoxify these compounds by means of a specifically evolved digestive system, such as a lumen, but do not select foods by their bitter/astringent taste so intensively.

Although primates are usually classified as omnivores, their main food items are plants. However, it is known that not only humans but all primates are more willing to eat bitter/astringent foods than to reject these (Koshimizu et al. 1993). Therefore, in contrast with the preference for sweetness studied by Simmen and Hladik, as discussed earlier, Ueno (in manuscript) compared the tolerance to bitter and astringent tastes among humans and 89 species of primates from prosimians to apes that were reared in the Japan Monkey Center, Sapporo City, Maruyama Zoo, or the Primate Research Institute, Kyoto University. Every animal species was fed fruits (apples, bananas, etc.), sweet potato, and monkey chow. An experiment was carried out with the following procedures at the home cage where the animals were housed.

Quinine chloride (QHCl) was used as the bitter taste and tannic acid as the astringent taste. Three different concentrations (QHCl: 10^{-5} M, 10^{-3} M, and 10^{-1} M; tannic acid: 2^{-5} M, 2^{-3} M, 2^{-1} M) were selected on the basis of three different thresholds of humans acquired by same procedure as this study: (1) detection threshold, (2) recognition threshold, and (3) rejection threshold. To minimize the effect of novelty and to increase the motivation of feeding, this test used apples as a basis for the stimuli. Apple slices were pierced to spread the solution into the inside and soaked for longer than 30 min before each session. Apple slices were presented to the subjects in order of ascending concentration. When animals rejected apples, the session ended. The amount and duration of feeding and responses during feeding (manipulation, facial expression, saliva, etc.) were directly observed and recorded by video. Each session used one stimulus condition, and only one session was carried out in a day to minimize the effect of the advancing aversive experience.

Every species altered their response to apples in accordance with the intensity of taste stimuli and also expressed a gustofacial reflex to the bitter taste (typically arched form of mouth; Steiner and Glaser 1984). Most species began to express hesitation of ingestion starting from the median concentration (10^{-3} M) of the bitter taste and rejected ingestion at the highest concentration (10^{-1} M). However, great apes, except for gorillas, and folivores such as the colobus monkey did not refuse to ingest bitter apples even at the highest concentration (10^{-1} M); that is, they showed great tolerance for the bitter taste. The tolerance of astringent taste was not so different among species, and most species could ingest acerbate apples at the highest concentration (2×10^{-1} M).

Similarities of the property of the sense of taste among 90 species including human, which were calculated by response to QHCl and tannic acid, were analyzed by the multidimensional scaling method. As a result, the species were roughly classified into two groups (Fig. 2). Group 1 consisted of gibbons, cercopithecines,

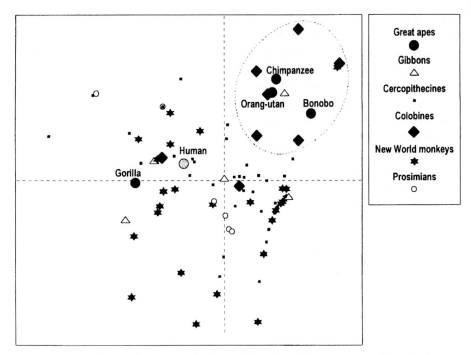

Fig. 2. Plot of similarity of sensitivity to bitter and astringent taste in primates by multidimensional scaling (MDS). Great apes, except gorilla, and colobines, are distributed at the upperright peripheral. The range of these two groups is roughly distinguished from that of the other groups

New World monkeys, and lemurs, and group 2 consisted of colobines and great apes except gorillas.

4 Hypothesis of Foraging Strategy to Bitter/Astringent Taste

Most primates were highly tolerant of astringent taste. It is known that browsers and some omnivores, which commonly ingest tannin-rich foods, are tolerant of astringent taste (Glendinning 1994; Hagerman and Robbins 1993). Further, the previous study reported that primates reject astringent foods on the basis not of the absolute intensity but of the relative intensity compared with sugar (Simmen 1994). Accordingly, primates probably do not use astringency independently as a cue for food selection.

On the other hand, the tolerance to bitter taste was different among groups. The rejection thresholds of group 1 were almost the same as that of omnivorous Rodentia (8×10^{-4} M; Glendinning 1994). Accordingly, the species in group 1, which have not evolved a physical detoxification system, must depend on bitter taste for food selection at the same level as omnivores. However, groups 2 and 3 did not reject the apples even at the highest concentration. This result suggested that although these species do not prefer bitter tastes, they do not employ bitter taste as a cue for food selection. For example, Hladik (1981) reported as well that chimpanzees include in their food items a proportion (15%) of plants likely to have a high alkaloid content similar to the proportion (14%) of plants that reacted positively to the alkaloid test among 382 species in the Gabon rain forest. Further, the recent comparison of chimpanzees with several sympatric monkeys elucidated that chimpanzees are more willing to ingest tannin-rich foods than are the monkeys (Reynolds et al. 1998).

The previous studies showed that folivorous species or colobines evolved a physical detoxification system such as a rumen and rumen microbes (Bauchop and Martucci 1968). Accordingly, these species evolved a physical construction and function and acquired the capacity to detoxify secondary compounds as did herbivores. On the other hand, great apes have not evolved the ability to detoxify as do colobines. Do they have any unknown strategy for coping with plant toxic substances? We propose here that logically they must avoid ingesting large quantities of a single toxic food item in any one day. Actually, it is known that the great apes, except gorillas, consume a variety of food items in 1 day; e.g., more than 15 for chimpanzees (Hladik 1977; Rodman 1977; Wrangham 1977). This food diversity in chimpanzees increases conspicuously in the dry season, when chimpanzees depend on plants other than fruits in comparison with the rainy season (Yamakoshi 1998). These observations can support the possibility that great apes, except gorillas, cope with unpredictable toxic risk by increasing their daily food diversity. On the contrary, gorillas, which alone of the great apes demonstrate a low tolerance to bitter taste, use fewer daily food items than the other great apes, although they have no fewer food items than the other great apes on an annual basis (Yamagiwa et al. 1996). Gorillas also eat more kinds of leaf and bark. Accordingly, gorillas must select foods at the same level as group 1.

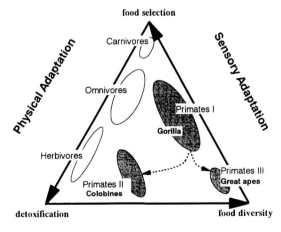

Behavioral Adaptation

Fig. 3. Hypothetical model of the evolution of the strategies for coping with plant toxins in mammals. The model shows that primates evolved a strategy such as omnivory intermediate to the behavioral strategy in great apes, which increased the daily dietary variety, or the physical strategy in colobines, which acquired a detoxification system

In summary, we hypothesize that primates evolved the following two different strategies for coping with unpredictable toxic risk (Fig. 3). The first is physical adaptation or the evolution of a detoxification system, and the second is behavioral adaptation or an increase in daily food diversity. In either case, one would expect a high tolerance to bitter tastes; in the first case, because they can detoxify poisons, and in the second case, because they are only eating small amounts of any foods. Also, the latter strategy, which uses various foods, does not contradict with that strategy for sweetness in primates with larger bodies as described earlier.

To support the latter hypothesis, first, it is important to examine whether primates crave variety in their diets. However, there is little knowledge concerning this phenomenon so far. It has been reported that humans crave different foods based on quality when they continue eating monotonous foods even though those are sufficient in nutrition (Pelchat and Schaefer 2000). Given that humans are a species of primates closely related to the chimpanzee, it is possible that the diversity of foods chosen by chimpanzees or other apes has also occurred intentionally to some extent. Second, primates must be able to remember the food items consumed in a day to eat a wide variety of items. Sawaguchi (1992) reported that frugivorous primates developed a larger cerebral cortex than folivorous primates. It has been hypothesized that this difference was caused by the need for spatial memory. That is, because fruits distribute more dispersively than leaves, frugivorous primates have to remember those food patches to forage efficiently. The development of spatial memory capacity ought to have affected other memory systems. Consequently, primates acquired the ability to increase this food diversity through the process of evolution.

5 Conclusion

Behavioral adaptation, which requires enlargement of the cerebral cortex, can be considered a specific strategy in primates. Primates, including humans, are sometimes classified as omnivores, although the feeding property of primates is probably quite different from that of other omnivores such as rats. Primates have evolved not only the ability to ingest various foods but also the motivation to crave varied foods. Therefore, the author proposes to relabel this foraging strategy of primates "variovorous" rather than "omnivorous." Additionally, craving a variety of foods in humans likely evolved with this biological background.

References

Altman ST (1998) Foraging for survival: yearling baboons in Africa. University of Chicago Press, Chicago
Bauchop T, Martucci RW (1968) Ruminant-like digestion of the langur monkey. Science 161:698–699
Bate-Smith EC (1972) Attractants and repellants in higher animals. In: Harbone JB (ed) Phytochemical ecology. Academic Press, London, pp 45–57

Glendinning JI (1994) Is the bitter rejection response always adaptive? Physiol Behav 56:1217-1227

Hagerman AE, Robbins CT (1993) Specificity of tannin-binding salivary proteins relative to diet selection by mammals. Can J Zool 71:628-633

Hladik CM (1977) Chimpanzees of Gabon and chimpanzees of Gombe: some comparative data on the diet. In: Clutton-Brock TH (ed) Primate ecology: studies of feeding and ranging behavior in lemurs, monkeys and apes. Academic Press, London, pp 481-501

Hladik CM (1981) Diet and the evolution of feeding strategies among forest primates. In: Harding RSO, Teleki G (eds) Omnivorous primates: gathering and hunting in human evolution. Columbia University Press, New York, pp 215-254

Hladik CM, Simmen B (1996) Taste perception and feeding behavior in nonhuman primates and human population. Evol Anthropol 5:58-71

Koshimizu K, Ohigashi H, Huffman MA (1993) Physiological activities and the active constituents of potentially medicinal plants used by wild chimpanzees of the Mahale Mountains, Tanzania. Int J Primatol 14:345-356

Laska M (1996) Taste preference thresholds for food-associated sugars in the squirrel monkey (*Saimiri sciureus*). Primates 37:91-95

Pelchat ML, Schaefer S (2000) Dietary monotony and food craving in young and elderly adults. Physiol Behav 68:353-359

Rodman PS (1977) Feeding behavior of orangutans of the Kutai Nature Reserve, East Kalimantan. In: Clutton-Brock TH (ed) Primate ecology: studies of feeding and ranging behavior in lemurs, monkeys and apes. Academic Press, London, pp 384-413

ReynoldsV, Plumptre AJ, Greenham J, Harborne J (1998) Condensed tannins and sugars in the diet of chimpanzees (*Pan troglodytes schweinfurthii*) in the Budongo Forest, Uganda. Oecologia (Berl) 115:331-336

Sawaguchi T (1992) The size of the neocortex in relation to ecology and social structure in monkeys and apes. Folia Primatol 58:131-145

Simmen B (1994) Taste discrimination and diet differentiation among New World primates. In: Chivers DJ, Langer P (eds) The digestive system in mammals: food, form and function. Cambridge University Press, Cambridge, pp 150-165

Simmen B, Hladik CM (1998) Sweet and bitter taste discrimination in primates: scaling effects across species. Folia Primatol 69:129-138

Steiner JE, Glaser D (1984) Differential behavioral responses to taste stimuli in nonhuman primates. J Hum Evol 13:709-723

Wrangham RW (1977) Feeding behaviour in chimpanzees in Gombe National Park, Tanzania. In: Clutton-Brock TH (ed) Primate ecology: studies of feeding and ranging behavior in lemurs, monkeys and apes. Academic Press, London, pp 503-538

Yamagiwa J, Maruhashi T, Yumoto T, Mwanza N (1996) Dietary and ranging overlap in sympatric gorillas and chimpanzees in Kahuzi-Biega National Park, Zaire. In: McGrew WC, Marchant LF, Nishida T (eds) Great ape society. Cambridge University Press, Cambridge, pp 82-98

Yamakoshi G (1998) Dietary response to fruit scarcity of wild chimpanzees at Bossou, Guinea: possible implication for ecological importance of tool use. Am J Physiol Anthropol 106:283-295

Part 3
Origin of Human Speech:
Auditory Perception and Vocalization

6
Lemur Vocal Communication and the Origin of Human Language

Ryo Oda

1 Introduction

To shed light on the evolution of human language, many studies have been made on the vocal communication of nonhuman primates. These studies have revealed that some rudimentary properties of human language can be seen in nonhuman primate vocal communication. In particular, much knowledge about the natural vocal communication of primates has accumulated since the method of playback experiments was established. For example, referential signaling (Zuberbuhler et al. 1999), categorical perception of vocalization (Masataka 1983), acoustic "rules" regulating vocal exchange (Sugiura 1993), and flexibility of vocal production (Sugiura 1998) have been found in some primate species in natural habitats. However, most of these studies have been conducted on the anthropoid primates. There have been relatively few studies on vocal communication in prosimians, including the lemurs of Madagascar, which are indigenous to the island and have evolved separately. Because of their uniqueness, the lemurs are important species as subjects for comparative studies of primate vocal communication.

Madagascar began its journey away from Africa as much as 165 million years ago, to reach its present separation of roughly 400 km from the continent. Having been separated from other ecological systems, many animals and plants on this island are unique. About 30 species of primates are now seen on the island. These primates, known as lemurs, belong to the infraorder Lemuriformes in the suborder Prosimii. Their ancestors are thought to have reached Madagascar in the Eocene epoch by floating across the Mozambique Channel from Africa on matted tangles of vegetation, and adaptively radiated to various environments of their "new world" (Tattersall 1982). Lemurs in Madagascar are of special interest since they have complex social structures and life styles in spite of their primitive physical characteristics and relatively small brains.

In this chapter, I describe several features of vocal communication by ring-tailed lemurs (*Lemur catta*) under natural conditions, and ask whether such features are the result of parallel evolution with anthropoids. Data are drawn from a popula-

Department of Humanities and Social Sciences, Nagoya Institute of Technology, Gokiso-cho, Showa-ku, Nagoya 466-8555, Japan

tion of ring-tails in Berenty Reserve, southern Madagascar. Additional data, supplementing those drawn from natural habitats, come from experiments conducted on a captive group of ring-tailed lemurs at the Izu Cactus Park, Japan. Based on these findings and other results of studies on primate cognition and behavior, I present an ethological view of the origin and evolution of human language.

2 Study Sites and Subjects

The ring-tailed lemur is found in the dry forests and bush of south and southwestern Madagascar. It is a medium-sized diurnal species which spends more time on the ground than any of the other lemurs. It is found in groups containing several adults of both sexes, with 15 individuals on average. The groups are female-bonded, and males migrate among the groups. The home range size varies from 5.7 ha to 34.6 ha. Its diet is principally fruits, leaves, and flowers. The ring-tailed lemur is one of the most studied of the lemurs, and is the most common lemur in zoos (Harcourt and Thornback 1990). The vocal repertoire of the ring-tailed lemur was described quantitatively by Macedonia (1993). There were up to 22 different adult call types, and the ring-tailed lemur has a moderately large vocal repertoire for a primate.

One of the study sites is the Berenty Reserve, a 200-ha forest reserve in the extreme south of Madagascar. Berenty is situated in an area of gallery forest surrounded by sisal plantations and the Mandrare River (Jolly 1966). There are large populations of three species of diurnal lemurs (*Lemur catta, Propithecus verreauxi verrauxi*, and *Eulemur fulvus*) as well as three species of nocturnal lemurs (*Cheirogaleus medius, Microcebus murinus*, and *Lepilemur mustelinus leucopus*). Several groups of ring-tailed lemurs have been identified and studied continuously since 1989 (Koyama 1991). Four of these groups (C1, C2, T1, and W) were chosen as study troops.

The colony group at the Izu Cactus Park, Shizuoka, Japan, inhabits a small island of about 400 m², composed of soil, rocks, grass, and trees. All individuals were born in Japan.

3 Interspecific Referential Signaling

Referential signaling, which is an important property of human language, was found in some vocal communications of nonhuman primates. Vervet monkeys (*Cercopithecus aethiops*) emit three different antipredator alarm calls which correspond to three kinds of predators (Cheney and Seyfarth 1990). Diana monkeys (*Cercopithecus diana diana*) have two types of alarm calls, which are used in response to eagles and leopards (Zuberbuhler et al. 1999). Agonistic screams emitted by rhesus macaques (*Macaca mulatta*) appear to refer to external objects and events, and serve as a possible source of information to recruit agonistic aid (Gouzoules et al. 1984). Trills given by spider monkeys (*Ateles geoffroyi*) function as referential signals to indicate fellow group members (Masataka 1986).

The ring-tailed lemur emits different vocalization in response to aerial (e.g., *Polyboroides radiatus*) and terrestrial (e.g., *Cyptoprocta ferox*) predators (Jolly 1966; Macedonia 1993). The response of semi-free-ranging ring-tailed lemurs in the Duke University Primate Center (DUPC), USA, to antipredator call playbacks suggested that these vocalizations denoted different classes of predators and functioned as referential signals (Macedonia 1990; Macedonia and Evans 1993). To rule out the response urgency hypothesis, which proposes that different antipredator calls denote the different levels of escape urgency that predators impose on their prey, Pereira and Macedonia (1991) presented models of aerial and terrestrial predators to the ring-tailed lemurs in DUPC. They investigated the response of the lemurs to each of four conditions (minimal/maximal urgency and aerial/terrestrial predator) and found that response urgency did not determine the antipredator call selection in ring-tailed lemurs. I did a series of playback experiments in the Berenty Reserve to test whether ring-tailed lemurs in their natural habitat were able to recognize their own alarm calls as referential signals.

The playback procedure basically followed that of Macedonia (1990). Stimulus alarm calls were recorded from several groups during encounters with actual predators in Berenty (Fig. 1a,b). The same call types were played to subjects at two locations (on the ground and in the trees). Subjects were tested once per playback type at a given location. Behaviors recorded for the subjects on the ground were: 1, looking skyward; 2, running into the branches of a tree; 3, running to any location

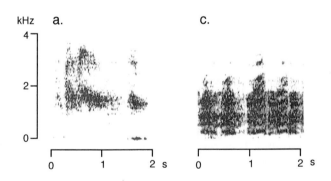

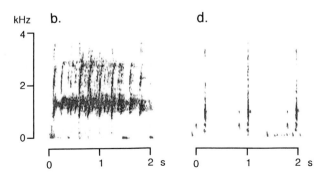

Fig. 1. Representative sound spectrograms of the four kinds of stimulus: **a** ring-tailed lemur antiraptor call; **b** ring-tailed lemur anticarnivore call; **c** Verreaux's sifaka antiraptor call; **d** Verreaux's sifaka anticarnivore call. (From Oda and Masataka 1996, with permission)

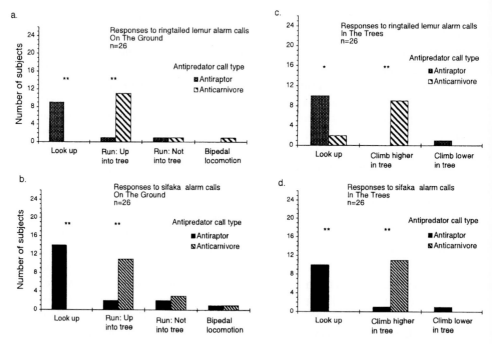

Fig. 2. Responses of free-ranging ring-tailed lemurs in Berenty on the ground (**a,b**) and in the trees (**c,d**) to playbacks of antipredator calls by their own species and by Verreaux's sifaka: n is the number of subjects; $*P < 0.05$; $**P < 0.01$. (From Oda and Masataka 1996, with permission)

except into a tree; 4, standing and walking or trotting bipedally. Behaviors recorded for the subjects in the trees were: 1, looking skyward; 2, moving upward; 3, moving downward. Detailed experimental procedures are given in Oda and Masataka (1996).

The alarm calls of ring-tailed lemurs were played back to 26 individuals (15 males and 11 females) on the ground. They responded appropriately to conspecific alarm calls (Fig. 2a). Nine lemurs looked skyward in response to the antiraptor calls, while none looked up in response to the anticarnivore calls, which is a statistically significant difference (Fisher's exact test, $P = 0.002$). On the other hand, only one individual ran into the trees in response to the antiraptor calls, whereas 11 animals did so in response to the anticarnivore calls ($P = 0.002$). Few subjects ran to any location except into the trees or responded by bipedal locomotion for either type of alarm call.

In the same way, alarm calls of ring-tailed lemurs were played back to 26 animals (14 males and 12 females) in the trees (Fig. 2c). Ten lemurs looked skyward in response to the antiraptor calls, whereas two animals did so in response to the anticarnivore calls, which is a statistically significant difference ($P = 0.019$). On the other hand, no individual climbed higher in response to the antiraptor calls, whereas seven animals did so in response to the anticarnivore calls ($P = 0.001$). Only one subject climbed lower.

Many animals communicate acoustically with members of other species as well as their own species. Some birds and primates form polyspecific groups. One possible benefit to forming mixed-species groups is to utilize other species as sentinels that give loud alarm calls (e.g., Munn 1986; Gautier-Hion et al. 1983). Interspecific communication is seen not only among species that form polyspecific groups, but also in many species living sympatrically. For example, vervet monkeys can respond to both their own and a sympatric bird species' alarm calls as if the calls denote specific types of danger (Cheney and Seyfarth 1990). Verreaux's sifaka (*Propithecus verreauxi verreauxi*) is another diurnal lemur in Berenty, which also gives different alarm calls for raptors and for terrestrial predators (Jolly 1966; Richard 1978; Fig. 1c,d). In southern Madagascar, ring-tailed lemurs and Verreaux's sifakas are sympatric and fall prey to the same predators. Sauther (1990) studied the antipredator responses of free-ranging ring-tailed lemurs in Beza–Mahafaly Special Reserve in Madagascar, and suggested that the lemurs respond both to alarm calls of their own species and to the alarm calls of sifakas as if the calls denote specific predators. I did another series of playback experiments to test whether ring-tailed lemurs in Berenty can recognize sifaka alarm calls as well as their own.

The alarm calls of Verreau's sifakas were recorded and played back to 26 ring-tailed lemurs (15 males and 11 females) on the ground. They also responded appropriately to sifaka alarm calls (Fig. 2b). Fourteen lemurs looked skyward in response to the sifaka antiraptor calls, although none looked up in response to the anticarnivore calls ($P = 0.000$). On the other hand, two individuals ran into the trees in response to the sifaka antiraptor calls, whereas 11 animals did so in response to the anticarnivore calls ($P = 0.009$). Few subjects ran anywhere except into the trees or responded by bipedal locomotion for either type of alarm call.

In the same way, alarm calls of Verreaux's sifaka were played back to 26 animals (14 males and 12 females) in the trees (Fig. 2d). Ten lemurs looked skyward in response to the antiraptor calls, whereas none did so in response to the anticarnivore calls ($P = 0.001$). On the other hand, only one individual climbed higher in response to the sifaka antiraptor calls, whereas 11 animals did so in response to the anticarnivore calls ($P = 0.002$). Only one subject climbed lower.

In Berenty, ring-tailed lemurs responded differently to the two types of sifaka alarm calls as well as their own calls. In the reserve, we could see ring-tailed lemurs and sifakas ranging closely at such a distance that they could see each other. They almost always ignored each other, even at a distance of 2 or 3 m. The diurnal activity pattern was different between these two species. Ring-tailed lemurs spent their daytime on the ground more often than Verreaux's sifakas (Howarth et al. 1986). In using different levels of the forest, sifakas could see areas that are blind spots for ring-tailed lemurs. Sifakas in the trees could be sentinels against raptors for ring-tailed lemurs on the ground. Indeed, the number of ring-tailed lemurs that responded appropriately was greatest when they heard sifaka antipredator calls on the ground.

These two diurnal species are threatened by the same predators and coexist successfully by niche separation. It would be advantageous for ring-tailed lemurs to recognize what sifaka alarm calls denote and to respond to them. In Berenty, I also conducted playback experiments of ring-tailed lemur alarm calls to 11 adult sifakas in trees (Oda 1998). The experiments are only preliminary, but they reveal the possibility that sifakas also perceive what types of predators the ring-tailed lemur calls denote. A mutual interspecific division of roles in warning behavior may exist in Berenty.

4 Learning Vocal Recognition?

The playback experiments indicate that ring-tailed lemurs in Berenty can perceive what type of predators the sifaka calls refer to. How do ring-tailed lemurs perceive the sifaka alarm calls? One possible way is that they find the same kind of acoustic features as their conspecific alarm calls in sifaka antipredator calls. That is, they might respond appropriately because the sifaka alarm calls have common acoustic features that elicit particular responses. Indeed, there are common characters in the sound structure of the alarm calls of both species. "Rasp" and "shriek," which ring-tailed lemurs emit against raptors, are atonal sounds, and that is also the case for "roar," which is the sifaka antiraptor call. On the other hand, the anticarnivore calls "yap" by ring-tailed lemurs and "fak" by sifakas are tonal sounds consisting of repeated, brief units (Fig. 1). An alternative possibility is that ring-tailed lemurs understand that sifaka alarm calls are other species' calls and respond to their "meaning" independently. I did playback experiments in the Izu Cactus Park to test whether ring-tailed lemurs that had never been exposed to sifakas were also able to recognize sifaka alarm calls as referential signals.

Ring-tailed lemurs in the Izu Cactus Park also emit different vocalizations in response to aerial and terrestrial predators just like wild lemurs. Figure 3a,c show the results of the playback experiments in which conspecific alarm calls were presented in the same way as in the previous study. Ten adults were chosen as subjects. On the ground, six lemurs looked skyward in response to the conspecific antiraptor calls, while none looked up in response to the anticarnivore calls ($P = 0.011$). On the other hand, no individual ran into the trees in response to the conspecific antiraptor calls, whereas five animals ran into the trees in response to the anticarnivore calls ($P = 0.047$). No subjects ran to any location except into the trees or responded by bipedal locomotion for either type of alarm call.

In the trees, five lemurs looked skyward in response to the conspecific antiraptor calls, whereas none looked up in response to the anticarnivore calls ($P = 0.033$). On the other hand, no individual climbed higher in response to the conspecific antiraptor alarm calls, whereas five animals climbed higher in response to the anticarnivore calls ($P = 0.033$). No subject was observed to climb lower.

In the same manner, alarm calls of sifakas in Berenty were played back to the same 10 ring-tailed lemurs. On the ground, three lemurs looked skyward in response to the antiraptor calls, and two animals did so in response to the anticarnivore calls. One subject ran into the trees in response to the anticarnivore calls, and no lemur ran anywhere except into the trees or responded by bipedal locomotion for either type of alarm call (Fig. 3b). In the trees, five lemurs looked skyward in re-

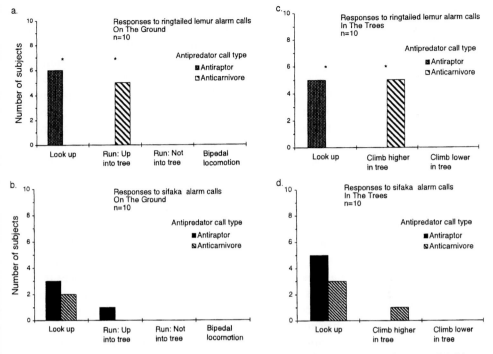

Fig. 3. Responses of a colony of ring-tailed lemurs in the Izu Cactus Park on the ground (**a,b**) and in the trees (**c,d**) to playbacks of antipredator calls by their own species and by Verreaux's sifaka: *n* is the number of subjects; *P < 0.05; **P < 0.01. (From Oda and Masataka 1996, with permission)

sponse to the sifaka antiraptor calls, and three animals did so in response to the sifaka anticarnivore call, revealing no statistically significant difference. Few subjects climbed higher or lower (Fig. 3d).

Ring-tailed lemurs in Berenty respond appropriately to heterospecific calls as much as to conspecific calls, whereas the lemurs in the Izu Cactus Park respond to conspecific calls more appropriately than to heterospecific calls. Although there are common acoustic features in the alarm calls of ring-tailed lemurs and sifakas, the results indicate that ring-tailed lemurs distinguish sifaka calls from their own calls. Acoustic experiences are important in order to recognize the referential signaling of alarm calls (Hauser 1988). Ring-tailed lemurs in the Izu Cactus Park, which had no contact with Verreaux's sifakas, had not learned the relationship between the sifaka calls and the things for which they stand.

5 Contact Calls as Social Communication

In group-living primates, members of a troop need to communicate with each other in order to maintain cohesion. This communication may be based on physical contact, e.g., grooming, as well as on the transfer of visual or acoustic information. Contact calls occur in many species of primates. They maintain the cohesiveness of the group (Macedonia 1986; Gautier and Gautier 1977; Snowdon 1989; Harcourt et al. 1993). In addition, contact calls given by some species function in a rudimentary representational manner (Cheney and Seyfarth 1990; Masataka 1986). Most contact calls share the properties of being tonal in structure and of relatively low frequency. Previous investigators noted that contact calls are often heard in rapid sequences involving two, or occasionally more, animals, and that group members do not vocalize at random (see the chapter by H. Sugiura in this volume).

In ring-tailed lemurs, vocalizations possessing the features of contact calls are often produced during daily activities, and these were named "meow calls" by Jolly (1966). Variants of this call have been described by Andrew (1963) and Macedonia (1993) as "moan" and "wail," and by Petter and Charles-Dominique (1979) as "cohesion miaouw." Macedonia (1986) recorded and spectrographically analyzed these vocalizations from eight semi-free-ranging individuals in the DUPC. The results of a discriminant function analysis indicated that statistical discrimination of individuals was possible on the basis of the acoustic differences in the calls. Thus, ring-tailed lemurs possess the acoustic basis for individual recognition. I conducted research on the contact calls of ring-tailed lemurs in Berenty Reserve to investigate the way they employ the calls in their social life.

I studied one troop of ring-tailed lemurs (C2 troop) between August and December 1993 (Oda 1996). The troop contained seven adult males (≥ 3 years), six adult females, one subadult female (≥ 2 year), and three juveniles. The group has been identified and studied since 1989 (Koyama 1991). The names of all the members are abbreviated to two letters of the alphabet. Sixteen animals in the C2 troop were the subjects of the focal animal sampling. I recorded all occurrences of vocalizations by a focal animal 2–4 m from the subject. In addition, I noted the focal animal's activity when calling and its proximity to other group members within a radius of 3 m. I also scored activity and proximity (within a radius of 3 m) to other group members at 5-min intervals using a point sampling method (Altmann 1974).

I numbered meow calls of high quality, and chose 30 calls per individual ($N = 480$) at random. I conducted acoustic analyses for those 480 calls. The measurements of acoustic features are restricted to the fundamental frequency (Fig. 4). The activities of focal animals are in one of three categories: resting, moving, or foraging. Ring-tailed lemurs emit meow calls more frequently while resting and moving than while foraging, and the frequency is not significantly different between resting and moving (Fig. 5). For acoustic parameters, the effect of context is not significant.

Significantly more calls occurred when other animals were absent than when others were present, despite the fact that the subjects had neighbors ≤ 3 m away significantly more often than not (Fig. 6). The effect of proximity on the acoustic parameters is significant for duration, maximum minus minimum frequency, and median frequency. For meow calls emitted when there were no other animals within a 3-m radius, the duration was shorter, frequency modulation was stronger, and median frequency was higher than those recorded when others were present

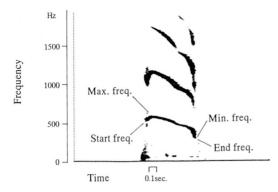

Fig. 4. Sound spectrogram of a meow call. Labels refer to the spectral measurements used in the analysis. (From Oda 1996, with permission)

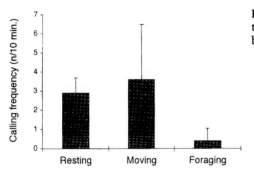

Fig. 5. Mean number of meow calls emitted per 10 min by 16 individuals in each behavioral context. T-bar indicates SD

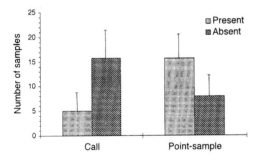

Fig. 6. Proximity of others to a focal animal emitting a meow call. Averages of 16 individuals are represented. T-bar indicates SD. (From Oda 1996, with permission)

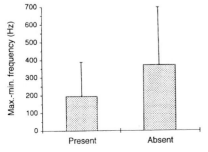

Fig. 7. Mean maximum minus minimum frequency of meow calls emitted by 16 individuals when others are proximate or not proximate. T-bar indicates SD

(Fig. 7). Frequency modulation is an important feature of acoustic communication in nonhuman primates. Sounds that are strongly modulated in frequency are much easier to localize than sounds with little frequency modulation. For example, Brown et al. (1978) found that Japanese macaques can localize the source of stimulus sounds more easily when frequency-modulated calls are played back than when nonmodulated sounds are broadcast. High-pitched calls may draw the attention of others, and frequency-modulated sounds may provide an important auditory cue for the location of group members.

The meow calls were frequently exchanged between troop members. Next I investigated how this exchange related to their social relationships in the troop. Social relationships among troop members were evaluated from grooming interactions. I divided each grooming session into sampling periods of 1 min duration. I did not consider the direction of grooming because most dyadic grooming is mutual in ring-tailed lemurs. The results are shown in Fig. 8a. I excluded three juveniles from the following analysis in order to simplify the expression of social relationships. For females, pairs for which grooming sessions exceeded the average +SD could be roughly divided into two groups. One included MI, MI90, and RH; the other, OD, OD90, SI, and KI. Grooming interactions for males occurred less frequently than among females.

a.

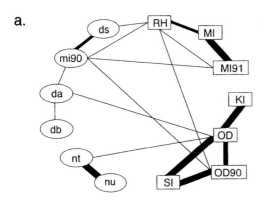

b.

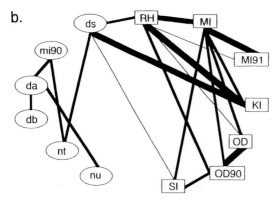

Fig. 8. a Sociogram of grooming interaction among the troop members. b Exchange network of the meow call among the troop members. Only frequency indices greater than the mean (X) are shown by lines. Scale: *Thin line, X – X+SD; medium-width line, X+SD – X+2SD; thick line, ≥X+2SD. Ovals* indicate males and *oblongs* indicate females. (From Oda 1996, with permission)

I analyzed call exchange data by noting the emitter and the responder. An individual emitting a call < 4 s before the focal animal's call, if any, is an emitter. An individual emitting a call < 4 s after the focal animal's call, if any, is a responder. Call exchange is one exchange of call emissions between two individuals. I excluded cases in which a third lemur vocalized < 4 s before the emitter's call. Figure 8b is the sociogram of call exchange in the troop. The pattern of call exchanges differed from that of grooming interactions, and I observed exchanges between the two subgroups of females illustrated in Fig. 8a. Ring-tailed lemurs sometimes divide into several subgroups during their routine daily activities (Jolly 1966), and they appear to use meow calls to communicate with other subgroups. I observed call exchanges between the two subgroups of females, as well as within each of them, which seemed connect the subgroup. Social communication among group-living primates involves nonphysical contact as well as physical contact. Acoustic contact and physical contact complement each other to maintain group cohesion.

Not only grooming but also olfactory communication relate to contact call production in ring-tailed lemurs. Lemurs are well equipped with scent-producing glands and use olfactory signals for communication. The ring-tailed lemur has sex-specific marking behavior. Although both sexes use genital marking, only males mark with their forearms (arm marking). Experimental studies suggested that male ring-tailed lemurs could identify the sex of the scent donor (Dugmore et al. 1984). Female marking was not affected by the breeding cycle, and male marking increased during the breeding season (Evans and Goy 1968). In the wild, there is individual variation in marking frequency and individuals have preferred marking sites, which are concentrated in the areas where the home ranges of troops overlap (Mertl-Millhollen 1988). Territory demarcation is the only function of scent marking that has been studied.

Although marking behavior has been studied as a means of intergroup information transfer, it is possible that scent marking functions in interindividual communication within a group (Kappeler 1998). I investigated the frequency of marking and contact call production in the C2 troop (Oda 1999). Six adult males and six adult females were the subjects of focal animal sampling. The mean frequency of genital marking by males was 1.2 per hour and by females it was 1.3 per hour. There was no sex difference in the frequency of genital marking (Fig. 9). While the frequency of genital marking was not significantly correlated with call frequency (Spearman rank correlation coefficient: $rs = -0.46$, $z = -1.51$, $n = 12$, ns), the frequency of arm marking by males was negatively correlated with call frequency ($rs = -0.93$, $z = -2.07$, $n = 6$, $P < 0.05$). Males that called frequently made relatively fewer arm markings and vice versa (Fig. 10). As I mentioned earlier, when group members were dispersed, the contact call was given frequently. This suggests that the call is used for long-distance communication. On the other hand, a scent on a branch cannot spread very far. Male lemurs could use calling when the troop was dispersed and use their arms to make scent marks when it was bunched together. This might explain the negative correlation between the frequency of calling and arm marking. Generally, primates rely on visual and acoustic signals to communicate with each other. In ring-tailed lemurs, however, which make active use of

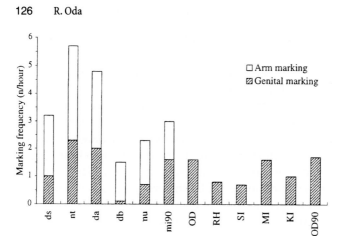

Fig. 9. The frequency of marking by each individual. *Capital letters* indicate females and *lower case letters* indicate males. (From Oda 1999, with permission)

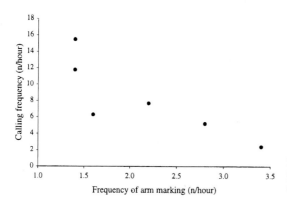

Fig. 10. The frequency of arm marking and calling by each male. (From Oda 1999, with permission)

olfactory communication, the acoustic and olfactory modes of communication might complement each other to facilitate contact between individuals.

6 Parallel Evolution of Acoustic Communication

Ring-tailed lemurs in natural habitats can respond to both their own and heterospecific sifakas alarm calls as if the calls denote a specific type of predator, and acoustic experience is important for perceiving the relationship between the calls and the predator. Moreover, the ring-tailed lemurs modify calling frequencies and acoustic features of their contact call according to their social and behavioral situations. They do not exchange the calls at random, and the pattern of interaction in call exchanges reflects their affiliate relationship. These findings have also been reported for many species of anthropoid primates. The ring-tailed lemur is indigenous to Madagascar and has evolved separately. Their vocal communication systems, however, represent some characteristics found in the anthropoids. Although their physical characteristics are primitive, it has become clear that terms such as "primitive" and "lower" are inappropriate for comparisons of the social organization of ring-tailed lemurs and anthropoid species (Taylor and Sussman 1985). The ring-tailed lemur forms multimale–multifemale groups comprising several matrilines. The groups are female-bonded, and males migrate among the groups. Such a social organization can be seen in some species of macaques and baboons. The highly developed social organization of ring-tailed lemurs seems to be the main factor affecting the evolution of vocal communication systems found in this study.

In general, primate social organization is thought to be determined by external factors such as predation pressure, and food competition within and between groups (van Schaik and van Hoof 1983). Matrilineal kin groups tend to be developed in the frugivorous primates such as macaques (Wrangham 1980), and that is the case for the ring-tailed lemur (Sussman 1991). Antipredator alarm calls are expected to develop in such a nepotistic group. Sherman (1977) found that when a Belding's ground squirrel (*Spermophilus beldingi*) gave an alarm call it was more likely to be attacked by a predator, but individuals nearby benefited from the early warning by retreating quickly down their burrows. In this case, the individuals that got the benefit were often relatives of the caller (sisters and sister's offspring), and so it is possible that alarm calling has spread by kin selection (Klump and Shalter 1984).

Moreover, group living is important to facilitate learning. I suggested that the perception of the alarm calls was learned through acoustic experiments. Socially transmitted information is important in learning novel behavior patterns in a wide range of species (Griffin 1992). Indeed, social reinforcement for perception of alarm calls was reported in vervet monkeys. If an infant gives an alarm call for a genuine predator (as opposed to a nonpredator), adults respond more strongly (Cheney and Seyfarth 1990). Ring-tailed lemurs live in groups of 15 individuals on average, and use relatively large home ranges compared with other lemurs. To maintain cohesion, members of a troop need to contact each other by the transfer of acoustic information. Under the social organization common to group-living anthropoids, vocal communication systems in ring-tailed lemurs might have undergone much parallel evolution with the anthropoids.

7 Differences Between Nonhuman Primate Vocal Communication and Human Language

Voluntary use of vocalization and referential signaling are parts of the foundation of human language. As mentioned above, such faculties could be evolved under some specific selective pressure for maintaining a complex society and escaping from predators. A possible story of language evolution is that the referential calls increased in variation and came to be produced in combination for complicated events. The combination then developed into a precursor of the universal grammar. However, this idea lacks the pragmatic view of language which asks how the human language is normally used and understood. True communication does not occur unless both the signal sender and the recipient take into account each other's states of mind (Grice 1989). Although we communicate through encoding meaning into sound and decoding it by a common language as a code, we manage to communicate much more than we encode and decode. When we communicate, we infer how the signals we send will be interpreted by the recipients, and predict what beliefs, desires, or knowledge the recipients will gain from the signals. From this point of view, could the vocal communication of nonhuman primates be described as true communication? Unfortunately, most studies of vocal communication in nonhuman primates have dealt with semanticity, development, or function. Do nonhuman primates have the faculty to attribute beliefs, knowledge, or desires to another individual? I review what kind of evidence we have about the mental attributions of nonhuman primates.

Through experiments using a trained chimpanzee, Premack and Woodruff (1978) presented an idea called the "theory of mind." To take account of others' mental states is to have a theory of mind. Unlike behavior, mental states are not directly observable; one needs a theory to infer another's state of mind. To test whether children have a theory of mind, developmental psychologists designed several patterns of "false belief task." The first such task was called the Maxi task, and was created by Wimmer and Perner (1983). Baron-Cohen (1985) and his colleague modified it and called it the Sally-Ann task. In this task, a child subject and another child watch as the experimenter places a toy in Location A. Then the other child leaves the room. After that, the experimenter moves the toy to Location B. The experimenter then asks the subject where the other child will look for the toy when they return. If the subject understands that others have beliefs independent of their own beliefs, they will answer that the returning child will believe the toy to be in the original Location A where they saw it hidden. On some occasions the subject child sees this acted out with puppets or in a cartoon. This task is generally known as the location change task, and usually relies on linguistic interactions, so it is difficult to use with nonhuman primates.

Call and Tomasello (1999) used a nonverbal version of the location change false belief task with chimpanzees and orangutans as well as with 4- and 5-year-old children. The task is a kind of hiding–finding game, in which an individual (the hider) hides a reward in one of two identical containers, and another individual (the communicator) observes the hiding process and attempts to help the subject by placing a marker on the container that is believed to hold the reward. The following three abilities must be acquired before this test is taken: (1) to track the reward to a new location, (2) to track the container marked as containing the reward to a new location, and (3) to ignore the communicator's marker when it is known to be incorrect. Assuming that these abilities had been acquired, the crucial false belief task was given. In this case, the communicator watched the hiding process and then left the area, at which time the hider switched the locations of the containers. When the communicator returned, she marked the container at the location where she had seen the reward hidden, which was incorrect. The hider then gave the subject the opportunity to find the reward. A verbal false belief task was also given to children in the same context. The results indicated that the children's performance in the verbal and nonverbal false belief tasks were highly correlated. They also gave the nonverbal tasks to two orangutans and seven chimpanzees. No ape succeeded in the nonverbal false belief task even though they succeeded in acquiring the prerequisites for the task.

The false belief task includes a presupposition that the subjects understand the psychological connection between seeing and knowing. Can nonhuman primates understand the relationship between perception and knowledge? Povinelli (1996) and his colleagues did some interesting experiments in this field. They suggested that chimpanzees can understand the perception–knowledge relationship, while rhesus macaques cannot. The hypothesis that macaques, unlike chimpanzees, do not understand the seeing–knowing relationship has also been supported by the experiments of Cheney and Seyfarth (1990). They reported that macaque mothers might not attend to the knowledge states of their infants when telling them about the presence of food or warning them about danger.

Joint attention is another presupposition for the theory of mind (Baron-Cohen 1995). An individual can understand what kind of information is shared between others by monitoring their gaze. For 11 species of nonhuman primates, including prosimians, macaques, and apes, Itakura (1996) investigated whether they could follow the experimenter's gaze, and whether they could find food by gaze-monitoring. The results indicate that while chimpanzees and orangutans were able to follow the experimenter's gaze and quickly learned to use the gaze as a key to finding food, other species had some difficulty in doing so. This study dealt with the joint attention between human and nonhuman primates. There is a possibility that chimpanzees and orangutans could follow the experimenter's gaze because of the closeness of the phylogeny of apes and humans. Tomasello et al. (1998) studied joint attention between individuals of the same species using captive chimpanzees, sooty mangabeys (*Cercocebus atys torquatus*), rhesus macaques, stump-tailed macaques (*M. arctoides*), and pigtail macaques (*M. nemestrina*). They reported that individuals from all species followed the gaze of conspecifics. These experimental studies indicate that some species of anthropoids have the ability of joint attention, and it has been suggested that chimpanzees are able to understand the relationship between visual perception and knowledge. However, their skill in mental attribution is not as well developed as that of humans.

Mental attribution is a part of social intelligence. Social interactions among group members result from the selective pressures of social intelligence, so such intelligence can be inferred by the social behavior of free-ranging individuals. Deception is a scale by which we can decide whether nonhuman primates in natural habitats have the ability of mental attribution. Field observations of tactical deception by many species of anthropoids have been reported (Byrne and Whiten 1988). However, the relevance of the data on deception has been questioned on the basis of two main arguments. First, these observations are anecdotal and ambiguous. Second, alternative nonmentalistic explanations, such as chance behavior or associative learning, could be proposed as more parsimonious explanations (Heyes 1998).

Deception at any level has never been reported among prosimians. Other scales for sophistication of social intelligence, such as "tripartite behavior" (Kummer 1967), reconciliation after conflict, and enlistment of others in attack are rare among ring-tailed lemurs (e.g., Kappeler 1993). Jolly (1998) said, "Numerically they range as kin groups, but in social relations they may still react as aggressively solitary animals, cozily grooming daughters and sisters but with little commitment to a wider troop nexus." We will probably not be able to find any evidence of scales among other species of prosimians which are less gregarious and have simpler societies than ring-tailed lemurs.

8 Evolution of the Language Circuit

Aitchison (1996) claimed that while Chomsky said that language consists of types of "mental organ," language uses numerous different parts of the brain, so it might be appropriate to say "language circuits" instead of "language organs." She said that language should be regarded as a London sightseeing bus which has its own specialized preordained routes. The results of ethological studies on nonhuman primate vocal communication support this claim. Several parts of the foundation of human language, such as the voluntary use of vocalization and referential signaling, can be seen among some species of anthropoids. These faculties are also found in ring-tailed lemurs, which are indigenous to Madagascar and have evolved separately from other primates. This fact means that such faculties could have evolved among primates under some specific selective pressure. However, nonhuman primates have not developed the attribution of mental states to others, which is an important factor of human language communication.

The evidence of social intelligence from the laboratory and the field indicates that prosimians might have few of the mental devices which indicate social intelligence. This seems curious, because the ring-tailed lemur forms large multimale-multifemale groups in which sophisticated social intelligence might be needed (Dunbar 1996). Jolly (1966) said, "Thus, the lemurs seem to have 'monkey-type' societies without having evolved monkey-level intelligence." One of the reasons why they lack monkey-level intelligence might be that their large social groups were established within a short period of time. In general, diurnal species form larger groups than nocturnal species. Most prosimians possess a tapetum lucidum within their eyes, i.e., a highly reflective layer lying between the choroid and the retina, which is associated with nocturnality. Even diurnal lemurs, such as ring-tailed lemurs, have the tapetum. This means that the ring-tailed lemurs have not spent a long enough time as a diurnal lemur species to atrophy the tapetum. It might not be long enough to evolve social intelligence devices either.

On the other hand, anthropoids have some other devices such as joint attention and perspective taking. However, these devices are not integrated and sophisticated enough to be the theory of mind mechanism. Moreover, the faculties of information transfer by vocalization and that of mental attribution might work separately (Fig. 11). The foundations of human language were established when these faculties were connected with each other as points in the language circuits. Since the foundation of human language, the voluntary use of vocalization has become

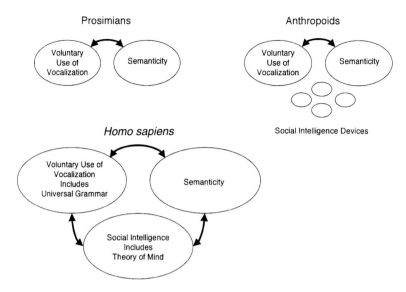

Fig. 11. A simple schema of vocal communication in nonhuman primates and human language. *Large ovals* represent domain-specific faculties. *Small ovals* represent modules for particular tasks, which compose the faculties

more complex in the need to follow the rule of universal grammar. Referential calls have increased explosively in their variation. Social intelligence has become more sophisticated to include the theory of mind, and been connected with the faculties of information transfer by vocalization (Fig. 11).

There are two views about the process of language evolution. One is that language emerged suddenly, and the other is that it developed slowly. Probably, each of the faculties that are now a part of human language evolved slowly. Then, these faculties were connected to each other and vocal communication developed suddenly into human language. While the combination of all faculties by connection might be a simple change, the results would provide remarkable benefits.

A discovery in neuroscience also supports this hypothesis. Single neurons in the premotor cortex of monkeys were found to be active when the monkeys make particular reaching gestures (see the chapter by S. Kojima, in this volume). Some of these neurons, which have been named mirror neurons, are also active when the monkeys watch an another individual making the same gesture. These cells are in an area of the monkey cortex that appears to be homologous to Broca's area in the human brain. The mirror neurons might be precursors of the ability to imagine the mental perspective of observed conspecifics, and this function might be a part of mind-reading ability (Gallese and Goldman 1998). The existence of such neurons in the area involved in the programming of human speech is highly suggestive of the evolutionary possibility of language circuits.

Acknowledgments

I would like to thank Profs. N. Koyama, K. Aoki, N. Masataka, A. Jolly, M. Hiraiwa-Hasegawa, C. Saito, T. Hasegawa, and N. Kondo, Drs. T. Tanaka, H. Sugiura, and T. Shima,and H. Tsutsumi and the Timbazaza Zoo for their advice and help. I am also grateful to the Government of the Republic of Madagascar, J. de Heaulme, and the Izu Cactus Park for allowing me to carry out this research. This study was supported by the Cooperative Research Fund of the PRI, Kyoto University, and a grant (No. 05041088) from the Ministry of Education, Science and Culture of Japan.

References

Aitchison J (1996) The seeds of speech. Cambridge University Press, New York

Altmann J (1974) Observational study of behavior: sampling methods. Behaviour 49:227–265

Andrew RJ (1963) The origins and evolution of calls and facial expressions of the primates. Behaviour 20:1–109

Baron-Cohen S (1995) Mindblindness. MIT Press, Cambridge

Brown CH, Michael DB, Moody DB, Stebbins WC (1978) Localization of primate calls by Old World monkeys. Science 201:753–754

Byrne RW, Whiten A (1988) Machiavellian intelligence. Oxford University Press, Oxford

Call J, Tomasello M (1999) A nonverbal false belief task: the performance of children and great apes. Child Dev 70:381–395

Cheney DL, Seyfarth RM (1990) How monkeys see the world. University of Chicago Press, Chicago

Dugmore SJ, Baily K, Evans CS (1984) Discrimination by male ring-tailed lemurs (*Lemur catta*) between the scent marks of male and those of female conspecifics. Int J Primatol 5:235–245

Dunbar R (1996) Grooming, gossip, and the evolution of language. Faber and Faber, London

Evans CS, Goy RW (1968) Social behaviour and reproductive cycles in captive ring-tailed lemurs (*Lemur catta*). J Zool 156:181–197

Gallese V, Goldman A (1998) Mirror neurons and the simulation theory of mind-reading. Trends Cogn Sci 2:493–501

Gautier JP, Gautier A (1977) Communication in Old World monkeys. In: Sebeok T (ed) How animals communicate. Indiana University Press, Bloomington, pp 890–964

Gautier-Hion A, Quris R, Gautier JP (1983) Monospecific vs polyspecific life: a comparative study of foraging and antipredatory tactics in a community of *Cercopithecus* monkeys. Behav Ecol Sociobiol 12:325–335

Gouzoules S, Gouzoules H, Marler P (1984) Rhesus monkey (*Macaca mulatta*) screams: representational signalling in the recruitment of agonistic aid. Anim Behav 32:182–193

Grice HP (1989) Studies in the way of words. Harvard University Press, Cambridge

Griffin DR (1992) Animal minds. University of Chicago Press, Chicago

Harcourt AH, Stewart KJ, Hauser M (1993) Functions of wild gorilla "close" calls. I. Repertoire, context, and interspecific comparison. Behaviour 124:89–122

Harcourt C, Thornback J (1990) Ring-tailed lemur. In: The IUCN Red Data Book. IUCN, Cambridge, pp 105–110

Hauser MD (1988) How infant vervet monkeys learn to recognize starling alarm calls: the role of experience. Behaviour 105:187–201

Heyes CM (1998) Theory of mind in nonhuman primates. Behav Brain Sci 21:101–148

Howarth CJ, Wilson JM, Adamson AP, Wilson ME, Boase MJ (1986) Population ecology of ring-tailed lemur, *Lemur catta*, and the white sifaka, *Propithecus verreauxi vereauxi*, at Berenty, Madagascar, 1981. Folia Primatol 47:39–48

Itakura S (1996) An exploratory study of gaze-monitoring in nonhuman primates. Jpn Psychol Res 38:174–180

Jolly A (1966) Lemur behavior. University of Chicago Press, Chicago

Jolly A (1998) Pair-bonding, female aggression and the evolution of lemur societies. Folia Primatol 69:1–13

Kappeler PM (1993) Reconciliation and post-conflict behavior in ring-tailed (*Lemur catta*) and redfronted (*Eulemur fulvus rufus*) lemurs. Anim Behav 45:901–915

Kappeler PM (1998) To whom it may concern: the transmission and function of chemical signals in *Lemur catta*. Behav Ecol Sociobiol 42:411–421

Klump GM, Shalter MD (1984) Acoustic behavior of birds and mammals in the predator context. Z Tierpsychol 66:189–226

Koyama N (1991) Troop division and inter-troop relationships of ring-tailed lemurs (*Lemur catta*) at Berenty, Madagascar. In: Ehara A, Kimura T, Takenaka O, Iwamoto M (eds) Primatology today. Elsevier, Amsterdam, pp 173–176

Kummer H (1967) Tripartite relations in hamadryas baboons. In: Altmann SA (ed) Social communication among primates. University of Chicago Press, Chicago, pp 63–73

Macedonia JM (1986) Individuality in the contact call of the ring-tailed lemur (*Lemur catta*). Am J Primatol 11:163–179

Macedonia JM (1990) What is communicated in the antipredator calls of lemurs: evidence from playback experiments with ring-tailed and ruffed lemurs. Ethology 86:177–190

Macedonia JM (1993) The vocal repertoire of the ring-tailed lemur (*Lemur catta*). Folia Primatol 61:186–217

Macedonia JM, Evans CS (1993) Variation among mammal alarm call systems and the problem of meaning in animal signals. Ehtology 93:177–197

Masataka N (1983) Categorical responses to natural and synthesized alarm calls in Goeldi's monkeys (*Callimico goeldii*). Primates 24:40–51

Masataka N (1986) Rudimentary representational vocal signalling of fellow group members in spider monkeys. Behaviour 96:49–61

Mertl-Millhollen AS (1988) Olfactory demarcation of territorial but not home range boundaries by *Lemur catta*. Folia Primatol 50:175–187

Munn CA (1986) Birds that "cry wolf." Nature 319:143–145

Oda R (1996) Effects of contextual and social variables on contact call production in free-ranging ring-tailed lemurs (*Lemur catta*). Int J Primatol 17:191–205

Oda R (1998) The responses of Verreaux's sifakas to anti-predator alarm calls given by sympatric ring-tailed lemurs. Folia Primatol 69:357–360

Oda R (1999) Scent marking and contact call production in ring-tailed lemurs (*Lemur catta*). Folia Primatol 70:121–124

Oda R, Masataka N (1996) Interspecific responses of ring-tailed lemurs to playback of antipredator alarm calls given by Verreaux's sifakas. Ethology 102:441–453

Pereira ME, Macedonia JM (1991) Response urgency does not determine anitipredator call selection by ring-tailed lemurs. Anim Behav 41:543–544

Petter JJ, Charles-Dominique P (1979) Vocal communication in prosimians. In: Doyle GA, Martin RD (eds) The study of prosimian behavior. Academic Press, New York, pp 247–305

Povinelli D (1996) Chimpanzee theory of mind?: the long road to strong inference. In: Carruthers P, Smith PK (eds) Theories of theories of mind. Cambridge University Press, New York, pp 293–329

Premack D, Woodruff G (1978) Does the chimpanzee have a "theory of mind?" Behav Brain Sci 4:515–526

Richard AF (1978) Behavioral variation: case study of a Malagasy lemur. Bucknell University Press, Lewisburg

Sauther ML (1990) Antipredator behavior in troops of free-ranging *Lemur catta* at Beza Mahafaly special reserve, Madagascar. Int J Primatol 10:595–606

van Schaik CP, van Hoof JARAM (1983) On the ultimate causes of primate social systems. Behaviour 85:91–117

Sherman PW (1977) Nepotism and the evolution of alarm calls. Science 197:1246–1253

Snowdon CT (1989) Vocal communication in New World monkeys. J Hum Evol 18:611–633

Sugiura H (1993) Temporal and acoustic correlates in vocal exchange of coo calls in Japanese macaques. Behaviour 124:207–225

Sugiura H (1998) Matching of acoustic features during the vocal exchange of coo calls by Japanese macaques. Anim Behav 55:673–687

Sussman RW (1991) Demography and social organization of free-ranging *Lemur catta* in the Beza Mahafaly reserve, Madagascar. Am J Phys Anthropol 84:43–58

Tattersall I (1982) The primates of Madagascar. Columbia University Press, New York

Taylor L, Sussman RW (1985) Preliminary study of kinship and social organization in a semi-free-ranging group of *Lemur catta*. Int J Primatol 6:601–614

Tomasello M, Call J, Hare B (1998) Five primate species follow the visual gaze of conspecifics. Anim Behav 55:1063–1069

Wimmer H, Perner J (1983) Beliefs about beliefs: representation and constraining function of wrong beliefs in young children's understanding of deception. Cognition 13:103–128

Wrangham RW (1980) An ecological model of female-bonded primate groups. Behaviour 75:262–300

Zuberbuhler K, Cheney DL, Seyfarth RM (1999) Conceptual semantics in a nonhuman primate. J Comp Psychol 113:33–42

7
Vocal Exchange of Coo Calls in Japanese Macaques

Hideki Sugiura

1 Introduction

The vocal exchange of contact calls in nonhuman primates (e.g., Snowdon and Cleveland 1984; Biben et al. 1986; Masataka and Biben 1987) may be the most similar form of communication to our own conversation. Vocal exchange can be expressed as a communication form in which a sender gives vocalization as an addressing signal and a respondent also gives vocalization as a response. Furthermore, responding vocalization can act as another addressing signal, which elicits further vocal response. Such vocal–vocal communication systems also appear in duetting (e.g., Farabaugh 1982; Cowlishaw 1992) and counter singing (e.g., Tenaza 1976; Maples et al. 1988) . In duetting, typically a pair of male and female sing a song alternately or simultaneously. In counter singing, typically a territorial animal sings a song against a rival's song. Among these, vocal exchange seems to be the most similar to human conversation, because it occurs among group members in affiliative contexts. As Dunbar (1996) argued, normal conversation may have played an important roll in the evolution of human language. Thus, it should be interesting to try to discover whether vocal communication parallels our own conversation.

A central issue in studies of vocalizations by nonhuman primates is the extent of their plasticity. Several studies have suggested that the acoustic features of vocalizations are inherited and little modified during development (e.g., Lieblich et al. 1980; Herzog and Hopf 1983) . The failure of researchers to find any evidence of learning is somewhat surprising, however, in view of the influence of learning on other behavior and the apparent importance of learning in human speech.

The evidence for plasticity in the way vocalizations are produced in nonhuman primates is limited. Masataka and Fujita (1989) showed, in cross-fostering experiments between Japanese macaques *(Macaca fuscata)* and rhesus macaques (*Macaca mulatta*), that the acoustic features of one of the species-specific calls is learned in these species. However, Owren et al. (1993) obtained no evidence of vocal learning by the same experimental paradigm. Elowson and Snowdon (1994) showed that pygmy marmosets, *Cebuella pygmaea*, modify the acoustic features of their con-

Primate Research Institute, Kyoto University, 41 Kanrin, Inuyama, Aichi 484-8506, Japan

tact calls: when they placed two unfamiliar populations together in a common acoustic environment, two parameters of the frequency of the calls changed in parallel in both populations after acoustic contact.

These earlier studies focused on changes in acoustic properties during development or over a relatively long period (more than a month). The acoustic changes during vocal exchange occur in the short term (within seconds), which is a somewhat different time scale. However, it would also provide evidence of flexibility in the production of calls. In Japanese macaques, some acoustic features of coo calls given as replies appeared to change to match those of preceding calls during successive call exchanges among group members.

Coo calls are uttered by most members of a troop of Japanese macaques in a variety of contexts, with the exception of agonistic interactions. Itani (1963) reported that coo calls are often given in dense vegetation. When macaques hear a coo call given by a group member, the most obvious and usual response is to utter another coo call. The basic function of vocal exchange appears to be to locate group members and to maintain within-group contact. In terms of its acoustics, the coo call has a basically tonal structure, with a high degree of variability in the fundamental frequency component (Green 1975).

In this paper, the timing and acoustic properties of coo call exchanges in Japanese macaques are described and analyzed. The temporal patterns of occurrence of consecutive coo calls were studied to determine whether or not following coos could be considered a response to preceding coos. I also examined whether the acoustic features of following coos were related to those of preceding coos. The observations suggested that some of the coos are uttered in response to preceding coos, and that in those cases the acoustic features of following coos were similar to those of preceding coos. Therefore, a playback experiment was performed with a group to eliminate unobserved factors which could affect the result, and to replicate the phenomenon experimentally.

2 Methods

2.1 Subject Groups

The subjects were wild and semi-free-ranging Japanese macaques (*Macaca fuscata yakui*): the Yakushima P, G, and B groups, and the Ohirayama group. The number of monkeys is shown in Table 1.

Generally speaking, in free-ranging Japanese macaques, the frequency of coo calls is much higher than in macaques under captive conditions. This may be because the basic function of coo calls is to locate group members vocally. Localization of group members should be needed more in wild monkeys, who often spread out and move over a large range. In addition, as ambient noise is usually less in the wild than in captive conditions, especially in the forest, we can record vocalizations more clearly. Thus, free-ranging Japanese macaques are good subjects for study of the vocal communication of coo calls.

Table 1. The number of monkeys in the subject groups

	Adult female (>= 5 years)	Adult male (>= 5 years)	Juvenile female (1–4 years)	Juvenile male (1–4 years)	Infant (< 1 year)	Total
Yakushima P group	6	6	2	3	1	18
Ohirayama group	18	13	7	19	7	64
Yakushima G group	9	9	2	2	1	23
Yakushima B group	9	11	2	2	1	25

For naturalistic observations conducted between 1990 and 1991, the Yakushima P group and the Ohirayama group were chosen as the subjects. The Yakushima P group is a wild group which ranges over approximately 40 ha of warm temperate forest in the western coastal region of Yakushima Island, south of Kyushu, Japan. The group has been studied since 1973 without provisioning (Maruhashi 1992). The Ohirayama group was originally a translocated group. The original members of the group were captured together in 1957 in Yakushima Island and immediately flown to Mt. Ohirayama, Aichi Prefecture, on the Japanese mainland (Tanaka 1995). Since then, they have been provisioned under semi-free-ranging conditions by the Japan Monkey Center. A description of the group and its habitat is given by Kawai (1960). The two groups can be regarded as genetically similar, but have been geographically separated without any contact for the past 33 years.

A playback experiment was conducted in the Yakushima G and B groups in 1994 and 1995. These groups inhabit the adjacent area to the home range of the Yakushima P group on Yakushima Island. These wild groups enabled me to conduct a playback experiment on an individual basis, in which I played back stimulus vocalization to a predetermined individual and attempted to evoke a vocal response from it. Although I had undertaken a preliminary playback experiment in the Ohirayama group (Sugiura 1993), such an individual-basis experiment was difficult. In that group, the number of potential respondents was so great that it was very difficult to evoke a vocal response from a single predetermined target animal. In the wild groups, in contrast, as the sizes of these groups were relatively small and the members are more likely to spread out over relatively large distances, such an experiment was possible.

2.2 Subject Animals

I chose adult females as my subjects. While adult females often exchange coo calls with other females, adult males seldom do so with females or with males (Mitani 1986). Therefore, I focused on vocal exchange among females.

For the naturalistic observation of coo call exchange, I chose five adult and juvenile females from the Yakushima P group and eight females from the Ohirayama group as target animals (all were age 3 years or older). For the playback experiment, I chose five adult females (Mdr, Ml, Tk, Yn, and Ts) in the G group and two adult females in the B group (Ks and St).

2.3 Acoustic Analysis

Vocalizations were tape-recorded and analyzed quantitatively, using a sound spectrograph (Kay DSP Sonagraph, model 5500 and model 4300, Kay Elemetrics, Lincoln Park, NJ, USA).

I defined coo calls as predominantly tonal calls basically following the classification of Green (1975). The calls have some harmonic overtones, and their fundamental frequency (F0) elements are spectrographically distinct and are usually the most dominant frequency components (Fig. 1). Therefore, for each call, the duration, the location of the maximum frequency in relation to the entire length of F0 elements, the start frequency, the end frequency, the minimum frequency, the maximum frequency, and the frequency ranges (maximum minus start frequency and maximum minus end frequency) were measured.

Green (1975) divided coo calls into seven "subtypes" from their acoustic parameters. He mainly used two quantitative and one qualitative parameters as the criteria for the classification of the subtypes: duration of call (≤ 0.19 s and ≥ 0.20 s), location of maximum frequency relative to total duration (<2/3 and $\geq$2/3), and the presence of a "dip" feature which was the local minimum frequency. In my study, however, the two quantitative parameters were distributed continuously rather than discretely, and coos with a dip feature were rare. Therefore, instead of dividing them into subtypes straight away, I analyzed all coo calls as a single category.

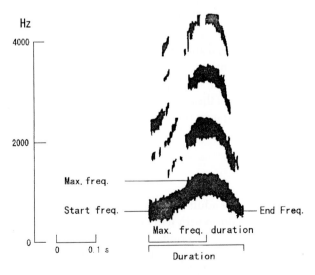

Fig. 1. A representative sound spectrogram of the coo calls and acoustic variables measured

I transformed the frequency parameters logarithmically before statistical analysis. Because the value of the frequency ranges could be zero, I added 10 Hz to these values before the logarithmic transformation. The time to maximum frequency and duration were transformed to square roots, and the location of the maximum frequency to arcsine.

3 Naturalistic Observations of Vocal Exchange of Coo Calls

3.1 Recording and Analysis

Data were collected by the focal animal sampling method, with a given observation session lasting for 60 min. In an observation session, I stood near the focal animal with a directional microphone aimed at it, and recorded target and other vocalizations, the identities of callers, and behavioral correlates with a tape recorder. The observations were conducted when most of the group members were located within visible or audible range. If I could not identify of the caller, I treated the caller as "unidentified" for the subsequent analysis. A total of 23 and 84 observation sessions were conducted with the Yakushima P group and the Ohirayama group, respectively.

3.2 Contextual and Individual Variability of Acoustic Features

I examined whether the acoustic features of coo calls vary depending on the contexts in which calls are given and the individuality of the caller. As to the context, Green (1975) argued that the acoustic features of coo calls co-vary with context. Individual differences in acoustic features have been reported in many animals, including Japanese macaques (Mitani 1986; Inoue 1988).

I conducted two-way analysis of variance with activity and caller as the two main factors. The ANOVA was a mixed model, with activity as the fixed factor and the caller as the random factor. The context was classified into four categories: feeding, moving, grooming, and resting. Because coos were rarely given while grooming in the Yakushima P group and while moving in the Ohirayama group, these calls were eliminated from the analysis.

Table 2 shows the results of the analysis. The acoustic features of coos did not differ among activity for any variable. In contrast, coo calls differed individually for many variables.

There were no salient differences in the acoustic properties of coos in the different contexts. Therefore, it may be safe to pool the coos given in different contexts for the following analysis. In contrast, acoustic features differed individually. Individual difference should be considered as a factor that can affect the acoustic variation of coo calls.

Table 2. Result of a two-way ANOVA of acoustic variables of coo calls, with "activity" as the fixed effect and the "individuality of callers" as the random effect in the Yakushima P group and the Ohirayama group. The result of two main factors are given

Yakushima P group	Activity		Caller	
Variables	F	df	F	df
Time to maximum frequency	0.5	(2, 9.8)	1.5	(4, 10.1)
Duration	0.2	(2, 9.0)	2.0	(4, 9.1)
Location of maximum frequency	1.5	(2, 12.0)	2.7	(4, 12.7)
Start frequency	0.2	(2, 9.7)	17.5	(4, 10.0)***
Maximum frequency	0.7	(2, 9.5)	5.2	(4, 9.8) *
End frequency	1.1	(2, 10.6)	10.0	(4, 11.1)**
Maximum minus start frequency	0.9	(2, 9.5)	3.3	(4, 9.8)
Maximum minus end frequency	0.5	(2, 11.4)	3.7	(4, 11.9)*

Ohirayama group	Activity		Caller	
Variables	F	df	F	df
Time to maximum frequency	0.1	(2, 16.5)	5.2	(7, 14.6)**
Duration	0.8	(2, 16.3)	3.6	(7, 14.5)*
Location of maximum frequency	0.0	(2, 17.0)	2.8	(7, 14.7)*
Start frequency	1.8	(2, 14.7)	6.1	(7, 13.8)**
Maximum frequency	0.8	(2, 15.1)	1.8	(7, 13.9)
End frequency	0.1	(2, 15.3)	2.3	(7, 14.0)
Maximum minus start frequency	0.5	(2, 15.6)	2.4	(7, 14.2)
Maximum minus end frequency	2.7	(2, 15.6)	1.1	(7, 14.1)

$*P < 0.05; **P < 0.01; ***P < 0.001.$

3.3 Timing of Vocal Exchange

An important issue for vocal exchanges is the timing of the utterance of calls. When a call was followed by an another call, we tended to regard the latter call as response. However, monkeys may give calls randomly, and the two consecutive calls (i.e., two calls separated by a short interval) may occur by chance alone. To examine whether the timing of calls follows a random process or not, I did a quantitative analysis of the timing of call utterances using the measure of intercall intervals.

When all the vocalizations were categorized, more than 85% of the total vocal sample consisted of coos. Because of their low incidence, call types other than coos were excluded from the subsequent analysis. Whenever two coos were uttered consecutively, and not interrupted by other call types, the intercall interval was defined as the time from the end of a call to the beginning of the next.

I classified sequences of two consecutive coos into two categories as follows: (1) a sequence of two consecutive coos where the second call was uttered by a different caller from the first (DC sequence), and (2) a sequence of two consecutive coos uttered by the same caller (SC sequence). A total of 244 and 140 intervals in DC sequences and 141 and 74 intervals in SC sequences were analyzed for the Yakushima

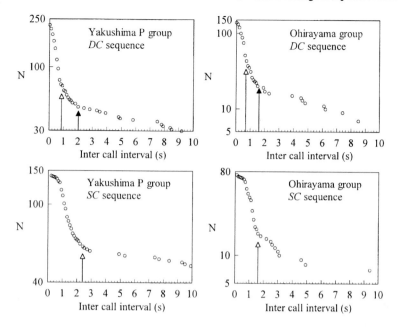

Fig. 2. Log-survival plots of intercall intervals in different caller (*DC*) sequences (*upper panels*) and same caller (*SC*) sequences (*lower panels*) in the Yakushima P group and the Ohirayama group. *Arrows* indicate abrupt change points, and *closed arrows indicate* the chosen criteria interval for the response calls

P group and the Ohirayama group, respectively. Because there were no significant individual differences in the intervals in DC sequences or in SC sequences within each population, I pooled the data together.

The overall distributions of intervals in DC sequences and SC sequences are shown by the log-survival plots in Fig. 2. If the calls occurred randomly, the distribution of time intervals between the first and second calls in DC sequences is expected to be distributed exponentially, and the graph should form a straight line with a negative slope (Sibly et al. 1990). However, these graphs do not show such a steady decrease. In both groups, the majority of calls in DC sequences were uttered after others' coo calls within a brief interval of approximately between 0 and 2 s in the Yakushima P group and between 0 and 1.5 s in the Ohirayama group. With intervals which are longer than these, the curves decline slowly.

These interval distributions indicate that two processes of coo call utterance may be involved in the DC sequence. One type of call occurred intensively shortly after the previous call, and the other type occurred less frequently after the previous call, with a relatively longer interval. The former seems to be elicited by the previous call, and thus it may be a "response call." The latter seems to occur independently of the previous call, and thus it may be a "spontaneous call."

For subsequent analysis, the criteria interval for the response and spontaneous calls was determined by the distribution of intervals using the "abrupt change point test for exponentially distributed bout length" (Haccou and Meelis 1992).

Between 0.3 and 10 s, two significant "abrupt change points" were found for both groups. Longer change points (2 s in the Yakushima P group and 1.6 s in the Ohirayama group) were used as criteria intervals. Thus, "response calls" were defined as those that followed the previous call after 0–2 s in the Yakushima P group and after 0–1.6 s in the Ohirayama group. "Spontaneous calls" were defined as those that occurred after 2 s in the Yakushima P group and 1.6 s in the Ohirayama group.

The distributions of SC sequences showed a different pattern from those of DC sequences. For approximately 1 s after their own calls, the curve decreased gradually. Thereafter, between approximately 1 and 3 s, the curve fell rapidly. Again, after that interval, the curve decreased slowly. Thus, the second coos in SC sequences rarely occurred during the periods in which the second coos in DC sequences generally occurred. Instead, the second call in SC sequences frequently occurred after the response calls had occurred.

Again, I tested the criteria interval in SC sequences, using the "abrupt change point test for exponentially distributed bout length" (Haccou and Meelis 1992). Between 0 and 10 s, one significant "abrupt change point" was found for each group. The change points were 2.4 s in the Yakushima P group and 1.6 s in the Ohirayama group. Thus, second calls in SC sequences within these intervals appear to occur dependently on the first call, and therefore can be regarded as "repetitious calls".

These results suggest that when an animal gives a coo spontaneously, she remains silent during the period when the group members are likely to respond, and if no animal makes any vocal response, she gives another coo, again addressing the other animals.

3.4 Similarity of First and Second Coos in DC Sequences

The question arises of whether acoustic properties differ between calls of DC sequences with short and long intervals. The acoustic features of the second coos given by a different caller with a short interval, which may represent responses, might be dependent on those of the first coos. On the other hand, second coos with long intervals, which may be spontaneous calls, might be acoustically independent of the first coos. For example, one might hypothesize a greater similarity between coos separated by short intervals than by those separated by longer intervals. Therefore, regression analysis was performed with each acoustic variable of the first calls being taken as an independent variable, and those of the second calls being taken as a dependent variable.

I examined the effects of the acoustic features of the first call on those of response calls by linear regression analysis. As shown above, the acoustic properties of coo calls were significantly different among individuals (Table 2). To control individual difference, I treated each subject individual who gave the second call as a "block," and calculated the variance and covariance by subject using the mean for each individual. The data obtained when the second callers were other than target animals or were unidentified were eliminated from the following analyses. Be-

Table 3. Results of regression analysis in response and spontaneous calls during the naturalistic observations

Variables	Response calls in Yakushima P group		Response calls in Ohirayama group		Spontaneous calls in Yakushima P group	
	b	$F_{1, 113}$	b	$F_{1, 65}$	b	$F_{1, 18}$
Time to maximum frequency	0.208	5.6*	1.130	0.0	0.220	0.7
Duration	8.660	1.3	0.204	3.8	0.830	7.9*
Location of maximum frequency	0.139	2.8	-0.146	1.4	0.139	0.4
Start frequency	4.990	1.5	-1.770	0.1	-0.397	29.1***
Maximum frequency	0.260	13.9***	0.351	11.6**	-0.347	2.2
End frequency	0.141	4.7*	-1.170	0.0	-0.340	14.3**
Maximum minus start frequency	0.256	11.5***	0.451	16.8***	3.240	0.0
Maximum minus end frequency	6.300	0.7	0.182	3.8	4.420	0.1

*$P < 0.05$; **$P < 0.01$; ***$P < 0.001$.

cause the sample size was not large enough, the data of the spontaneous calls in the Ohirayama group were not analyzed.

Table 3 shows the overall regression coefficient (b) and goodness of fit to the regression. In the response calls in the Yakushima P group, regressions of four variables (time to maximum frequency, maximum frequency, end frequency, and maximum minus start frequency) showed significantly positive effects. In the Ohirayama group, two variables (maximum frequency and maximum minus start frequency) showed positive effects. In spontaneous calls in the Yakushima P group, in contrast, one variable (duration) showed a significant positive coefficient and two variables (start frequency and end frequency) showed significant negative coefficients.

In both groups, two interrelated variables (maximum frequency and maximum minus start frequency) of response calls were similar to those of the previous calls. These acoustic similarities between the first and second calls support the probability that second calls with a short interval in DC sequences were given as responses.

The similarity suggests that the second caller might alter the acoustic features (e.g., frequency range) of their response call according to the same features of the first call. In this analysis, however, the identity of the first caller was not considered. Thus, the identity of the first caller might affect the acoustic features of the second call. For example, the caller might have a tendency to respond to juveniles with a high-pitched coo call and to adults with low-pitched calls. (The fundamental frequency of coo calls becomes lower as the caller gets older, Inoue 1988.) If we can control the acoustic features of the first call, including the caller's identity, these questions can be answered. Therefore, I then conducted a playback experiment.

4 Playback Experiment

The observations of vocal exchanges of coo calls suggest that respondents alter some acoustic features of their own coos, and match them to the preceding coos. If the subject animals indeed match the acoustic features of their calls with those of playback vocalizations, the acoustic parameters of the subjects' coos should be varied depending on the acoustic quality of the stimuli. To explore this possibility, I carried out the following playback experiment.

4.1 Stimulus and Procedures

I presented each subject female with a set of stimuli that consisted of six or eight coo calls that had been recorded from one of the adult females in the same group ("stimulus female"). I chose as stimulus animals adult females that were habituated and gave coo calls relatively frequently: four adult females (Mdr, Ml, Tk, and Yn) in the G group, and two adult females (Ks and St) in the B group (Mdr's calls were played back to the female Ml, Ml's to the females Mdr and Tk, Tk's to the female Yn, Yn's to the female Ts, Ks's to the female St, and St's to the female Ks). All the six stimulus females were taken from the seven subject animals, except for one subject (Ts). For each subject female, I chose as the stimulus animal the female which had exchanged coo calls most often with the current subject female during the recording of the calls of the stimulus females.

The results of the previous section, as well as those of preliminary experiments (Sugiura 1993; Sugiura and Masataka 1995), suggested that the maximum frequency and the frequency range of responses were the acoustic features most likely to be matched to those of preceding calls. On the basis of these results, I chose as stimuli coo calls that had been recorded clearly and that had different maximum frequencies and frequency ranges (maximum frequency minus starting frequency) of the fundamental frequency component, as well as a similar starting frequency in the fundamental frequency.

For each experimental trial, I hid a speaker approximately 15–20 m away from the subject female. I stood about 10 m from the speaker and waited until all the monkeys were silent for at least 5 s before the playback to prevent any other vocalizations having any effect on the subsequent calls of the females. To avoid an effect from the presence of the stimulus animal, I also confirmed that the females whose calls were used as playback stimuli in the trial were not close to the subject. Thereafter, I played one of the set of stimuli. In a trial, only a single stimulus was played back. Stimuli were played back in random order. I recorded these stimuli, as well as vocalizations, from the subjects during the 10-s period after the playback.

If the subject female gave coo calls during the 10 s that followed the playback, the trial was regarded as successful. If any vocalization occurred simultaneously with a stimulus, or if the first vocalization that followed the stimulus was not a coo from the subject animal, I regarded the trial as a failure and excluded it from subsequent analysis (a further discription can be found in Sugiura 1998).

4.2 Response Calls and Spontaneous Calls

Of 2047 trials for the seven subjects, 549 (27%) were successful. The average numbers of trials and of successful trials per subject were 292 and 78 (27%), respectively. In natural conditions, 24.2–31.3% of coos were responded to within 1.5 s by the other group members (H. Sugiura, unpublished data). Thus, the percentage of successful trials was not very different from that in natural conditions.

To examine whether subject animals responded to stimulus calls, I plotted the interval between the end of a stimulus and the beginning of the subject's subsequent coo calls using a log-survival plot, as shown in Fig. 3. Similarly to the results of the natural habitat observations, the majority of calls were uttered within a brief interval of approximately 0.3–1.5 s. Few were uttered within intervals of less than approximately 0.3 s or more than approximately 1.5 s. This distribution of intervals indicates that two types of coo call were involved in the successful trials: the

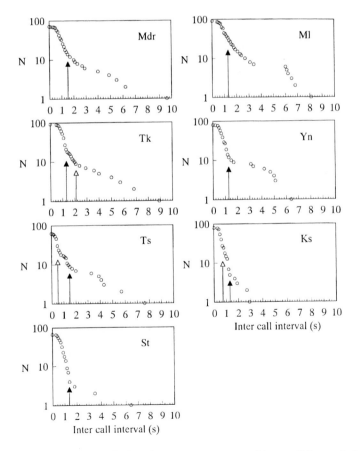

Fig. 3. Log–survival plots for seven macaques of intercall intervals between playback of a stimulus coo call from an adult female and the subject's response coos. *Arrows* indicate the "abrupt change points," and *closed arrows* indicate the end of the response calls

response call, elicited by the stimulus, and the spontaneous call, which occurred independently of presentation of the stimulus. Since a certain length of time is necessary for a response, response calls should occur with only a brief delay after presentation of the stimulus. Then, within a certain time interval after the brief delay, response calls should occur frequently. Calls that were uttered a very short time after the end of the stimulus presentation, and those uttered after a long interval, must be given independently of the stimulus, and should thus be spontaneous calls.

I classified two types of coo call according to the interval between the stimulus and the call. First, I determined the shortest interval before a response call, considering the minimum reaction time and the delay in sound transmission. Considering the reaction time from the stimulus presentation to a motor reaction (Saslow 1972), and a delay between the actual occurrence of the subject's call and the recording, I defined the shortest interval as 0.3 s. Thus, I classified calls that occurred within 0.3 s of the end of the stimulus as spontaneous calls, and those at 0.3 s or later as response calls.

I determined the longest intervals before response calls from the distribution of intervals. To determine the appropriate criteria of intervals, I used the "abrupt change point test for exponentially distributed bout length" (Haccou and Meelis 1992). As the distribution of intervals differed significantly between subject animals (Kruskal–Wallis test: $\chi^2 = 89.8, P < 0.001$), these criteria were determined for each subject. Between 0.3 and 10 s, one or two significant abrupt change points were found for all subjects (Fig. 3). For four subjects (Mdr, Ml, Yn, St), for whom only one change point each was found, I chose the point that was the longest interval before the response calls (mean =1.43 s). For the other three subjects (Tk, Ts, Ks), for whom two change points were found, I chose one change point that was closest to the mean change point for the other four subjects (1.43 s) as the longest interval for the response call. Thus, calls that occurred at or after 0.3 s but before the longest interval were classified as response calls, and those that occurred within 0.3 s or after the longest interval were classified as spontaneous calls.

4.3 Acoustic Difference Between Response and Spontaneous Calls

I analyzed the variation in acoustic parameters of the coo calls given in the successful trials ($N = 549$) to describe their basic structures. Two-way ANOVA was conducted for each of eight parameters, using the subject (df = 6) and two types of coos (response calls and spontaneous calls, df = 1) as two independent factors. For all acoustic parameters, the differences between subjects were highly significant ($P < 0.001$), but the differences between response calls and spontaneous calls were not significant ($P > 0.05$).

Table 4. Results of regression analysis in response and spontaneous calls, and difference of goodness of fit of regressions between the two types of calls (From Sugiura 1998, with permission)

Parameter	Response calls			Spontaneous calls			Between response and spontaneous calls
	Pooled regression		Between subject	Pooled regression		Between subject	
	b	$F_{1,38}$	$F_{6,32}$	b	$F_{1,30}$	$F_{6,24}$	$F_{7,70}$
Time to maximum frequency	0.238	10.1**	1.5	-0.003	0.0	0.3	0.9
Duration	0.511	16.6***	1.2	0.489	13.5***	1.4	1.1
Location of maximum frequency	0.001	0.0	1.0	-0.170	1.8	0.7	1.2
Maximum frequency	0.347	39.0***	2.1	0.222	4.7*	3.1*	3.8**
End frequency	0.161	2.1	1.8	-0.075	0.2	0.9	2.0
Maximum minus start frequency	0.359	66.9***	2.7*	0.057	0.3	2.3	43.5***
Maximum minus end frequency	0.191	12.0**	1.5	0.167	2.3	0.7	9.0***

$*P < 0.05; **P < 0.01; ***P < 0.001.$

4.4 Acoustic Properties of Stimuli and Calls

I examined the effects of the acoustic properties of stimuli on the acoustic properties of response calls by linear regression analysis. All the data from all seven subjects were analyzed together, with each individual treated as a block (Table 4). In the response calls, pooled regressions of five parameters (time to maximum frequency, duration, maximum frequency, maximum minus start frequency, and maximum minus end frequency) showed significantly positive effects. In the spontaneous calls, pooled regressions of duration and maximum frequency were significant.

The differences in the goodness of fit of regressions between all seven subjects were tested in response calls and spontaneous calls separately. Maximum minus start frequency in response calls and maximum frequency in spontaneous calls showed significant individual differences in the goodness of fit of regressions.

The differences of goodness of fit of regressions between response calls and spontaneous calls were also examined. Maximum frequency and two frequency ranges of the response calls had significantly better goodness of fit than those of spontaneous calls. For the other four parameters, there was no significant difference between the response and spontaneous calls.

Next, I conducted a regression analysis for each of the seven subjects, to see the effect of the acoustic properties of the stimuli on those of coos at the individual level. For response calls, five subjects (Mdr, Ml, Tk, Yn, and Ts) had significant positive regressions with stimulus calls at least in maximum minus start frequency (Table 5, Fig. 4). St had a significant regression in duration and a nearly significant

Table 5. Results of regression analysis in response and random calls for each subject (From Sugiura 1998, with permission)

Parameter	Mdr		Ml		Tk		Yn		Ts		Ks		St	
	b	F	b	F	b	F	b	F	b	F	b	F	b	F
Response calls														
Time to maximum frequency	0.33	2.3	0.64	7.2†	0.24	11.5*	0.24	0.7	0.09	0.1	-0.49	6.5*	0.16	0.5
Duration	0.73	2.3	1.12	14.6*	0.35	2.0	0.44	0.7	0.60	2.1	-0.02	0.0	0.23	8.6*
Location of maximum frequency	0.04	0.1	0.29	4.6	0.02	0.1	0.23	1.7	-0.12	0.3	-0.04	0.1	-0.18	3.4
Maximum frequency	0.58	35.7**	0.57	8.5*	0.43	30.1**	0.48	6.5†	0.23	3.5	0.09	0.6	0.35	4.4
End frequency	0.53	1.5	1.13	4.7	0.04	0.0	1.29	6.3†	0.08	1.3	0.01	0.0	-0.23	0.1
Maximum minus start frequency	0.42	19.8*	0.43	12.4*	0.45	17.9*	0.60	52.4**	0.42	11.1*	0.06	0.3	0.22	7.1†
Maximum minus end frequency	0.36	15.8*	0.24	2.8	0.33	4.8	0.34	2.3	-0.22	0.7	0.02	0.0	0.05	0.3
df	1, 4		1, 4		1, 4		1, 4		1, 6		1, 6		1, 4	
Spontaneous calls														
Time to maximum frequency	0.13	0.1	-0.06	0.0	0.06	0.2	-0.20	0.2	-0.33	1.0	0.43	0.2	-0.71	1.2
Duration	0.53	5.0	0.48	2.7	0.67	6.9†	-0.06	0.0	-0.30	0.6	1.62	3.9	-0.25	2.6
Location of maximum frequency	-0.38	1.9	-0.03	0.0	0.03	0.0	-0.55	0.7	-0.44	1.8	0.37	0.5	-0.03	2.6
Maximum frequency	0.40	2.5	-0.06	0.4	0.35	7.5†	-0.53	2.5	0.01	0.0	0.80	4.4	-0.20	1.1
End frequency	0.08	0.0	-0.15	0.0	0.22	0.2	-1.78	2.7	-0.12	1.5	0.40	0.3	-0.50	1.0
Maximum minus start frequency	0.13	0.2	-0.08	1.8	0.32	9.8*	-0.93	10.7*	-0.04	0.0	0.67	2.1	-0.11	0.3
Maximum minus end frequency	0.24	0.6	-0.11	0.8	0.16	4.4	0.36	0.7	0.01	0.0	0.88	1.4	-0.02	0.0
df	1, 4		1, 3		1, 4		1, 3		1, 5		1, 3		1, 2	

†$P < 0.1$; *$P < 0.05$; **$P < 0.01$.

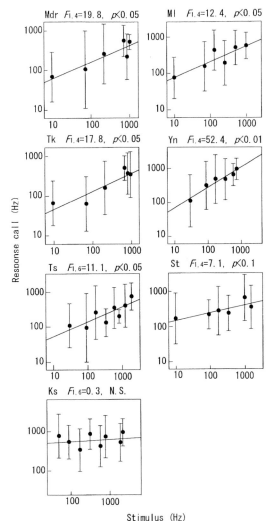

Fig. 4. Regressions of the maximum minus start frequency of the response coos of seven macaques on that of the playback stimuli (coo calls from an adult female). Means and standard deviations for coos in response to each stimulus and regression lines are indicated

regression in maximum minus start frequency. Only one animal, Ks, had no significant positive regression. For spontaneous calls, only maximum minus start frequency showed significant regressions and only in two subjects (Tk and Yn), and one of these (Yn) was negative (Table 4). Therefore, only one subject, Tk, showed a positive regression in any parameter which could have occurred by chance. In contrast to response calls, the acoustic features of stimulus coo calls had no significant positive effect on the acoustic features of spontaneous calls in the majority of the subjects.

Table 6. Kendall's rank correlation coefficient between response rate, median interval, and determinant coefficient of regression analysis for response calls and spontaneous calls (From Sugiura 1998, with permission)

	Response call		Spontaneous call	
Parameter	Response rate	Median interval	Response rate	Median interval
Time to maximum frequency	0.524	-0.619†	-	-
Duration	-0.429	0.333	-0.524	0.238
Location of maximum frequency	-	-	-	-
Maximum frequency	-0.810*	0.714*	-0.333	0.238
End frequency	-	-	-	-
Maximum minus start frequency	-0.619†	0.333	-	-
Maximum minus end frequency	-0.905**	0.619†	-	-

-, Not analyzed.
†$P < 0.10$; *$P < 0.05$; **$P < 0.01$.

4.5 Call Rates, Intervals, and Goodness of Fit of Regression

In the regression analysis, the goodness of fit of the regression varied between subjects. I conducted post hoc analysis to explore the possible factors that affected that variation.

Matching should require more careful listening to the acoustic features of preceding calls and more time to utter calls with controlled acoustic features than nonmatching. Thus, it would be more difficult for the animals to match to the majority of the calls they hear or to match with shorter intervals. The correlation between the call rates and the determinant coefficients (R^2) of the regression analysis and that between the median interval and R^2 was thus examined by Kendall's rank correlation test (Table 6). This analysis was performed only for the parameters that showed significant pooled regression (Table 4).

For the response calls, significant negative correlations were found between call rates and the determinant coefficients in the maximum frequency and the maximum minus end frequency. A significant positive correlation was also found between R^2 and the median interval in maximum frequency. Figure 5 shows the R^2 of the maximum frequency plotted against the response rate and median interval. In contrast, for spontaneous calls, no significant correlations were found between the call rates and the determinant coefficients, or between the median interval and determinant coefficients.

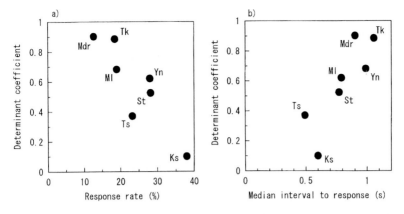

Fig. 5. Relationship between (**a**) the call rate and (**b**) the median interval and the degree of the goodness of fit of the regression analysis at maximum frequency. The name of the subject is indicated next to each point. The call rate is the percentage of response calls for the number of trials. Median interval is the median of the intervals between the end of the playback stimulus (coo call of an adult female) and the start of the subject's coo call. (From Sugiura 1998, with permission)

5 Discussion

The distribution of intervals between two consecutive calls in DC sequences in the naturalistic observations as well as those between the playback stimuli and subsequent calls was not a simple exponential distribution. It appeared, however, that calls were given more frequently approximately before 1–2 s and less frequently after that interval. In fact, statistically significant abrupt change points were found between 1 and 2 s.

Although calls occurred frequently just after 0 s from the previous calls in the observations, calls were rarely given before 0.3–0.5 s from the stimulus in the playback experiment. This discrepancy between the observation and the playback experiment was probably caused by the difference in the situations. In a natural situation, the first call can be preceded by another call, and this may allow the second caller to respond more rapidly. In addition, some of the second calls may be given not as responses to the first call, but as responses to the call before the first call. In the playback experiment, in contrast, no calls preceded the stimulus, and that is probably why the delay appeared.

Thus, in summary, the results indicate that the monkeys gave response coo calls after a delay of approximately 0.3–2 s.

The second coos in SC sequences rarely occurred during the periods in which most of the second coos in DC sequences occurred. These results suggest that when an animal gives a coo spontaneously, she remains silent during the period when the group members are likely to respond, and when no animal makes any vocal response, she gives another coo, again addressing the other animals.

These results suggest that Japanese macaques are able to alter the timing of emission of coo calls according to the group members' vocal response. Furthermore, when Japanese macaques call repeatedly, they may able to alter the timing of the production of coo calls according to the rapidity of the response (H. Sugiura, un-published data). When a monkey stayed close to other group members, response calls tended to return more rapidly than when a monkey stayed further from the group. In the former situation, calls in SC sequences were repeated more rapidly than in the latter situation. This suggests that callers adjust the time they wait for a response depending on the expected latency in response.

The playback experiment demonstrated that the basic hypothesis that monkeys alter the acoustic features of their coos to match those of previous coos was valid. When all seven subjects were pooled, the acoustic quality of the response calls was positively correlated with that of the playback stimuli, while spontaneous calls were acoustically independent of the stimuli.

At the individual level, five of the seven subjects showed significant positive regressions in maximum minus start frequency. However, the degree of goodness of fit in regressions differed between subjects, i.e., matching was better in some subjects than others, and two of the subjects showed no evidence of matching. Although this individual difference needs further examination, I can suggest two possible factors which could affect these differences. The subjects who responded with higher rates or with short intervals after the playback tended to match acoustic properties with those of stimuli to a lesser extent. Frequent or fast responses might need a trade-off with the degree of vocal matching.

The frequency range of response calls from the start to the maximum was significantly influenced by that of the playback stimuli for most subjects. The results indicate that Japanese macaques matched the acoustic features of their coo calls with those of the stimulus calls to which they responded, and they could modify the acoustic features of their calls, in particular the range of frequency modulation of the fundamental frequency component, under natural conditions.

There is no general agreement about whether nonhuman primates learn the acoustic features of their vocalization (e.g., Snowdon 1990). Although the ability to match might not be sufficient, it would be necessary for vocal learning. Thus, my results support the possibility that the acoustic features of the coo calls of Japanese macaques are learned.

The basic function of vocal exchange of coo calls appears to be to locate group members and maintain contact with them (Itani 1963; Okayasu 1987). A possible functional reason why macaques match the acoustic features of their vocal response with those of the preceding call is to indicate that the call is given as a response to the preceding caller. Occasionally, two or more callers give coo calls successively or simultaneously during a vocal exchange (Mitani 1986; Sugiura 1993). In such situations, the acoustic cues of responses might help the preceding callers to discriminate between a true response and an independent call. Another possible explanation for matching is related to the ease of location of coo calls, which depends on the range of their frequency modulation (Brown et al. 1979). Japanese macaques exchanging coo calls might use a similar range of frequency modulation because it is the most suitable one for the distance between the callers and ambient noise levels at the time. In fact, Japanese macaques give coo calls with a greater frequency range when they stay far away from group members than when they stay close to them (H. Sugiura, unpublished data).

The present study described the fundamental pattern of vocal exchange in Japanese macaques. The temporal pattern of this vocal exchange is similar to that that found in squirrel monkeys (Masataka and Biben 1987) and that between a human mother and her infant (Masataka 1993). Vocal matching during vocal exchange appears to be comparable with the vocal behavior of human infants, who match their pitch patterns to presented song tones (Kessen et al. 1979) or to those of their mother's vocalization (Masataka 1992). These similarities may suggest that vocal exchanges of contact calls in nonhuman primates parallels human conversation. In addition, vocal exchanges have been reported in several other nonhuman primates (Snowdon and Cleveland 1984; Biben et al. 1986; Mitani 1986), and it may be a common form of communication in primates. However, there still remains the possibility that such vocal exchanges may not be restricted to primate species, but that they are a common form of communication among social species that make contact with other group members. A recent study of dolphins suggested that they exchange "whistles" among group members in a similar temporal pattern to that found in nonhuman primates (Nakahara 1998) and match the acoustic features of the preceding whistle when they respond (Janik 2000). Comparative studies over a broad phylogenetic range may be necessary to further clarify the evolution of vocal exchanges in nonhuman animals and its relation to human conversation.

Acknowledgments

I am grateful to Nobuo Masataka for his advice and comments throughout this study, and to the Japan Monkey Center and the Yakushima Forest Environment Conservation Center for permission to perform field research. I also thank the Field Research Center of the Primate Research Institute, Kyoto University, for affording the full use of the facilities in the Yakushima Field Station. This work was partially supported by a Research Fellowship of the Japan Society for the Promotion of Science for Young Scientists and the Cooperative Research Fund of the Primate Research Institute, Kyoto University.

References

Biben M, Symmes D, Masataka N (1986) Temporal and structural analysis of affiliative vocal exchanges in squirrel monkeys (*Saimiri sciureus*). Behaviour 98:259–273
Brown CH, Beecher MD, Moody DB, Stebbins WC (1979) Locatability of vocal signals in Old World monkeys: design features for the communication of position. J Comp Physiol Psychol 93:806–819
Cowlishaw G (1992) Song function in gibbons. Behaviour 121:131–153
Dunbar RIM (1996) Grooming, gossip, and the evolution of language. Harvard University Press, Cambridge, pp 230
Elowson AM, Snowdon CT (1994) Pygmy marmosets, *Cebuella pygmaea*, modify vocal structure in response to changed social environment. Anim Behav 47:1267–1277
Farabaugh SM (1982) The ecological and social significance of duetting. In: Kroodsma DE, Miller ED (ed) Acoustic communication in birds. Acdemic Press, New York, pp 85–124

Green S (1975) Variation of vocal pattern with social situation in the Japanese monkey (*Macaca fuscata*): a field study. In: Rosenblum LA (ed) Primate behavior. Academic Press, New York, pp 1–102

Haccou P, Meelis E (1992) Statistical analysis of behavioural data: an approach based on time-structured models. Oxford University Press, New York, 396 pp

Herzog M, Hopf S (1983) Effects of species-specific vocalizations on the behaviour of surrogate-reared squirrel monkeys. Behaviour 86:197–214

Inoue M (1988) Age gradation in vocalization and body weight in Japanese monkeys (*Macaca fuscata*). Folia Primatol 51:76–86

Itani J (1963) Vocal communication of the wild Japanese monkey. Primates 4(2):11–66

Janik VM (2000) Whistle matching in wild bottlenose dolphins (*Tursiops truncatus*). Science 289:1355–1357

Kawai M (1960) A field experiment on the process of group formation in the Japanese monkey (*Macaca fuscata*), and the releasing of the group at Ohirayama. Primates 2:181–255

Kessen W, Levine J, Wendrich KA (1979) The imitation of pitch in infants. Infant Behav Dev 2:93–99

Lieblich AK, Symmes D, Newman JD, Shapiro M (1980) Development of the isolation peep in laboratory-bred squirrel monkeys. Anim Behav 28:1–9

Maples EG, Haraway MM, Collie L (1988) Interactive singing of a male Mueller's gibbon with a simulated neighbor. Zoo Biol 7:115–122

Maruhashi T (1992) Fission, takeover, and extinction of a troop of Japanese monkeys (*Macaca fuscata yakui*) on Yakushima Island, Japan. In: Itoigawa N, Sugiyama Y, Sackett GP, Thompson RKR (eds) Topics in primatology. University of Tokyo Press, Tokyo, pp 47–56

Masataka N (1992) Pitch characteristics of Japanese maternal speech to infants. J Child Lang 19:213–223

Masataka N (1993) Effects of contingent and noncontingent maternal stimulation on the vocal behaviour of three- to four-month-old Japanese infants. J Child Lang 20:303–312

Masataka N, Biben M (1987) Temporal rules regulating affiliative vocal exchanges of squirrel monkeys. Behaviour 101:311–319

Masataka N, Fujita K (1989) Vocal learning of Japanese and rhesus monkeys. Behaviour 109:191–199

Mitani M (1986) Voiceprint identification and its application to sociological studies of wild Japanese monkeys (*Macaca fuscata yakui*). Primates 27:397–412

Nakahara F (1998) Vocal development and vocal behavior of dolphins (in Japanese with English abstract). IBI Reports 8:43–51

Okayasu N (1987) Coo sound communication (in Japanese). Quaternary Anthropol 19:12–30

Owren MJ, Dieter JA, Seyfarth RM, Cheney DL (1993) Vocalizations of rhesus (*Macaca mulatta*) and Japanese (*M. fuscata*) macaques cross-fostered between species show evidence of only limited modification. Dev Psychobiol 26:389–406

Saslow CA (1972) Behavioral definition of minimal reaction time in monkeys. J Exp Anal Behav 18:87–106

Sibly RM, Nott HM, Fletcher DJ (1990) Splitting behaviour into bouts. Anim Behav 39:63–69

Snowdon CT (1990) Language capacities of nonhuman animals. Yearb Phys Anthropol 33:215–243

Snowdon CT, Cleveland J (1984) "Conversations" among pygmy marmosets. Am J Primatol 7:15–20

Sugiura H (1993) Temporal and acoustic correlates in vocal exchange of coo calls in Japanese macaques. Behaviour 124:207–225

Sugiura H (1998) Matching of acoustic features during the vocal exchange of coo calls by Japanese macaques. Anim Behav 55:673–687

Sugiura H, Masataka N (1995) Temporal and acoustic flexibility in vocal exchanges of coo calls in Japanese macaques (*Macaca fuscata*). In: Zimmermann E, Newman JD, Jeurgens U (eds) Current topics in primate vocal communication. Plenum Press, New York, pp 121–140

Tanaka T (1995) Populational differences of vocal behavior in Japanese monkeys (in Japanese with English abstract). Jpn J Psychol 66:176–183

Tenaza RR (1976) Songs, choruses and countersinging of Kloss' gibbons (*Hylobates klossii*) in Siberut Island, Indonesia. Z Tierpsychol 40:37–52

8
Hearing and Auditory–Visual Intermodal Recognition in the Chimpanzee

KAZUHIDE HASHIYA[1] and SHOZO KOJIMA[2]

1 Introduction

In contrast to the large body of studies on chimpanzee vision, few studies have examined the function of other sensory modalities in chimpanzees. In 1929, Yerkes and Yerkes noted a lack of studies on chimpanzee perception "except for the sense of sight". The situation has remained largely unchanged for 70 years, though there are a few exceptions. Vocal–auditory functions are still the least understood behavioral pattern of chimpanzees (Mitani 1994; Nishida 1994; Tomosello and Call 1996) despite many researchers having pointed out that vocal–auditory channels play particularly important roles in behavior. Observations under completely natural conditions are not always practicable (though desirable) to evaluate the cognitive abilities of chimpanzees because of the difficulty in excluding extraneous factors. Behavioral experimentation is also needed in this sense, but there are only a limited number of studies at this point. This is probably because of the fact that training nonhuman primates in auditory-related tasks is very difficult, compared to the corresponding visual tasks.

The first part of this chapter reviews studies on auditory perception of the chimpanzee, mainly using the reaction time (RT) task. The second part reviews studies examining how the chimpanzee recognizes perceived auditory signals. To study auditory recognition of chimpanzees, the authors propose that studies on the correspondence between auditory recognition and visual recognition, that is, auditory–visual intermodal recognition is a good approach.

2 Auditory Perception

2.1 Basic Auditory Function of the Chimpanzee

In order to evaluate an animal's cognitive ability properly, it is important to measure the basic function of each sensory channel. Except for the pioneering studies

[1] Department of Cognitive Psychology, Graduate School of Education, Kyoto University, Yoshida Honmachi, Sakyo-ku, Kyoto 606-8501, Japan
[2] Primate Research Institute, Kyoto University, 41 Kanrin, Inuyama, Aichi 484-8506, Japan

by Elder (1934; 1935), few studies have examined auditory acuity in chimpanzees (Stebbins 1971; Tomonaga 1990).

2.1.1 Absolute Threshold

The absolute threshold and loudness of pure tones at various frequencies were measured in 2 female chimpanzees (Kojima, 1990). A reaction time (RT) task was employed, on the basis of the suggestion that reaction times for detection of tones reflect loudness (Pfingst et al.1975a, b).

A trial of the RT task consisted of the events as follows. The subject wore earphones and faced a panel containing a lamp and a telegraph key in a double-walled, sound attenuating room (Fig. 1). Following illumination of the lamp, the subject was required to hold the key down until a pure tone was presented. On detecting the tone, the subject had to release the key as quickly as possible. The duration between the key press and the onset of the tone was varied from 2 to 6 s. The latency between the onset of the tone and key release was measured as RT. Key-release during the 1-s presentation of the tone was followed by the delivery of a food reward, a small piece of fruit, and the termination of the lamp. If the subject did not release the key within 1 s after the onset of the tone, or the subject released the key before the onset of the tone, the trial terminated without a food reward. Tone signals were presented to the subject through the earphone. (Note: the setup of the task was basically common among the RT tasks described hereafter, but differed in

Fig. 1. The chimpanzee Pan performing a reaction time (RT) task

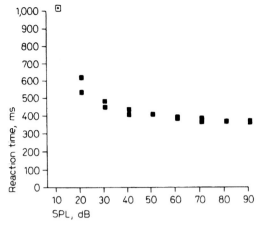

Fig. 2. Intensity-reaction time functions at 1 kHz for one of the chimpanzees (Pen). The results for two sessions for one subject are shown in this figure. *Filled squares* indicate the mean reaction time of 8 trials. The *open square with a dot* indicates that the mean reaction time was the same for the two sessions

some parts. For details of the procedures of each experiment, refer to the original papers). The frequency of the tones used as stimuli varied from 0.125 kHz to 32 kHz in 1-octave intervals as well as 24 kHz, with 15 ms rise-fall time. Intensities were in 10 dB steps from −10 dB to 90 dB SPL and changed randomly for every 8 trials except for the first 8 trials, where a 90 dB tone was always presented. A single frequency was used in a daily session of 96 trials.

As a source of comparison, one human was tested using the same apparatus and procedures, except for lack of delivery of the food reward. The frequencies of pure tones used for the human subject were from 0.125 kHz to 16 kHz in 1-octave intervals.

The results showed that RT became shorter as the intensity of the tone increased (Fig. 2). A curvilinear regression line was fitted to intensity-RT functions and equal reaction time contours were derived from them for the chimpanzee and human subjects, respectively (Fig. 3). The absolute threshold at a particular frequency was defined as the sound intensity associated with a mean RT of 800 ms. The chimpanzees were sensitive to 1 kHz and 8 kHz tones; the best frequency was 8 kHz. Both of the chimpanzee subjects showed a 10 to 20 dB SPL loss in auditory sensitivity at 2 to 4 kHz tones for the threshold level. In contrast, the human subject was most sensitive to middle range frequencies of 0.5 to 4 kHz. Auditory sensitivity was better in chimpanzees than in the human at higher frequencies (> 8 kHz), and vice versa at lower frequencies (< 0.25 kHz). Chimpanzees showed a W-shaped auditory sensitivity function whereas humans showed a U-shaped function. Elder (1934) reported a "4-kHz dip" and high sensitivity to 8 kHz in chimpanzees. Though there are some differences, the present results basically support the finding by Elder. A W-shaped audibility function had been observed with experiments with using earphones in Old and New World monkeys (Beecher 1974; Stebbins 1973 [See Brown and Waser (1984) for a case where subjects were tested in a free field situation]). In this sense, the results obtained in chimpanzees were similar to those for monkeys and differed from those for humans.

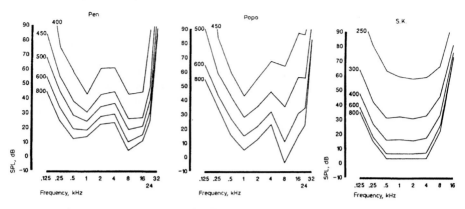

Fig. 3. Equal reaction time contours for the chimpanzees (Pen and Popo) and the human subject (SK). Numbers within each graph are reaction times in milliseconds

2.1.2 Resonance of External Auditory Meatus

What then are the origins of the loss of sensitivity at middle range frequencies in chimpanzees? The auditory system is divided into four parts: the external ear, the eardrum and middle ear, the inner ear and the auditory nervous system. Kojima (1990) measured the resonance of external auditory meatus in the formaldehyde-preserved cadaver of a chimpanzee and three living humans. In contrast to the results from behavioral experiments, the resonance frequency of the external ear of the chimpanzee was 2.5 kHz. The decrease in resonance at 2–4 kHz, frequencies at which the chimpanzees show a loss in sensitivity, was not observed. The frequency and the magnitude of the gain were approximately the same for humans and the chimpanzee (Fig. 4). This means that the resonance of the external ear does not account for the difference in auditory sensitivity between chimpanzees and humans. Rather, it is possible that resonance of the external ear may itself be a cause of the loss of sensitivity. Chimpanzees (as well as other species of nonhuman primates) often use loud calls, such as pant-hoots and screams (e.g., Nishida et al. 1999) in everyday life. These loud calls, which are amplified by the external ear, may affect auditory transmission in the middle and the inner ear. Histological and electrophysiological examination of these structures are necessary to test this possibility. There has been a suggestion that human speech might have evolved from vocal repertoires such as the short-distance, relatively soft calls of monkeys (Itani 1963). The evolution of human hearing may accordingly have some relation to the evolution of human speech.

2.1.3 Difference Threshold

Difference thresholds for frequency and intensity were examined in 2 chimpanzees and a human, using the RT task. In a departure from the aforementioned case of measuring the absolute threshold, pressing of the telegraph key in the presence

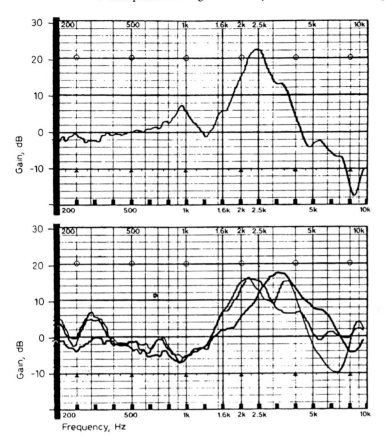

Fig. 4. Resonance of the external meatus of the formaldehyde-preserved cadaver of a chimpanzee (*upper panel*) and of three humans (*lower panel*)

of the lamp illumination initiated repeated presentation of a background stimulus with a 1.512-s inter-stimulus interval. Following two to eight times (varied among the trials) of repetition of the background stimulus, the target stimulus was presented; the subject had to release the key as quickly as possible when he/she detected the change in stimulus. In measuring the frequency difference threshold, the background stimulus varied from 0.5 to 4 kHz in 1-octave intervals and its intensity was fixed to 70 dB SPL. The differences between the background and the target stimulus were from ± 0.005 to ± 0.08 kHz (in 6 steps) for the chimpanzee and from ± 0.0025 to ± 0.04 kHz (in 6 steps) for the human subject, respectively. In measuring the intensity difference threshold, the background stimulus varied from 50 to 90 dB SPL in 10 dB SPL intervals and the frequency was fixed to 1 kHz. The differences between the background and the target stimulus were from ± 1 to ± 8 dB SPL (in 6 steps) for the chimpanzee and from ± 0.25 to ± 4 dB SPL (in 6 steps) for the human subjects, respectively.

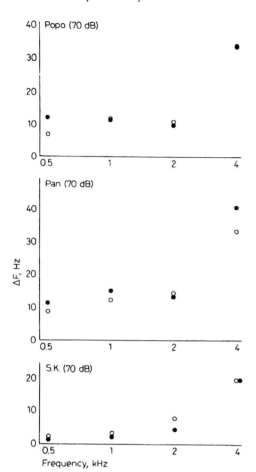

Fig. 5. Difference thresholds for frequency (DF). *Filled circles* indicate thresholds for frequency increments and *open circles* those for frequency decrements for the chimpanzees (Popo and Pan) and the human subject (SK)

The difference threshold was defined as the difference in frequencies or in intensities between the background and the target stimuli that was accompanied by a detection rate of 50%. Frequency difference thresholds at each frequency of background stimuli are shown in Fig. 5. In chimpanzees, frequency difference thresholds were about 0.01–0.015 kHz at the background stimuli of 0.5, 1 and 2 kHz. However, they rose abruptly to 0.03–0.035 kHz at the background stimulus of 4 kHz. The results for human subject showed a similar tendency, except that the thresholds were lower than those of the chimpanzees.

The intensity difference thresholds at each stimulus intensity are shown in Fig. 6. Intensity difference thresholds of chimpanzees became lower as the stimuli became louder. However, at the highest stimulus intensity (90 dB SPL of background stimuli), difference thresholds rose again, especially for frequency increment thresholds. In humans, intensity difference thresholds were lower than those of the chimpanzees and they decreased with the increase of stimulus intensity; the lowest thresholds were observed at the highest intensity.

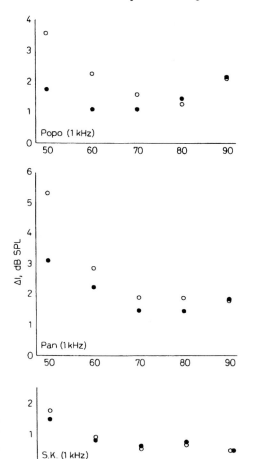

Fig. 6. Difference thresholds for intensity (DI). *Filled circles* indicate thresholds for intensity increments and *open circles* those for intensity decrements for the chimpanzees (Popo and Pan) and the human (SK)

2.2 Perception of Complex Signals

How does the perception of complex auditory signals reflect basic auditory functions? Chimpanzee perception of complex auditory signals was examined, using natural and synthetic signals.

2.2.1 Human Speech Signals: Consonant Perception

In humans, the structure of consonant perception and its relation to the feature system has been investigated in terms of the similarity or confusion among consonants (Miller and Nicely 1955; Wilson 1963). The perception of 20 French consonants was examined in chimpanzees, using the RT task (Kojima et al. 1989). Each consonant was followed by the vowel [a] pronounced by a native female speaker. Within the set of 20 consonant-vowel (CV) syllables, all the possible combinations of the background and the target syllable were tested. Following the repeated pre-

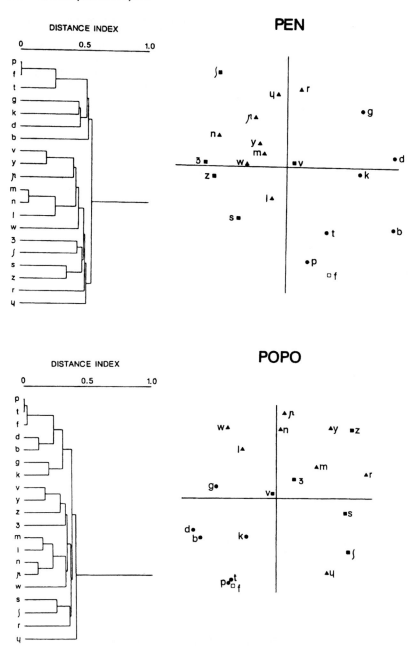

Fig. 7. The perceptual structure of the 20 French consonants by the chimpanzees (Pen and Popo). *Left panels* are the results of a cluster analysis (dendrograms) and *right panels* are those of MDSCAL. *Circles* are stop consonants, *squares* are fricatives, and *triangles* are other consonants. Because of a mistake in digitizing in the experiment, the consonant [f] became very similar in sound to [p]

sentation of the background CV syllable from 2 to 6 times (varied among the trials) with 1-s inter-stimulus interval, the subjects had to detect the change in syllables to the target. A nonmetric, multidimensional scaling method (MDS) and cluster analyses were applied to data on RTs, which were taken as an index of similarity between CV syllables (Fig. 7).

Results showed that in both of the chimpanzees stop consonants constituted a cluster. Results of clustering in chimpanzees roughly corresponded to the classification of consonants according to the manner of articulation by humans, though differentiation of consonants other than stops was not very clear. Also, the place-of-articulation and voicing features seemed to determine the structure of stop consonant perception by chimpanzees. That is, stop consonants seemed to be distributed on the two-dimensional MDS space, reflecting the voiced/unvoiced dichotomy: voiced stop consonants ([b], [d] and [g]) located peripherally, whereas unvoiced stop consonants ([p], [t] and [k]) located relatively apart from the voiced stop consonants.

The perception of stop consonants was examined in detail in a subsequent experiment, using the same RT task. The CV syllables used were natural and synthetic stop consonants ([p], [b], [t], [d], [k] and [g]) articulated by male voices, followed by the vowel [a].

The results of MDS analysis, taking the RT for discrimination between CV syllables as an index of similarity (Fig. 8), revealed that for chimpanzees, one of the two dimensions in MDS space was that for the voicing feature. In natural CV syllables, unvoiced stop consonants [p, t, k] were located peripherally in one dimension whereas the voiced stop consonants [b, d, g] were located on the opposite side of the same dimension. On the other hand, the place-of-articulation feature was not clearly reproduced in MDS space. The human showed a similar tendency. These results were reproduced for discrimination between synthetic CV syllables.

For the synthetic CV syllables, RTs for discrimination between consonants differing in voicing features (e.g., [pa] vs. [ba]) and in place-of-articulation features (e.g., [pa] vs. [ta]) were compared (Fig 9). Longer RTs were required for discrimina-

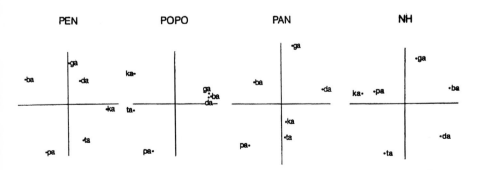

Fig. 8. The perception of the synthetic Japanese stop consonants by the chimpanzees (Pen, Popo, and Pan) and a human (NH)

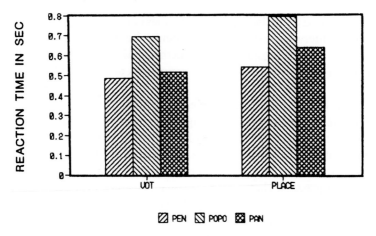

Fig. 9. Mean reaction time for discrimination of voicing (VOT) and the place-of-articulation (*PLACE*) features of the synthetic stop consonants by the three chimpanzees (Pen, Popo and Pan)

tion based on the place feature than on the voicing feature. That is, chimpanzees had difficulty in discriminating the place-of-articulation feature of stop consonants (e.g., in discriminating [ba] from [da]), as compared to discriminating the voicing feature.

These findings suggest that the major grouping of the stop consonants in chimpanzees is based on voicing features. Voicing is the main determinant for the perception of stop consonants in humans (Miller and Nicely 1955; Wilson 1963). The basic mechanism for the identification of stop consonants might be similar for chimpanzees and humans.

Humans show enhanced discrimination ability at the boundary between phonetic categories. This phoneme-boundary effect had once been regarded as one of the prerequisites for human speech and to be unique to humans. However, the same effect was found in perception in chinchillas (Kuhl 1981; Kuhl and Miller 1975, 1978) and in macaque monkeys (Kuhl and Padden 1982, 1983; Morse and Snowdon 1975; Sinnot et al. 1976; Waters and Wilson 1976). These findings have had an impact on the theories of speech perception. Kojima et al. (1989) examined the phoneme-boundary effect in chimpanzees.

Eight stop consonants on a [ga] (voiced consonant) - [ka] (unvoiced consonant) continuum with a discrimination of voice onset time (VOT) of 8 ms step were synthesized to test the effect based on VOT differences. Also, eight stop consonants on a [ba] (voiced, bilabial stop consonant) - [da](voiced, alveolar stop consonant) continuum were synthesized to test the effect between stops with different place-of-articulation. The starting frequencies of F2 and F3 transitions were 1.1 kHz and 2.25 kHz for [ba], respectively, and 1.8 kHz and 2.819 kHz for [da], respectively. Each stimulus the on [ba] - [da] continuum was synthesized at 0.1 kHz and at 0.0813 kHz steps for F2 and F3 frequencies, respectively (see Fig 10).

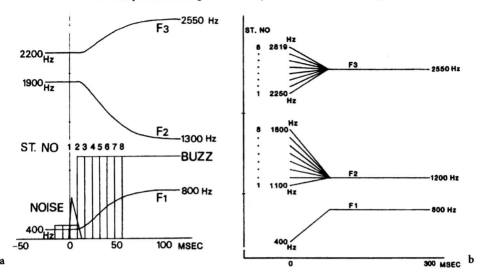

Fig. 10. a Eight synthetic stop consonants on the [ga]-[ka] continuum with VOT of 8-ms steps used to study the phoneme-boundary effect between voiced and unvoiced stop consonants. **b** Eight synthetic stop consonants on the [ba]-[da] continuum used to study the phoneme-boundary effect between stops with different places of articulation

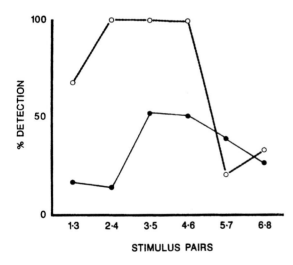

Fig. 11. Discrimination of stimulus pairs based on VOT differences. *Filled circles* with *solid lines* are the results by the chimpanzee subject and *open circles* with *broken lines* are those by the human subject

For each continuum, stimuli that differed at 2 steps were paired and the discrimination within the pair was tested in the RT task. That is, stimulus no. 1 was paired with no. 3, while stimulus no. 2 was paired with no. 4, and so on. For both VOT different continuum and for place-of-articulation different continuum, the phoneme-boundary effect was evident (Figs. 11 and 12).

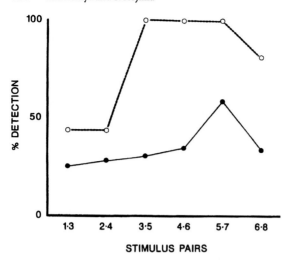

Fig. 12. Discrimination of stimulus pairs based on the difference in the starting frequencies of the second and third formants, i.e., place of articulation. *Filled circles* with *solid lines* are the results of the chimpanzee subject and *open circles* with *broken lines* are those by the human subject

2.2.2 Human Speech Signals: Vowel Perception

In contrast to consonants, vowels seems to have "fuzzy edges". It is difficult to determine the exact point at which, for example, an [a] sound becomes an [o] sound. Three chimpanzees and three humans (Japanese males) were tested in the RT task (Kojima and Kiritani 1989). Three sets of vowels were used as stimuli: five synthetic and natural Japanese vowels pronounced by a native male speaker and eight basic natural French vowels pronounced by a native female speaker (see Table 1). The subject had to detect the change in vowels from the background to the target after 2–6 times (varied) of repeated presentation of the background vowel. Within a single stimulus set, all the possible combinations of the background and the target vowel were tested. An MDS and cluster analyses were applied to data on RTs, which were taken as an index of similarity between vowels.

Table 1. The frequencies of the First (F1) and Second (F2) formant of the vowels used in the experiments

	Synthetic Japanese vowels				
	[i]	[e]	[a]	[o]	[u]
F1	230	500	750	550	300
F2	2100	1900	1300	950	1500
	Natural Japanese vowels				
	[i]	[e]	[a]	[o]	[u]
F1	390	651	911	520	390
F1	2604	2213	1367	911	1302

	Natural French vowels							
	[i]	[e]	[ɛ]	[a]	[α]	[ɔ]	[o]	[u]
F1	260	520	716	911	911	911	455	325
F2	2734	2437	2213	1367	1236	2148	846	781

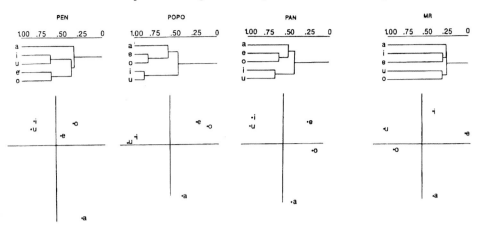

Fig. 13. Discrimination between synthetic Japanese vowels by the chimpanzees (Pen, Popo and Pan) and a human (MR). *Upper panels* are the results of a cluster analysis (dendrograms) and *lower panels* are those of MDSCAL. Numbers above each dendrogram are similarity indices (reaction time in seconds)

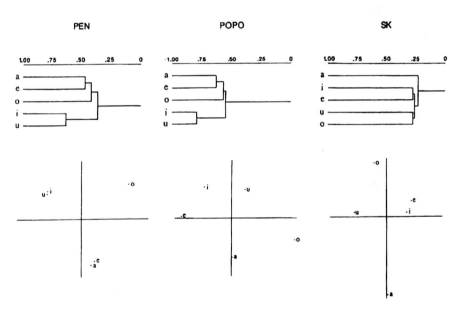

Fig. 14. Discrimination between natural Japanese vowels by chimpanzees (Pen and Popo) and a human (SK)

Figures 13, 14 and 15 show the perception of synthetic Japanese vowels, natural Japanese vowels and French vowels by the chimpanzee and the human, respectively. For the synthetic Japanese vowels, the longer RTs were necessary for discrimination of [i] from [u] and of [e] from [o] by all the chimpanzee subjects, whereas the humans required longer RTs for discrimination of [u] from [o] and of

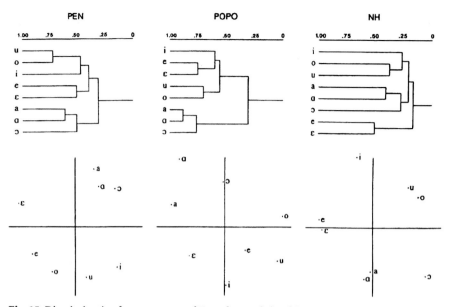

Fig. 15. Discrimination between natural French vowels by chimpanzees (Pen and Popo) and a human (NH)

[i] from [e]. For the natural Japanese vowels, three chimpanzees discriminated [i] from [u] with the longest RT, whereas the humans required the longest RT for discrimination of [i] from [e]. For the natural French vowels, the chimpanzees classified vowels into two groups, that is the group of [i], [e] and [u] and the group of [a] and [o]. The human subject, a Japanese, classified the French vowels into five Japanese vowels [u, o, a, e, and i].

The differences in auditory sensitivity functions might explain the difference in vowel perception between chimpanzees and humans. As described above, chimpanzees were less sensitive to the frequencies lower than 0.25 kHz, compared to humans. This might lead the chimpanzees to have a difficulty in hearing the first formants of [i] and [u]. Also, chimpanzees were less sensitive to middle-range frequencies of 2–4 kHz than to 1 and 8 kHz, and thus had a W-shaped auditory sensitivity function, whereas humans had a U-shaped auditory sensitivity function. This might be a factor in explaining the difficulty for the chimpanzee in hearing the second formants of [i] and [e], since the second formants of these vowels fall into the range between 2 to 4 kHz.

From another perspective, the repertoires of vocalization in chimpanzees may have some relation to their vowel perception. Chimpanzees usually produce vowel-like sounds, heard as [a], [o] and [u], but rarely produce vowels heard as [i] or [e] (Kojima and Nagumo, unpublished). A set of eight vowels on the [i]-[u] continuum were synthesized and discrimination among them was tested in a chimpanzee using the RT task. The background vowel was a typical [i] or [u] and one of the other members within the continuum was randomly allotted to the target

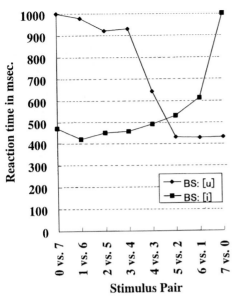

Fig. 16. Perceptual magnet effect. Eight vowels of the [i]-[u] continuum were synthesized and the discrimination between the vowels was examined. Stimulus No. 0 represents typical [u] and No. 7 represents typical [i]. *Solid black circles* with *solid lines* show the results when typical [u] was used as background stimulus (BS). *Shaded squares* with *solid lines* show the results when typical [i] was used as BS. The subject required a longer reaction time in the discrimination of stimulus numbers 0 vs. 7, 1 vs. 6, 2 vs. 5, and 3 vs. 4 when typical [u] was used as standard stimulus. Such tendencies were not observed when typical [i] was used as standard stimulus. The perceptual magnet effect was observed in vowel [u] but not in [i]

vowel. The chimpanzee seemed to have a "confusion" in discrimination even between typical [u] and relatively acoustically distant vowels, as compared with the case of using typical [i] as background vowel (Fig. 16). This seems to suggest that the vowel [u] had a "perceptual magnet effect" (Kuhl 1991) on the vowel perception of the chimpanzee, but the vowel [i] did not. The lack of the vowel-like sound [i] in the chimpanzee vocal repertoires might be another factor in explaining the difference in perception of the vowels.

2.2.3 Vocal Tract Normalization

The ability to perceive the same vowel from different speakers and the capacity for vocal tract normalization was examined in chimpanzees (Kojima and Kiritani 1989). Humans can perceive the same vowel regardless of the differences among speakers in voice. The property of the vowel is defined mainly by the first and the second formants (F1 and F2) distribution. However, even in a single vowel, the formant frequencies vary among speakers, reflecting differences in size of the vocal tract of the speaker. These facts lead to the phenomenon whereby the vowels of the same F1-F2 distribution uttered by different speakers are sometimes heard as different vowels, such as [a] and [o]. The listeners seem to perform a computation that adjusts or "normalizes" for these types of variations in the speech signals. It is of interest whether chimpanzees are endowed with such abilities of normalization.

First, Kojima and Kiritani examined whether the chimpanzee could ignore the difference in speakers who pronounced the same vowel. In the RT task, three Japanese vowels [a], [i], and [u] uttered by a female and a male speaker, respectively, were used as stimuli. As background stimuli in a single trial, the same vowels uttered by the 2 speakers were used and presented in random order. The background vowels were repeated for 3 to 6 times and changed to the different, target

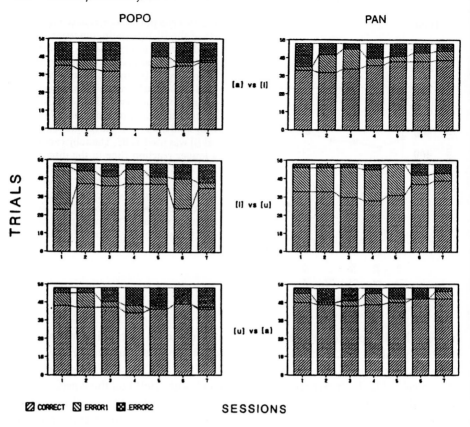

Fig. 17. Discrimination of the three vowel pairs ([a] vs. [i], [i] vs. [u], and [u] vs. [a]) uttered by different speakers, i.e., the results of the perceptual constancy experiment with the chimpanzees. *Error 1* indicates errors in which there was no response to change in vowels (miss) and *Error 2* indicates errors in the response to changes in the sex of the speaker of the background vowel (false alarm). The data for Popo at the fourth session of the [a]-[i] pair were accidentally lost

vowel. Only the response to the change in vowels produced food reward: the subjects had to ignore the change in speakers of the background vowel. The results showed that the chimpanzee detected the change in vowels with 70%–83% accuracy and responses to the change in speakers occurred only in 15%–16% of the trials in the last two sessions (Fig. 17). This indicated that the chimpanzee could perceive vowels as the same vowel irrespective of the speaker.

Using synthesized vowels as stimuli, a subsequent experiment examined the difference in perception of [o]-[a] boundaries of male and female voices, that is, ability for vocal tract normalization, in chimpanzees and a human. Two sets of eight vowels on the [o]-[a] continuum were synthesized (Fig. 18). Each vowel from one set was the same in F1 and F2 frequencies as one of the vowels from the other set, but differed in pitch, F3 and F4 frequencies. The stimuli were prepared considering the suggestion that pitch, F3 and F3 frequencies serve as cues for normalization (Fujisaki and Kawashima 1968). The set of vowels with higher pitch, F3,

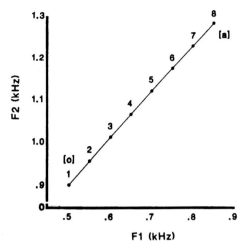

Fig. 18. Synthetic [o]-[a] continuum in the first-second formant (F1-F2) plane. Frequencies of F1 and F2 were determined by the following equations: $F1 = 500+50(i-1)$; $F2 = 1.1 \times F1+350$. The fundamental frequency of one of the sets is 180 Hz and it is heard as female voices. The other set has a fundamental frequency of 100 Hz and is heard as male voices. The frequency of the third and fourth formants of the female voices were 3 and 4 kHz, respectively, while those of the male voices were 2.5 and 3.5 kHz, respectively. Numbers printed beside *filled black circles* in this figure represent stimulus numbers. The stimulus duration was 300 ms and the rise and fall times are 10 and 20 ms, respectively

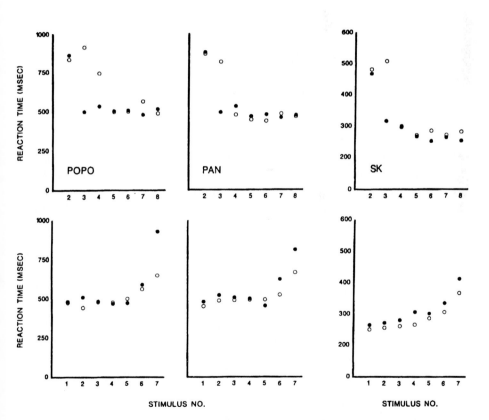

Fig. 19. Reaction times for discrimination of vowels in the synthetic [o]-[a] continuum by the chimpanzees (Popo and Pan) and a human (SK). The *upper panel* in each pair indicates the results of trials in which the background vowel was a typical [o] (Stimulus No. 1) and the *lower panel* indicates those of trials in which the background vowel was a typical [a] (Stimulus No. 8). *Open circles* correspond to female voices, and *filled circles* to male voices

and F4 frequencies (0.18, 3, and 4.2 kHz, respectively) were heard as female voices and the other (0.1, 2.5, and 3.5 kHz of pitch, F3, and F4 frequencies, respectively) as male voices. After confirming the adequacy of stimuli in line with the suggestion by Fujisaki and Kawashima for the human subject, two chimpanzees and a human were tested in the RT task. The background vowel was a typical [o] or [a] (stimulus No. 1 or 8 in Fig. 17, respectively) in each vowel set and one of the other members within the same set was randomly allotted to the target vowel.

Both the chimpanzees and the human subjects showed similar tendencies (Fig. 19): longer RT was required for female than for male voices in discrimination between Stimulus Nos. 1 and 3 (and between Nos. 1 and 4). In contrast, longer RT was required for male than for female voices in discrimination between Stimulus Nos. 8 and 6 (and between Nos. 8 and 7).

The results suggest that chimpanzees have a capacity for vocal tract normalization. The ability of vocal tract normalization has been found from early infancy in humans (Kuhl 1979). It seems to be evident that the chimpanzee has a potential ability for grouping the same conspecific signals regardless of the age and/or sex of the vocalizer.

3 Auditory–Visual Intermodal Recognition

The first part of this chapter has described the manner of auditory perception in the chimpanzee. The next question is how the perceived signals were understood and used in the natural lives of chimpanzees. How do they process meaningful auditory information? It is still unclear how chimpanzees understand and use complex auditory signals including vocal signals (Tomasello and Call 1996). Primatologists working with wild chimpanzees have suggested that some of the chimpanzees' social behavior relies on vocal recognition (Marler and Tenaza 1977; Goodall 1986; Mitani and Nishida 1993). However, it is difficult to observe overt behavior to evidence vocal recognition. Few playback experiments have been conducted (Kajikawa and Hasegawa 1996).

Laboratory studies may be able to contribute in this sense. To study chimpanzees' recognition of auditory signals examining the correspondence between vocal recognition and visual recognition is a good approach. From other perspectives, laboratory studies on visual recognition and auditory recognition have been conducted separately in most cases. Little is known about auditory–visual intermodal information processing in nonhuman primates. How are these separate and qualitatively distinct modalities coordinated and put together? This "binding problem" is still unanswered and the behavioral consequences of combining signals from different sensory channels remain poorly understood (Hauser 1996; Partan and Marler 1999).

There are only a limited number of studies on auditory–visual intermodal recognition in nonhuman primates (Davenport et al. 1975; Bauer and Philip 1983; Colombo and D'Amato 1986; Savage-Rumbaugh et al. 1986; Gaffan and Harrison 1991; Boysen 1994). It is difficult to train nonhuman primate including chimpanzees to perform auditory-related tasks through standard conditional discrimination paradigms. There have been claims that the limitation in auditory–visual intermodal recognition might be one factor explaining the lack of language in nonhuman primates (Ettlinger 1960, 1977; Geschwind 1965).

3.1 Auditory–Visual Matching-to-Sample Task (AVMTS)

A female chimpanzee was trained in the computerized task named auditory–visual matching-to-sample (AVMTS). In AVMTS, though the detailed procedures vary among experiments, the subject had to choose from 2 alternatives a photograph that was associated with the auditory sample stimulus previously presented (Fig. 20). For example, following the sound of castanets, the subject had to choose a photograph of castanets which was pitted against a photograph of a bell. The sample auditory stimulus was manually produced by the experimenter in the acquisition process, though this was substituted by automatic presentation of digitally recorded sound in later experiments. The choice alternatives always consisted of two photographs presented on a monitor side by side. The subject's touch response to the monitor was automatically detected by touch screen and recorded by the computer. A response to the photograph associated with the sample auditory stimulus was defined as a correct response and was followed by a delivery of food reward. Figure 21 shows a schematic illustration of one trial of AVMTS used in Hashiya and Kojima (1997).

3.1.1 Acquisition

The acquisition training started using 4 objects as sources of stimuli. A photograph of each object was taken and was used as a comparison stimulus. The object itself

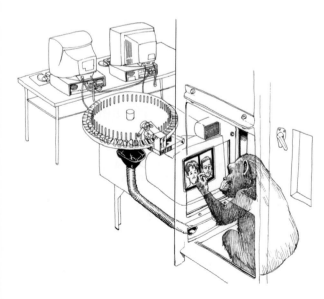

Fig. 20. The experimental setup and the chimpanzee performing auditory–visual matching-to-sample (AVMTS)

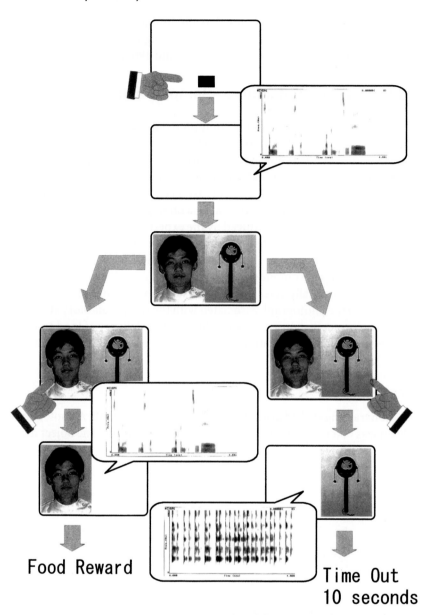

Fig. 21. Schematic illustration of one trial from AVMTS (Hashiya and Kojima 1997). The example shows a set of unexperienced stimuli in the test session: a human and a traditional Japanese toy. Color photographs were used in the actual trials. Sound spectrograms (CSL model 4300B, Kay Elemetrics, NJ, USA) represent the auditory stimuli in this illustration. A human utterance was presented as the sample stimulus in this trial

was handled by the experimenter to produce an auditory sample stimulus or was shown to the subject as a visual sample stimulus. After 79 sessions of training, the subject acquired the AVMTS. The major findings in the process of training were as follows (Hashiya and Kojima 1997).

1. In the process of training, auditory–visual bimodal presentation of the sample stimulus was effective. That is, the subject was shown the object producing the same sound as the auditory sample in the course of training AVMTS. Alternation of AV bimodal and auditory unimodal exposure of the sample stimulus in every other trial seemed to facilitate the subject's learning of AV intermodal relation.
2. Once AVMTS was acquired, rapid learning of AV association for novel stimuli was observed: the performance was generalized to the novel stimulus.

Does the acquired AVMTS performance effectively assess the chimpanzee's auditory and AV intermodal recognition? In other words, what was learned by the subject through the training of AVMTS? There are some steps to evaluate the generality of AVMTS performance.

3.1.2 Memory Retention

Before testing the generality of the AVMTS performance, we examined the effect of sample modalities (visual and auditory) on the subject's matching performance based on working memory. A delay interval was inserted between the termination of the sample presentation and the onset of comparison stimuli and changes in matching performance as a function of the delay interval were compared between auditory sample and visual sample conditions. Except for the modalities of sample stimuli, procedures were the same between the conditions (Fig. 22).

When visual stimuli were used as samples, matching performance remained at more than 90% correct. Significant change in matching performance as a function of delay intervals was not observed (Pearson's product-moment correlation coefficient; $r = -0.54$, n.s.). When auditory stimuli were used as samples, matching performance deteriorated seriously when the delay interval was lengthened ($r = -0.93, P < 0.05$). These results formed a clear contrast to that of the visual sample condition.

Previous studies have suggested that performance based on auditory working memory decays easily or is fragile in monkeys (Kojima 1985). The present results were comparable to previous studies of auditory memory in monkeys. On the other hand, Davenport et al. (1975) reported at least 20 s of memory retention in a visual (sample)–haptic (choice) intermodal matching-to-sample task in chimpanzees. Fujita and Matsuzawa (1990) reported more than 90 s of memory retention by a chimpanzee in a constructive matching-to-sample task, which can be regarded as a form of conditional matching-to-sample task in visual modality. These findings suggest that the AVMTS performance does not decay due to the general property of conditional / symbolic matching-to-sample tasks or of intermodal matching-to-sample tasks. The sharp decay of the subject's matching performance seemed to reflect the limited ability to store auditory information.

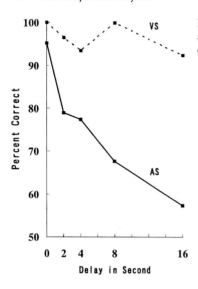

Fig. 22. Percentage correct in auditory sample (*AS*) and visual sample (*VS*) conditions as a function of delay interval

3.1.3 Categorical Transfer

In addition to the object stimuli used in the previous experiment, human stimuli (recorded voice and photographs) were introduced (Hashiya and Kojima 1997). To form choice alternatives, a human stimulus was always pitted against an object stimulus: the choice alternatives never consisted of two stimuli from the same category. The chimpanzee received extensive training with a limited number of stimuli (6 humans and 6 objects) to match a recorded human voice to the speaker's photograph, and to match a recorded object sound to the photograph of the sound source, respectively. Within a session of the training, the particular pairing of a human stimulus and an object stimulus was fixed but changed among the sessions.

After 20 training sessions, the novel human and object stimuli, which the subject had not experienced before, were presented. A novel human stimulus was always pitted against a novel object stimulus. There were 26 pairs of the novel stimuli to be tested. The subject succeeded in matching novel human voices and object sounds to the corresponding novel photographs, respectively, on the first trials even though she had no prior experience for the particular stimulus [79% (41/52), $Z = 4.02$, $P < 0.001$, Table 2]. No obvious positional bias was observed. No consistent preference to human or object comparison stimuli was observed. The total percent of correct responses on the first session was also significantly higher [71% (223/312), $Z = 7.53$, $P < 0.001$] than by chance alone (50%). Matching performances to unexperienced stimuli in 9 out of 26 first sessions were significantly higher than the by-chance level ($Z = 2.02$ at minimum, $P < 0.05$). However, there were 17 sessions in which the subject's performance to unexperienced stimuli did not reach a significant level.

The mean response time to unexperienced stimuli (5.8 s, SD = 2.6) was significantly longer than that to training stimuli (2.6 s, SD = 0.8) ($P < .05$, *t*-test). It took

Table 2. Response in the first trial and percent correct in the first session for each stimulus

Pair	First trial Human	First trial Object	First session Human	First session Object	Total	Binomial test
1	○	●	6/6	4/6	10/12	*
2	●	○	2/6	5/6	7/12	n.s.
3	○	○	5/6	3/6	8/12	n.s.
4	○	●	6/6	2/6	8/12	n.s.
5	●	○	3/6	4/6	7/12	n.s.
6	●	○	1/6	2/6	3/12	n.s.
7	○	○	4/6	6/6	10/12	*
8	○	○	4/6	6/6	10/12	*
9	○	○	3/6	3/6	6/12	n.s.
10	○	○	3/6	3/6	6/12	n.s.
11	○	○	5/6	4/6	9/12	n.s.
12	●	○	1/6	4/6	5/12	n.s.
13	○	○	6/6	5/6	11/12	**
14	○	○	4/6	5/6	9/12	n.s.
15	○	○	3/6	6/6	9/12	n.s.
16	●	○	4/6	6/6	10/12	*
17	○	○	6/6	6/6	12/12	**
18	○	○	3/6	6/6	9/12	n.s.
19	●	○	5/6	6/6	11/12	**
20	●	○	3/6	5/6	8/12	n.s.
21	●	○	3/6	6/6	9/12	n.s.
22	○	○	5/6	5/6	10/12	*
23	○	○	4/6	5/6	9/12	n.s.
24	○	○	5/6	4/6	9/12	n.s.
25	○	○	6/6	6/6	12/12	**
26	○	●	4/6	2/6	6/12	n.s.
Total	18/26	23/26	102/156	121/156	223/312	
% Correct	69	88	67	76	71	
Z1.77	3.73	3.76	6.81	7.53		
Statistics	*	**	**	**	**	

A white unfilled circle means that the subject's choice at the first trial was correct. A black filled circle means that the subject's choice was wrong.
$*p < .05; **p < .01$.

much longer for the subject to respond to unexperienced stimuli. This means that the subject discriminated unexperienced stimuli from training stimuli.

In sum, the chimpanzee generalized the intermodal matching skill in the first trials of unexperienced stimulus presentation. However, those results are valid only to the extent that the stimulus is classified into the previously-trained categories. However, the performance remained relatively low and only slightly above the by-chance level.

3.1.4 Recognition of Conspecific Signals

In addition to the human and object stimuli used in the previous experiments, the subject was presented with auditory and visual stimuli from two novel categories, the chimpanzee and bird (Hashiya 1999). The subject had never experienced the stimuli from these two categories before. The author tested whether the matching performance can be generalized to novel stimuli whose auditory and visual characteristics are completely different from the stimuli the subject had been trained to discriminate.

The chimpanzee stimuli were used to determine if AVMTS performance reflects the subject's natural cognition beyond the specific learning of the particular auditory and visual stimuli. The conspecific chimpanzee vocalizations and visual images are experienced daily by the subject and recognition of such auditory–visual signals should be important as a basis for communication. The bird stimuli were used, in addition to the chimpanzee stimuli, as a control condition. This was done to examine the possibility that the matching of the chimpanzee vocalization and the photograph might be based only on the novelty of the stimuli: the novel auditory and visual stimuli might be matched only because both stimuli were new to the subject. Additionally, differences in previous experience and differences in the ecological importance of the natural information, which are derived from the stimuli, may also have an effect on the subject's AVMTS performance. See Fig. 23.

Throughout the experiment, the subject responded correctly (Z = 2.46 at minimum, $P < 0.05$, two-tailed binomial test with Yates' correction for continuity) for four, 11, 11 and 13 out of 16 stimuli in the bird, chimpanzee, human and object category, respectively. The percent correct responses for the remaining 25 stimuli were in the range of the by-chance level. There was no stimulus for which the percent correct responses were significantly lower than the by-chance level. The overall percent correct was 54 (242/448, Z = 1.65, n.s.), 79.5 (356/448, Z = 12.4, $P <$ 0.001), 82.4 (369/448, Z = 13.7, $P < 0.001$) and 79.6 (357/448, Z = 12.5, $P < 0.001$)

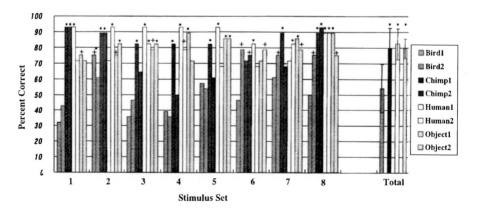

Fig. 23. Percentage correct responses to each stimuli for each session ($^{+}P < 0.05$, $^{*}P < 0.01$, significance of percentage correct differing from the chance level, 50%). See text for more information

in the bird, chimpanzee, human and object category of stimuli, respectively. For the chimpanzee category, the subject performed as well as for the human and object categories which had already been experienced, despite the fact that the subject experienced the particular stimuli from the chimpanzee category for the first time. In contrast, the subject's performance for the bird category remained in the range of the by-chance level.

3.1.5 Human Individual Recognition

For captive chimpanzees, evidence for visual recognition of conspecific and human individuals was reported. Matsuzawa (1990) found very rapid matching of letters with pictures of individual conspecifics and humans by a human-raised chimpanzee. Boysen and Berntson (1986) reported a captive chimpanzee's selective response to the pictures of familiar and unfamiliar human individuals by measuring the subject's heart rate.

Face–voice matching of familiar human or conspecific individuals by chimpanzees has been reported by two laboratory studies. However, transfer tests of the acquired performance using novel stimuli were not conducted by Bauer and Philip (1983) and thus the generality of performance was unclear. Though Boysen (1994) conducted a transfer test and demonstrated the generality of performance, it is still unclear whether the performance of matching particular face–voice was acquired as a result of training or if it reflected the chimpanzees' natural cognition. It is important to eliminate the chance that the particular association to be learned by the subject was learned from already-trained variants.

We tested face–voice recognition of familiar humans, instead of conspecifics. The advantage of using humans as stimuli in the experiment was being able to collect various recorded samples and photographs under well-controlled conditions, compared to the case of using chimpanzees as stimuli. It would also be interesting to ask whether a human-raised chimpanzee recognizes familiar human voices without visual information and can match them to visual images. The subject had received no previous training to discriminate those variants of which the author tested the subject's recognition in the following experiments.

Familiar Individuals

Ten human individuals, eight males and two females were selected from among the researchers who constantly worked with the subject, or staff of the institute who always took care of the subject. They participated in the experiment as a source of auditory and visual stimuli as explained below.

Sample stimuli were digitally recorded voices. As stimuli, each participant in the experiment was requested to read 10 different Japanese phrases of conversation from a book. The phrases requested were different for each participant. The duration of each recorded sound was edited to 4 s after the recording. Sound intensity of each auditory stimulus ranged from 50 to 70 dB SPL in the experimental booth. Visual stimuli were color photographs of human faces from the front. They served as comparison stimuli in the experiment and also served as sample stimuli in visual–visual matching-to-sample task as mentioned below. Thirty pictures per participant were taken with a digital video camera on different backgrounds. Each auditory sample stimulus was randomly paired with one of the photographs of its speaker.

Eight male participants were randomly separated into four pairs. Two female participants formed one pair. The subject was tested with one of the five stimulus pairs in one session. To confirm the discrimination of human faces by the subject, visual–visual matching-to-sample task (VVMTS) was tested before AVMTS. In VVMTS, a picture presented on the monitor served as sample stimulus. Only when the subject's performance at the first session of VVMTS was significantly better than the by-chance level was AVMTS tested for the particular stimulus pair. A schematic illustration of a trial from each of AVMTS and VVMTS are presented in Fig. 24.

A trial consisted of the following events. *VVMTS*: When the subject touched the start key presented at the center of the monitor, the key disappeared and a picture of a human was presented on the monitor as the sample stimulus. The subject had to touch the picture three times. This procedure was introduced to increase the chance for the subject to observe the sample stimulus. The position of the picture was randomized to the left or the right side of the monitor for the first two responses. The position of the picture was fixed to the center only at the third response. Every response to the picture was followed by a beep sound and the third response to the stimulus was followed by the termination of the sample and the presentation of the two choice alternatives. The chimpanzee had to respond three times to the picture of the same human as the sample stimulus to get a reward, although the picture itself was different from the sample. Multiple response to the comparison stimulus was set to decrease the probability of reinforcing the subject's random response. The position of the choice alternatives was randomized to the left or the right side of the monitor for each response. Each correct response was followed by a beep sound. The third correct response was followed by a food reward and a presentation of a chime sound. An incorrect response was followed by the termination of the trial after a 50-s timeout. The first response to the comparison stimuli in each trial was taken to represent the subject's response.

AVMTS: As in VVMTS, when the subject touched the start key the key disappeared and a recorded human voice was presented for 4 s. Immediately after the termination of the auditory stimulus, the two choice alternatives were presented on the monitor. The chimpanzee had to respond three times to the picture of the speaker of the sample auditory stimulus to get a reward. The procedure after the onset of choice alternatives was the same as described in VVMTS except that the third correct response in AVMTS was followed by the playback of sample stimulus instead of the presentation of the chime sound.

Figure 25 shows the results of VVMTS and AVMTS for each session. For all of the five pairs, the subject's performance of VVMTS was significantly better than the by-chance level (Z = 2.69 at minimum, $P < 0.01$, Binomial test). The total percent correct was 87.5% (SD = 6.61). The subject's performance of AVMTS was significantly better than the by-chance level in four of five pairs (Z = 3.64 at minimum, P

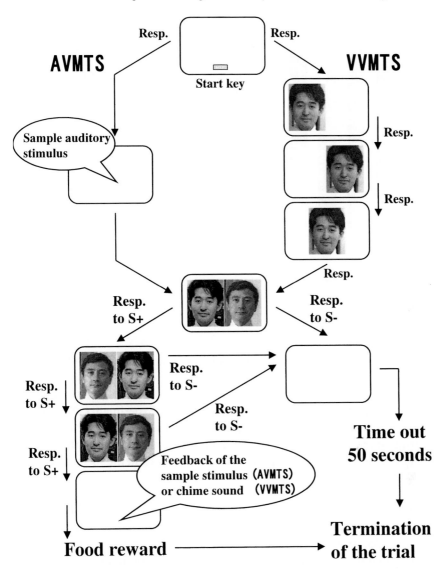

Fig. 24. Schematic illustration of one trial from VVMTS and AVMTS. Color photographs were used in the actual trials

< 0.01, Binomial test) but not in the other pair (Z = 1.74, n.s.). Total percent correct was 80% (SD = 8.84) and was significantly better than the chance level (Z = 8.4, P < 0.01,).

The chimpanzee matched the photographs and recorded voices of familiar humans, despite the subject not having been previously trained to discriminate or categorize such variants. The results suggested auditory–visual intermodal recognition of familiar individuals by chimpanzees.

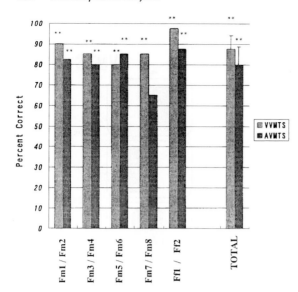

Fig. 25. Percentage correct of each set of stimuli in VVMTS and in AVMTS in Experiment 1. *Fm*, Familiar male; *Ff*, familiar female. **$P < 0.01$; *$P < 0.05$; n.s., no significance. (Results of binomial test). Chance level is 50% correct

Control Experiment

As a control experiment, matching performance to unfamiliar voices and faces was tested. This experiment was conducted to determine whether the subject could rapidly acquire discriminative performance based on some cues extraneous to individual identification. All methods were the same as Experiment (1) except for the following points: Fourteen people, eight males and six females, participated in the experiment as the stimuli. They were all unfamiliar individuals to the subject. The eight males were randomly separated into four pairs and the six females were randomly separated into three pairs.

Figure 26 shows the results of the experiment. The subject's performances in VVMTS for three of seven pairs were not significantly better than the by-chance level ($Z = 1.74$, n.s., for two pairs; significantly worse performance than chance for one pair, $Z = 2.06$, $P < 0.05$). Thus, only the four pairs where the subject performed significantly better than the by-chance level ($Z = 3.6$ at minimum, $P < 0.01$) were tested in AVMTS. The average percent correct for the 4 pairs was 85.6% (SD = 5.54). For all the tested pairs, the subject's performance of AVMTS remained in the range of the by-chance level ($Z = 1.74$ at maximum, n.s.). The average percentage correct was 43.8% (SD = 7.22). The subject could match familiar individual faces and voices but could not match those of unfamiliar individuals.

Speaker Identification Based on Exclusion

Abilities to recognize more general cues than individuality that lie in auditory and visual information were examined, since they are assumed to be essential for speaker identification and as a basis of intermodal individual recognition.

The ability to sort acoustic variants of the vocal signals into two categories, mother and others, or group members and strangers has been reported in monkeys. This

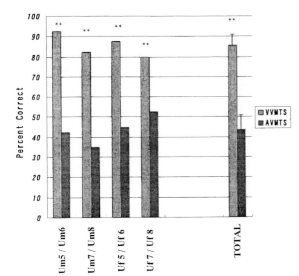

Fig. 26. Percentage correct of each set of stimuli in VVMTS and in AVMTS in the Control Experiment. Results of the pairs that did not pass the criterion in VVMTS are not presented here. The total percentage correct in VVMTS was calculated as the average percentage correct of the sets of stimuli that passed the criterion. *Um*, Unfamiliar male; *Uf*, unfamiliar female. **$P < 0.01$; *$P < 0.05$; n.s., no significance. (Results of binomial test). Chance level is 50% correct

ability is assumed to be important to avoid substantial danger, as well as to form mother–infant or intra group social relations. Such an ability might be tied to the detection or recognition of an unfamiliar pattern of auditory signals. We tested whether the chimpanzee recognized voices of unfamiliar humans and matched the voice to unfamiliar human faces, when the unfamiliar face was paired with a familiar human face. This kind of matching is called exclusion. Previous studies have reported the difficulty of chimpanzees in acquiring exclusion in arbitrary visual matching to sample task (Tomonaga et al. 1991; Yamamoto and Asano 1991; but see Tomonaga 1993 for an exception).

In this experiment, the sources of the stimuli were familiar individuals and unfamiliar individuals. To form choice alternatives, a portrait of a familiar individual was always pitted against a portrait of an unfamiliar individual. The subject's performances in VVMTS were better than the by-chance level ($Z = 3.00$ at minimum, $P < 0.01$) only in four of seven pairs. Thus, AVMTS was tested for the four pairs and the performance for all the tested pairs was significantly better than the by-chance level ($Z = 3.32$ at minimum, $P < 0.01$, see Fig. 27). The total percent correct was 90% (SD = 0) when familiar individual voices served as sample stimuli, and 75% (SD = 10.8) when unfamiliar individual voices served as sample stimuli. There was no significant difference in the subject's performance dependent upon the familiarity/unfamiliarity dichotomy of sample stimuli (chi square test, n.s.).

The subject could distinguish strangers' voices from those of familiar individuals and match them to the corresponding individual's pictures. This means that the subject matched unfamiliar individuals' voices and faces based on a form of reasoning; "this voice should belong to this face, since it does not belong to the other face", that is, exclusion. Tomonaga (1993) suggested that the lack of previous experience being reinforced through the response to the unfamiliar stimulus might

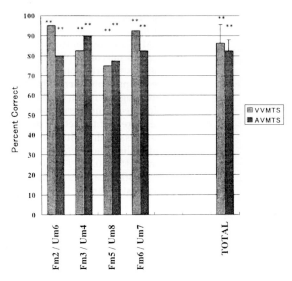

Fig. 27. Percentage correct of each set of stimuli in VVMTS and in AVMTS in Experiment 2. *Fm*, Familiar male; *Um*, unfamiliar male. **P < 0.01; *P < 0.05; n.s., no significance. (Results of binomial test). Chance level is 50% correct

explain the difficulty in acquiring exclusion in an arbitrary visual matching task. In the present experiment, however, the previous experience of reinforcement was almost the same for each stimulus. This might explain the different results from previous studies that have reported the difficulty of chimpanzees to perform on the basis of exclusion.

Speaker Identification Based on Sex Difference

This experiment was conducted to determine whether the chimpanzee's ability for face–voice matching of human individuals could be generalized to the task that requires the recognition of cues other than individuality, such as sex differences. In visual modality, Itakura (1992) reported sex discrimination of photographs of humans by a chimpanzee. It is still unclear at this point whether there are overt sex differences in voice throughout the vocal repertoires of chimpanzees (see Marler and Hobbett 1975 for sex differences in pant-hoot). However, there are clear sex differences in adult human voices, which Jusczyk et al. (1992) reported that 2-month-old infants can discriminate. Do these features "naturally" serve as a cue for categorization by the chimpanzee in a human environment?

The sources of the stimuli were unfamiliar males and females who had no contact with the subject. Ten people, five males and five females, participated in the experiment as the stimuli. To form choice alternatives, a portrait of a male was always pitted against a portrait of a female. The performance of VVMTS was significantly better than the by-chance level for all of five pairs ($Z = 3.95$ at minimum, $P < 0.01$). The percentage correct for all ten participants was significantly better than the by-chance level ($Z = 2.46$ at minimum, $P < 0.05$). The total percent correct was 90.5% (SD = 5.7). No increase in percentage correct was observed between the

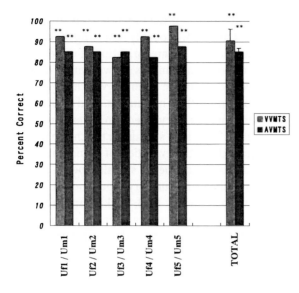

Fig. 28. Percentage correct of each set of stimuli in VVMTS and in AVMTS in Experiment 3. *Um*, Unfamiliar male; *Uf*, unfamiliar female. **$P < 0.01$; *$P < 0.05$; n.s., no significance. (Results of binomial test). Chance level is 50% correct

first exposure and the second exposure of the same sample stimuli (93%, SD = 7.58 for the first exposure; 88%, SD = 10.37 for the second exposure of the stimuli). No increase in percentage correct was observed between the first half of the session and the second half of the session (93%, SD = 4.47 in the first; 88%, SD = 9.75 in the second half of the session).

The performance of AVMTS were significantly better than the by-chance level for all five pairs (Z = 3.95 at minimum, $P < 0.01$, see Fig. 28). The total percentage correct was 85% (SD = 1.77). No increase of percent correct was observed between the first exposure and the second exposure of the same sample stimuli (83%, SD = 5.70 in the first; 87%, SD = 4.47 in the second half of the session).

The results showed that some of the differences that exist between the vocal signals of human females and males are naturally attributed to the different features in their faces, respectively, though it is still inconclusive whether the chimpanzee recognizes such differences as a sex difference. The chimpanzee seemed to recognize auditory–visual intermodal relations other than individuality and made use of this ability in speaker identification.

4 Discussion and Conclusions

These studies have clarified the manner of chimpanzee audition, which had been one of the least understood aspects of chimpanzee cognition. They have described auditory acuity and the manner of understanding of perceived auditory signals in correspondence to visual recognition.

Studies on auditory perception showed constraints of "effective" auditory signals for the chimpanzee. Chimpanzees shared the capacity for categorical perception of consonants and of vocal tract normalization with humans. On the other hand, the W-shaped auditory sensitivity function of the chimpanzee was similar to that of monkeys and differed from that of humans. Probably reflecting the difference in auditory sensitivity function, the manner of categorization of vowels was different between chimpanzees and humans. Chimpanzees showed larger difference thresholds than humans both in frequency and in intensity, which suggested a lower resolution of processing auditory input than in humans. And also, there were limitations for chimpanzees in auditory memory retention. These findings are important in knowing the constraints of "effective" signals for the chimpanzee. In addition, they might hold information about the evolution of human hearing systems, especially in relation to the adaptation of speech signals. Speech signal processing requires rapid computation of changing auditory input and sufficient memory capacity to store verbal information (Hockey 1973; Watkins and Watkins 1980; Kojima 1985). Chimpanzees are not endowed with a capacity in such aspects, in contrast to their ability for consonant perception and for vocal tract normalization. It is worth studying how these chimpanzee hearing systems adapt to perception of conspecific vocal signals (Kojima and Nagumo 1999).

We also reported the successful case of AV intermodal matching by a chimpanzee. The acquired AVMTS performance worked as an effective tool to investigate natural cognition in the chimpanzee. The findings obtained through AVMTS showed that the chimpanzee naturally recognizes auditory–visual intermodal relations between conspecific face and voice, as well as individual (human) face and voice. They further suggested an ability of speaker identification based on estimation, which seems to be essential to form and expand knowledge in naturalistic situations. The possible importance of exclusion in the expansion of vocabulary has been noted in human development (Hayes 1994). The ability to focus on more general or more abstract features such as sexuality, which is evident in humans (Walker-Andrews et al. 1991), plays an important role in affording flexibility of AV intermodal recognition. In the process of learning intermodal relations through experience, the chimpanzee might make use of such abilities.

Using AVMTS, intermodal recognition of conspecific signals such as individuality or the affective states should be assessed in the future. We should also clarify the basis of acquiring or learning the intermodal relation of auditory and visual signals by the chimpanzee. It is interesting that some of the previous reports of AV intermodal recognition in apes (Premack 1976; Savage-Rumbaugh et al. 1986) are based on the subject's spontaneous learning of the intermodal relation, not on the result of active teaching through the standard conditional discrimination paradigm. The gap between the standard conditional discrimination paradigm and spontaneous learning is still difficult to explain. However, the implication from previous studies is that some factors involved in face-to-face communication, which cannot be found in the standard procedure of training, might form a basis of learning auditory–visual intermodal relation. Considering that in the present experiment the simultaneous presentation of auditory and visual stimuli did facilitate the subject's learning of AVMTS, temporally corresponding bimodal experience might be crucial to form chimpanzee intermodal recognition.

Premack and Schwartz (1966) once described a phonetic system that might be used as a basis for an artificial language with chimpanzees, based on Jakobson's (Jakobson et al. 1952) distinctive feature model. When considering the recent findings of chimpanzee hearing and by using the AVMTS paradigm, an attempt such as that by Premack and Schwartz (1966) might be revisited.

Acknowledgments

The authors thank Sumiharu Nagumo for his help in programming and the technical assistance to construct the apparatus, to the staff of the Laboratory Primate Center of the Primate Research Institute, especially Kiyonori Kumazaki and Norihiko Maeda, for their daily care of the chimpanzees, and to the staff of the Primate Research Institute, especially Michael A. Huffman and Tetsuro Matsuzawa, for their help and support. Thanks are also due to Miso Suzuki for the illustration in Fig. 20. This study was supported by a fellowship (#2827 and #6649) to Kazuhide Hashiya from the Japan Society for the Promotion of Science and a Grant-in-aid for Scientific Research from the Ministry of Education, Science, Culture, and Sports grants "The Emergence of Human Cognition and Language" (Shozo Kojima, director), #04610053 to Shozo Kojima, #04551002, #04651017, #05044006, and #06260222 to Tetsuro Matsuzawa.

References

Bauer HR, Philip MM (1983). Facial and vocal individual recognition in the common chimpanzee. Psychol Rec 33:161–170

Beecher MD (1974). Hearing in the owl monkey (*Aotus trivirgatus*). I. Auditory sensitivity. J Comp Psychol 86:898–901

Boysen ST (1994) Individual differences in the cognitive abilities of chimpanzees. In: Wrangham RW, McGrew WC, de Waal FBM, Heltne PG (eds) Chimpanzee cultures. Harvard University Press, Cambridge, pp 335–350

Boysen ST, Berntson GG (1986) Cardiac correlates of individual recognition in the chimpanzees (*Pan troglodytes*). J Comp Psychol 100:321–324

Brown CH, Waser PM (1984) Hearing and communication in blue monkeys (*Cercopithecus mitis*). Anim Behav 32:66–75

Colombo M, D'Amato MR (1986) A comparison of visual and auditory short-term memory in Monkeys (*Cebus apella*). Q J Exp Psychol B 38:425–448

Davenport RK, Rogers CM, Russell IS (1975) Cross-modal perception in apes: altered visual cues and delay. Neuropsychologia 13:229–235

Elder JH (1934) Auditory acuity of the chimpanzee. J Comp Psychol 17:157–183

Elder JH (1935) The upper limit of hearing in chimpanzee. Am J Physiol 112:109–115

Ettlinger G (1960) Crossmodal transfer of training in monkeys. Behaviour 16:56–64

Ettlinger G (1977) Interactions between sensory modalities in nonhuman primates. In: Schrier AM (ed) Behavioral primatology. Advances in research and theory, Vol 1. Lawrence Erbaum, Hillsdale, pp 71–104

Fujisaki H, Kawashima T (1968) The roles of pitch and higher formants in the perception of vowels. IEEE Trans Audio Electroacoust AU-16:73–77

Fujita K, Matsuzawa T (1990) Delayed figure construction in a chimpanzee (*Pan troglodytes*). J Comp Psychol 104:345–351

Gaffan D, Harrison S (1991) Auditory–visual associations, hemispheric specialization and temporal-frontal interaction in the rhesus monkey. Brain 114:2133–2144

Geschwind N (1965) Disconnexion syndromes in animal and man. Part 1. Brain 88:237–294

Goodall J (1986) The chimpanzees of Gombe/ patterns of behavior. Harvard University Press, Cambridge, MA

Hashiya K (1999) Auditory–visual intermodal recognition of conspecifics by a chimpanzee (*Pan troglodytes*). Primate Res 15:333–342

Hashiya K, Kojima S (1997) Auditory–visual intermodal matching by a chimpanzee (*Pan troglodytes*). Jpn Psychol Res 39:182–190

Hauser M (1996) The evolution of communication. MIT Press/Bradford Books, Cambridge USA

Hayes SC (1994) Relational frame theory: a functional approach to verbal events. In: Hayes SC, Hayes LJ, Sato M, Ono K (eds) Behavior analysis of language and cognition. Context, Reno, pp 9–30

Hockey R (1973) Rate of presentation in running memory and direct manipulation of input processing strategies. Q J Exp Psychol 25:104–111

Itakura S (1992) Sex discrimination of photographs of humans by a chimpanzee. Percept Motor Skills 42(2):157–172

Itani J (1963) Vocal communication of the wild Japanese monkeys. Primates 4:11–66

Jakobson R, Fant CGM, Halle M (1952) Preliminaries of speech analysis. MIT Press Cambridge, USA

Jusczyk PW, Pisoni, DB, Mullenix J (1992) Some consequences of stimulus variability on speech processing by 2-month-old infants. Cognition 43:253–291

Kajikawa S, Hasegawa J (1996) How chimpanzees exchange information by pant-hoots: a playback experiment. In: The emergence of human cognition and language (Vol. 3, pp 123–128). Grant-in-aid for Scientific Research, Ministry of Education, Science, Sports and Culture, Japan. Annual Report

Kojima S (1985) Auditory short-term memory in the Japanese monkey. Int J Neurosci 25:255–262

Kojima S (1990) Comparisons of auditory functions in the chimpanzee and human. Folia Primatol 55:62–72

Kojima S, Kiritani S (1989) Vocal–auditory functions in the chimpanzee: vowel perception. Int J Primatol 10:199–213

Kojima S, Tatsumi IF, Kiritani S, Hirose H (1989) Vocal–auditory functions in the chimpanzee: consonant perception. Hum Evol 4:403–416

Kuhl PK (1979) Speech perception in early infancy: Perceptual constancy for spectrally dissimilar vowel categories. J Acoust Soc Am 66:1668–1679

Kuhl PK (1981) Discrimination of speech by nonhuman animals: basic auditory sensitivities conductive to the perception of speech-sound categories. J Acoust Soc Am 70:340–349

Kuhl PK (1991) Human adults and human infants show a "perceptual magnet effect" for prototypes of speech categories, monkeys do not. Percept Psychophys 50:93–107

Kuhl PK, Miller JD (1975) Speech perception by the chinchilla: Voiced-voiceless distinction in alveolar-plosive consonants. Science 190:69–72

Kuhl PK, Padden DM (1982) Enhanced discriminability at the phonetic boundaries for the voicing feature in macaques. Percept Psychophys 32:542–550

Kuhl PK, Padden DM (1983) Enhanced discriminability at the phonetic boundaries for the place feature in macaques. J Acoust Soc Am 73:1003–1010

Marler P, Hobbett L (1975) Individuality in long-range vocalizations of wild chimpanzees. Z Tierpsychol 38:97–109

Marler P, Tenaza R (1977) Signaling behavior of apes with special reference to vocalization. In: Sebeok T (ed) How animals communicate. Indiana University Press, pp 965–1003

Matsuzawa T (1990) Form perception and visual acuity in a chimpanzee. Folia Primatol 55:24–32

Miller GA, Nicely PE (1955) An analysis of perceptual confusion among some English consonants. J Acoust Soc Am 27:338–352

Mitani JC, Nishida T (1993) Contexts and social correlates of long-distance calling by male chimpanzees. Anim behav 45:735–746

Mitani JC (1994) Ethological studies of chimpanzee vocal behavior. In: Wrangham RW, McGrew WC, de Waal FBM, Heltne PG (eds) Chimpanzee cultures. Harvard University Press, pp 195–210

Morse PA, Snowdon CT (1975) An investigation of categorical speech discrimination by rhesus monkeys. Percept Psychophy 18:9–16

Nishida T (1994) Afterward-review of recent findings of Mahale chimpanzees: implications and future research direction. In: Wrangham RW, McGrew WC, de Waal FBM, Heltne PG (eds) Chimpanzee cultures. Harvard University Press, pp 373–396

Nishida T, Kano T, Goodall J, McGrew WC, Nakamura M (1999) Ethogram and ethnography of Mahale chimpanzees. Anthropol Sci 107:141–188

Partan S, Marler P (1999) Communication goes multimodal. Science 283:1272–1273

Pfingst BE, Hienz R, Miller J (1975a) Reaction time procedure for measurement of hearing: II Threshold function. J Acoust Soc Am 57:431–436

Pfingst BE, Hienz R, Kimm J, Miller J (1975b) Reaction time procedure for measurement of hearing: I Suprathreshold function. J Acoust Soc Am 57:421–430

Premack D (1976) Intelligence in ape and man. Lawrence Erlbaum Assoc

Premack D, Schwartz A (1966) Preparation for discussing behaviorism with chimpanzee. In: Smith FL, Miller GA (eds) The genesis of language. MIT Press

Savage-Rumbaugh ES, McDonald K, Sevcik RA, Hopkins WD, Rubert E (1986) Spontaneous symbol acquisition and communicative use by pygmy chimpanzees (Pan paniscus). J Exp Psychol [Gen] 115:211–235

Sinnot JM, Beecher MD, Moody DB, Stebbins WC (1976) Speech-sound discrimination by monkeys and humans. J Acoust Soc Am 60:687–695

Stebbins WC (1971) Hearing. In: Schrier AM, Stollniz F (eds) Behavior of nonhuman primates: modern research trends. Academic Press, Vol 3, pp 159–192

Stebbins WC (1973) Hearing of old world monkeys (Cercopitecinae). Am J Phys Anthropol 38:357–364

Tomasello M, Call J (1996) Primate cognition. Oxford University Press

Tomonaga M (1990) Multidimensional auditory stimulus control in a chimpanzee (Pan troglodytes). Primates 31:545–553

Tomonaga M (1993) Tests for control by exclusion and negative stimulus relations of arbitrary matching to sample in a "symmetry-emergent" chimpanzee. J Exp Anal Behav 59:215–229

Tomonaga M, Matsuzawa T, Fujita K, Yamamoto J (1991) Emergence of symmetry in a visual conditional discrimination by chimpanzees (Pan troglodytes). Psychol Rep 6:51–60

Walker-Andrews AS, Bahrick LE, Raglioni SS, Diaz I (1991) Infant's bimodal perception of gender. Ecol Psychol 3:55–75

Waters RA, Wilson WA Jr (1976) Speech perception by rhesus monkeys: the voicing distinction in synthesized labial and velar stop consonants. Percept Psychophys 19:285–289

Watkins OC, Watkins MJ (1980) The modality effect and echoic persistence. J Exp Psychol [Gen] 109:251–278

Wilson KV (1963) Multidimensional analyses of confusion of English consonants. Am J Psychol 76:86–95

Yamamoto J, Asano T (1991) Formation of stimulus equivalences in a chimpanzee. In: Ehara A, Kimura T, Takenaka O, Iwamoto M (eds) Primatology today. Elsevier, pp 321–324

Yerkes RM, Yerkes AW (1929) The great apes—a study of anthropoid life. Yale University Press

9
Early Vocal Development in a Chimpanzee Infant

Shozo Kojima

1 Introduction

The vocal behaviors of human infants develop in stages. At the age of about 1 year, the typical infant acquires the first word of spoken language (Oller 1980, 1986; Stark 1980). Although there have been many studies of vocal development in human infants, few studies have systematically reported on the vocal development of chimpanzee infants (Heyes 1951; Marler and Tenaza 1977; Plooij 1984). In the present study, the early development of vocal behaviors in a chimpanzee infant was investigated.

Although human infants usually do not utter words during the first year of life, they do develop a capability to produce the kinds of sounds that are found in words during this period (Oller 1981). Oller (1980) and Stark (1980) have proposed similar stage theories on the development of vocal behavior of human infants leading to the first word of spoken language. According to Oller's infraphonological analyses of non-crying utterances, there are five stages of vocal development in human infants. These stages include Phonation (0–1 months of life), Goo (2–3 months), Expansion (4–6 months), Canonical babbling (7–10 months), and Variegated babbling (11–12 months) stages. There are characteristic nonreflexive vocalization types for each stage; for example, quasi-resonant nuclei in the Phonation stage, goo in the Goo stage, fully resonant nucleus, raspberry, and others in the Expansion stage, canonical babbling in the Canonical stage, and variegated babbling and gibberish in the Variegated Babbling stage. In the present experiment, the development of vocal behaviors in a chimpanzee infant was investigated. Many differences in vocal development between chimpanzees and humans are expected. Does the chimpanzee infant reach the Goo, the Expansion, and the Canonical babbling stages?

Primate Research Institute, Kyoto University, 41 Kanrin, Inuyama, Aichi 484-8506, Japan

2 Methods

The subject was a female chimpanzee named Pan born in the Primate Research Institute, Kyoto University. Because the mother of the subject rejected her soon after birth, she was raised by human caretakers. The use of the subjects adhered to the Guide for the Care and Use of Laboratory Primates (Primate Research Institute, Kyoto University, 1986). The vocal behaviors of the chimpanzee infant were observed about 10 h a day, 7 days a week, from her first day of life. Vocal sounds and situations in which the chimpanzee vocalized were described in notebooks, and a total of 2340 min of tape recording (TC-D5M, Sony) was conducted. Vocalizations recorded on the tape recorder were analyzed by sound spectrographs (Digital Sona-Graph, model 7800, Kay and SA-70, Rion, Tokyo, Japan) and by an LPC (linear predictive coding) program. The order of prediction of the LPC analysis was 10. In the present study, the vocal behaviors of the chimpanzee infant during the first 18 weeks of life are reported. Because the subject was extensively trained to vocalize after the 18th week, data after this week were not analyzed.

3 Results

3.1 Classification of Non-crying Vocalizations

The following vocalizations were non-crying vocalizations unrelated to aversive emotional states. Rather, they were related to excitement or attention. Staccato sounds were short and breathy, produced in series (Fig. 1A). This vocal sound was sometimes accompanied by phonations with frequency modulations (Fig. 1B). Situations in which staccatos were produced and their developmental changes are described later. Grunts were vowel-like sounds, heard as [u], [o], or [a] by the author (Fig. 1C,D,E). Although the vowel-like sound [oe] was produced several times, the chimpanzee did not produce [i]- and [e]-like sounds. Laughter sounds were breathy and repetitive, elicited when the chimpanzee infant was tickled.

3.2 Development of Non-crying Vocal Behaviors

There were four stages in the early development of non-crying vocalizations in the infant. Three of the four phases were included in the observation period of this chapter. These vocal sounds were almost always elicited by environmental stimuli and were occasionally emitted throughout the observation period.

   In the first stage (stage I, 0–3 weeks after her birth), the chimpanzee infant produced about 14 staccatos or grunts per day in response to various stimuli. One of the non-crying vocalizations (staccatos) was observed within 24 h of birth (see Fig. 1A). These vocal sounds were elicited when the caretaker or the author presented low-pitched, loud sounds to her. The infant responded with vocalization when she was satiated, alert, and gazing at the caretaker or the author. The infant produced more staccatos than grunts in this stage.

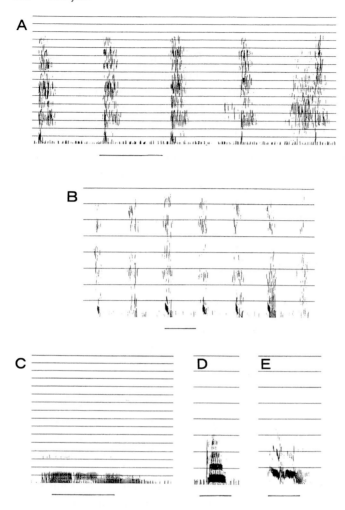

Fig. 1A–E. Sound spectrograms of staccatos. **A** This staccato was observed within 24 h after birth. **B** A staccato accompanied by phonation with frequency modulation. **C** An example of a grunt. **D** A coo (CV syllable) uttered by the chimpanzee infant. **E** An example of a grunt with frequency modulation. In this and the following sonograms, *horizontal lines* indicate 1 kHz and *time scales* indicate 0.2 s

In the second stage (stage II, 4–6 weeks), non-crying vocalizations of the infant decreased. The author did not hear these vocal sounds on the 41st day. Crying vocalizations also decreased in this stage.

In the third stage (stage III, 7–18 weeks), the infant's non-crying vocalizations were readily elicited. These vocal sounds were observed about 85 times a day. Various visual and auditory stimuli elicited staccatos and grunts. For example, when the caretaker approached the infant or when the author spoke to her, she produced grunts. The high level of elicitation was maintained throughout this

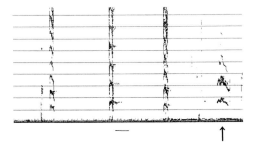

Fig. 2. An example of vocal imitation of an external sound by the chimpanzee infant. The chimpanzee infant imitated the same sound just before this episode. The first three sounds were those of a toy presented by the author. The last sound with an *arrow* is the voice of the chimpanzee infant. The fundamental frequency of this vocal sound was unusually high as a non-crying utterance

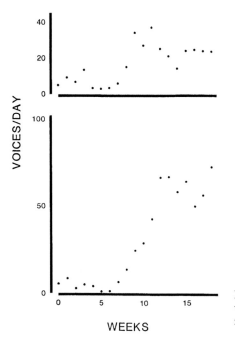

Fig. 3. Early developmental changes of staccatos (*upper*) and grunts (*lower*) in the chimpanzee infant

stage. Her first laughter was observed at 10 weeks after her birth. The infant appeared to imitate environmental sounds (Fig. 2).

Although it did not occur during the observation period of this study, the infant decreased vocalizations from the 18th week of life. In this stage IV (19 weeks), the infant seemed to habituate to environmental stimuli. For example, although the infant produced grunts in response to an approaching stranger, she did not respond to the caretaker or the author. Figure 3 shows the frequency of staccatos and grunts in the first three stages.

3.3 Infraphonological Analyses of Non-crying Vocalizations

Preliminary infraphonological analysis of the infant's vocalizations was performed by visual inspection of sonograms because vocal resonance patterns were insufficient. As shown in Fig. 1(C-E), there were no energies at higher frequencies for these grunts. Although grunts were accompanied by normal phonation in the first phase, they often became breathy in the second phase. Frequency modulations of both crying and non-crying vocalizations gradually increased throughout the observation period (see Fig. 1D). Goos or coos, which were a kind of consonant-vowel (CV) syllable, also increased (see Fig. 1E).

4 Discussion

4.1 Developmental Changes in Elicitation of Vocalizations

The chimpanzee infant rarely emitted non-crying vocalizations spontaneously. Rather, her vocalizations were almost always elicited by environmental stimuli throughout the observation period. The author could easily elicit vocalizations of the infant in the third stage, but it was difficult in the second and fourth stages. The first two stages of the infant correspond approximately to the human Phonation stage. The chimpanzee's third stage has common features with the human Goo stage. A similar time course for the elicitability of vocalization was observed in human infants by Kaye and Fogel (1980). That is, they observed vocal behaviors of human infants at 6, 13, and 26 weeks, ages that correspond to the Phonation, Goo, and Expansion stages, respectively. The elicitation of vocal behaviors increased from 6 to 13 weeks, that is, from the Phonation to the Goo stage, but it decreased from 13 to 26 weeks, that is, from the Goo to the Expansion stage. Thus, this aspect of vocal behavior in the chimpanzee infant and human infants may have a common base. However, there were differences between chimpanzee and human infants. Although human infants increased spontaneous vocalizations throughout from the Phonation to the Expansion stage (Kaye and Fogel 1980), the chimpanzee infant rarely vocalized spontaneously. The spontaneity of vocalization is an important factor for the development of babbling, which is crucial for the development of spoken language. Another difference is early appearance of non-crying vocalizations in the chimpanzee infant. In fact, in the chimpanzee first stage, which corresponds to the early human Phonation stage, it was easier for the author to elicit vocalizations from the chimpanzee infant than from human infants. This observation indicates that some parts of non-crying vocalizations of the chimpanzee infant may be genetically determined.

4.2 Contexts That Favor the Elicitation of Vocalizations

Throughout these three developmental stages, vocalizations were elicited by the author when the chimpanzee infant was alert and attentive to the author or the caretaker or to various stimuli presented by us. A similar condition was reported for elicitation of vocalizations in human infants (Kaye and Fogel 1980; Locke 1993). This similarity indicates that active interactions between infants and caretakers are important for vocalizations in both species. Chimpanzee mothers usually do not try to elicit vocalizations from their infants. This absence of interaction may account in part for the limited vocal development of the chimpanzee. The chimpanzee infant imitated an environmental sound. Acceleration of vocal imitation around the Goo stage has been reported in human infants (Kuhl and Meltzoff 1982).

4.3 Infraphonological Analyses of Vocalizations

Analyses of vocalizations suggested that infraphonological features of chimpanzee vocal sounds were similar to those of human infants in the Phonation and the Goo stages. As with quasi-resonant nuclei, the resonance of the vocal tract was concentrated in low frequencies in the chimpanzee infant's vocalization. The high larynx may be responsible for this phenomenon. The Goo, a kind of CV syllable generally including very limited formant transitions, increases in both chimpanzee and human infants. Unlike human infants, however, the chimpanzee infant's vocalizations became breathy early in the second stage. This result indicates that voluntary control over the vocal folds, which is critical to the development of vocal behavior, is weak in the chimpanzee infant.

5 Conclusion

In conclusion, chimpanzee and human infants have similar vocal tendencies, especially when elicited before about 20 weeks, that is, in the Phonation and Goo stages. The larynx of the human infant moves downward in the Expansion stage, permitting greater vocal resonance and expanded vocal repertoires. The larynx of the chimpanzee remains high. Thus, it may be difficult for the chimpanzee to reach the human Expansion stage and therefore to achieve Canonical babbling.

Acknowledgment

The author thanks Drs. D. Kimbrough Oller and John L. Locke for their comments on the manuscript. This research was supported by a grant-in-aid for scientific research, Ministry of Education, Science, Sports and Culture of Japan (no. 05206109). Oller updated his stage model of early vocal development of human infants, based on summaries of longitudinal research from around the world; see Oller (1995).

References

Heyes C (1951) The ape in our house. Harper, New York
Kaye K, Fogel A (1980) The temporal structure of face-to-face communication between mothers and infants. Dev Psychol 16:454–464
Kuhl PK, Meltzoff AN (1982) The bimodal perception of speech in infancy. Science 218:1138–1140
Locke JL (1993) The child's path to spoken language. Cambridge University Press, Cambridge
Marler P, Tenaza R (1977) Signalling behavior of wild apes with special reference to vocalization. In: Sebeok T (ed) How animals communicate. Indiana University Press, Bloomington, pp 965–1033

Oller DK (1980) The emergence of the sounds of speech in infancy. In: Yeni-Komshian GH, Kavanagh JF, Ferguson CA (eds) Child phonology, Vol 1, Production. Academic Press, New York, pp 93–112

Oller DK (1981) Infant vocalizations: exploration and reflexivity. In: Stark RE (ed) Language behavior in infancy and early childhood. Elsevier, New York, pp 85–103

Oller DK (1986) Metaphonology and infant vocalizations. In: Lindblom B, Zetterstrom R (eds) Precursors of early speech. Stockton Press, New York, pp 21–35

Oller DK (1995) Development of vocalizations in infancy. In: Winitz H (ed) Human communication and its disorders: a review, Vol 4. York Press, Timonium, MD, pp 1–30

Plooij FX (1984) The behavioral development of free-living chimpanzee babies and infants. Ablex, Norwood

Stark RE (1980) Stages of speech development in the first year of life. In: Yeni-Komshian GH, Kavanagh JF, Ferguson CA (eds) Child phonology, Vol 1, Production. Academic Press, New York, pp 73–92

Part 4
Learning and Memory

10
Chimpanzee Numerical Competence: Cardinal and Ordinal Skills

Dora Biro[1] and Tetsuro Matsuzawa[2]

Imagine two 22-year-olds side by side: Candidate A and Candidate B. Both can label collections of up to at least nine objects with the corresponding number; both use symbols to represent these numbers; both can arrange their symbols in ascending order to indicate the sequence in which they denote increasing quantities. However, Candidate A is incapable of representing 'absence' or 'nothing' symbolically, and is hence incapable of higher mathematics. Candidate B has a symbol for 'zero' and can use it in the ordinal as well as the cardinal sense. Candidate A is a citizen of Ancient Rome. Candidate B is a chimpanzee.

The chimpanzee in question is Ai: a 22-year-old female at the Primate Research Institute of Kyoto University, who from the age of one-and-a-half has learned a symbolic language with the aid of which she has been able to communicate her visual world to experimenters. She can describe how she perceives colors, objects, or even other individuals by indicating her choice of symbol on a computer panel. How she perceives a perhaps more abstract aspect of the environment—number— can be investigated within the same paradigm. This chapter outlines Ai's early training in the use of symbols as labels—including Arabic numerals to label sets of items such as real-life objects and stimuli on a computer screen—followed by a series of experiments designed to test her ability in tasks requiring the use of some form of numerical skill and to elucidate the cognitive processes that underlie numerical competence in the chimpanzee.

1 Introduction

The ability to count and to perform various forms of mathematical reasoning is an uncontested attribute of adult *Homo sapiens*. Recent evidence suggests that even infants as young as 4.5 months old are able to individuate and recognize numerical distinctness of everyday items (Spelke et al. 1995) as well as to perform simple mathematical operations, such as addition and subtraction on them (Wynn 1992). For non-human subjects, investigations of numerical abilities have addressed a far

[1] Animal Behaviour Research Group, Department of Zoology, University of Oxford, Oxford OX1 3PS, UK
[2] Primate Research Institute, Kyoto University, 41 Kanrin, Inuyama, Aichi 484-8506, Japan

more contentious issue. Discrimination based on quantities has been demonstrated in a variety of species: non-primates such rats (Davis and Bradford 1986), raccoons (Davis 1984), and a parrot (Pepperberg 1987), and primates such as squirrel monkeys (Olthof et al.1997; Terrell and Thomas 1990; Thomas et al. 1980), macaques (Washburn and Rumbaugh 1991), a gorilla (MacDonald 1994), and chimpanzees (Boysen and Berntson 1989; Matsuzawa 1985; Rumbaugh et al. 1987; Woodruff and Premack 1981). To what extent the skills demonstrated by the subjects in these studies resemble human performance is hotly debated. One of the main stumbling blocks has been the lack of agreement on functional definitions for the different kinds of processes through which numerical discrimination is accomplished (see Davis and Perusse 1988, and Boysen and Capaldi 1993, for a review of efforts to construct definitions). For the purposes of relating numerical competence studies in non-human subjects to human performance, Gelman and Gallistel's (1978) "counting model" has been among the most widely used criteria.

The counting model outlines five main principles which define numerical labeling behavior as "true" counting. Among these are the "one-to-one principle" which involves the sequential tagging of all items in a set (with each item receiving exactly one tag), the "stable order principle" which entails that these tags be assigned in a particular order that does not change between instances of counting, and the "cardinal principle" stating that the last tag assigned in a set then serves to label the total number of items in that set. In addition, the model also requires that numerical labeling be abstract—anything can be counted using the same tagging procedure.

Researchers have often relied on one or more of the above principles to demonstrate the presence of numerical abilities in *Homo sapiens'* closest relative, the chimpanzee. Subjects have been asked to make numerical judgments by selecting from among available alternative sets according to some criterion (such as choosing the larger of two arrays), while in so-called cardinal tasks, subjects were required to label sets of items by symbols representing the number corresponding to their total. Furthermore, some proto-mathematical skills have been considered by a variety of authors. Woodruff and Premack (1981) examined the chimpanzee's ability to handle proportions by matching quarters, halves, three quarters, and wholes of apples to corresponding proportions of liquid in a jar. A mathematical operation—summation—was tested by Rumbaugh et al. (1987) by asking their subject to choose between two pairs of food wells containing variable numbers of chocolate pieces. Boysen and Berntson (1989) also examined summation in a spatially larger scale task, using both food items and Arabic numerals as the stimuli to be summed. These experiments have demonstrated the presence of rudimentary mathematical abilities in the chimpanzee—exhibited spontaneously to the astonishment of the researchers themselves.

Work carried out at the Primate Research Institute of Kyoto University has, for the most part, focused on cardinal and ordinal meanings of number. The principal subject, a female chimpanzee named Ai, arrived at the institute at the age of approximately one, and began her training with a computer-controlled apparatus soon after her arrival. The following sections shall describe, after a brief introduction to the experimental paradigm and Ai's early training, the emergence of cardinal and ordinal applications of the same numerical symbols (currently ranging from 0 to 9) and their productive as well as receptive use by the subject: a constellation of abilities which at present make her unique in the chimpanzee world.

2 Early Training

Ai began her training in the use of numerical symbols in October 1981 (at the age of approximately 5 years) using a computer-controlled apparatus with which she had by that time become thoroughly familiar. The apparatus consisted of a small display window and a computer console located in a training room (190 × 220 × 180 cm). This set-up, or slight variations thereof, had been used from the age of 2 onwards as the medium through which the chimpanzee was taught to label a range of colors and everyday objects using a specially created artificial visual language comprised of so-called lexigrams (Fig. 1).

2.1 Object and Color Naming

In her earliest encounter with the computer operated apparatus (Asano et al. 1982), Ai was trained in an identity matching-to-sample task first using colored rectangles then geometric figures. Samples were presented in random order on a small panel, while a matrix of keys with a set of alternatives, containing one exact replica of the sample, were available on the response panel, also in a pattern randomized across trials. Pressing the matching alternative 1–5 times in consecutive trials resulted in the subject receiving a food reward, such as a raisin or a small piece of apple. Once Ai had mastered this matching, a symbolic labeling task followed. Everyday objects were placed in the display window by the experimenter, and the chimpanzee had available a set of the previously encountered geometric figures,

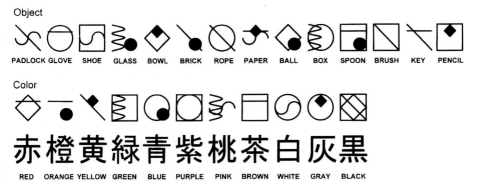

Fig. 1. The list of lexigrams, representing 14 objects and 11 colors, that formed Ai's vocabulary at the start of number training. Also shown are *kanji* characters, introduced later, that correspond to the same 11 colors

each of which had been arbitrarily assigned to represent one type of object. Discrimination was initially trained between two objects (glove and padlock); others—along with the symbols associated with them—were added successively one by one once the labeling of all previously introduced objects had reached a certain criterion of accuracy (90% correct in two consecutive sessions). The labeling of colors was taught following a similar procedure: starting with just two (red and blue), more colors were added successively as the subject's performance reached the criterion.

To combine object and color naming, Asano et al. (1982) then selected five objects (glove, bowl, brick, rope, and paper) and five colors (red, green, blue, yellow, and black) from Ai's vocabulary, preparing a total of 25 sample items in all possible color/object combination. Starting with discrimination between samples of one object in two colors and adding on more objects and colors as performance improved, Ai was now required to describe both the identity of the object presented and its color from among the sets of lexigrams presented to her. In the final stages of training, her accuracy in object and color naming reached 94% and 93% accuracy, respectively.

2.2 Numerical Labeling

At the start of number training (Matsuzawa 1985; Matsuzawa et al. 1986), Ai's vocabulary included 14 object names and 11 color names, each associated with a single unique lexigram. For the purposes of numerical labeling, five objects (pencil, paper, brick, spoon, and toothbrush) and five colors (red, yellow, green, blue, and black) were selected from this list. All 25 possible sample items were made available in multiple copies, for presentation as sets of homogenous items in the display window. A 5 × 5 matrix of keys on the subject's computer console was used to present Arabic numerals (in random patterns across trials) as the symbols to which the numerical attribute of the sample sets needed to be matched.

Initial training involved discrimination between 1 and 2 using a single type of object, one or two red pencils. Ai was required to respond by pressing the key showing the Arabic numeral "1" when one red pencil was presented, and "2" when the experimenter held up two. At this stage, only two keys (1 and 2) on the console were operative, their positions changing randomly between trials to prevent the learning of positional cues rather than the symbols. In later stages, once this first discrimination was mastered, sample sets of objects as well as the set of console keys were expanded one by one, until the use of numbers 1 through 6 reached the learning criterion (at least 90% correct for all numerical labeling in two consecutive sessions). At each stage, training consisted of various phases designed to test the extent to which the use of numerals already learned would generalize to sample sets containing novel colors and/or novel objects. Probe trials were inserted among background trials: For example, after training discrimination using 1 or 2 red pencils, approximately every tenth trial in a session with an otherwise unaltered procedure would consist of an unrewarded probe with samples of 1 or 2 blue pencils (novel color blue), 1 or 2 red papers (novel object paper), or 1 or 2 blue

Table 1. Generalization of numerical labeling to a new color, new object, or both in probe trials inserted among trained samples

	Proportion correct on probe trials			
Numbers	New color, trained object	Trained color, new object	New color, new object	Accuracy by chance
1, 2	0.83	0.50	0.50	0.50
1, 2, 3	0.60	0.33	0.33	0.33
1, 2, 3, 4	0.71	0.54	0.56	0.25
1, 2, 3, 4, 5	0.79	0.57	0.56	0.20

Rows represent successive stages of training in which Ai's numerical repertoire was gradually extended.
Adapted from Matsuzawa 1985.

papers (novel object/color blue paper). This was followed by training with reward on the newly introduced object-color combinations, before moving on to sets of $n+1$ items once accuracy in labeling n of all previously encountered sample items reached the criterion.

Hence, with new colors and objects being introduced at every stage of training, by the time Ai learned to label 1, 2, 3, 4, 5, and 6, she was familiar with 25 kinds of sample items (5 objects in each of 5 possible colors). Results at each stage (see Table 1) showed that generalization of numerical labeling for newly introduced colors and objects was incomplete, although often significantly above chance levels of accuracy, especially in later stages. Novel colored but familiar objects disturbed performance less than novel objects in familiar colors, which in turn lowered accuracy to a lesser extent than novel objects in novel colors. Overall, accuracy in numerical labeling of novel samples improved as Ai's training progressed to include larger numbers and more objects and colors.

The start of training with novel numbers was at each stage associated with a characteristic pattern of errors. By far the majority of mistakes were scored between neighboring numbers, where the new number being introduced was identified as the previously acquired highest. Towards the end of a total of 95 sessions, however, errors all but disappeared and Ai scored over 98.5% correct in labeling 1-5 items. Nine sessions later she also passed the criterion for labeling 1-6.

Matsuzawa (1985) then extended Asano et al.'s (1982) object and color naming study by introducing a further attribute, number. Presenting Ai with 1 to 6 of one of five different trained objects each painted one of five familiar colors (as listed above)—such that items shown simultaneously as samples were always identical—the experiment required her to name all three attributes in any order she liked (Fig. 2). Reaching more than 90% accuracy after just one block of three sessions, results also showed a clear preference for two of the six possible "word orders": color-object-number and object-color-number. This tendency remained even when new objects or familiar objects in new colors were presented as samples (to which Ai spontaneously transferred the 3-term naming skill): number was always

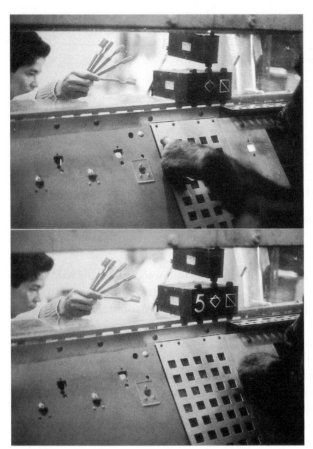

Fig. 2. a By pressing keys with the appropriate lexigrams, Ai performs a symbolic matching task, labeling the items shown to her in the display window. Here, she already has selected "red" and "toothbrush". **b** Ai selects the Arabic numeral "5" to describe another attribute of the sample besides color and object names. Note that number was the last of the three attributes assigned, even though the order of naming was entirely up to Ai

identified as the last of the three attributes. Matsuzawa cites the difficulty with which each attribute is identified as one of the potential factors contributing to preference exhibited towards particular sequences. The first number discrimination (1 vs. 2) took almost twice as many trials to acquire as the first object discrimination (glove vs. padlock) and more than three times as many trials as the first color discrimination (red vs. green). Furthermore, numerical naming was generally accomplished with lower accuracy than object or color naming. Nevertheless, Ai's extensive prior training with object and color naming only (Asano et al. 1982) may also have played a role in these two attributes being named after one another as the first two responses in 3-term naming. The level of reliability with which she was able to apply these labels clearly may have prompted Ai to preferentially respond with the "easier" two of the three required attributes first.

3 Further Cardinal Tasks: "Dot-Counting"

3.1 Transfer to Computer-Generated Stimuli

To assess further the extent to which Ai's labeling skills represented some form of rudimentary numerical ability, a variety of additional studies were conducted. Murofushi (1997) reports an experiment testing the transfer of skills acquired in labeling real life objects to computer-generated stimuli. The original apparatus was adapted by the introduction of a computer monitor that was to present stimuli instead of objects appearing in a display window. The subject continued to make her responses on a separate response panel. This set-up allowed for more controlled stimulus presentation as well as the accurate recording of variables such as response latency. The pace of experimental sessions remained fully under the control of the chimpanzee, with a start key present on the panel.

Stimuli presented on the computer monitor were green dots positioned randomly as determined by computer-generated sequence files. Dots numbered 1 or 2 initially, the maximum number increasing progressively as Ai attained the criterion (two consecutive sessions with 90% correct overall and at least 80% correct for each individual number) at each stage. Results showed that the numerical labeling skills acquired by Ai in previous years did not transfer easily to this new task. The number of sessions required to reach the criterion increased along with the maximum number of dots presented at each stage. The types of errors encountered in Ai's performance were comparable to those seen in her earlier training in that the largest proportion of mistakes were scored between neighboring numbers. Another commonly recurring response involved the selection of the largest numeral when the largest as well as when the second largest number of dots were presented, which resulted in higher accuracies for the labeling of the largest of the stimulus sets.

Sixteen, eighteen, twenty-four, forty-five, and fifty-five sessions were required, respectively, for Ai to attain the criterion when stimulus sets consisted of 1-2, 1-3, 1-4, 1-5, and 1-6 dots. At this stage her accuracy in labeling was over 80% for all numbers from 1 through 6. Generalization tests, where newly-generated random patterns of dots were used (i.e., those which the chimpanzee had never before encountered) did not interfere with accuracy, revealing that Ai's performance was not dependent upon familiarity with a finite set of on-screen dot arrangements.

To further test the effects of dot patterns, various non-random arrangements were presented. Results showed that Ai's performance in numerical labeling was affected only to a slight extent. Accuracy dropped to 79.2% in labeling 4, 5, and 6 dots when these were aligned vertically, horizontally, or diagonally thereby forming a line, in contrast to 83.3% for patterns resembling the faces of a die, 82.4% for patterns on the faces of a die plus a randomly positioned extra dot, and 83.6% for randomly positioned dots.

Thus, Ai's performance in the "dot-counting" task was highly accurate and comparable to that obtained with real-life objects. But to what extent can we say that Ai was "counting"? Humans accomplish tasks of this sort by one of three principal mechanisms (Mandler and Shebo 1982). Subitizing, a perceptual process, entails making a rapid but accurate judgment about the size of a (relatively small) set through viewing all items in it simultaneously; true counting relies on sequential tagging of all items in turn, using the last tag assigned as a label for the whole set; and estimation involves quick matching of the perceived magnitude of the set to a memorized magnitude associated with a label. Which of these strategies did the chimpanzee make use of? Analysis of patterns apparent in Ai's response latency may be instrumental in revealing the mechanism underlying the matching behavior.

Response latency varied as a function of dot number during sessions of labeling 1 to 5, 6, or 7 dots, revealing an interesting pattern. The function was flat for sample sets of one, two, and three, then increased gradually for numbers four and above, up until the second largest number. In turn, for the largest number of dots within a session, response latency dropped again. The shape of the latency curve thus does not provide a simple explanation. While the initial flatness may indicate subitizing where sets containing small numbers of items are perceived and identified at similar rates as human data suggests (Mandler and Shebo 1982), the gradual increase for numbers 4 and above may suggest the appearance of counting-like behavior where items are in some way tagged sequentially, with more items requiring more time to process. However, the final drop in response latency when the largest number is encountered indicates that other processes may instead be at work.

It is interesting to note at this point that the introduction of eight and nine dots in following years (Matsuzawa et al. 1991) was accompanied by a similar trend, suggesting in each case that the largest number was dealt with differently from others. In sessions where stimuli included all dot numbers between one and nine—and Ai's accuracy was as high as 95%—response latency data revealed a flat function up to four, then began monotonically to increase up to eight, followed by a drop when responding to nine. Matsuzawa et al. (1991) also tested a human subject following exactly the same procedure as that used for Ai, and found a response latency curve with several similarities as well as differences to the chimpanzee's (Fig. 3). The human RT curve was also flat up to 4 followed by monotonic increase, however, there was no drop in the case of the largest number, 9. In addition, the chimpanzee also seemed to be making her responses at a higher speed overall. To account for the shape of the chimpanzee's response latency curve, Murofushi (1997) suggests relative-magnitude estimation as a possible mechanism (after Dehaene 1992) responsible for the final drop on the largest number.

This conclusion was supported by the results of a follow-up study by Tomonaga and Matsuzawa (in manuscript). Here, Ai as well as four human subjects were tested in two alternative versions of the "dot-counting" task using 1–9 dots. In one, the unlimited exposure condition, subjects were allowed to view the dot patterns displayed until they made their selection of Arabic numeral to label the set. In the other, brief-exposure condition, dot patterns were displayed for 100 ms only, and subjects were required to select numerals without the dots being displayed at the same time. Results showed some clear differences between the two species. In the unlimited exposure condition, the previous findings of the largest-number effect (i.e., increased accuracy and a drop in latency to respond when 9 dots were pre-

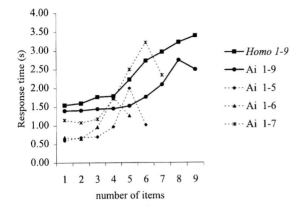

Fig. 3. Response latency as a function of the number of dots presented in the "Dot-counting" task for Ai and for a human subject tested under identical conditions. Note the similarities in the two curves (despite the chimpanzee's having accomplished the task somewhat more quickly overall), as well as the prominent largest-number effect in Ai. Nine dots were labeled significantly more quickly than eight; no such trend was found in the human subject. (Adapted from Matsuzawa et al. 1991.) Also included are data from stages of training when the upper limit of Ai's repertoire was 5, 6, and 7 (*dashed lines*); again, the largest-number effect is clearly illustrated. (Adapted from Murofushi 1997)

sented) were replicated for Ai, while humans showed the expected monotonic increase in response latency between 4 and 9. In the 100 ms condition, response latency was again flat in the subitizing range (1–4) for both species, but also between 6 and 9 in Ai's case, whereas humans showed monotonic increase after 4.

Tomonaga and Matsuzawa (in manuscript) interpret this difference under the brief-exposure condition with reference to "mental counting" by humans (Mandler and Shebo 1982): A representation of the sample items is held in consciousness in the form of an image and the same counting process is used to enumerate them as under unlimited exposure. The chimpanzee, on the other hand, resorted to estimation in the brief-exposure condition, and the largest-number effect disappeared. Evidence to show that Ai was also using a form of estimation during unlimited exposure—despite the superficial similarities with human data—came from her behavior during that condition. She often looked back at the sample once the numeral alternatives became available—a behavior whose frequency increased along with the number of dots in the sample, leading to an overall increase in average response time. Furthermore, analysis of the distribution of response times for each number revealed a peak shift with increasing dot number in humans, as expected in the case of true counting. Ai's data showed no such peak shift, instead, the distribution became skewed, average response latency increasing due to the appearance of instances of looking back at the sample. Tomonaga and Matsuzawa (in manuscript) conclude that the single peak that was maintained by Ai for all numbers therefore probably reflected the time required for a single estimation.

Thus, the kind of sequential tagging which is evident in humans and is often referred to as "true" counting did not seem to underlie the chimpanzee's performance. Nevertheless, high levels of accuracy seemed to suggest that Ai was responding to some aspect of the sample stimuli which corresponded to numerosity. Various manipulations of the on-screen stimuli attempted to answer this question.

3.2 Heterogeneous Stimuli

As the appearance of individual stimuli in "dot-counting" had so far been identical between trials, one possible alternative to using the number of dots to solve this task was for Ai to attend to dot density or the total area of screen surface covered by dots instead. Therefore, experimenters introduced novel types of stimuli: dots of different sizes as well as video images of real-life objects in various colors presented either as homogeneous or heterogeneous stimulus sets. At the same time as beginning testing in this phase, an adjustment was made to the apparatus, with the response panel being replaced by a touch-sensitive monitor which displayed Arabic numerals as alternatives at random locations within an imaginary 6 × 3 matrix. Touching this screen at a location over the numerals displayed constituted a response.

When testing with heterogeneous dot patterns, the original green dots were replaced by white stimuli in three distinct sizes: large (2 cm diameter), medium (1.5 cm diameter), and small (1 cm diameter). Sample sets consisted of combinations of large and small, large and medium, or medium and small dots. By varying the composition of sets with respect to the number of small, medium, or large dots present, the correspondence between total dot area and total dot number was disrupted, and an estimation of the former by the subject was no longer sufficient to accurately label the number of items in a set. Ai was initially trained on homogeneous sets of either large or small dots, maintaining around 90% accuracy from the start. This was followed by the introduction of sets of mixed dot sizes for two to three sessions. Results showed only a small drop in accuracy (approximately 5%) when samples included small dots in combination with large or medium ones, while heterogeneous large-medium sets were labeled as accurately as homogeneous sets after a single session. The distribution of errors as a function of the proportion of large dots in the sample showed that with more large dots being present, Ai was more likely to make the mistake of matching the sample to a larger numeral than to a smaller one, suggesting that dot density played at least some part in controlling the behavior. Overall, however, accuracy remained high, and numerical labeling of stimuli differing in size posed little extra difficulty for the chimpanzee.

How heterogeneity in shape as well as color may affect Ai's behavior was investigated in an additional study. Instead of white dots as stimuli, a laser-disc player was used to project images of real-life objects on a video monitor. The objects were blocks, pencils, and padlocks, colored either red or green, and numbering between one and seven. Stimulus sets displayed were of three types: homogeneous, where all items were identical (same object, all either red or green), color-heterogeneous (same object, some red, some green), and object-heterogeneous (same color, some blocks, some pencils).

Ai was first trained on homogeneous sets. Sets containing the already familiar dots were used as baseline, then green blocks were introduced. Red blocks followed when her accuracy reached 90%, then red pencils, green pencils, green padlocks, and red padlocks in turn, once the criterion had been attained at each stage. When all possible color/object combinations were thus exhausted, transfer tests to color-heterogeneous sets were conducted in sessions where half the trials consisted of homogeneous displays of one to seven identical objects of the same color, and half consisted of one to seven identical objects, some red, some green. Object-heterogeneous sets were tested in the same way, except that half the trials in transfer tests consisted of displays showing a mixture of pencils and blocks, totaling one to seven, all either green or red.

While changing the identity of objects displayed in the homogeneous set condition caused a slight drop in accuracy (which nevertheless recovered after a single session), changing the colors of already familiar items had no such effect, and accuracy was maintained at over 90%. In heterogeneous trials, dealing with sets comprised of two different objects of the same color upset performance to a greater extent than sets with identical objects in different colors. The latter was associated only with a relatively small decrease in accuracy, which then quickly recovered to levels attained with homogeneous sets. Accuracy under the object-heterogeneity condition dropped to below 60% in the first session, and although there was gradual improvement over the sessions that followed, it remained at around 85%.

Although encountering for the first time heterogeneous stimulus sets—dots of varying sizes, objects in different colors, and different objects in the same set—generally led to Ai being less accurate in numerical labeling, her performance tended to remain at high levels, recovering to baseline within a few sessions. It seemed that the skills she had acquired over the years in quantifying real-life objects as well as stimuli displayed on a computer monitor could also be transferred to a variety of novel stimuli as well as to heterogeneous sets.

3.3 Introduction of "Zero" in Cardinal Tasks

Throughout the studies described in the previous paragraphs, Ai's range of sample sets and associated numerical symbols had one as the lowest number, with seven, eight, and eventually the current maximum of nine being introduced over years of occasionally interrupted training. The challenge was now to extend Ai's range in the other direction: Could she handle the concept of zero?

Rumbaugh et al.'s (1987) study of summation in chimpanzees already successfully introduced the concept of "absence of food" (numerical equivalent 0), while Boysen and Berntson's (1989) study of labeling food items showed that the chimpanzee Sheba had no difficulty in learning to match an empty food tray to the numeral 0. Furthermore, when tested on summation using first food items then cards showing Arabic numerals hidden in various target locations around a room, Sheba successfully solved the addition task when "absence of food" or "zero" was included in the required operation. More recently, Olthof et al. (1997) trained squirrel monkeys on a similar symbolic counting task, first by having them choose one of a pair of Arabic numerals from among 0, 1, 3, 5, 7, and 9, yielding a corresponding number of peanuts as reward. Their subjects then tended to correctly choose the larger of two sums (displayed on two cards, each with one, two, or three Arabic numerals printed on it) when 0 was included among the stimuli used.

Instead of beginning Ai's training in the use of the numeral 0 with a food related task, researchers opted for the dot-counting paradigm familiar to her from previous years (Biro and Matsuzawa 2001). The apparatus and task were adapted in the following ways. The separate video screen and response panel were united into a single touch-sensitive monitor, the lower half of which displayed a start key followed by a sample, while alternatives, two at a time, appeared in the upper half in two distinct locations (left and right). Ai was first allowed to grow accustomed to the new set-up through training in sessions involving dot numbers and numerals she was already familiar with. Samples were presented as 1–9 dots inside a thin white frame (10 cm × 11 cm). Once her accuracy in labeling these reached levels attained in previous experiments, the numeral zero was introduced. The sample was shown as an empty white frame, and the numeral 0 was displayed as one of two alternatives at the top of the screen (Fig. 4). Within a session, it was presented 18 times—interspersed between background trials of labeling 1–9 dots—pitted against all numerals 1–9 with presentation as the left or right alternative counterbalanced over all trials.

Results of previous "dot-counting" experiments with regards to 1–9 dots were replicated in this study, referred to as dot-to-numeral matching (DNMAT), including the largest-number effect: nine dots were labeled more accurately and more quickly than were eight. A notable difference concerned the distribution of errors: instead of the vast majority being scored between neighboring numbers, confusions now occurred between more widely separated numbers. However, this was probably an artefact of the two-alternative condition in that it may have encouraged some slight positional bias in Ai. Overall accuracy averaged 86.5% per session. Meanwhile, accuracy in labeling 0 dots exceeded 80% correct after 7 sessions, and continued to increase rapidly, leveling at 99.4% during the last ten sessions.

Examining the process of acquisition leading up to this highly accurate performance, we encounter four distinct phases of learning:

(I)   The symbol '0' is completely avoided. The empty frame is matched to every alternative between 1 and 9; incorrect responses are scored.
(II)   Correct responses appear in trials where large numbers serve as alternatives.
(III) Upsurge in so-called "false alarms" where zero is chosen erroneously in response to the presentation of other dot numbers.
(IV)   "False alarms" disappear almost completely, and accuracy in correctly identifying zero dots approaches 100%.

Phase (II) is probably the most interesting of the four, and illustrates a phenomenon which was encountered in other numerical tasks involving zero. We will return to it in more detail in Section 4.3.

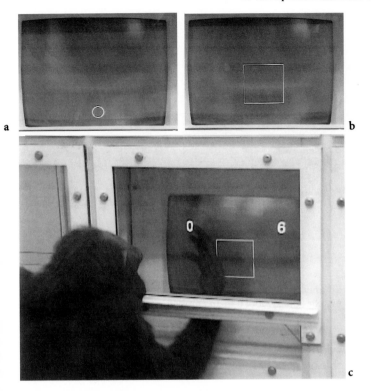

Fig. 4. The two-alternative dot-counting task in the context of which the numeral 0 was first introduced. The start key (**a**) is followed by the presentation of the sample (**b**) in the lower half of the screen, then two alternatives pitted against each other in the upper half. (**c**) Here, Ai correctly matches an empty frame to the symbol 0

3.4 Receptive Versus Productive Use of Numerals

At the same time as Ai was being introduced to the numeral zero by matching dots to one of two alternative numerals, she was also trained on an additional task which involved the receptive use of the same numerals.

Numerical labeling of real-life objects as well as the "Dot-counting" or DNMAT tasks are said to involve the productive use of numerals: The subject is presented with a set of items and is required to respond by selecting a numeral describing their total. Receptive use, on the other hand, is a reversal of this scenario, where the subject is to select the correct number of items when shown an Arabic numeral as the sample. We may argue that to describe the former sort of matching behavior as true symbol use would entail the subject showing transfer between the two types of usage.

Rumbaugh et al. (1989) report the receptive use of numerals in a chimpanzee. Lana was presented with an Arabic numeral and was required to make a corresponding number of responses, such as using a joystick to successively select boxes on a computer monitor. In Boysen and Berntson's (1989) study, three chimpanzee

Fig. 5. Receptive use of the numeral 0. The task is the reverse of the dot-to-numeral matching task (productive use). Ai matches the symbol 0 to an empty frame

subjects learned to select placards showing the correct number of markers (one, two, or three) when Arabic numerals as samples were presented to them on a video monitor.

Ai's training in the receptive use of numerals used the same apparatus as described above for the introduction of zero. The procedure was altered such that an Arabic numeral between 0 and 9 was shown in the lower half of the screen, and two thin white frames containing sets with alternative numbers of dots were displayed in the top. To score a correct response, the subject was required to select the frame containing dots whose number corresponded to the value of the numeral (Fig. 5).

Acquisition of this task (numeral-to-dot matching, NDMAT) did not show perfect transfer from DNMAT, with overall accuracy exceeding 80% only after 34 sessions. Average accuracy was 76.6% during the last ten sessions of training, significantly lower than DNMAT, suggesting that Arabic numerals were not used as "symbols" in the strictest sense by the chimpanzee.

The introduction of zero was accompanied by a similar set of four learning phases as listed in the previous section, with the exception that in Phase IV "false alarms" did not disappear completely, but remained frequent when numerals 1 or 2 were the sample items. Nevertheless, accuracy in trials where 0 served as the sample leveled out at 96.7% in the last ten sessions.

4 Ordinal Tasks

Gelman and Gallistel's (1978) counting model—conceived for human children—emphasizes the role of comprehension of the ordinal relationship among cardinal meanings of numbers. In other words, having an understanding of the nature of numbers entails that one is able to order numerals into an ascending or descending series.

How Ai's knowledge of cardinals translated to the use of these symbols as items in an ordinal series was investigated in the following set of experiments, addressing issues such as transitive inference, the representation of linear sequences, and perceptual/cognitive aspects of numerical ordering.

4.1 Initial Training with Pairs of Adjacent Numerals

Ai's earliest encounter with ordinal tasks (Tomonaga et al. 1993; see also Tomonaga and Matsuzawa 2000) occurred after she had already mastered the use of numerals 1–9 in the "dot-counting" task, and she was therefore thoroughly familiar with the apparatus (touch screen monitor) as well as the Arabic numerals as stimuli. To familiarize her with sequential responses, training began with three sessions of presenting seven homogeneous white circles on screen, each of which disappeared following a touch by the subject. Ai was required to touch all circles to conclude a trial. This was followed by the presentation of adjacent pairs of numerals (1-2, 3-4, 5-6, 7-8), where a touch to one numeral caused it to disappear, a touch then to the second caused it also to disappear, such that the subject cleared the screen in stages to conclude each trial. Reinforcement was given irrespective of the order in which the stimuli were touched, as an attempt to examine the possibility that cardinal training allowed Ai to spontaneously construct the ordinal scale. Results of this early test showed that she responded more or less randomly: no consistent ascending or descending patterns were detected in her responses.

Individual adjacent pairs were then trained in turn with differential reinforcement of responses from small to large, starting with 1-2. When accuracy reached 90% or more, probe trials of the as yet untrained pairs were inserted randomly into sessions, under non-differential reinforcement, but Ai showed no evidence of having transferred the rule of responding in ascending order to these pairs. As a result, pairs 1-2, 3-4, 5-6, and 7-8 had to be trained explicitly, in successive blocks of sessions. With differential reinforcement, she quickly attained highly accurate performance, reaching 90% correct after one or two sessions only.

When she had learned to respond correctly to all pairs listed above, new adjacent pairs were tested as probes against a background of baseline trained pairs. These new combinations included 2-3, 4-5, and 6-7, none of which seemed to have been acquired spontaneously by the chimpanzee. In fact, she performed significantly below chance, indicating that she had used a rule other than responding according to the ordinal scale. 2 was touched after 3, for example, based on her training history where 2 (in the 1-2 pair) was always touched second while 3 (in the 3-4 pair) was always first. This tendency, however, quickly disappeared once training with differential reinforcement of the new pairs started. 2-3, 4-5, and 6-7 were mastered within one to two sessions each.

In the final phase of training with adjacent pairs of numerals, all possible combinations between 1 and 8 were included in a session in a random sequence of trials, constituting the baseline maintained at close to 100% correct. A single type of probe trial was included, namely 8-9. In this case, Ai responded with an accuracy of 87.5%, showing that this was a relationship that she had spontaneously acquired.

4.2 Beyond Pairs and Adjacent Numerals

What did Ai learn from her training with adjacent pairs about the nature of the ordinal scale 1–9? This question was examined further by Tomonaga et al. (1993).

Carrying on with baseline trials of adjacent pairs, the researchers introduced all possible non-adjacent pairs as probes over several sessions. In contrast to the chance level performance noted with probe trials involving adjacent pairs in the initial phases of ordinal training, Ai's accuracy on non-adjacent probes averaged 80.4%. This was considered evidence that Ai had acquired some form of an ordinal scale from adjacent pair training alone, and was able to apply this knowledge to pairs of items taken from anywhere along this scale. This skill is referred to as transitive inference, or transitivity, where the subject is thought to build, from a limited number of trained comparisons between items, a representation of a linear sequence incorporating individual items, and is able to make inferences about the relative "values" of pairs of items that had never explicitly been trained before. Boysen et al. (1993) reported a comparable finding in their language trained chimpanzee, Sheba, as did D'Amato and Colombo (1990) using capuchin monkeys. Worthy of note is the fact that tests of transitivity have so far failed with pigeons (Straub and Terrace 1981; Terrace and McGonigle 1994), while the case for rats is still under consideration (see Roberts and Phelps 1994). Human children at the age of 6 have no trouble with this task (Chalmers and McGonigle 1984).

The next step following extensive training on all possible pairs taken from the 1–9 range (again reaching close to 100% accuracy on average) was to introduce 3-item sequences. Again, trials were of the "disappearing" type where Ai's response to each numeral caused it to disappear in turn. Initial probe trials containing only adjacent sequences revealed imperfect transfer to this condition, although accuracy was higher than expected by chance. Then, after nine sessions of training with reinforcement, Ai reached over 90% correct, and was given probe trials of non-adjacent 3-item sequences. Here, accuracy was already around 80% without differential reinforcement, clearly above chance (16.7%).

In subsequent years, Ai was also trained on the ordering of 4, 5, …, 9 adjacent numerals (Fig. 6a), and transfer between conditions (sequences gradually increasing in length) was evident at each stage. In the last stage, ordering all nine numerals, Ai's accuracy averaged 82.5% over the final ten sessions. Response latencies (Fig. 6b) under this condition reveal a striking pattern that was evident during all stages of the experiment, irrespective of the length of the sequence dealt with. The longest time to respond to a numeral was always taken when selecting the first item in the sequence, followed by significantly shorter response latencies for all other numerals. There was no significant difference in the speeds with which the latter were chosen. What this finding might tell us about the way in which Ai solved the numerical ordering task is addressed further in Section 4.4.

Tomonaga et al. (1993) also examined other aspects of the response latency data

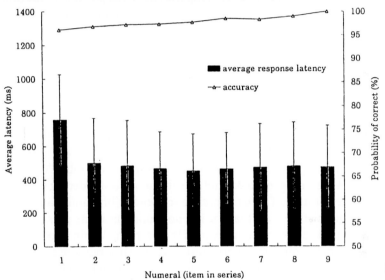

Fig. 6. a Ai performing the 9-unit ordering task. She is required to touch all nine numerals between 1 and 9 in ascending order. Each numeral disappears, if selected in the correct position in the sequence, such that scoring a correct trial entails the clearing of the screen in stages. b Response latencies associated with selecting each of the nine numerals. Initial latency in selecting the first numeral is significantly longer than all other latencies, which in turn do not differ significantly from each other. *Triangles* show probabilities of correctly identifying a numeral as the next in the sequence when all previous numbers have already been selected

from 3-item sequences, specifically the effect that the identity of the first item and the distance between items on an ordinal scale had on Ai's performance. Looking at all 84 possible 3-unit combinations from the range 1–9, they found that latency increased with the position of the first item on the scale, that is, higher numbers were chosen more slowly as first items than lower numbers. In addition, responses to the first item were made faster if the second item was a number separated from the first by a larger gap.

Both of these phenomena have been investigated by various authors in a variety of species, and have been thought to reveal something about the nature of subjects' cognitive representation of the sequence learned. While pigeons show no evidence of the same kind of positional or symbolic distance effects (e.g., Terrace and McGonigle 1994), D'Amato and Colombo (1988) and Swartz et al. (1991) found comparable increases in response latency along with the serial position of the first item, as well as with the increasing "gap" between the first and second item in capuchins and rhesus monkeys. Monkeys and apes may therefore solve serial recognition tasks in a qualitatively different way from other taxa, likely involving mental representations of the sequences in question. Ai's performance also pointed towards the latter possibility. What kinds of characteristics her representation possessed was investigated by experiments described in the following sections.

4.3 Introduction of "Zero" in Ordinal Tasks

Subsequent to Ai's training on the labeling of zero dots with the symbol "0" in the DNMAT and NDMAT tasks (Sections 3.3 and 3.4), the numerical ordering task was also extended by the addition of zero to possible sequences (Biro and Matsuzawa 2000). Training began with a 2-unit task using non-adjacent items presented at random locations within a 5 × 8 imaginary matrix on screen. Within a session, all possible pairings from the range 0–9 were presented once in a random sequence, such that 0 appeared in combination with all numerals 1 through 9. The subject's task was once again to touch the two numerals in ascending order.

Accuracy in baseline trials 1-2, 1-3, 1-4, …, 8-9 was maintained at near perfect levels (99%) throughout. In contrast, the introduction of 0 initially resulted in frequent errors, suggesting that the knowledge of "zero" learned in cardinal tasks did not transfer spontaneously to the ordinal task. Instead, 0 was treated as a "wildcard" (see also next section; cf. Tomonaga and Matsuzawa 2000): when presented in combination with large numbers, Ai accurately chose "0" first as the lower of the pair, whereas being pitted against small numbers tended to yield incorrect answers. This is less than surprising considering Ai's training history where numbers toward the higher end of the scale were selected more often as the second item of a pair, while those near the lower end were reinforced if chosen first. This tendency disappeared gradually with training, with the "virtual" position of zero on the continuous numerical scale shifting steadily towards the lower end (Fig. 7). In other words, while in the very first session Ai identified zero as being positioned somewhere between 6 and 7, in subsequent sessions correct responses began to appear when 0 was presented with 6, then 5, 4, and so forth. Eventually, errors with

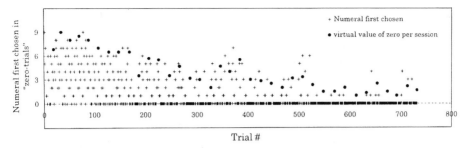

Fig. 7. The introduction of zero in a two-item ordering task. The subject is shown two Arabic numerals and is required to select the lower of the two first, followed by the higher. Trials shown are those where one of the numerals presented was 0. *Crosses* represent the numeral first chosen in individual trials; *filled circles* show the changes from session to session in the "virtual" value of zero along the continuous numerical scale between 0 and 9, calculated using the pattern of correct and incorrect answers given in particular sessions to trials involving zero

comparison numerals higher than 1 disappeared almost completely. However, the confusion between 0 and 1 continued to be a prominent feature of Ai's performance, with accuracy on these trials hovering just above 50%.

At this stage, the task was extended to 3-unit sequences, sessions consisting of all 120 possible consecutive and successive but non-consecutive sequences from 0-1-2 to 7-8-9 presented in a random order. Ai easily adapted to this new condition: her overall accuracy was 90% in the first session, increasing to an average of 94% correct after a few more sessions. Trials involving 0 also quickly reached 89% accuracy on average, with those of the 0-2-X, 0-3-X, ..., 0-8-9 type (where X was any number larger than the previously listed one) scoring over 90% in all sessions. However, 0-1-X trials remained a problem—average accuracy over all sessions was 58%.

Thus, the persistent problem of confusions between 0 and 1 we encountered in the cardinal "dot" tasks DNMAT and NDMAT (Sections 3.3 and 3.4) was also evident in the 2-unit and 3-unit ordinal tasks. Errors where zero dots were labeled as "1", one dot was selected upon presentation of the symbol "0", zero dots were selected upon presentation of "1", or "1" was ordered before "0" continued to be scored until the end of the testing period. Taken together with the evidence for incomplete transfer of the use of zero between cardinal and ordinal tasks, these errors suggest that Ai's understanding of the meaning of zero in relation to the rest of the number symbols in her repertoire was not consistent with an "absence of items vs. presence of items" scheme which excludes zero from the set of natural numbers in human arithmetic. Instead, in her representation of the number system, zero appeared to shift along the length of a continuous numerical scale, gradually approaching the low end—yet ultimately still continuous with, adjacent to, and often confused with one.

Nevertheless, as mentioned briefly above, another potential explanation for the particular problems we experienced in trying to introduce a chimpanzee to the meaning of zero may also be provided by a consideration of Ai's history of number training. For all numbers between 1 and 9, each new number was added on to her already existing repertoire successively, in ascending order. This is analogous to a child's learning of numbers (Gelman and Gallistel 1978). The training of zero therefore represented a sharp contrast, as it was the first to join the series at the low end of the scale, severely disturbing the sovereignty of "one" as the lowest number. This fact may help to explain why comparisons of zero and one caused so much difficulty for the chimpanzee, both in cardinal and in ordinal number problems. The receptive cardinal task, NDMAT, where this problem was most apparent, was probably solved by Ai using a form of relative magnitude estimation (relative numerousness judgment, RJF; Thomas et al. 1980) rather than "true" counting, such that a sample of the numeral "1" may have elicited selection of the least number of dots available, i.e., none. Similarly, in the ordinal task, presentation of the numeral "1" in combination with any other alternative may have been a strong stimulus for selecting 1 as the first—and lowest—numeral. These studies thus suggest that the chimpanzee's competence in numerical tasks is a function not only of an underlying understanding of abstract numerical concepts, but also of the training history of the subject.

4.4 Mechanisms of Ordering

Aside from the slight difficulties caused by zero, Ai showed very high levels of accuracy in ordering consecutive as well as successive but non-consecutive numerical sequences of various lengths in the range 0–9. What was the nature of the mental representation she constructed and what kinds of cognitive processes facilitated her performance? Subtle alterations to the now familiar ordering task attempted to answer these questions.

Tomonaga and Matsuzawa (2000) adapted the procedure introduced by D'Amato and Colombo (1989) of including a "wild-card" item in a serial recognition task, changing the original design such that the wild-card did not substitute for one of the items in the sequence learned, but appeared in addition to them. Numerals in groups of three were presented on each trial in a baseline condition, followed by sessions where trials consisted of three numbers plus one or two white squares that served as wild-cards. Two conditions were tested. In the first, referred to as the "terminator" condition, touches to the white square could be made at any point in the sequence, without it disappearing. Multiple responses could be made, up until the conclusion of the trial (touching the highest numeral after both lower numerals had disappeared), at which point the screen was cleared and the next trial could commence. The computer recorded all touches to the square. In the second, "wild-card" condition, either one or two (varied randomly between sessions) white squares were presented along with the three numerals. Responses to these wild-cards resulted in their disappearance, and the trial continued as under baseline conditions.

The terminator condition clearly revealed a serial position effect (where the identity of the first item influenced performance irrespective of the other items present) as well as a symbolic distance effect (where the ordinal distance between items controlled accuracy and/or response latency). First, responses to the white square item increased in frequency on trials with larger first items, and second, touches to the square between the first and second and between the second and third numerals increased as a function of increasing ordinal distance between the individual items. In the wild-card condition where squares disappeared when touched, the majority of responses were made before the first item, and furthermore, the higher the first item was, the more likely that the wild-card was pressed initially. Ordinal distances between first and second and between second and third numerals had no effect on choice of the wild-card in this condition.

Tomonaga and Matsuzawa (2000) interpret this result proposing that Ai understood the ordinal positions of the numerical symbols she had been taught to handle, and had formed an integrated 9-item linear representation of these items from her training with only (at that time) 2- or 3-item sequences. The patterns noted in selecting wild-cards suggested that these were used to fill the "gaps" between numerals, the likelihood of a response depending on the absolute as well as relative positions of the items along the numerical scale. While the authors concede that both serial position and symbolic distance effects can be explained not only by a cognitive but also a reinforcement theory account (see Wynne 1995), they suggest that at least response latency data does point toward the possibility that Ai solved the serial learning task in a way quite different from non-primate subjects. Cognitive explanations are therefore to be considered.

The characteristic pattern in Ai's response latency in ordinal tasks (long latency on the first response followed by short latencies on all ensuing responses) prompted an investigation of the perceptual and cognitive aspects of solving such tasks (Biro and Matsuzawa 1999). We conducted sessions with baseline trials of the standard 3-unit ordering described previously, this time using all numbers between 0 and 9. All 120 possible 3-unit sequences were presented in a random order in a session, and interspersed among them were twelve probe trials of a particular type, designed to shed light on the kinds of mental processes Ai made use of during different stages of solving this task. These trials were referred to as "switch" trials and entailed the following manipulation: Immediately after the subject correctly identified the lowest of the three numerals, the on-screen positions of the remaining two were exchanged by the computer (Fig. 8a). Over a series of ten sessions all 120 3-unit sequences were tested as switch trials, and we compared accuracy and response latency data as well as the subject's behavioral topography under baseline and switch conditions.

A drop in accuracy from 94% in baseline trials to 45% in switch trials showed that merely exchanging the locations of second and third items following the start of a trial (or, more precisely, the start of Ai's response) had a profound effect on the subject's performance. Three kinds of errors were possible in 3-unit ordering: pressing the second item first, pressing the third item first, and pressing the first item followed directly by the third. Of these, the single one responsible for the lowered accuracy was the third type; the other two kinds of errors were scored with the same very low frequencies under both conditions (Fig. 8b). Analysis of response

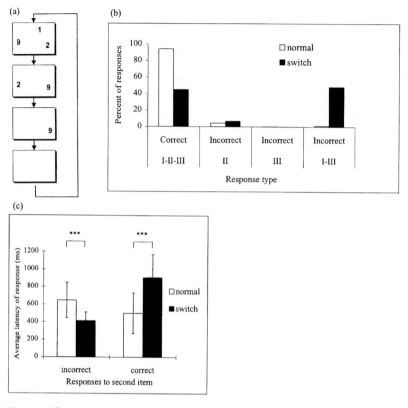

Fig. 8. a Schematic representation of sequence of events on the touch-screen monitor during a correctly performed "Switch-trial" in a three-item ordering task. **b** Effects on accuracy of switching. Responses are classified as "Correct" when numerals are touched in the appropriate sequence (lowest-intermediate-highest, i.e., I-II-III), or "Incorrect" if there is any deviation from this sequence. **c** Response times in selecting the second item of each sequence

latency revealed two interesting differences. First, in trials where Ai performed correctly, responses to the second item were made significantly more slowly under the switch condition. Second, the opposite was true for incorrect responses: normal trials involved significantly longer latencies to respond to the second item (Fig. 8c). To explain this result, we considered Ai's behavioral topography during both conditions.

All sessions were videotaped, and Ai's behavior during switch trials was categorized according to the movement her finger made over the screen following the occurrence of the switch. We noted the following main patterns. In the vast majority of cases where Ai eventually scored a correct response, her finger initially moved towards the highest numeral (now occupying the pre-switch position of the intermediate numeral), then changed course somewhere along the way and followed the path to the intermediate item. Rarely did she select the intermediate numeral correctly without such course correction. In approximately half of the total number of incorrect switch trials, where the highest numeral was erroneously selected in the second position resulting in the automatic clearing of the screen of the one remaining stimulus, Ai's finger continued on a trajectory towards the now blank position of the third item (the location, pre-switch, of the highest and post-switch of the intermediate numeral) before being withdrawn.

From the distribution of these finger movements, we drew the following conclusions: Ai had, at the onset of each trial, inspected all three available numerals, deciding on the order in which they were to be responded to. In addition, she also planned the appropriate motor sequence that would lead to a correct response. Switching the positions of the numerals during the execution of this motor sequence therefore interfered with the pre-planned response and only on approximately half of all switch trials was Ai able to disengage and alter the movement of her finger in time to avoid scoring incorrectly. When we examined the distribution of errors as a function of on-screen distances between numerals, we found that correct responses were more likely to be scored if the first and second items were located far apart. In other words, when Ai's finger had a relatively long way to travel from the first to the second item (which became the first and third items after switching), there was just enough time available for her to notice she was about to make a mistake. This also helped to explain the second of our observations on differences in response latency under baseline and switch conditions: incorrect switch trials were associated with shorter latencies of response to the second item than baseline trials, because it was in those cases that Ai's finger had a short distance to travel—hence the lower likelihood of being able to correct the path of movement in time.

In contrast with this account is the step-by-step serial search process proposed, for example, for macaques (Ohshiba 1997), in which the subject at each stage identifies only the target immediately to follow, and only after—or perhaps during—the selection of this target is there an attempt to identify and locate the next item. Ai's performance seemed to suggest that she solved the 3-unit ordering task using a more sophisticated strategy involving pre-planning followed by the execution of the appropriate motor sequence. In addition, we also found evidence of a monitoring phase that allowed for quick corrections to be made when the numeral about to be chosen did not correspond to the numeral next in the pre-planned sequence. How humans would solve such serial recognition tasks has not been investigated thoroughly, yet we suspect their strategy would be highly similar to Ai's.

What kinds of limits were imposed on Ai's performance in the ordering task? In other words, how far was she able to "plan ahead"? Kawai and Matsuzawa (2000) investigated these issues in an extension of the switch experiment with fascinating results. The experiment once again entailed the ordering of numerals, where in special "masking trials" after the selection of the lowest item, all remaining numerals were replaced in their original positions by white squares. The fact that the subject was able to complete trials of this sort with over 90% accuracy up to four items and with 65% accuracy at the five-item stage (well above chance level) clearly reinforced the pre-planning hypothesis. Ai's performance suggested that she could remember sequences of at least five numbers—a result that has prompted the authors to refer to the phenomenon as the "magic number 5 in the chimpanzee", drawing parallels with the magic number 7 effect in adult human information processing (Miller 1956). The number of items that can be handled simultaneously by the chimpanzee's brain in a numerical task thus appears to be at least five, corresponding to levels attained by preschool children.

5 Conclusions

Over the course of more than 15 years of training, Ai's ability to handle abstract concepts and symbols in the form of Arabic numerals has progressed from learning to communicate her ability to differentiate between one red pencil and two red pencils to a variety of skills once thought unique to humans. Beginning by acquiring the skill of cardinal labeling of homogeneous sets of real-life objects, she later learned to extend this skill to stimuli displayed on a computer monitor as well as to sets containing heterogeneous items. Subsequent training allowed her to build a linear representation of the same symbols she used in the aforementioned cardinal tasks, yielding ordinal knowledge of numbers and showing that she was capable of handling both of these abstract meanings of number. At the age of 22 in 1998, her number repertoire included all numerals from 0 to 9, applied in both cardinal and ordinal tasks with consistently high accuracy.

What Ai's abilities might tell us about the chimpanzee's numerical competence is a complex issue. Ai's accuracy in cardinal tasks involving sets of up to nine items has reached high levels through training, and has in addition acquired a certain degree of abstraction—perhaps the hallmark of human-like counting skills—where the application of the symbols no longer depends on the identity of the items counted. Yet a variety of experiments have shown that the way Ai solves such tasks is not necessarily analogous to human "counting" (Murofushi 1997; Tomonaga and Matsuzawa in manuscript). While the phenomenon of subitizing seems to be shared by the two species for up to four or five items as shown by response latency analysis, higher numbers are treated differently. Ai was likely using some form of estimation (such as relative magnitude estimation) to match the size of sets to numerals, whereas humans under similar circumstances perform sequential tagging of all items leading to more accurate but often slower performance. Bearing in mind the experimental setting, which from its earliest days involved differential reinforcement and large numbers of consecutive trials, it may be reasonable to suppose that Ai's estimating was at least in part a result of a speed-accuracy trade-off. The computer-controlled apparatus that was adopted in the majority of the studies described in this chapter allowed for rapid responding in individual trials and may thus have encouraged the emergence of Ai's own unique strategies for solving numerical tasks. Nevertheless, the level of objectivity offered by a fully automated experimental paradigm must not be underestimated. The complete exclusion of the experimenter from testing sessions has allowed for the elimination of any form of social cueing that may contaminate data. In addition, exact measurements of response latency have proved instrumental in shedding light on the cognitive processes underlying performance on both cardinal and ordinal tasks. Strictly controlled stimulus presentation consistent between tasks examined in parallel (productive vs. receptive use of numerals; ordering of numerals) also seems to be of fundamental importance when testing for transfer between the use of the same symbols/concepts in different contexts.

An issue encountered at several places in this chapter is the role of previous training in influencing Ai's performance in numerical tasks. Tracking this train of thought, we may argue from a developmental perspective that comparisons with human children rather than adults have more relevance to studies of numerical competence in non-human subjects. The emergence of counting in children, however, is in most cultures preceded by the memorization and recital of number words in a stable order, even before they are applied in one-to-one correspondence to items to be counted. This familiarity with labels may well facilitate counting performance once the cognitive apparatus reaches the maturity required to proceed to true counting and, eventually, mathematical reasoning. In a somewhat similar vein, Ai's prior training in the use of symbols as labels for real-life objects may have been instrumental in allowing her to handle numerical labeling tasks with the competence that we observed. The abilities that have already come to light as a result of experience in the use of numerical symbols thus seem to reveal sophisticated mechanisms for dealing with quantities. Further studies will doubtless probe more deeply into the ways in which Ai, as well as other chimpanzees, see the world of numbers.

References

Asano T, Kojima T, Matsuzawa T, Kubota K, Murofushi K (1982) Object and color naming in chimpanzees (*Pan troglodytes*). Proceedings of the Japan Academy 58(B):118–122

Biro D, Matsuzawa T (1999) Numerical ordering in a chimpanzee (*Pan troglodytes*): Planning, executing, and monitoring. J Comp Psychol 113:178–185

Biro D, Matsuzawa T (2001) The concept of number in a chimpanzee (*Pan troglodytes*): Cardinals, ordinals, and introduction of zero. Anim Cogn 4(2) (in press)

Boysen ST, Berntson GG (1989) Numerical competence in a chimpanzee (*Pan troglodytes*). J Comp Psychol 103:23–31

Boysen ST, Capaldi EJ (eds) (1993) The development of numerical competence: Animal and human models. Lawrence Erlbaum Associates, Hillsdale, NJ

Boysen ST, Berntson GG, Shreyer TA, Quigley KS (1993) Processing of ordinality and transitivity by chimpanzees (*Pan troglodytes*). J Comp Psychol 107:208–215

Boysen ST, Berntson GG, Shreyer TA, Hannan MB (1995) Indicating acts during counting by a chimpanzee (*Pan troglodytes*). J Comp Psychol 109:47–51

Chalmers M, McGonigle B (1984) Are children any more logical than monkeys on the five-item series problem? J Exp Child Psychol 37:355–377

D'Amato MR, Colombo M (1988) Representation of serial order in monkeys (*Cebus apella*). J Exp Psychol [Anim Behav] 14:131–139

D'Amato MR, Colombo M (1989) Serial learning with wild card items by monkeys (*Cebus apella*): Implications for knowledge of ordinal position. J Comp Psychol 103:252–261

D'Amato MR, Colombo M (1990) The symbolic distance effect in monkeys (*Cebus apella*). Anim Learn Behav 18:133–140

Davis H (1984) Discrimination of the number three by a raccoon (*Procyon lotor*). Anim Learn Behav 12:409–413

Davis H, Bradford SA (1986) Counting behavior by rats in a simulated natural environment. Ethology 73:265–280

Davis H, Pérusse R (1988) Numerical competence in animals: Definitional issues, current evidence and a new research agenda. Behav Brain Sci 11:561–615

Dehaene S (1992) Varieties of numerical abilities. Cognition 44:1–42

Gelman CR, Gallistel CR (1978) The child's understanding of number. Harvard University Press, Cambridge, MA

Kawai N, Matsuzawa T (2000) Numerical memory span in a chimpanzee. Nature 403:39–40

MacDonald SE (1994) Gorillas' (*Gorilla gorilla gorilla*) spatial memory in a foraging task. J Comp Psychol 108:107–113

Mandler G, Shebo BJ (1982) Subitizing: An analysis of its component processes. J Exp Psychol [Gen] 111:1–22

Matsuzawa T (1985) Use of numbers by a chimpanzee. Nature 315:57–59

Matsuzawa T, Asano T, Kubota K, Murofushi K (1986) Acquisition and generalization of numerical labeling by a chimpanzee. In: Taub DM, King FA (Eds) Current perspectives in primate social dynamics. Van Nostrand Reinhold, New York, pp 416–430

Matsuzawa T, Itakura S, Tomonaga M (1991) Use of numbers by a chimpanzee: A further study. In: Ehara A, Kimura T, Takenaka O, Iwamoto M (Eds) Primatology today. Elsevier, Amsterdam, pp 317–320

Miller GA (1956) The magical number seven plus or minus two: Some limits on our capacity for processing information. Psychol Rev 63:81–97

Murofushi K (1997) Numerical matching behavior by a chimpanzee (*Pan troglodytes*): Subitizing and analogue magnitude estimation. Jpn Psychol Res 39:140–153

Ohshiba N (1997) Memorization of serial items by Japanese monkeys, a chimpanzee, and humans. Jpn Psychol Res 39:236–252

Olthof A, Iden CM, Roberts WA (1997) Judgements of ordinality and summation of number symbols by squirrel monkeys (*Saimiri sciureus*). J Exp Psychol [Anim Behav] 23:325–339

Pepperberg IM (1987) Evidence for conceptual quantitative abilities in the African Grey parrot: Labeling of cardinal sets. Ethology 75:37–61

Roberts W, Phelps M (1994) Transitive inference in rats: a test of the spatial coding hypothesis. Psychol Sci 5:368–374

Rumbaugh DM, Savage-Rumbaugh ES, Hegel M (1987) Summation in a chimpanzee (*Pan troglodytes*). J Exp Psychol [Anim Behav] 13:107–115

Rumbaugh DM, Hopkins WD, Washburn DA, Savage-Rumbaugh ES (1989) Lana chimpanzee learns to count by "Numath": a summary of videotaped experimental report. Psychol Rec 39:459–470

Spelke ES, Vishton P, von Hofsten C (1995) Object perception, object directed action, and physical knowledge in infancy. In: Gazzaniga M (Ed) The cognitive neurosciences MIT Press, Cambridge, MA, pp 165–179

Sternberg S (1969) Memory-scanning: Mental processes revealed by reaction-time experiments. Am Scientist 57:421–457

Straub R, Terrace H (1981). Generalization of serial learning in pigeons. Anim Learn Behav 9:454–468

Swartz K, Chen S, Terrace H (1991) Serial learning by rhesus monkeys I: acquisition and retention of multiple four-item lists. Anim Behav Proc 17:396–410

Terrace HS, McGonigle B (1994) Memory and representation of serial order by children, monkeys, and pigeons. Curr Dir Psychol Sci 3:180–185

Terrell DF, Thomas RK (1990) Number-related discrimination and summation by squirrel monkeys (*Saimiri sciureus* and *S. boliviensus*) on the basis of the number of sides of polygons. J Comp Psychol 104:238–247

Thomas RK, Fowlkes D, Vickery JD (1980) Conceptual numerousness judgements by squirrel monkeys. Am J Psychol 93:247–257

Tomonaga M, Matsuzawa T (2000) Sequential responding to Arabic numerals with wild cards by the chimpanzee (*Pan troglodytes*). Anim Cogn 3:1–11

Tomonaga M, Matsuzawa T, Itakura S (1993) Teaching ordinals to a cardinal-trained chimpanzee (in Japanese with English summary). Primate Res 9:67–77

Washburn DA, Rumbaugh DM (1991) Ordinal judgments of numerical symbols by macaques (*Macaca mulatta*). Psychol Sci 2:190–193

Woodruff G, Premack D (1981) Primitive mathematical concepts in the chimpanzee: proportionality and numerosity. Nature 293:568–570

Wynn K (1992) Addition and subtraction by human infants. Nature 358:749–750

Wynne CDL (1995) Reinforcement accounts for transitive inference performance. Anim Learn Behav 23:207–217

11
Reproductive Memory Processes in Chimpanzees: Homologous Approaches to Research on Human Working Memory

Nobuyuki Kawai and Tetsuro Matsuzawa

1 Introduction

Memory is extremely important to our everyday activities. We cannot recognize, plan, or decide even a trivial issue unless we can hold information in mind for periods of time and retrieve it from our memory. This is also the case in nonhuman animals. Thus, it is not surprising that research on animal memory has a long history (Hunter 1913; Tinklepaugh 1928), as well as research on human memory (Ebbinghaus 1885).

Memory is usually divided into "long-term memory," which can retain information for months or even years, and "short-term memory" or "working memory," which can hold information for only a short period of time. Cognitive scientists have mainly concentrated on the latter, which has a flexible capacity and can store and manipulate information within certain limits which have yet to be assessed. Comparative cognitive scientists have also explored working memory in animals.

Many studies on working memory have shown similarities between humans and nonhuman animals (Wright et al. 1985). By using delays of different durations in the delayed matching to sample (DMTS) procedure, D'Amato (1973) investigated how long three capuchin monkeys could hold information about the sample stimulus. It was found that they were able to perform well above the chance level even with 60 s delay, but that this retaining ability showed a marked decrease in accuracy as the length of the delay increased.

Other similarities between humans and nonhuman animals were also found in the more complex mechanisms of working memory. For instance, a series of studies by Wright (e.g., Wright 1999) revealed that animals also demonstrated the dual interference effects in a list memory task, the "primacy effect" and the "recency effect," which are known to affect the human list memory. The primacy effect refers to a better memory for the first items of a list of stimuli, whereas the recency effect refers to a better memory for the last items of a list. Therefore, list memory in humans is characterized as a U-shaped function with poor performance in the middle of a list. Such U-shaped functions (i.e., serial position effect) are demon-

Department of Behavior Brain Science, Primate Research Institute, Kyoto University, 41 Kanrin, Inuyama, Aichi 484-8506, Japan

strated by using serial probe recognition tasks with pigeons, capuchin monkeys, rhesus monkeys, and humans (Wright 1999; Wright et al. 1985). The other mechanisms which are equivalent to human working memory, such as rehearsal (Grant 1984), prospective coding (Roitblat 1980), retrospective coding (Urcuioli and Zentall 1986), and directed forgetting (Maki and Hegvik 1980), were also demonstrated. These studies suggest that information processes in working memory are very similar between humans and nonhuman animals.

2 Another Approach to Examining Working Memory in Nonhuman Animals

Research with pigeons and monkeys has substantially increased our understanding of animal working memory, which has many similarities with that of humans. It is noticeable, however, that there have been few studies of working memory in chimpanzees, despite the fact that the cognitive abilities of chimpanzees have been intensively examined in areas such as visual perception (e.g., Chapter 3 by Tomonaga, this volume), numerical competence (e.g., Matsuzawa 1985; Woodruf and Premack 1981), analogies (Gillan et al. 1981), and linguistic abilities (Asano et al. 1982; Gardner and Gardner 1969).

Although working memory is essential for such "higher" cognitive processes, there are only two studies of chimpanzees' working memory (Kawai and Matsuzawa 2000b; Fujita and Matsuzawa 1990), an outline of which will be described in the following sections. These two research projects are unique in that a chimpanzee was required to retrieve all the items or components which had to be memorized during a trial. In other words, the chimpanzee was investigated for reproductive memory processes, not for recognition.

Apart from research into long-term memory (e.g., serial learning), most previous studies on memory in nonhuman animals have analyzed only recognition memory processes. Nevertheless, most studies on human memory have examined reproductive memory processes with verbal reports as a response. Bahrick (1970) suggested that the reproduction of memorized items is more difficult than recognizing a reproduction of them, as the reproduction of items requires both recognition and production of the items, while recognition of the items requires only the former. This suggests that reproduction and recognition may require different ways of information processing in working memory.

To understand the underlying processes of a primate's working memory in more detail, it will be necessary to devise new tasks in a fresh approach to investigating reproductive memory processes in nonhuman primates. The following sections describe a method of investigating a chimpanzee's working memory using a reproductive procedure.

There are two conventional ways to assess human working memory. These test how long the subject can retain memorized information, and how much information the subject can store at one time. Both of these approaches have been applied to testing the linguistic and numerical competence of a chimpanzee.

3 Delay Interval

3.1 Constructive Matching to Sample (MTS)

Although chimpanzees cannot generate a verbal report as a response on reproductive memory processes, a language-trained chimpanzee can represent "words" by using symbols such as geometric figures (i.e., a lexigram), Arabic numerals, or letters of the alphabet. A female chimpanzee, Ai, had learned a number of language-like skills with lexigrams as symbols (Asano et al. 1982).

Ai showed an ability to construct the "words" of a lexigram from nine fundamental elements called design elements (graphemes) (Matsuzawa 1989). The task, called "constructive matching-to-sample," was to create copies of samples from these elements. The samples were any possible two- or three-element compounds of the nine elements (Fig. 1).

Sample stimuli were presented on a CRT monitor. Ai had to select the elements from the keyboard attached on the left of the CRT. A correct choice led to the appearance of the selected element on the CRT. Further correct choices composed figures on the CRT made of combined elements. A completed copy of the sample led to the delivery of a piece of food. Any mistaken choice led to an immediate suspension of the trial and imposed a 5-s time-out interval as a light penalty.

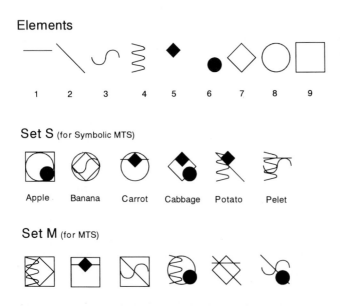

Fig. 1. Stimuli used in Fujita and Matsuzawa (1990) for matching to sample (MTS). *Top*, The nine figure elements. *Middle*, The six three-element figures used in naming training for the chimpanzee in Experiment 1 of Fujita and Matsuzawa (1990). *Bottom*, Six three-element figures used in matching training for the chimpanzee in Experiment 2 of Fujita and Matsuzawa (1990)

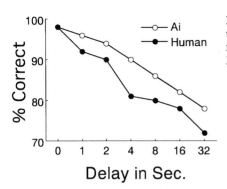

Fig. 2. Accuracy of delayed figure reconstruction by a chimpanzee (Ai) and by humans as a function of retention interval (in Fujita and Matsuzawa 1990)

3.2 Delayed Constructive Matching to Sample

Fujita and Matsuzawa (1990) imposed intervals of 0, 1, 2, 4, 8, 16, and 32 s between the presentation of a sample and the illumination of the keyboard in the constructive MTS task. In order to construct the samples correctly, Ai had to remember them during these intervals. The samples were all the possible three-element compounds from the nine elements shown at the top of Fig. 1: a total of 84 samples.

In Ai's previous training history, four of these lexigrams had represented certain objects or colors. Others were new to her and had no symbolic meaning.

For the purposes of comparison, four human subjects also participated in this experiment with the same apparatus and the same stimuli. They were required to add or subtract two-digit numbers on a hand-held computer during delay periods, in order to prevent verbal rehearsal.

Figure 2 shows the accuracy of the reconstructions after each delay interval by Ai, and the average of the accuracy of humans. Both species showed a decrease in accuracy as a function of delay duration. Ai reconstructed the lexigrams with an accuracy as high as 80% correct for a 32-s delay, which was much higher than the chance level (1.2%). Her ability to reconstruct lexigrams was slightly better than that of humans.

3.3 Semantic vs. Nonsemantic Items

In human memory processing, meaningful items are memorized better than meaningless ones (Craik and Tulving 1975). Ai learned to construct lexigrams associated with actual objects and colors more quickly than meaningless ones (Matsuzawa 1989), suggesting that any "verbal rehearsal" might improve her retention. Fujita and Matsuzawa (1990), in their Experiment 2, tested whether such meaningfulness would help in reconstructing the lexigrams after the delays.

Two sets of lexigrams were used as samples. Six lexigrams represented foods, such as an apple, a banana, etc. (Fig. 1, Set S). The other six figures did not symbolize anything (Fig. 1, Set M). The accuracy of reconstruction decreased as a function of delay duration in both species, as it did in their Experiment 1. However, only humans showed a better memory for meaningful items than for meaningless ones, whereas Ai did not show any difference in reconstruction ability between the two groups of samples. Currently, it is uncertain what conditions can serve as a "verbal rehearsal" in linguistically competent chimpanzees. Further research will be required to examine the effects of "verbal rehearsal" upon the working memory of chimpanzees.

4 Memory Span

4.1 Numerical Ordering and a Switch Trial Experiment

Biro and Matsuzawa (1999) reported indirect evidence that a chimpanzee could memorize three items at once. In their research, a single chimpanzee, Ai, participated in the experiment. Before describing their study, we will briefly summarize Ai's numerical competence.

The chimpanzee Ai has learned to use Arabic numerals to represent numbers (Matsuzawa 1985). She can correctly label 0 through 9 items shown to her by touching the corresponding numeral on a touch-sensitive monitor (Murofushi 1997). She can also choose the numerals 0 through 9 in succession, and arrange them in ascending order (Tomonaga et al. 1993).

Utilizing Ai's abilities, Biro and Matsuzawa (1999) conducted a unique experiment to study her perceptual and cognitive mechanisms in solving a numerical ordering task. In their experiment, a random set of three numerals, ranging from 0 to 9, was presented (e.g., 0, 5, 9), spatially distributed on a touch-sensitive monitor. Ai was required to choose the numerals in ascending order. Occasionally, probe trials were inserted during this task. In probe trials, immediately after the correct selection of the lowest numeral in a stimulus set ("0" in the above example), the positions of the remaining two on-screen numerals were exchanged by the computer. If Ai selected the second number incorrectly (i.e., the position now occupied by the highest number), the screen was cleared automatically. Her accuracy dropped to 45% in these "switch" trials, in contrast to 95% correct in the normal background trials (see also Chapter 10 by Biro and Matsuzawa, this volume). These results suggest that Ai had already memorized all the numerals and their positions before the initiation of the first response. If this is the case, another question arises: how many numerals can she memorize at once ?

4.2 Masking Experiment

As explained above, Ai had learned the relations between ordinal numbers. When a sequence of numbers was presented on a touch-sensitive monitor, and despite inter–integer differences among the numbers, she could touch each number from the lowest to the highest and with remarkable speed and accuracy. Taking advantage of this ability, Kawai and Matsuzawa (2000b) set up a memory span task.

A set of numbers was presented on a CRT: say, 1, 3, 4, 6, and 9. Immediately after

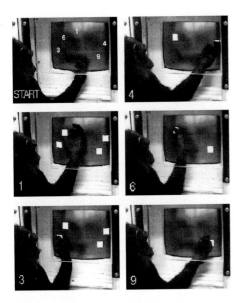

Fig. 3. The chimpanzee Ai performing the five-number ordering task in the "masking" trial. Five numbers (1, 3, 4, 6, and 9) were presented on the touch-sensitive monitor. Immediately after Ai had correctly chosen the lowest number (1), the remaining numbers were automatically masked. Ai continued to identify the numbers one by one in ascending order, ending with 9

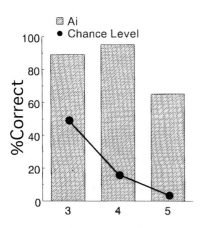

Fig. 4. Accuracy and chance level of masking trials by the chimpanzee Ai (in Kawai and Matsuzawa 2000b)

Ai had selected the lowest number in the sequence (i.e., 1), all the remaining numbers were masked by white squares (Fig. 3). In order to score correctly in a trial, Ai had to memorize all the numbers as well as their respective positions before making the first response. Ai was tested with three different set sizes, the 3-item, 4-item, and 5-item conditions, of which the chance levels were 50%, 17%, and 4%, respectively.

As shown in Fig. 4, Ai reached high levels of accuracy in these masking trials, with over 90% correct in a set size of four, and about 65% correct in a set size of five. Thus, her performance in each case was significantly above the chance level. This is comparable to human preschool children.

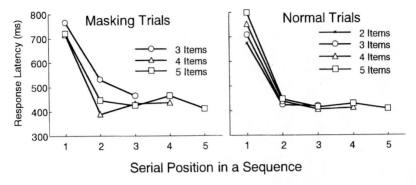

Fig. 5. Response latency in selecting the first, second, third, fourth, and fifth numerals in masking trials (*left*) and normal trials (*right*) (in Kawai 2000)

The reaction time of her response was of considerable interest. Independently of set size, Ai took the longest reaction time to select the lowest number, with her reaction time remaining constant for all subsequent responses. For instance, her mean reaction time for the first response to a set size of five was 721 ms, with 446 ms, 426 ms, 466 ms, and 411 ms for the last four masked responses, respectively. These reaction times were the same as in normal trials (Fig. 5). This suggests that Ai first explored the number space, calculating the ordinal relations and spatial locations of each number, and then used this stored information to give the lead correct responses (Kawai in manuscript).

An interesting episode worthy of note occurred during one of the testing sessions. After the correct selection of the lowest number, all the remaining numbers were masked by white squares. Right at that moment, a fight broke out among a group of other chimpanzees outside, accompanied by loud screaming by one chimpanzee. Ai stopped her responses and paid attention to the fight while looking away from the CRT for about 20 s. After she returned to the CRT, she selected the remaining four masked stimuli without error.

5 Conclusions

The chimpanzee Ai could hold information about a sample stimulus, and reconstruct lexigrams even with a 32-s retention interval, if the sample stimulus consisted of three fundamental elements (Fujita and Matsuzawa 1990). Ai could also memorize a random set of five numbers (Kawai and Matsuzawa 2000b). Our recent study revealed that Ai could memorize even six numbers (Kawai and Matsuzawa 2001). Taken together, Ai could memorize at least five items at one time, and retain the memory for some period of time, whatever the stimulus type (number or figure).

In the literature on human cognition, the capacity of working memory, in which information can be processed over a period of time, has been investigated in various ways such as digit span, pattern recall, listening span, and so on (see Gathercole 1999 for a summary). Adult humans can memorize strings of codes such as phone numbers if they consist of up to seven items. Beyond seven items, humans have great difficulty in memorizing them. This effect has been called "the magical number seven" (Miller 1956), and represents a limit imposed on the number of items that can be handled simultaneously by the brain. This rule has been believed and has influenced human cognition research for over 40 years (Baddeley 1994). If we follow this idea, we can talk about "the magical number five in a chimpanzee."

Recently, however, "the magical number" in humans has caused controversy, and now it is being reconsidered. Luck and Vogel (1997) suggested that the capacity of the visual working memory in humans might be four items. Cowan (2001) reviewed the previous research on the capacity of human working memory and concluded that "the magical number" in humans should be four. If this is the case, the chimpanzee's memory span exceeds that of humans.

We clearly need more information about memory processes in chimpanzees as well as in other primates. Ai is the only chimpanzee to have had her working memory assessed in relation to that of humans by means of a homologous procedure. Other chimpanzees should be examined in the same way, and Ai should also be tested in a variety of other ways which can be used to compare her working memory with that of humans (Hauser 2000; Kawai and Matsuzawa 2000a; 2001).

Working memory is essential to almost all the cognitive processes of both humans and nonhuman animals. Without it, we cannot engage in any daily activities such as calculations, constructing a sentence, or planning any sequential event. Ai's performance tells us that a chimpanzee has a memory capacity that is comparable to that of humans. This suggests that the cognitive gap between humans and chimpanzees may be smaller than has been thought. Further research on chimpanzee cognition promises to increase our understanding of the origins of human intelligence.

References

Asano T, Kojima T, Matsuzawa T, Kubota K, Murofushi K (1982) Object and color naming in chimpanzees (Pan troglodytes). Proc Jpn Acad 58(B):118–122

Baddeley A (1994) The magical number seven: still magic after all these years? Psychol Rev 101:353–356

Bahrick HP (1970) A two-phase model for prompted recall. Psychol Rev 77:215–222

Biro D, Matsuzawa T (1999) Numerical ordering in a chimpanzee (Pan troglodytes): planning, executing, and monitoring. J Comp Psychol 113:178–185

Cowan N (2001) The magical number 4 in short-term memory: a reconsideration of mental storage capacity. Behav Brain Sci 24:87–185

Craik FIM, Tulving E (1975) Depth of processing and the retention of words in episodic memory. J Exp Psychol Gen 104:268–294

D'Amato MR (1973) Delayed matching and short-term memory in monkeys. In:Bower GH (ed) The psychology of learning and motivation, vol 7. Academic Press, New York

Ebbinghaus H (1885) Memory. Duncker, Leipzig

Fujita K, Matsuzawa T (1990) Delayed figure reconstruction by a chimpanzee (*Pan troglo-dytes*) and humans (*Homo sapiens*). J Comp Psychol 104:345–351

Gardner RA, Gardner BT (1969) Teaching sign language to a young chimpanzee. Science 165:664–672

Gathercole SE (1999) Cognitive approaches to the development of short-term memory. Trends Cogn Sci 3:410–419

Gillan DJ, Premack D, Woodruff G (1981) Reasoning in the chimpanzee. I. Analogical reasoning. J Exp Psychol Anim Behav Process 7:1–17

Grant DS (1984) Rehearsal in pigeon short-term memory. In: Roitblat HL, Bever TL, Terrace HS (eds) Animal behavior. Erlbaum, Hillsdale

Hauser MD (2000) Homologies for numerical memory span? Trends Cogn Sci 4:127–128

Hunter WS (1913) The delayed reaction in animals and children. Behavior Monographs, 2, No. 1

Kawai N (2000) Relationship between transitive inference and comprehension of ordinality by chimpanzees (original in Japanese with English summary). Cognitive Studies: Bulletin of the Japanese Cognitive Science Society 7:202–209

Kawai N, Matsuzawa T (2000a) A conventional approach to chimpanzee cognition. Response to M. D. Hauser. Trends Cogn Sci 4:128–129

Kawai N, Matsuzawa T (2000b) Numerical memory span in a chimpanzee. Nature 403:39–40

Kawai N, Matsuzawa (2001) "Magical number 5" in a chimpanzee. Behav Brain Sci 24 (in press)

Luck SJ, Vogel EK (1997) The capacity of visual working memory for features and conjunctions. Nature 390:279–281

Maki WS, Hegvik DK (1980) Directed forgetting in pigeons. Anim Learn Behav 8:567–574

Matsuzawa T (1985). Use of numbers by a chimpanzee. Nature 315:57–59

Matsuzawa T (1989) Spontaneous pattern construction in a chimpanzee. In:Heltne PG, Marquardt LA (eds) Understanding chimpanzees. Harvard University Press, Cambridge, pp 252–265

Miller GA (1956) The magical number seven plus or minus two: some limits on our capacity for processing information. Psychol Rev 101:343–352

Murofushi K (1997) Numerical matching behavior by a chimpanzee (*Pan troglodytes*): subitizing and analogue magnitude estimation. Jpn Psychol Res 39:140–153

Roitblat HL (1980) Codes and coding processes in pigeon short-term memory. Anim Learn Behav 8:341–351

Tinklepaugh OL (1928) An experimental study of representative factors in monkeys. J Comp Psychol 8:197–236

Tomonaga M, Matsuzawa T, Itakura S (1993) Teaching ordinals to a cardinal-trained chimpanzee (original in Japanese with English summary). Primate Res 9:67–77

Urcuioli PJ Zentall TR (1986) Retrospective coding in pigeons' delayed matching-to-sample. J Exp Psychol Anim Behav Process 12:69–77

Woodruff G, Premack D (1981) Primitive mathematical concepts in the chimpanzee: proportionality and numerosity. Nature 293:568–570

Wright AA (1999) Visual list memory in capuchin monkeys (*Cebus apella*). J Comp Psychol 113:74–80

Wright AA, Santiago HC, Sands SF, Kendrick DF, Cook RG (1985) Memory processing of serial lists by pigeons, monkeys, and people. Science 229:287–289

12
Establishing Line Tracing on a Touch Monitor as a Basic Drawing Skill in Chimpanzees (*Pan troglodytes*)

Iver H. Iversen[1] and Tetsuro Matsuzawa[2]

1 Introduction

1.1 Written Communication

Using written symbols is essential for human communication. Effective written communication requires that each subject can read and write or produce the visual symbols. Therefore, considerable worldwide educational efforts are expended to teach children symbols such as letters, signs, and numbers. Several studies of human–animal communication have established that nonhuman subjects can also be taught to discriminate visual symbols presented by humans. Of particular interest is the fact that several studies have shown that chimpanzees can learn to discriminate complex lexigrams (Rumbaugh 1977; Savage-Rumbaugh 1986; Tomonaga and Matsuzawa 1992) by pointing to them or moving them about (Premack 1976) in an appropriate manner. Chimpanzees have also been trained to compose symbols from their elements (Fujita and Matsuzawa 1990) and to produce signs with their fingers (Gardner and Gardner 1978).

1.2 Drawing in Apes

While the ability to write symbols is prevalent among humans, the production of symbols by actually writing them by hand apparently has not been attempted with nonhuman animals. In a few studies, apes and monkeys have been shown to be able to scribble with paint material. For example, when Kluver (1933) presented a piece of chalk to cebus monkeys, they would spontaneously use it to scribble on the floor of their cage. Other studies with chimpanzees have established scribbling on paper with crayons or paintbrushes (Boysen et al. 1987; Morris 1962; Schiller 1951; Smith 1973). While the scribbles may on occasion be directed toward stimuli presented on the paper, one cannot predict what each stroke will look like, when it will occur, which direction it will take, or where it will be located.

[1] Department of Psychology, University of North Florida, Jacksonville, FL 32224, USA
[2] Primate Research Institute, Kyoto University, 41 Kanrin, Inuyama, Aichi 484-8506, Japan

1.2.1 Anecdotes of Drawing and Writing in Apes

In a short letter to *Nature* in 1942, the famed British zoologist Julian S. Huxley related an incident involving a young captive mountain gorilla named Meng. Huxley had observed one day that when Meng stood in a certain position he cast a shadow on a white wall in his cage. "Seeing his shadow before him at one moment, he stopped, looked at it, and proceeded to trace its outline with his forefinger" (Huxley 1942, p. 637). When Huxley later tried to reproduce this behavior by placing a lamp so that Meng cast a shadow on the wall, "the gorilla refused to take any interest, and was never seen to repeat the original performance" (p. 637). Huxley hypothesized that the origin of human graphic art might be found in simple tracing of shadows cast by a low sun against a cave wall. Other anecdotes of alleged structured ape drawing appear briefly in Garner (1900), Witmer (1909), Hoyt (1941), and Gardner and Gardner (1978). The literature on ape drawing was reviewed by Davis (1986) and Morris (1962) (see also Mitchell 1999).

1.2.2 Reliability of Ape Drawing

These anecdotes provide interesting evidence of the possibilities for ape writing. Yet they also highlight an important problem in examining such behavior. When the behavior cannot be reproduced regularly under similar conditions, one cannot possibly study it in a scientific manner. Obviously, an important element in human education in drawing and writing is repeated production of a given symbol under appropriate instruction, such as "write an A" or "draw a circle." Thus, even before attempting a scientific study of written communication in apes, one must first establish whether they can in fact produce and reproduce even the simplest form of drawing, such as a straight line that connects two points.

1.2.3 Previous Training of Drawing in Nonhuman Primates

Morris (1962), who studied scribbling in several chimpanzees, also attempted to train one subject to draw. In the words of Morris: " … a chimpanzee was once subjected to bribery with a food reward to encourage it to draw more intensely. The outcome of this experiment was most revealing. The ape quickly learned to associate drawing with getting the reward but as soon as this condition had been established the animal took less and less interest in the lines it was drawing. Any old scribble would do and then it would immediately hold out its hand for the reward. The careful attention the animal had paid previously to design, rhythm, balance and composition was gone and the worst kind of commercial art was born!" (pp. 158–159). This apparent inability of a chimpanzee to learn to draw on command was later highlighted by Gardner and Gardner (1988) and Litt (1973) as an illustration of the alleged general failure of operant conditioning to capture and maintain or generate such spontaneous and complex behavior (see Iversen 1988 for a comment on this argument). Brewster and Siegel (1976) attempted to train two *Macaca mulatta* to draw and reported that drawing with crayons could be selected and maintained with intermittent reinforcement. However, the authors reported that the monkeys often made drawing movements without looking at the crayon or tried to "draw" even without the crayon. To study the neuronal components of motor learning, Georgopoulos (1990) trained rhesus monkeys to move a handle to a target in a planar surface; the resulting behavior was said to resemble drawing movements although it produced no visual feedback. Schwartz (1994) similarly demonstrated that rhesus monkeys could be trained to track with a finger an object that moved in a spiral on a touch monitor. This tracking behavior was called drawing although the response did not produce visual feedback. Thus, the literature indicates that continuous hand movement on a flat surface can be trained in nonhuman subjects. However, the reliable production of signs directly by drawing apparently has not been attempted previously in apes or monkeys.

1.2.4 Can Apes Learn to Draw Reliably?

The purpose of our experiments, which began in 1992, was to determine whether captive chimpanzees can be taught to draw in a controlled manner. We used a fully automated, computer-controlled system with precise recording of the location of drawing. The subject faces a touch monitor. When the subject touches the screen, the location of the touch is recorded as a pair of x, y coordinates. A graphical symbol, such as a small filled circle, is presented immediately at the touched location. As the finger moves over the screen, it leaves a trace of computer-generated electronic ink that provides immediate visual feedback of the movement.

1.3 Trial and Error Versus Shaping Methods in Training of Drawing

When the desire is to modify behavior, one must distinguish between two fundamentally different methods; trial-and-error learning and shaping.

1.3.1 Trial-and-Error Learning

The trainer waits for the appropriate target behavior and then reinforces it. Applying this method with many species and response forms, Thorndike (1911) originally formulated the well-known "law of effect," which states that any such reinforced behavior will increase in strength. Trial-and-error learning is the method of choice when one wants to maintain or strengthen an already existing behavior. Thus, with trial-and-error learning, the experimenter selects behavior but does not generate it. One problem encountered with this method is that reinforcement may not strengthen the entire movement the experimenter selects, especially if it lasts for several seconds and consists of many components. Instead, the selected behavior may break down and be replaced with only one of the components. Experiments with reinforcement of grooming have illustrated this issue. Reinforcement of grooming may result in a breakdown of the behavior (e.g., Shettleworth 1975) unless the entire bout of grooming is reinforced (Iversen et al. 1984). It is likely that something similar happened when Morris (1962) tried to reinforce drawing and observed that drawing deteriorated to brief perfunctory scribbles (i.e., Sect. 1.2.3). An additional problem encountered with the trial-and-error method is that one must wait for the existing behavior before giving reinforcement. If the selected behavior is infrequent, then reinforcement becomes infrequent. With animal subjects, infrequent reinforcement reduces general activity and thus lowers the frequency of the target behavior.

1.3.2 Shaping (Operant Conditioning)

The method of shaping by successive approximation is a powerful tool that can be used to generate novel (previously nonexistent or rare) behavior. Skinner (1953, 1971, 1974) repeatedly emphasized the usefulness of shaping (see also Gleeson 1991; Iversen 1992). Shaping can be used to develop a variety of new skills in human subjects (e.g., Cooper et al. 1987; Martin and Pear 1996). The method of shaping changes behavior along a continuum from existing to novel forms of behavior. By selectively reinforcing existing forms of behavior, which deviate only slightly from the norm but in the direction of the target novel form, the behavioral repertoire will change with the emergence of novel responses. With explicit and careful application of the shaping method, entirely new forms of behavior, which never occurred before, can be generated in relatively short time periods (Skinner 1953). Shaping may be preferred over the trial-and-error method in a systematic study of drawing because frequent reinforcement will maintain a high level of activity, and because one can explicitly construct the target behavior, piece by piece. The method of shaping also allows one to "tag" the behavior to specific stimuli while it is being generated, so that the behavior will later occur when the tags are presented and not at other times.

1.4 Perceptual and Motor Components of Drawing

1.4.1 Seeing and Touching

Before one begins to train drawing, it is worth pausing to examine exactly what it is one wants the subject to be able to do. Consider the example of drawing a line between two dots with finger-painting on a touch monitor. The perceptual component consists of seeing the two dots and orienting to the dot where drawing is to begin (hereafter, startdot). The motor component consists of lifting the arm and aiming the finger at, and eventually touching, the startdot. This touch produces perceptual visual feedback in the form of ink on the drawing surface. Next, the subject must locate the other dot (hereafter, stopdot) and move the finger over the screen without lifting it to this location. This movement produces instantaneous visual feedback in the form of a trace of ink that forms on the drawing surface exactly when and where the finger moves. When the finger reaches the stopdot, the subject must lift the finger, and the visual feedback of the movement will stop at the same time. It is important to notice this last element of drawing. If the finger lifts before it reaches the stopdot, the trace will become too short; similarly, if the finger continues beyond the stopdot, the trace will become too long. Thus, aiming the finger toward the startdot and lifting it when it reaches the stopdot are essential components of drawing. In the vocabulary of stimulus control, one must make the startdot a stimulus for touching it without immediately releasing the finger and the stopdot a stimulus for moving the finger toward it and also a stimulus for withdrawing the finger when it is reached.

1.4.2 Trial-Termination Response

An additional aspect of drawing, which is not fully recognized, is that drawing must be terminated by the subject and not by the experimenter. For example, if the experimenter presented reinforcement when the finger reached the stopdot, the subject would probably learn to aim at the startdot and drag the finger (without lifting it) over the screen toward the stopdot until the movement produced reinforcement. Thus, the sound from reinforcement delivery would be the stimulus that made drawing stop; the subject would probably not learn to use the visual stopdot on the screen as a signal for stopping. A little reflection reveals that without control of where to stop, human drawing and writing would in fact be almost meaningless. That stopping the movement and lifting the finger (or pen) are customarily unheeded aspects of drawing has been illustrated to us several times in questions from the audience at scientific meetings where we have presented material on our project. Thus, a common question has been why we did not use a joystick for our experiments on drawing. This question rests on the assumption that because chimpanzees and other primates are very adept at using a joystick to select stimuli (e.g., Hopkins et al. 1993; Rumbaugh et al. 1993; Washburn et al. 1994), they might also learn to draw using a joystick. Indeed, the cursor could be made to leave a trace so as to "draw" on the screen in accordance with joystick movement. However, if the joystick method were to be used for drawing, it would not afford the subject an opportunity to select where to start and where to finish drawing because any movement of the stick would produce ink on the screen. Hence, a clearly identifiable start and end of each stroke, which is the hallmark of any drawing or writing, would be lost entirely with the joystick method. This method was therefore not suitable for our project. If the method were to be changed such that moving the stick produced ink on the monitor only when one pressed a switch on top of the stick, as seen on some joysticks used with video games, then it could of course be used for drawing. However, teaching a nonhuman subject to use such a switch along with stick movement has apparently not been tried before. We considered such a response combination unnecessarily complex compared with the simpler idea of using a touch monitor on which drawing begins when and where the finger touches the screen and ends when and where the subject lifts the finger.

1.4.3 Controlled Drawing

Evidence of controlled drawing is obtained when one can predict in advance, trial by trial, what the subject will draw. A schematic example of controlled drawing is illustrated in Fig. 1. At the onset of a trial, two dots are presented in one of four orientations (A–D). The subject places the finger on the startdot and moves it in one sweep to the stopdot while producing ink on the screen. The subject stops

Start   Aim   Sweep   Lift   End

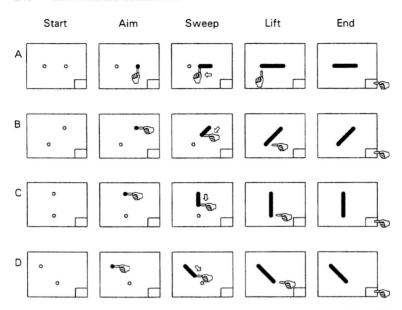

Fig. 1. Schematic diagram of drawing a line that connects two dots. *"Start"* shows what the screen looks like at the beginning of the trial. The subject aims at one dot (startdot) and sweeps a finger across the screen toward the other dot (stopdot) producing electronic ink on the screen, lifts the finger when it reaches the stopdot, and ends the trial by pressing the trial-termination key. The guide dot pattern varied in each trial among four types (*A–D*). (From Iversen and Matsuzawa 1996, with permission)

drawing by lifting the finger when it reaches the stopdot. Appropriate drawing would be evident if the subject indeed draws a line between two points. If the subject's drawing depends on the two-dot configuration presented at the start of a trial, then one will be able to predict, for each trial, what type of line the subject will draw based on the dot pattern. Thus, for two dots aligned vertically at a distance of 10 cm, one would predict a drawing of a vertical line about 10 cm-long; for two dots aligned horizontally at a similar distance, one would predict a drawing of a horizontal line of the same length, etc.

2 Research Goal and General Outline

The goal of the present research was to establish such a perceptual–motor pattern of controlled line drawing in the chimpanzee. If such controlled drawing could be established, an experimental basis would exist for subsequent teaching of writing to a nonhuman subject. This chapter outlines the main steps of training two chimpanzees to draw reliably on a touch monitor (Iversen and Matsuzawa 1996). We also present some examples from tests to illustrate how drawing is controlled by the stimuli on the screen. Particular emphasis is placed on the control of lifting the finger at the correct place. First, the subjects were trained to connect two dots at a fixed distance apart but at different angles. Second, we tested how the subjects would draw when the dots were presented at new distances apart. Would the subjects draw in the previous fixed distance or would they follow the dot pattern? Third, we attempted to teach the subjects to draw a copy of a model. A bar was presented on the screen and the subjects were to draw a line parallel to the bar. Last, the broader usefulness of the research method of moving a finger on a touch monitor is illustrated with additional examples from work in progress with the same subjects on tracing multiline figures, sorting objects into categories, and solving finger mazes.

2.1 General Method

The details of the apparatus used and the subjects have been described in Iversen and Matsuzawa (1996, 1997). Only the most important features are mentioned here.

2.1.1 Apparatus

Figure 2 shows a video print of one subject (Pendesa) sitting in front of a touch monitor while drawing a line that connects two points. The monitor was situated in an opening on one wall of an experimental booth. The surface of the monitor was 47 cm wide and 28 cm high. An opening under the monitor provided access to bits of seasonal fruit, peanuts, raisins, or candy delivered as reinforcement from a universal feeder. To provide an immediate indication of reinforcement delivery, chimes sounded at the onset of the delivery cycle. A computer recorded touches on the screen and controlled all stimuli on the screen as well as reinforcement deliv-

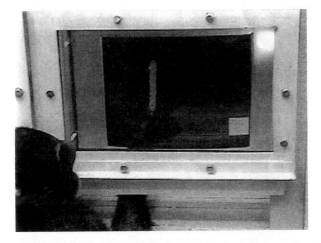

Fig. 2. Video print of the drawing behavior. The chimpanzee Pendesa connected the two dots by drawing on the screen in a sweeping motion. The finger movement generated instantaneous electronic ink in the form of fingertip-sized blue disks on the screen

ery. Stimulus sizes are expressed in pixels; for the monitor used, the conversion between pixels and metric units is 100:6 (i.e., a line 100 pixels long equals 6 cm).

2.1.2 Subjects

Two captive, adult, female chimpanzees (*Pan troglodytes*) served as subjects. Ai was 16 years old and Pendesa was 15 years old at the start of the experiment. Both subjects lived in an outdoor enclosure with a troop of six other chimpanzees. The chimpanzees were housed and cared for according to the *Guide for the Care and Use of Laboratory Primates* of the Primate Research Institute, Kyoto University. Subjects were neither food- nor water-deprived during the experiment. Subjects were usually trained for 4 or 5 days each week, with several consecutive sessions daily. The number of trials each session varied with the procedure and the subject's degree of cooperation. Commonly 100–200 reinforcements were presented on a training day.

2.1.3 Subjects' Previous Skills

Both subjects had considerable experience with pressing stimuli on touch monitors in a variety of tasks, such as matching to sample (e.g., Matsuzawa 1985a, b; Fujita and Matsuzawa 1990; Tomonaga and Matsuzawa 1992). Thus, it is important to note that the subjects brought specific skills to the experiment. First, they readily touched stimuli on the screen and collected the reinforcements. Second, they lifted their finger from the screen as soon as they touched a stimulus. The first skill of aiming at and touching a specific stimulus is compatible with the first elements of the drawing task of visually orienting to, aiming at, and then touching the startdot. However, the second skill of lifting the finger immediately after it touches the screen is directly incompatible with the drawing motion, which obviously entails continuous contact with the screen during movement of the finger over the screen. Thus, our intended training benefited from the already well-learned aim-and-touch behavior, but had to overcome the equally well engrained lifting of the finger from the screen immediately after a touch.

2.2 Overall Experimental Plan

The continuous motion of the finger over the screen was a nonexisting response in these subjects. We surmised that such a motion might be generated by gradually reducing the time spent lifting the finger between successive presses on several consecutive spatially displayed stimuli. We began by reinforcing existing pointing behavior. First, we reinforced pressing a small unfilled circle presented on the monitor; a press on the circle immediately filled the circle so as to produce visual feedback. Next, several nonoverlapping circles were presented in one of four linear patterns (vertical, horizontal, or two diagonals), and the subject had to press all circles to produce reinforcement. When pressing became efficient, we reduced the distance between circles to shorten the time spent lifting the finger between individual presses. We also introduced a directional pointing requirement to facilitate acquisition of continuous finger motion. The circles had to be pressed from top to bottom on trials with vertically or diagonally arranged circle patterns, or from right to left for horizontal patterns. To reduce finger lifting further, the circles were eventually made to overlap. When continuous motion emerged, the circles were faded from the screen in several steps and replaced by two dots that the subject had to connect by drawing a line between them.

2.2.1 Specific Procedural Details

Training progressed in 10 steps. The number of training sessions for each step depended on each subject's performance. Each session had 50–100 trials separated by 2-s intertrial intervals with a dark screen. Figure 3 shows a schematic of the training steps showing the visual stimuli presented on the screen. For illustrative purposes, the stimulus patterns are enlarged relative to the screen size. To save space, the stimuli are shown here only for trial types with a vertical orientation. The four stimulus configurations alternated randomly from trial to trial, as exemplified for Step 2 below. Steps 1–10 had 9, 4, 9, 2, 2, 4, 4, 6, 31, and 35 sessions, respectively, for Ai, and 7, 4, 6, 3, 7, 3, 4, 4, 27, and 35 sessions, respectively, for Pendesa.

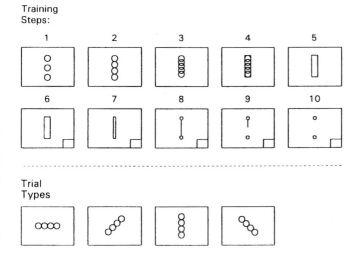

Fig. 3. Schematic diagrams showing the 10 steps in training the subjects to draw. Only the vertical stimulus display is shown; the stimulus display in each trial was one of four types, as shown below for training Step 2. Each frame exemplifies what the screen looks like at the start of the trial. The subject had to fill the circles in Steps *1–4* by pressing them. Steps *5–8* changed the stimulus display from a series of circles to a line with two small endpoints. In Step *9*, the line was gradually faded out, and in Step *10* only two dots appeared on the screen. In Steps *1–9*, ink could be produced only in the target area, but in Step *10* the subject could draw freely at any place on the screen. From Step *6*, a white rectangle appeared in the lower right-hand corner of the screen to enable the subject to make a trial-termination response after completion of the drawing. This response had no effect until drawing was completed. (From Iversen and Matsuzawa 1996, with permission)

After each subject had reliably pressed a single open circle, we presented several open circles on the screen in Step 1. All circles had to be pressed to produce reinforcement. Each open circle became a blue disk when pressed. Each repeated press on a blue disk produced a brief sound, but no penalties were incurred for this incorrect response. Reinforcement was delivered on each trial as soon as the last circle was filled. Note that an important contingency embedded in this procedure is the time from first screen contact to reinforcement. A trial takes longer to complete if, say, five circles have to be pressed individually as compared with sweeping the finger across the five circles without lifting it. Therefore, the time to reinforcement will shorten as repeated pressing diminishes and the movement eventually becomes a single sweep across the circles. We surmised that this small gain in time would facilitate the development of a sweeping motion. As pressing became very efficient, with short lifts between individual presses on circles, we moved the circles closer to each other in Step 2 (see Fig. 3). So, the circles first had gaps between them (Step 1), then they touched (Step 2), and then they overlapped in Step 3. This gradual change in stimuli was intended to facilitate development of continuous finger movement because of the reduced finger lifting from circle to circle.

We thought that a sweeping movement across circles might develop more rapidly if circles had to be pressed in a certain order as opposed to randomly. Therefore, once pressing had become reliable, we introduced a directional requirement in Step 2. We reasoned that movement from top to bottom would be appropriate for our task because the trace left by the finger would be visible to the subject, as opposed to moving the finger upwards where the hand would cover the trace. Therefore, in trials with a vertical or diagonal circle pattern, pressing was from top to bottom. On trials with a horizontal circle pattern, pressing was from right to left; we chose this direction because both chimpanzees appeared to press the outer right circle first on most trials with a horizontal display.

2.3 Development of a Sweeping Motion

To illustrate the development of a sweeping motion, Fig. 4 shows all the successive trials for Pendesa for sessions 11–16 (training Steps 2 and 3). The number of circles was gradually increased and the distance between circles was reduced, as indicated for each session. The time to complete each trial from first touch to last lift is given as a vertical line above the trial data. Up to session 13, Pendesa had pressed all the circles individually; that is, she pressed a circle and then lifted her finger to press the next circle, etc. A press followed by lifting the finger is indicated by a small disk inside each circle (on the screen, the disk covered the circle). In session 13 (Step 3), Pendesa began to connect successive circles without lifting her finger, as indicated by a diagonal bar from one circle to the next. Several trials with a full sweep emerged in session 14. In such trials, the time taken to complete the drawing was considerably reduced, as the vertical lines above each trial show. By session 16, Pendesa rapidly swept her finger over the stimulus display in each trial without lifting. For the remaining trial types, vertical sweeps began in session 15, and horizontal sweeps as well as sweeps along the other diagonal began in session 16. For Ai, the first

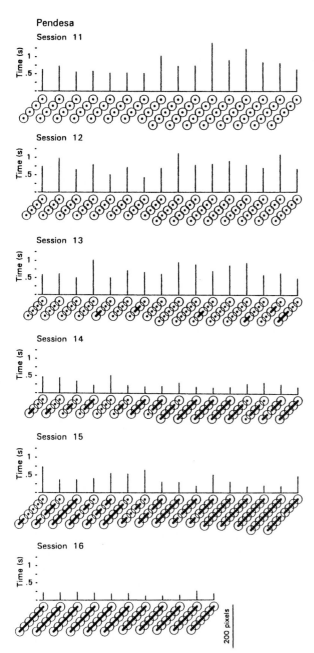

Pendesa

Fig. 4. Illustration of the development of a sweeping drawing motion by Pendesa for one trial type: the 45° diagonal. All trials of this type are shown for six successive sessions, covering training Steps 2 and 3. A *small dot inside a circle* represents a single press followed by a lift at the same location (when pressed, the circle on the screen became a blue disk). A *bar* connecting two or more successive circles represents a continuous movement from one circle to the next without lifting the finger; thus, a bar going through all circles indicates a continuous sweeping motion without lifting. Each trial also shows, as a *vertical line above the top circle*, the time taken to complete the drawing from the first touch to the last finger lift from the screen. (From Iversen and Matsuzawa 1996, with permission)

sweep occurred for the 45° diagonal in session 19, followed by vertical and horizontal sweeps in session 20 and the remaining diagonal in session 21. For both subjects, the trial completion time shortened considerably once sweeping occurred. Before sweeps had developed, the completion time was generally 0.5–1.5 s. Once sweeps had developed, the completion time was generally 0.15–0.5 s. Notice that although sweeping developed within a few sessions for all four stimulus patterns, sweeping did not immediately transfer from one circle pattern to another. Hence, there was no evidence of instantaneous, "insightful" learning (e.g., Kohler 1925). Thus, for both subjects, the behavior topography changed within a few sessions from repeated pressing and lifting to continuous sweeping once the subject made contact with the embedded contingency of finishing the trial sooner by moving their finger over the screen as opposed to repeated pressing and lifting.

2.4 Tracing and Trial Termination

Once the sweeping motion across the circles had developed, we gradually replaced the circles with a straight line (Steps 5–8, see Fig. 3) and introduced the trial-termination response (from Step 6). First we enclosed the circles with a rectangle. Then the circles were removed and the rectangle was reduced in width. Eventually, a line with a small dot at each end was the only stimulus presented on the screen. By Step 8, the task was thus tracing a line. As visual feedback, a blue disk would still be presented at each touched (invisible) target area although circles were no longer seen on the screen. Because the invisible target areas overlapped considerably, a thick blue trace formed and covered the stimulus line as the subject moved finger over the line. The width of the trace was the diameter of the blue disk (2 cm). As before, the reinforced movement was from top to bottom or from right to left (on horizontal lines). No visual feedback (ink) was produced at any screen location other than over the line (except for Step 10, see Sect. 2.6).

2.5 Development of Appropriate Finger Lifting from the Screen

Up to Step 5, reinforcement would appear when the subject reached the last target circle in the display. Hence lifting of the finger from the screen was prompted by the sound associated with reinforcement delivery. To make the visual stimuli on the screen control cessation of drawing, we therefore introduced a trial-termination response from Step 6. Lifting of the finger would no longer be prompted by a sound. Instead, the visual configuration at the end of the trace should come to control cessation of drawing. From Step 6, a white key appeared at the lower right-hand corner of the screen (see Fig. 3). Pressing this trial-termination key after the completion of drawing produced reinforcement. Both subjects acquired the trial-termination response promptly without any training. Because both subjects have had extensive training in a variety of tasks, they usually press new stimuli presented on the screen and therefore quickly make contact with procedural changes.

2.6 Length of the Drawn Line

By the end of Step 8 we had developed a program that detected the actual path of movement over the screen so that we could analyze how accurately the subjects were drawing. The program detected screen contact as points, and also detected a

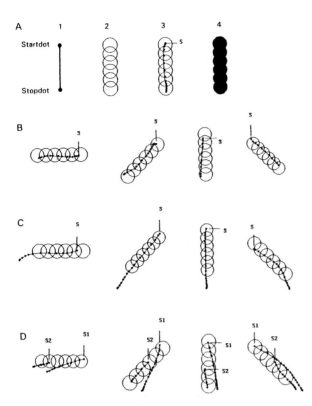

Fig. 5. Illustration of a movement–path analysis. **A** Schematic diagram of the stimulus and movement detection (shown for the vertical stimulus only). *1,* line stimulus shown on the screen; *2,* invisible target areas around the line; *3,* movement path of detected screen contacts within the invisible target areas (*S* indicates the start of the path); *4,* electronic ink appearing on the screen at the target locations after screen contact with the targets. **B** Examples of one-stroke movement paths that end in the last target area (correct drawing). **C** Examples of one-stroke movement paths that extend beyond the last target area. **D** Examples of two-stroke movement paths where the first stroke moves away from the target areas and the second stroke fills in the remaining target areas; stroke order is indicated by the labels *S1* and *S2.* All examples were drawn by Ai

lift from the screen. Figure 5 illustrates how the movement path related to the stimuli and shows some examples of actual paths. Panel A illustrates the relationship between stimulus and path (for the vertical stimulus only): 1, the stimulus line on the screen; 2, the invisible overlapping circular targets "under" the line; 3, a series of successive invisible recorded contact points (beginning at S); 4, the resulting formation of visible ink on the screen as the circular targets are contacted (from top to bottom in accordance with the drawing motion). Panel B shows examples of trials of each orientation where the subject correctly starts drawing (S) within the top circle and stops drawing by lifting the finger within the bottom circle. However, both subjects often moved considerably past the end of the stimulus line before they lifted their finger from the screen, as panel C shows. This extra movement did not produce visual feedback on the screen. We defined a path as too long if its endpoint was more than 1.5 cm (25 pixels) from the endpoint of the stimulus line on the screen. We analyzed the data for the last four sessions of Step 8 for each subject. The percentage of paths for Ai that were too long for the 0°, 45°, 90°, and 135° patterns were 62%, 94%, 89%, and 91%, respectively; for Pendesa they were 88%, 70%, 69%, and 87%, respectively. The average excessive length for all patterns was 35 and 25 pixels for Ai and Pendesa, respectively. Thus, both subjects had problems ending the drawing movement at the correct location; most often they went too far. In an attempt to sharpen the precision of drawing, reinforcement was withheld in Step 9 if the finger moved beyond the stopdot by more than 50 pixels. An additional problem was that the subjects sometimes moved their finger away from the vicinity of the line, as shown in panel D in Fig. 5, and therefore did not produce visual feedback. The procedure allowed for the subjects to make more than one stroke at each trial. Both subjects learned to correct such trials by adding a second stroke beginning roughly where the last blue disk appeared. The percentage of trials for Ai with two or more strokes for the 0°, 45°, 90°, and 135° patterns were 24%, 17%, 45%, and 37%, respectively; for Pendesa they were 52%, 36%, 15%, and 14%, respectively. Trials with multiple strokes rapidly dropped in frequency, and were nearly absent in Step 10.

2.7 Drawing Freely on the Screen

In Steps 1–9, drawing was confined to the target areas on the screen. The visual feedback was produced only at fixed locations when the finger contacted those locations. Touching outside the target areas produced no ink. This almost physical guidance of finger movement was completely removed in Step 10, where the subjects had to connect two dots 200 pixels apart with no line connecting the dots. Thus, we asked whether the drawing movement trained in Steps 1–9 would transfer to a new situation where a touch would produce ink at any screen location. In Step 10, the subjects were therefore drawing completely freely with no constraints on where ink would appear on the screen. The criterion for correct drawing was changed from filling invisible target circles to a different algorithm based on an on-line analysis of the screen contact points. Briefly, the path length from the first touch to the last lift had to be 200±40 pixels, the angle of an imaginary straight line connecting the first touch and the last lift had to be within ±15° of the angle of an invisible straight line connecting the two stimulus dots, and the variability of the trace had to be below a certain value (see Iversen and Matsuzawa 1996).

Remarkably, both subjects continued to draw in the previously trained directions by connecting the two dots with a trace of ink. Quite simply, the subjects were now drawing exactly as indicated in the diagram shown in Fig. 1. Thus, the previous training (Steps 1–9) of directional drawing in a confined area transferred to the new task of connecting two points by free drawing (Step 10). To analyze the precision of drawing generated by the training method, data were sampled for both reinforced and nonreinforced trials for five sessions in Step 10 ($n = 72$ trials for each trial type for each subject). The four boxes in the upper part of Fig. 6 show printouts of individual trials showing what the screen looked like when drawing was completed (the two white dots that appear on the screen at the start of a trial were covered by ink, but are shown here to indicate the precision of drawing; S indicates the start of the drawing). The next two panels show the frequency distributions of the angle of the drawn trace to the horizontal for each subject for each dot orientation, as indicated by the individual trials above. The clear separation of the distributions reveals how effectively the two-dot patterns guided the angular component of drawing. For Ai, all trials fell within the ±15° reinforcement criterion. For Pendesa, all but one trial fell within this criterion. For each dot pattern, the mean angle of the drawn trace deviated by not more than 3° from the angle formed by an invisible ideal line between the two dots. The mean (SD) angles for the 0°, 45°, 90°, and 135° patterns for Pendesa were –2° (4°), 44° (3°), 87° (3°), and 133° (3°); for Ai the values were –2° (5°), 42° (4°), 89 (3°), and 132° (5°). The last two panels in Fig. 6 show the frequency distributions of trace length for each dot orientation. The distributions appropriately peaked at a value near the distance between the two dots (200 pixels). For Pendesa, in 3.5% of all trials (10/288) the length of the drawn trace was outside the reinforcement criterion of ±40 pixels. For Ai, the length of the drawn trace fell outside this criterion in 15.3% (44/288) of the trials. For Pendesa, the mean (SD) lengths for the 0°, 45°, 90°, and 135° patterns were 203 (21), 203 (18), 202 (14), and 191, (14), respectively (units are one pixel). For Ai, the values were 231 (27), 201 (23), 209 (18), and 216 (26), respectively. The percentages of reinforced trials were 74 for Ai and 92 for Pendesa. For both subjects, unreinforced trials primarily resulted from a trace drawn at an appropriate angle but with a length which fell outside the reinforcement criterion of 200±40 pixels. Occasional unreinforced trials were drawings with an appropriate angle and length but with a curvature such that the criterion for variability was exceeded (see Iversen and Matsuzawa 1996).

2.8 Summary of Experiment on Drawing a Line Between Two Points

The drawing precision shown in Fig. 6 reveals that both subjects had indeed learned to draw in a controlled manner, as was the purpose of the experiment. The gradual modification of the stimuli on the screen changed the topography of screen contact from a repeated touch–lift mode to a touch–drag–lift mode. In early trials in the touch–drag–lift mode, the ape's finger often went beyond the last stimulus element on the screen, but by the end of Step 10, both subjects lifted their finger within about 20 pixels of the stopdot in most trials. By the end of Step 9, the subjects connected the two dots, as they would also do later in Step 10. However, the ink on the screen was confined to the target area. An important step in training was the change to the "free" screen in Step 10, where any screen contact produced ink on the screen. Thus, had the subjects drawn too far or touched outside the target area it would show on the screen as blue "ink." At the time of the experiment, we were quite worried that the control of drawing in Step 9 might not transfer to the free-screen situation in Step 10. Our worries turned out to be unjustified, as the subjects

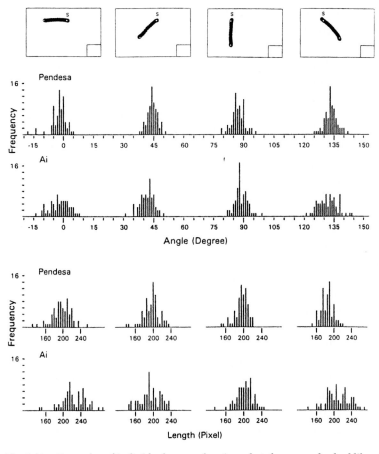

Fig. 6. *Top,* Examples of individual traces showing what the screen looked like after the subject had moved her finger from one dot to the other (*S* indicates where drawing starts). The white guide dots that appeared at the start of the trial were actually covered by the electronic ink, but are shown here to indicate the precision of the drawing (the dot distance was 200 pixels). *Middle,* Quantitative analysis of the angular component of drawing in Step 10. Frequency distributions of the angle of the drawn trace to the horizontal are shown for each trial type, as indicated by the top display. The angular criterion for reinforcement of ±15° is indicated for each distribution. *Bottom,* Quantitative analysis of the length component of drawing in Step 10. Frequency distributions of the length of the drawn trace are shown for each trial type, as indicated at the top. The length criterion for reinforcement of 200±40 pixels is indicated for each distribution. For both analyses, data cover both reinforced and nonreinforced trials and are sampled from the last five sessions in Step 10. Six nonreinforced trials for Pendesa and eight for Ai were excluded from analysis because of extreme curvature. (From Iversen and Matsuzawa 1996, with permission)

immediately drew correctly on the first session in Step 10. Thus, the stepwise training proved sufficient to generate drawing guided by the visual stimuli on the screen. The subjects had acquired the basic elements of drawing outlined earlier in this chapter. In addition, the resulting drawing could readily be reproduced, which is necessary for a scientific analysis. The experiment demonstrates that chimpanzee drawing can be brought forward from scribbling to precise line drawing guided by visual elements on the drawing surface.

3 When to Lift the Finger?

Both subjects accurately connected two dots, separated always by 200 pixels, in Step 10. Earlier in training, the subjects had problems learning to lift their finger at the correct location, so we wondered whether the subjects had in fact learned that the stopdot was a signal for them to lift their finger. Maybe they had learned instead that the hand should move a fixed distance of 200 pixels from the startdot. We tested this possible difference in the reason for lifting their finger by inserting probe trials where the distance between the dots was 120 or 300 pixels. Regular trials maintained a 200-pixel dot distance. If the stopdot exerted full control over lifting the finger from the screen, then the subjects should have no problems with the probe trials. If, on the other hand, the subjects had learned to draw at a fixed length of about 200 pixels, then they should draw too far on probe trials with a 120-pixel dot distance and not far enough on probe trials with a 300-pixel dot distance. Each of eight test sessions with free drawing, as in Step10, had 40 trials with the familiar 200-pixel dot distance (baseline trials) and eight probe trials.

Figure 7 shows dot locations (large open circles), the points where drawing began (small open circles), and the points where the finger was lifted and drawing ended (small filled circles) in individual trials. These data are presented merely to

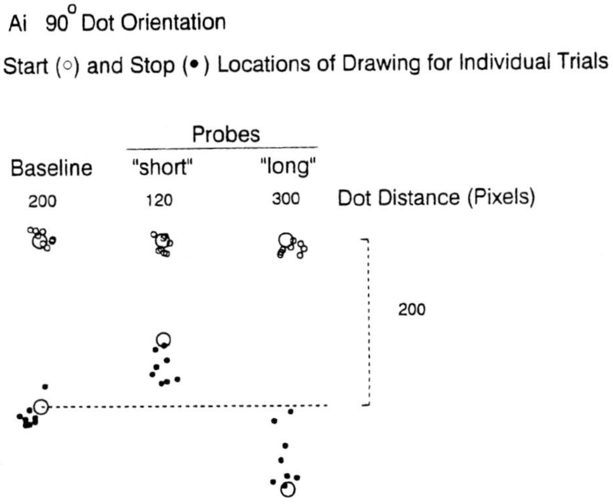

Ai 90° Dot Orientation

Start (○) and Stop (●) Locations of Drawing for Individual Trials

Fig. 7. Analysis of drawing in probe trials with altered guide dot distances. The starting and stopping points of drawings are shown for individual trials for the 90° stimulus display for Ai. Two *large open circles* show the guide dots. *Small open circles* indicate the start of drawing and *small closed circles* indicate where the drawing stopped. Data are shown for all probe trials of 120- and 300-pixel distances between the guide dots and for baseline trials that preceded 120-pixel probe trials

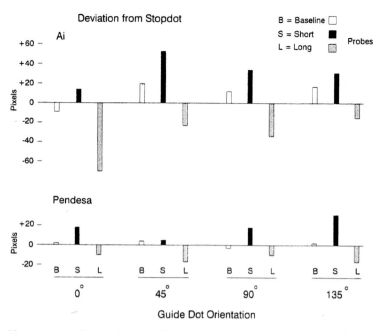

Fig. 8. Average distance between the stopdot and the point of lifting the finger for both subjects for all four guide dot orientations. Positive values indicate that the finger moved beyond the stopdot (trace is too long), and negative values indicate that the finger did not reach the stopdot (trace is too short)

illustrate the control problem, and are for vertical dot orientation for Ai only. On "short" probes, Ai usually did not lift her finger before it had passed beyond the stopdot. Conversely, on "long" probes, Ai usually lifted her finger before it had reached the stopdot. Figure 8 shows the average distance from the stopdot to the place where the finger was lifted for all four dot orientations for both subjects. Moving the finger too far is indicated by positive values (the drawn trace was too long), and not moving the finger far enough by negative values (the drawn trace was too short). Ai went too far on "short" probes and not far enough on "long" probes. Pendesa also went too far on most short probes and not far enough on most long probes; however, she did not deviate as much from the stopdot location as did Ai. The variation around the stopdot on baseline trials was from –10 to +20 pixels for Ai; for Pendesa, this variation was only from –5 to +5 pixels. In the novel situation with varying dot distances, the subjects correctly aimed at the startdot and drew at a correct angle, but the length of the trace was not as well controlled as with a fixed dot distance because of the weak control by the stopdot. Thus, with novel dot distances in probe trials, the component of drawing that broke down was the one that they had experienced problems with before, namely lifting the finger at the correct location. That both subjects went too far in short probes and not far enough in long probes suggests that moving the finger (or hand) a fixed distance of about 200 pixels in training must have exerted some degree of control over the drawing. The resulting drawing in probe trials is thus a mixture of trying to move the finger a fixed length (200 pixels as in baseline trials) and aiming toward the stopdot. Both subjects later received considerable training with varying dot distances and learned to lift the finger within 5–15 pixels of the stopdot, thereby illustrating direct guidance by the dot pattern.

4 Simple Copying

Now that the subjects had learned to draw a line that connected two dots, the next question we asked was whether it might be possible to teach the subjects to draw a line parallel to an already existing line (Iversen and Matsuzawa 1997). Establishing such a performance would amount to a very simple form of copying.

4.1 Pretraining and Probe Testing

Because the subjects could already connect two dots at a distance of 200 pixels, we built upon this existing performance. We simply added a model (a bar) on the screen next to the two guide dots, as illustrated for the 0° orientation in the top display in Fig. 9 (baseline trials). At the start of each training trial the screen showed the bar and two dots. The two dots were distant from, and aligned with, the endpoints of the bar. The task for the subject was to aim at the startdot, sweep the finger across the screen surface to the stopdot, lift the finger, and then press a white key on the screen to end the trial. When the trace produced by the subject was correctly drawn, this trial-termination response produced reinforcement. In each trial, the bar and the two dots appeared in one of four orientations, 0°, 45°, 90°, or 135° with the horizontal. First, we wanted to know whether the bar would serve as a model for drawing after we removed the stopdot in probe trials, as illustrated in the second panel in Fig. 9. What would the subjects do in such trials? If the subjects were able to draw a line parallel to the model, then the training of connecting two dots next to the model would have proven sufficient to establish the bar as a model for drawing; the subjects would be making a copy of the model. After 15 sessions each of 96 trials of pretraining with the same reinforcement criteria as described for Step 10 above (Sect. 2.7), we mixed baseline trials (model and both startdot and stopdot) with probe trials (model and startdot only). One test session had 72 baseline trials and 6 probe trials for each model orientation; drawing in baseline trials was reinforced automatically, while drawing in probe trials was reinforced manually. (Some additional probe tests are described in Iversen and Matsuzawa 1997.) In baseline trials, both subjects connected the dots with a trace at a length that was within ±10 pixels of the distance between the dots and at an angle that was within 3° of the angle of an imaginary line between the dots, just as they had done in Step 10. Drawing in the first three probe trials of each type is shown for both subjects in the bottom panel in Fig. 9. In probe trials, the model did not reliably guide the drawing by Ai. Pendesa drew a trace at an angle that roughly followed the angle of the model, but the trace was too short in all cases. Thus, the method of simply present-

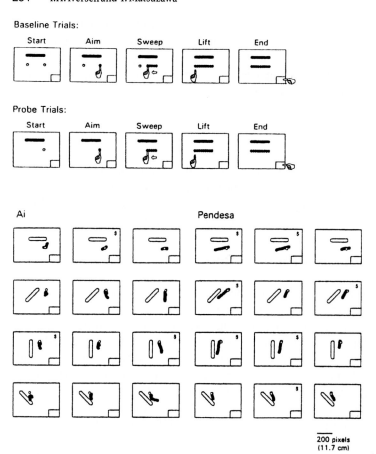

Fig. 9. *Top,* Baseline trials and probe trials illustrated for the 0° display only; four trial types with the model at 0°, 45°, 90°, or 135° relative to the horizontal occurred within each session. Each frame represents an outline of the screen. For baseline trials, a model (*dark bar*) and two guide dots (*small open circles*) appeared on the screen at the start of the trial. The subject aims at one dot, sweeps the finger over the screen, lifts the finger at the next dot, and ends the trial by pressing the trial-termination key in the lower right-hand corner of the screen. For probe trials, the stopdot was removed from the display to test how the model guided the drawing. Probe trials were mixed with baseline trials. The subject had to draw a line parallel to the model and of the same length as the model. *Bottom,* The first three probe trials (no stopdot) for each model are presented for Ai and Pendesa. Each frame shows the drawing for one trial. The model bar is indicated in outline to avoid confusion with the drawn trace, which appears in black; for the subjects, the model was solid blue. The startdot is shown here to indicate the precision of drawing although it was actually covered by the drawn trace. Reinforcement is marked by a $ sign in the upper right-hand corner; this sign did not appear on the screen. (From Iversen and Matsuzawa 1997, with permission)

ing a model on the screen is not sufficient to establish guidance by the model. Because the subjects could already draw a line that connected the two dots, the model was not a necessary stimulus for correct drawing, and therefore the subjects did not have to look at it. Nonetheless, some rudimentary form of model control was evident for Pendesa.

4.2 Training to Draw a Line Parallel to a Model

The method of merely presenting the model on the screen next to the guide dots proved an ineffective teaching strategy. Therefore, we next attempted to teach model guidance by different methods. Because Ai did not show reliable drawing in probe trials, she was given additional training that faded out the existing control by the stopdot. This training passed through various steps that cannot be described briefly (see Experiment 2 in Iversen and Matsuzawa 1997). By the end of this training, Ai drew as Pendesa had done in the early probe trials (see Fig. 9).

4.2.1 Simplification of the Stimulus Display

For both subjects, the display from the probe trials was used in all trials. Thus, each trial showed the model and the startdot only. The criteria for reinforcement were widened to increase the likelihood of reinforcement. Thus, a trace was reinforced if its length was 200±50 pixels, and its angle to the horizontal was ±15° (the variability criterion was unchanged). To simplify the display, the 0° model was removed, and the three remaining models shared the same startdot location. In previous sessions, both subjects customarily drew a trace that was shorter than the model and then pressed the trial-termination key, which ended the trial. The procedure was changed to a multiple-stroke contingency. If the drawn trace had an appropriate angle and variability but was shorter than the length criterion for reinforcement, then the trial did not end at the first trial-termination response. Instead, the model and the drawn trace remained on the screen, and the subject could now add another stroke to the screen to make the already drawn trace longer. When the subject then made a second trial-termination response, the computer reevaluated the combined trace, and the trial ended in reinforcement when all three criteria were met. Ai had 16 and Pendesa had 7 sessions each of 48 trials.

4.3 Model Control

The best evidence of model control obtained with this teaching method is shown in the top panel in Fig. 10, which presents all the trials from one session for Ai. Trials were scheduled in mixed order, but are arranged sequentially here for each model orientation. Note how the trace angles are clearly differentiated by the model orientation. Only the traces on the first two trials for the 45° model resemble traces for some of the 90°-model trials. Although many trials went unreinforced because the trace fell short of the angle or length criterion, the data show that Ai could draw a line parallel to the model in most trials.

Ai   All Trials From One Session

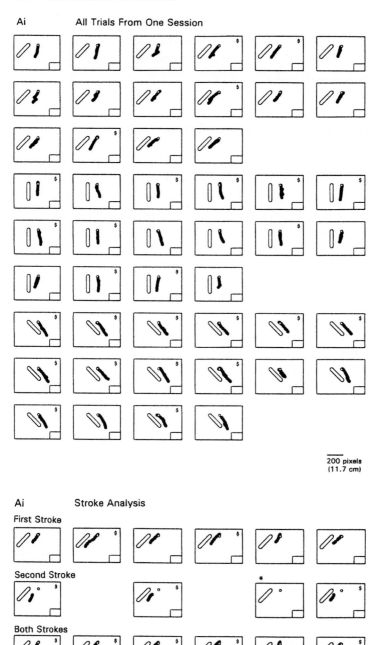

200 pixels
(11.7 cm)

Ai   Stroke Analysis

First Stroke

Second Stroke

Both Strokes

4.4 Multiple-Stroke Contingency

For both subjects, the multiple-stroke contingency led to more effective drawing. The lower panel in Fig. 10 shows selected individual trials for the 45° model for Ai. Six trials show the first stroke, the second stroke if one occurred, and the cumulative trace. Only two trials were reinforced after one stroke. When a stroke was not reinforced because it was too short, the stimuli remained on the screen, and Ai could supplement the drawing by adding a second stroke. In three trials for the 45° model, Ai thus added a second stroke that started roughly where the first had ended. Note how the model also guides the angle of the second stroke. A similar pattern was seen for the 90° and 135° model. Based on an analysis of all sessions for Ai, 60% of the trials had a second stroke and 78% of these strokes had the appropriate angle; 46% of the trials with a second stroke improved the cumulative trace enough to result in reinforcement. For Pendesa, 31% trials had a second stroke and 55% of these strokes had the appropriate angle; 25% of the second-stroke trials were reinforced.

4.5 Quality of Copying

The relationships between the angle and length of the drawn line are presented as scatterplots for each model orientation in Fig. 11 for both subjects. The x-axis presents the trace angle and the y-axis the trace length for each single trial. Data are coded for each model orientation, as shown under the x-axis. The boxes represent the reinforcement criteria for length and angle; hence, a data point falling inside the appropriate box indicates that both length and angle were correct in that trial. Note that no model was presented at the 0° angle in any trial. Ai achieved 76%, 64%, and 43% reinforced trials for the 45°, 90°, and 135° models, respectively. In

---

←

Fig. 10. *Top,* Drawing in all 48 trials from one session for Ai. All trials featured the model and the stopdot only; no 0°-model orientation was presented. Trials are presented sequentially from left to right for each model although trials actually occurred in mixed order within the session. Each frame shows the drawing for one trial. The model bar is indicated in outline to avoid confusion with the drawn trace, which appears in black; for the subjects, the model was solid blue. The startdot is shown here to indicate the precision of the drawing although it was actually covered by the drawn trace. Reinforcement is marked by a $ sign in the upper right-hand corner; this sign did not appear on the screen. *Bottom,* An illustration of the multiple-stroke contingency shown as drawing in six individual trials selected from the 45° model for Ai. Data show the first stroke, the second stroke if one occurred, and the cumulative trace. Thus, each trial reads from top to bottom. A stroke was defined as cessation of drawing followed by lifting the finger and drawing at a new location or followed by a trial-termination response. If the first stroke was reinforced there was no opportunity for a second stroke in that trial, hence the blanks under reinforced first strokes. If the first stroke had a correct angle but was too short, then the trial-termination response did not end the trial and a second stroke was allowed. On the trial marked by *, Ai pressed the trial-termination key twice without adding a second stroke. The combined stroke was analyzed by the reinforcement criteria of 200±50 pixels for length and ±15° for angle. (From Iversen and Matsuzawa 1997, with permission)

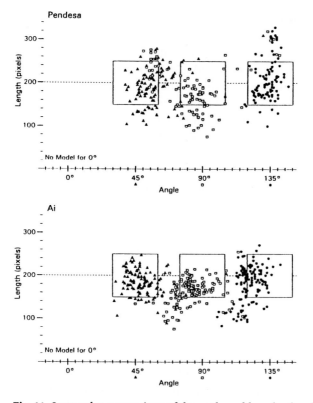

Fig. 11. Scatterplot comparison of the angle and length of each drawn trace for each subject during the line-copy task with three model types (the 0° model was not presented during sessions). All trials in a session displayed the model and the startdot only. Data are from eight sessions for Ai and seven sessions for Pendesa; each session had 48 trials. Data points are marked for each model orientation, as indicated by the symbols under the *x*-axis. *Boxes* indicate the criteria for reinforcement of ±15° for angle and ±50 pixels for length; data points inside an appropriate box thus represent reinforced trials. Nineteen trials for Pendesa and four for Ai were rejected from analysis because of an extreme variability score that prevented a linear analysis. (From Iversen and Matsuzawa 1997, with permission)

total, 61% of all traces were within the reinforcement criteria of 200±50 pixels and ±15°. Pendesa achieved 48%, 38%, and 71% reinforced trials for the 45°, 90°, and 135° models, respectively. In total, 52.3% of all traces were within the reinforcement criteria of 200±50 pixels for length and ±15° for angle. However, Pendesa also drew traces that overlapped with the angle criterion for one of the other models. For example, 19.8% of the trace angles for the 90° model were within 45°±15°, and 6.5% of the trace angles were within 135°±15°. In total, in 12.9% of all trials, the trace was drawn at an angle that was inappropriate for the model.

4.6 Summary of the Copying Experiment

This experiment shows that a model that is adjacent to the target area can give some degree of drawing control. Thus, both Ai and Pendesa could draw a line that was parallel to a model presented a few centimeters to the left of the target area. However, the drawn trace was often too short, indicating that control over lifting the finger from the screen was not well established. Simplifying the stimuli on the screen by removing the 0° model and having all models share the same startdot, as well as introducing the "multiple-stroke contingency," improved drawing for both subjects compared with early probe trials (i.e., Figs. 9 and 10). The results suggest that the subjects do not lack the necessary perceptual and motor components to draw accurately from a model. However, the performances were not stable, indicating that the stimuli we used did not guide performance optimally. A detailed behavior-pattern analysis of individual trials (see Iversen and Matsuzawa 1997) indicated that the instability of the performance was not random, but instead reflected control by the prevailing stimuli and the immediate pretraining. For example, both subjects drew toward the trial-termination key in some trials and toward the model in other trials, revealing competing sources of drawing control. Our research was exploratory, and we cannot yet pinpoint the conditions that are necessary to establish reliable single-line copying. The multiple-stroke contingency improved drawing considerably and may be of general use in teaching of drawing. In subsequent work on the tracing of complex stimuli with the same subjects, we have found a similar multiple-stroke contingency to be of great use. Other improvements have been to locate the trial-termination key outside the screen and to use more varied stimulus exemplars than we did here. For example, in later work the model appeared in different lengths as well as different orientations. These changes further improved the length of the drawn trace. We suspect that an important aspect of drawing a line parallel to a model is to learn that when the stopdot is removed, then the end of the model must "correspond" spatially to the location of the now absent stopdot. However, it is not immediately apparent how one would go about teaching this relationship to the subjects. Unfortunately, the vast literature on children's drawing (e.g., Cox 1992; Goodnow 1977; Harris 1963) does not provide much guidance about how to establish accurate drawing or copying other than to give verbal instructions to subjects, who have already acquired verbal behavior and some drawing skills. Although still incomplete, the teaching methods developed in the present experiments for establishing model-guided, independent drawing in the chimpanzee without the use of verbal instruction may offer a possible way to establish basic drawing and copying skills in human subjects who may lack such skills. In summary, the precise recording and immediate visual feedback offered by the touch monitor, combined with the automated, on-line analysis of the performance, may form an ideal setting for establishing drawing in both human and nonhuman primates.

5 Using a Touch Monitor for Multiline Tracing, Object Sorting, and Fingermazes

The touch-monitor technique has proved very useful for the study of drawing. The subject works with a finger on the screen in a manner that is very similar to drawing with a pen on paper. The drawing traces are given immediate and automated visual feedback just as a pen leaves immediate feedback (marks) on paper. In addition, the movement path is recorded electronically and can be analyzed precisely, which is not typically the case with pen movements on paper. The touch monitor also proved very useful in subsequent studies with the same subjects in tasks that involved more complex drawing and transport of an object with the finger over the screen surface. These tasks make use of the same perceptual motor relations that were established in the basic drawing task described above, namely aiming, sweeping, and stopping. Figure 12 illustrates these new tasks in displays of single trials that show what the screen looks like when a given trial begins and what it looks like after the subject has completed that trial (presentation by Iversen and Matsuzawa 1998).

5.1 Multiline Drawing

The tracing task was expanded to complex figures. Thus, on each trial the monitor presents a stimulus composed of several thin lines, for example the outline of a square, as shown in the top left-hand panel in Fig. 12. The task is to trace this outline with the finger. The subject can start at any place on the square. Finger movement leaves electronic ink on the screen as in the previous drawing task. The square shown in the top right-hand panel shows the drawing in a single trial; Pendesa composed the square in two strokes, each forming a 90° angle. To end the trial after drawing is completed, the subject presses the trial-termination key in the lower right-hand corner of the screen. This response causes the computer to analyze the drawn figure according to two criteria; first, the proportion of the stimulus covered by ink, and second, the proportion of ink that covers the stimulus. When both criteria are satisfied, usually to 95% accuracy or higher, the subject receives reinforcement. The task has no restriction on the number of strokes used to complete the drawing. The subjects can now trace very accurately a variety of multiline stimuli, such as the letters of the alphabet and complex geometrical figures. Considerable training, progressing from one line, to two lines, to several lines, was necessary to establish this elaborate tracing performance.

5.2 Object Sorting

In a somewhat different task, the subjects sort randomly arranged objects into categories by moving the objects about on the screen surface with a "click and drag" method. The subject moves an object by first placing the finger on it and then dragging it by moving the finger over the screen. When the finger movement stops

Multi-Line Tracing

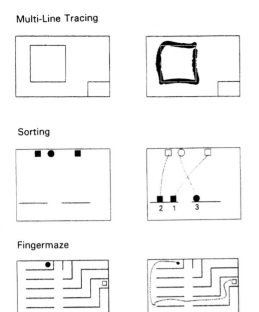

Sorting

Fingermaze

Fig. 12. Illustration of three tasks that use finger movement on the touch screen as the response. Each frame pair (left and right) shows a single trial from each task. The panels on the *left* show what the screen looks like at the beginning of a trial. The panels on the *right* show what the screen looks like after the subject has finished the trial. In the multiline tracing task, the screen shows four thin lines that form a square and the subject traces the outline of this square with a finger, thereby producing electronic ink over the square. The stimulus lines are covered by ink during drawing, but are superimposed here (in white) on the frame to the right to show the precision of drawing. The *square* in the lower right-hand corner shows the trial-termination key. In the sorting task, the left-hand frame shows three objects at the top of the screen that are to be moved to the two trays at the bottom of the screen. To form two categories, the subject must drag the objects with a finger one at a time to the trays. The frame on the right shows the movement paths and their sequential order; the paths and numbers are not shown on the screen during the task, and the objects disappear from the top during movement. In the fingermaze task, the screen shows a movable dark disk, a target (*small white square*), and obstacles (*lines*) that the disk cannot move through. The subject moves the disk to the target by dragging it with a finger over the screen. The movement path is shown in the frame on the right but did not appear on the screen during the task. Data are from sessions with Pendesa

or the finger is lifted from the screen, the object stops moving. Thus, the object moves in accordance with the finger. The task is very similar to aiming at, clicking on, and then dragging an object to a different location in modern human computer-mouse applications. First, we trained the subjects to move objects from one location to another. Then we introduced objects of different shape and color. The task for the subject was to sort the objects into categories. Previous studies have demonstrated that chimpanzees can, spontaneously or with training, sort "real" three-dimensional objects on trays in a face-to-face situation where an experimenter is present with the subject at all times (Matsuzawa 1990; Spinozzi 1996; Tanaka 1995). Our objective was to develop a task that presents objects automatically, keeps a record of all object movements, and delivers reinforcement automatically. Because the computer screen is flat, our subjects by necessity were sorting two-dimensional objects and not real objects. The left frame in the second panel in Fig. 12 shows what the screen looked like at the onset of a given trial. The objects to be sorted appear at the top of the screen. The two lines at the bottom of the screen are the receptacles or "trays" that the objects have to be sorted onto. The subject must form two categories of objects by dragging the objects to the trays one by one. When all objects have been brought down to the trays, the subject presses a switch next to the screen as a trial-termination response. As in the drawing task, this response starts the automated evaluation process by the computer. Reinforcement is presented when the objects are sorted correctly; if the objects are sorted incorrectly, a tone sounds and the subject can rearrange the objects. The frame center right shows how the subject sorted these objects. The movement path of each object is indicated here but does not appear on the screen during the task. The objects did actually move from the top of the screen to the bottom, but for the purpose of illustration, the initial top locations are outlined in the frame. In this task, each trial had different arrangements of different objects that could be sorted into one or two categories; with one category, only one tray should be filled. Both subjects learned this sorting task with great accuracy after considerable training. The aiming, dragging, and stopping developed in the drawing task carried over directly to the sorting task. Although the subjects were not drawing on the screen, they immediately acquired the task of moving an object from one location to another. The difficulty of the sorting task was not the object movement per se, but the formation of categories.

5.3 Solving Fingermazes

The fingermaze has a long tradition in experimental psychology and provides a way to examine the acquisition of spatial navigation skills (Woodworth and Schlosberg 1954). An electronic fingermaze was previously developed for rhesus monkeys where the subjects controlled a cursor with a joystick (Washburn 1992); the cursor moved at a constant speed, and the subject could control the direction of movement by the direction of engagement of the joystick. In our task, the subject moved a small object on the screen with the finger. The left-hand frame of the lower panel in Fig. 12 shows what the screen looked like at the start of a given trial. The trial features a "ball" that can be moved, a stationary square goal, and several obstacles that the ball cannot pass through. In each trial, there is at least one open path from the ball to the goal. Each trial in a session shows a different maze pattern. Movement of the ball leaves no trace on the screen while the subject moves the ball. However, a path can be added to the maze pattern later for analysis. Thus, the frame on the right shows the path that the subject moved the ball in this trial. The ball stops if it is moved into an obstacle, and it can only be moved again when the finger moves away from the wall the ball faces. Training progressed from very simple mazes with just one obstacle to mazes with several obstacles of increasing complexity. At the present stage of the research, the subjects learn to negotiate each maze pattern by trial-and-error; they literally move the ball into blind alleys on the first trials of each maze pattern. One of the subjects, Pendesa, has solved more than 100 different maze patterns.

6 Conclusion

6.1 Learning to Draw

Theories about children's drawing development range widely (see Cox 1992; Harris 1963). At one extreme, drawing is considered a reflection of how the child understands or interprets the world. At the other, drawing is considered to be an acquired skill that results from complex perceptual–motor learning. Previous work with drawing by chimpanzees had demonstrated that they would readily scribble with markers on paper, but their drawing does not develop spontaneously beyond this stage. This is in contrast to children's drawing, which develops from scribbling to more complex forms to reproductive drawing. If one considers the view that drawing reflects how the subject "understands" the world, it is a bit hard to accept the implication that chimpanzees' understanding of the world only matches the rather primitive scribbles that they may make spontaneously. A multitude of studies has shown that chimpanzees can make very detailed discriminations of complex visual stimuli that far surpass the primitive scribbles that they can produce. Considering the alternative view that drawing is an acquired skill, one may suggest that perhaps chimpanzees could in fact learn to draw in a structured manner. Our research was based on this latter premise. After training, the two chimpanzees in our study had learned to draw lines that very accurately connected two guide dots. Drawing was built piece by piece from existing looking and pointing behavior. The learning consisted of motor and perceptual components. Continuous finger movement while keeping contact with the drawing surface, which is necessary for drawing, was a motor component that was lacking in the subjects' repertoire. Once this movement had been learned, the next step was to guide it by the stimulus elements on the screen, which is the perceptual component of drawing. Apparently, one of the aspects of drawing that our subjects found most difficult to learn was to lift their finger at the correct location. Even when drawing was seemingly well established, new stimulus configurations, such as presenting the guide dots at new distances, revealed that the subjects had probably not learned a generalized stop response. A similar problem appeared when we attempted to teach the subjects to make a simple copy of a line. When the stopdot was removed, the subjects often lifted their finger from the screen too soon and therefore made traces that were too short. Considerable training was necessary before some degree of transfer of control of finger lifting from the stopdot to the model was evident, and the copy performance never became stable.

The literature on children's drawing is particularly silent about what happens during the transitions from scribbling to line drawing and from line drawing to tracing and copying. In tests of abilities to trace and copy, children are instructed verbally to trace or copy a model, and deviations between the model and the drawing are commonly interpreted as reflective of perceptual and cognitive development or deficits (see, for example, Cox 1992; Hulstijn and Mulder 1986). Maybe one reason for the poor showing of studies of learning to trace and copy in children is that learning to copy has been considered detrimental to artistic development (see, for example, Cox 1992; Duncum 1988). Morris (1962) used a similar argument when reinforcement of drawing in one chimpanzee resulted in repetitive strokes (see Sect. 1.2.3). Disregarding the issues of artistry and creativity, our studies suggest, nonetheless, that the skills of drawing and copying can be learned by subjects who at first can only scribble, and that this learning process can be examined systematically.

6.2 The Chimpanzee as a Model for Teaching Without Verbal Instruction

Because our chimpanzee subjects do not have the prerequisite language behavior that enables the teacher to use verbal instruction with success, we instead used "contingency-shaped" training. Thus, without any kind of "demonstration," "imitation," "assistance," "molding," or verbal instruction, the subjects eventually reached a moderate degree of success in drawing a line parallel to a model. Our work with the chimpanzees thus serves as a model for teaching without the use of verbal instructions. The vast literature on drawing in children is primarily diagnostic of underlying perceptual and cognitive processes or disorders in these processes. When emphasis is placed on learning how to draw, the method used is primarily that of verbal instruction (e.g., Abercrombie 1970; Kephart 1971; Tuch and Judy 1975; Witsen 1967). We have shown that considerable learning can take place without the use of verbal instructions. The individual feedback produced both visually and in terms of reinforcement are sufficient elements to promote advances in drawing. Certain disabilities may deprive humans of the opportunity to learn to follow verbal instructions; such subjects might learn to draw with feedback-based training methods similar to those presented here.

6.3 Drawing as a Measure of Intelligence or an Acquired Skill

Using drawing as a measurement of intelligence in children has had a long tradition in psychology since the book by Goodenough (1926) on that topic. The more general inference of mental processes from overt behavior has had an even longer life in both animal and human psychology. For example, Romanes (1882) suggested that overt behavior is the "ambassador of the mind." In the case of animal subjects, the inference of hidden cognitive or mental processes from overt behavior has fostered an "epidemic of anthropomorphism" in the 20th century (Davis 1997). Previously, Henton and Iversen (1978) referred to the same problem as a special case of phrenology, where movements (behavior instead of skull features) are used as indicants of underlying brain or cognitive structures. Davis (1997) asks, as did Skinner (1953) before him, what cognitive inferences and descriptions add to our understanding of behavior and why they are so compelling. Davis attempts the following answer: "… I believe it is largely rooted in misguided folk wisdom surrounding *human* behavior. If we did not hold a needlessly complex view of the importance of conscious (verbal) thought in the genesis of our own behavior, we would not even be tempted by the rampant analogizing … [between humans and animals]" (p. 342; italics by Davis). So, according to folk wisdom, because children's drawing skills are commonly used to infer the presence of underlying cognitive processes, one might by analogy be tempted to use the drawing skills of chimpanzees to infer something about their cognitive apparatus. In fact, some years ago a scholarly visitor to the laboratory where our experiments took place saw one session with one of our subjects drawing lines between the two guide dots. It so happened that day that the drawing was incorrect (too variable) on a few trials with the diagonal patterns. The visitor immediately drew the inference that chimpanzees have the same perceptual or conceptual problems that young children are reported to have with diagonal lines. In our studies we have found no empirical basis for a distinction in the accuracy of the drawing between lines of different orientation (i.e., Fig. 6). Nevertheless, the teaching methods we used to build drawing are quite different from the assessment-directed methods used to study children's drawings. Hence, a comparison of drawing abilities between children and chimpanzees is premature.

We do not find it useful to explain the progress in drawing of our subjects as a result of enhancement of their intelligence. Rather, the step-by-step changes in drawing ability and the emerging problems in drawing in certain test situations can be dealt with entirely from the perspective of how the subjects learn to interact with the stimuli on the screen. Our view is consistent with at least some theories of children's drawing, which account for drawing development in terms of the organism's repeated contacts with its environment as sources of individual feedback of perceptual–motor interactions (e.g., Harris 1963). Similarly, Cox (1992), in her book on children's drawings, concludes that acquisition of skill is essential for drawing development, and encourages teachers to put emphasis on giving feedback to children for their drawings. By focusing on building up the drawing performance step by step by careful manipulation of stimuli in accordance with the existing skills, one knows how to construct the resulting performance as opposed to merely admiring it or ascribing it to internal processes.

6.4 Objective Studies of Behavior Using a Touch Monitor

The touch monitor offers a unique opportunity to study a variety of complex performances in human as well nonhuman subjects. A computer can present stimuli, and touching the screen can be recorded with great precision. In particular, the development of a sweeping finger motion on the screen by the chimpanzee subjects sets the stage for subsequent developments such as object movement in sorting and fingermaze tasks. The novel ability to move the finger from one point to another was acquired quite readily in both subjects. This motor component is generic to all the tasks we have designed. The tasks then differ in how the stimuli shown on the monitor guide the behavior. A combination of thin lines can result in the drawing of complex figures consisting of multiple strokes. In the sorting task, thin lines serve as trays that objects can be dragged onto from other screen locations, and in the fingermaze task, thick lines serve as obstacles that the subject should avoid by moving a single object around them to a target. Thus, the tasks differ in terms of the perceptual skills that the subject has to acquire. For example, in the task of drawing a line parallel to a model, the subject must learn that what previously controlled lifting of the finger (the stopdot) is now represented in a more complex form by the model. The issue for the experimenter is to find an efficient way to teach the subject that a model at one location can in fact serve as a guide for behavior at an adjacent location (a few centimeters to the right of the model).

More broadly, the literature on acquisition of behavior customarily focuses on shaping of behavior as a modification of the topography of components of a particular final response. McIlvane and Dube (1992) have articulated the need for precision in the development of terminology and methodology for the shaping of stimulus–control relations. Thus, in the various tasks described here, it was not so much a response that had to be changed as the degree to which the new stimulus configurations could come to guide already established responses. For example, in the initial stages of training to sort objects, we first presented just one object at the top of the screen that had to be moved to one tray located at the bottom of the screen. The subjects had no problems with this, but when we next presented two identical objects that should be placed on the same tray, both subjects invariably and without hesitation tried to move the second object on top of the first object and not toward the tray. We then changed the situation so that one object appeared first; when it was placed on the tray the second object would appear. The subjects still tried to move the second object on top of the first object and not next to it on the tray. Considerable training was necessary to teach the subjects to place two or more objects next to each other. There was little doubt that the subjects could identify the objects because they correctly aimed at them and moved them. Similarly, there was little doubt that the response of moving one object from one location to another was well established, but one object next to another on the tray was not a familiar stimulus and did not control lifting the finger. The subjects had to be taught that the movement should stop when one object is near another object and on the tray. Thus, the complexity of the skills that the chimpanzees proved capable of acquiring depended very much on the ability of the experimenters to design teaching methods that would allow the subjects to come into contact with the relevant controlling stimuli.

References

Abercrombie MLJ (1970) Learning to draw. In: Connolly K (ed) Mechanisms of motor skill development. Academic Press, London, pp 307–335

Boysen ST, Berntson GG, Prentice J (1987) Simian scribbles: a reappraisal of drawing in the chimpanzee (*Pan troglodytes*). J Comp Psychol 101:82–89

Brewster JM, Siegel RK (1976) Reinforced drawing in *Macaca mulatta*. J Hum Evol 5:345–347

Cooper JO, Heron TE, Heward WL (1987) Applied behavior analysis. Merrill, Toronto

Cox M (1992) Children's drawings. Penguin, London

Davis H (1997) Animal cognition versus animal thinking: the anthropomorphic error. In: Mitchell RW, Thompson NS, Miles HL (eds) Anthropomorphism, anecdotes, and animals. State University of New York Press, New York, pp 335–347

Davis W (1986) The origins of image making. Curr Anthropol 27:193–215

Duncum P (1988) To copy or not to copy: a review. Stud Art Educ 29:203–210

Fujita K, Matsuzawa T (1990) Delayed figure reconstruction in a chimpanzee (*Pan troglodytes*) and humans (*Homo sapiens*). J Comp Psychol 104:345–351

Gardner RA, Gardner BT (1978) Comparative psychology and language acquisition. In: Salzinger K, Denmark FL (eds) Psychology: the state of the art. New York Academy of Sciences, New York, pp 37–76

Gardner RA, Gardner BT (1988) Feedforward versus feedbackward: an ethological alternative to the law of effect. Behav Brain Sci 11:429–493

Garner RL (1900) Apes and monkeys: their life and language. Ginn, Boston

Georgopoulos AP (1990) Eye–hand coordination and visual control of movement: studies in behaving animals. In: Berkley MA, Stebbins WC (eds) Comparative perception, vol 1. Wiley, New York, pp 375–403

Gleeson S (1991) Response acquisition. In: Iversen IH, Lattal KA (eds) Techniques in the behavioral and neural sciences. Experimental Analysis of Behavior, Part 1. Elsevier, Amsterdam, pp 63–86

Goodenough FL (1926) Measurement of intelligence by drawings. World Books, Chicago

Goodnow J (1977) Children drawing. Harvard University Press, Cambridge

Harris DB (1963) Children's drawings as measures of intellectual maturity. Harcourt, Brace & World, New York

Henton WW, Iversen IH (1978) Classical conditioning and operant conditioning: a response pattern analysis. Springer, New York

Hopkins WD, Fagot J, Vauclair J (1993) Mirror-image matching and mental rotation problem solving by baboons (*Papio papio*): unilateral input enhances performance. J Exp Psychol G 112:61–72

Hoyt AM (1941) Toto and I. JB Lippincott, New York

Hulstijn W, Mulder T (1986) Motor dysfunction in children: toward a process-oriented diagnosis. In: Whiting HTA, Wade MG (eds) Themes in motor development. Martinus Nijhoff, Dordrecht, pp 109–126

Huxley JS (1942) Origins of human graphic art. Nature 149:637

Iversen IH (1988) How to change behavior? Behav Brain Sci 11:457–458

Iversen IH (1992) Skinner's early research: from reflexology to operant conditioning. Am Psychol 47:1318–1328

Iversen IH, Matsuzawa T (1996) Visually guided drawing in the chimpanzee (*Pan troglodytes*). Jpn Psychol Res 38:126–135

Iversen IH, Matsuzawa T (1997) Model-guided line drawing in the chimpanzee (*Pan troglodytes*). Jpn Psychol Res 39:154–181

Iversen IH, Matsuzawa T (1998) Automated training of line drawing and object movement (in fingermaze and sorting tasks) on a touch-sensitive monitor in captive chimpanzees (*Pan troglodytes*). Presentation at Measuring Behavior 1998, Groeningen, The Netherlands, August (abstract available at http://www.noldus.com/events/mb98/abstracts/iversenl.html)

Iversen IH, Ragnarsdottir GA, Randrup KI (1984) Operant conditioning of autogrooming in vervet monkeys (*Cercopithecus aethiops*). J Exp Anal Behav 42:171–189

Kephart NC (1971) The slow learner in the classroom. 2nd edn. Merrill, Columbus

Kluver H (1933) Behavior mechanisms in monkeys. University of Chicago Press, Chicago

Kohler W (1925) The mentality of apes. Harcourt Brace, New York

Litt S (1973) Shaping up or self-shaping: a look at modern educational theory. J Hum Psychol 13:69–73

Martin C, Pear MJ (1996) Behavior modification. Prentice Hall, Upper Saddle River

Matsuzawa T (1985a) Use of numbers by a chimpanzee. Nature 315:57–59

Matsuzawa T (1985b) Color naming and classification in a chimpanzee (*Pan troglodytes*). J Hum Evol 14:283–291

Matsuzawa T (1990) Spontaneous sorting in human and chimpanzee. In: Parker ST, Gibson KR (eds) "Language" and intelligence in monkeys and apes: comparative developmental perspectives. Cambridge University Press, Cambridge, pp 451–468

McIlvane WJ, Dube WV (1992). Stimulus control shaping and stimulus control topographies. Behav Anal 15:89–94

Mitchell RW (1999) Scientific and popular conceptions of the psychology of great apes from the 1790s to the 1970s: déjà vu all over again. Primate Rep 53:3–76

Morris D (1962) The biology of art. Methuen, London

Premack D (1976) Intelligence in ape and man. Elrbaum, Hillsdale

Romanes GJ (1882) Animal intelligence. Kegan, Paul, Trench, London

Rumbaugh DM (1977) Language learning by a chimpanzee. Academic Press, New York

Rumbaugh DM, Hopkins W, Washburn DA, Savage-Rumbaugh ES (1993) Chimpanzee competence for counting in a video-formatted task situation. In: Roitblat HL, Herman LM, Nachtigall PE (eds) Language and communication: comparative perspectives. Erlbaum, Hillsdale, pp 329–346

Savage-Rumbaugh ES (1986) Ape language: from conditioned response to symbol. Columbia University Press, New York

Schiller PH (1951) Figural preferences in the drawings of a chimpanzee. J Comp Physiol Psychol 4:101–111

Schwartz AB (1994) Direct cortical representation of drawing. Science 265:540–542

Shettleworth SJ (1975) Reinforcement and the organization of behavior in golden hamsters: hunger, environment, and food reinforcement. J Exp Psychol Anim Behav Process 1:56–87

Skinner BF (1953) Science and human behavior. Macmillan, New York

Skinner BF (1971) Beyond freedom and dignity. Knopf, New York

Skinner BF (1974) About behaviorism. Knopf, New York

Smith DA (1973) Systematic study of chimpanzee drawing. J Comp Physiol Psychol 82:406–414

Spinozzi G (1996) Categorization in monkeys and chimpanzees. Behav Brain Res 74:17–24

Tanaka M (1995) Object sorting in chimpanzees (*Pan troglodytes*): classification based on physical identity, complementarity, and familiarity. J Comp Psychol 109:151–161

Thorndike EL (1911) Animal intelligence. Macmillan, New York

Tomonaga M, Matsuzawa T (1992) Perception of complex geometric figures in chimpanzees (*Pan troglodytes*) and humans (*Homo sapiens*): analysis of visual similarity on the basis of choice reaction time. J Comp Psychol 106:43–52

Tuch B, Judy H (1975) How to teach children to draw, paint, and use color. Parker, West Nyack

Washburn DA (1992) Analyzing the path of responding in maze-solving and other tasks. Behav Res Meth Instrum Comput 24:248–252

Washburn DA, Harper S, Rumbaugh DM (1994) Computer-task testing of rhesus monkeys (*Macaca mulatta*) in the social milieu. Primates 35:343–351

Witmer L (1909) A monkey with a mind. Psychol Clin 3:179–205

Witsen BV (1967) Perceptual training activities handbook. Teachers College Press, New York

Woodworth RS, Schlosberg H (1954) Experimental psychology (rev edn). Holt, Rinehart & Winston, New York

13
Object Recognition and Object Categorization in Animals

Masako Jitsumori[1] and Juan D. Delius[2]

1 Introduction

One of the most important attributes of cognitive activities in both human and nonhuman animals is the ability to recognize individual objects and to categorize a variety of objects that share some properties. Wild-living spider monkeys, for example, individually recognize their partners and a large number of other conspecifics quickly and accurately regardless of their highly variable spatial attitudes and also discriminate them from other species (J. Delius, personal observation). Object recognition and object categorization are both equally vital for most of the advanced animals.

The retinal image of an object varies as a function of orientations, distances, lighting conditions, background scenes, and so forth at the time of viewing. Invariance operations are required for animals to identify objects despite variability of retinal stimulation. Observers may recognize the differences in orientation, location, size, and other aspects, but these differences do not obscure the identity of the objects. On the other hand, animals categorize a large number of individual objects into the same classes. There is little or no doubt that animals categorize natural objects sharing some properties, such as edible, water-offering, mateable, threatening, and so forth. The ability for animals to categorize objects enables them to learn about their environments economically with a drastic decrease in the stimulus information that they have to cope with. The processes that underlie object recognition and object categorization may well differ, but both require a common response to a variety of visual inputs. Pooling of a plurality of sensory inputs into fewer but more comprehensive signals is of great ecological relevance for animals. How and to what extent is such information pooling accomplished by animals?

[1] Department of Cognitive and Information Sciences, Chiba University, 1-33 Yayoi-cho, Inage-ku, Chiba 263-8522, Japan
[2] Allgemeine Psychologie, Universität Konstanz, D-78434 Konstanz, Germany

2 Object Recognition in Animals

First, we review recent studies on object recognition in animals. An overview of object recognition is an essential basis for a wide range of empirical and theoretical research focusing on object categorization in animals that we wish to present later in this chapter.

2.1 Orientation Invariance

Orientation invariance in animals has been studied mostly by employing tasks similar to the "mental rotation" problem used with human subjects (see Shepard and Metzler 1971). Hollard and Delius (1982) taught pigeons a matching-to-sample (MTS) task that involved geometric figures as the samples and the same figures and their mirror images as the comparisons. Humans generated a typical reaction-time function that increased monotonically with the angular disparity between the sample and comparison stimuli. Pigeons responded more quickly than humans and produced essentially flat speed functions. It thus seems that pigeons are more efficient than humans at discriminating mirror images, regardless of their orientations in the frontal plane. Hollard and Delius (1982) argued that the excellence of orientation invariance in pigeons may have arisen phylogenetically because of the special demands that the typical avian lifestyle makes on them. Pigeons operate visually predominantly on the horizontal plane where the orientation of objects is largely arbitrary, being relative to the position of the observer. Humans, because of their ground-bound upright stance, mainly operate visually on the vertical plane, where they themselves and most objects have standardized orientations determined by gravity. According to this argument, the deficit of humans is attributable to bioevolutionary adaptation. Since then, a number of studies have been conducted with diverse species: pigeons (Delius and Hollard 1995; Lohmann et al. 1988), baboons (Hopkins et al. 1993; Vauclair et al. 1993), and dolphins (Herman et al. 1993). Ample evidence has been accumulated showing that animals can spontaneously identify images of an object in different orientations.

By using arbitrary different odd stimuli rather than mirror images as comparisons, Delius and Hollard (1995) revealed that neither humans nor pigeons showed a rotation effect. They argued that the absence of rotation effect in pigeons in the tasks where mirror images were served as comparisons is best explained by assuming that pigeons, unlike humans, do not experience any difficulties in discriminating mirror images. It is likely that pigeons may concentrate on local features rather than global shapes, which enables them to immediately discriminate mirror images much in the same manner as humans quickly discriminate arbitrary different odd stimuli. This easy-to-discriminate mirror-images hypothesis predicts that comparison stimuli that are sufficiently difficult for pigeons to discriminate before any orientation disparities may force pigeons to show a mental rotation effect. This notion awaits further research in which local features are arranged to precisely control the difficulty in discriminating paired comparisons. Studies by using tasks that require pigeons to attend only to global shapes but not to any local features are also needed.

2.2 Rotation Invariance

The consequences of rotation in depth are more profound than rotation in the frontal plane. When an object moves, as is often the case if it is a live animal such as conspecifics and predators, the retinal image changes drastically. Different features of the object come into or move out of view for an observer. Similarly, the observer's locomotion often changes the retinal images of an object. Laboratory experimenters have begun to examine rotation invariance or viewpoint invariance in pigeons by using static (Cerella 1977, 1990a, 1990b; Wasserman et al. 1996) and dynamic (Cook and Katz 1999) two-dimensional (2-D) representations of the 3-D objects as stimuli.

Jitsumori, Zhang, and Makino (in preparation) trained pigeons to discriminate frontal views of human faces. As far as we know, there have been no studies that examine rotation invariance with pigeons by using pictures of natural objects. As it is the case for most natural objects, a rotated human face drastically changes not only its global shape but also local features such as eyes, nose, and mouth. It was revealed that pigeons successfully discriminated the faces over untrained depth orientations, but discrimination accuracy decreased when the faces rotated far away from the training angle. Such performance decay, or generalization decrement, is often shown by pigeons and humans when they are trained to discriminate complex geometric patterns rotated in depth. The finding appears to be consistent with the viewer-centered models of object recognition (cf. Edelman and Bülthoff 1992; Ullman 1989), which assume that an object is represented by stored memory view(s) determined by the perspective of the viewer. Jitsumori et al. further trained the same pigeons to discriminate the faces rotating to left and to right. It was expected that training with the dynamic stimuli would possibly broaden the range of invariance with the static stimuli, by increasing collections of stored views that may belong together to a structured single human face rotated in depth. However, the dynamic training failed to broaden the range of invariance. This finding was rather surprising because we have already obtained data clearly indicating that the pigeons having been trained to discriminate dynamic video images of conspecifics showed transfer to a variety of corresponding static scenes (Jitsumori et al. 1999). How can we explain the discrepancy between the findings? Do pigeons see real images of conspecifics but not human faces in dynamic 2-D video representations?

If animals show rotation invariance with a variety of 2-D pictures, then the performance may suggest that they recognize the 2-D pictures as depicting a 3-D object. It is yet not clear whether pigeons see a 2-D video image as a snapshot depicting a particular object moving in a 3-D space. The so-called inverse optics problem assumes that the 3-D shape is inferred based on metrics of the 2-D retinal images and the estimated rotation in depth. So far, most findings with laboratory animals, specifically with pigeons, are better described simply by viewpoint-dependent 2-D mechanisms or stimulus generalization, making it unnecessary to posit a more complex mechanism of object recognition. It is apparent, however, that animals including pigeons in their natural environments readily identify their conspecifics, predators, and other natural objects despite of changes in viewing angle. A natural object, say a live pigeon, changes its 2-D images continuously when it moves in a natural setting. Moreover, particular features, such as gray color, plumage texture, round smooth body, and many others are left unchanged. A large collection of stored views seen in quick succession in the natural setting as well as the orientation-invariant features may provide animals with excellent object recognition in the natural environment. What remains unclear is whether and how laboratory animals recognize unfamiliar 3-D objects based on their restricted 2-D representations.

2.3 Object-Picture Equivalence

We humans do not truly experience a 2-D image of an object as being really 3-D. Nevertheless, we can recognize real objects in 2-D photographs and videos. Do animals recognize and treat pictures as real objects? This issue is directly explored in apes (Savage Rumbaugh et al. 1980), monkeys (Bovet and Vauclair 1998; Tolan et al. 1981; Winner and Ettlinger 1978), pigeons (Cabe and Healey 1979; Delius 1992; Lumsden 1977; Ryan and Lea 1994; Watanabe 1993), and chickens (Bradshaw and Dawkins 1993; Evans et al. 1993; McQuoid and Galef 1993; Patterson-Kane et al. 1997). The basic experimental paradigm mostly used was that animals were first trained to discriminate real objects and then tested for transfer to corresponding pictures (object-picture transfer). The findings are controversial, except for those obtained with laboratory apes that have some previous experiences with picture recognition.

Animals, including humans, may have to learn to "see" real objects in pictures through interaction with the objects and their pictures (Gibson 1986). It has been reported that humans without prior knowledge of pictures (Miller 1973), chimpanzees (Winner and Ettlinger 1978), and monkeys (Bovet and Vauclair 1998) have difficulties in spontaneously recognizing pictures as corresponding to real objects. Jitsumori and Matsuzawa (1991) trained monkeys and pigeons to discriminate orientations (upright versus upside-down) of slides of a wide variety of frontal views of humans. Only monkeys with prior experience of pictures successfully transferred the discrimination to novel slides depicting natural objects such as apes, monkeys, and other animals but not to the slides of artificial objects. Both monkeys with no prior experience of pictures of natural objects and pigeons failed to show transfer to novel slides of animals other than humans.

Delius (1992) and Watanabe (1993) showed substantial object-picture transfer in pigeons. The objects depicted in these experiments were rather simple. The latter study used a very small number of seedlike (edible) objects and nonseedlike objects (nonedible), and 2-D cues may thus be sufficient to support transfer from real objects. This may be the case even with a relatively large number of spherical and nonspherical junk objects used by Delius (1992; see also Delius et al. 1999).

The findings in these studies clearly demonstrated that pigeons recognize consistency between real objects and their 2-D representations to some extent, but perceptual similarities suffice to account for the findings. This similarity-based generalization account agrees with the findings that animals often have more difficulties in picture-object transfer than object-picture transfer (Cole and Honig 1994). If bidirectional object-picture transfers are not fulfilled, it is premature to conclude that animals have an innate command over object-picture equivalence.

We do not yet know how animals, particularly avian species, perceive static and dynamic video images. Videos are designed for the human eye having a particular scan rate, color mixing, and pixel density. It is well known that avian color vision is very different from that of humans. The color video images that are adequately realistic for humans are, therefore, most likely chromatically false for birds. Also, pigeons have a higher flicker-fusion threshold than humans (Emmerton 1983; Hendricks 1966; Powell 1967). The video monitor (60-Hz or 50-Hz scan), designed for the human eye (for which the frequency is about 30 Hz at the approximate brightness of video monitors), possibly breaks up the image for pigeons (see also Jitsumori et al. 1999). So far, there is no clear evidence that monkeys and pigeons recognize pictures as being more than only partially equivalent to real objects.

3 Categorization in Animals

Information pooling of a variety of stimuli of a similar nature has been studied in humans under the heading of concept formation. The responses of interest in the human studies are often verbal labels, i.e., words that through suitable experience come to stand semantically for the particular collections of stimulus items. Humans learn to classify items into the same classes at several different levels of categorization. For example, we recognize a carrot as a member of the named categories "carrot," "vegetables," "plants," "nonanimals," "natural objects," and so forth. Members of a given category thus can be members of other categories. In this sense, naming may constitute a sort of one-to-many MTS, with a given object as a sample and its category names as correct comparisons. Humans may learn about the conceptual classification and the precise meaning of a given object, by integrating new information into knowledge representations through experiences at different levels of categorization. Because of the very existence of language competence of humans, concept formation has been long assumed to be a cognitive competence uniquely restricted to humans.

Early reports about the classification of objects or pictures thereof by animals were often headed with "concept discrimination" or "categorization." Fersen and Lea (1990) correctly pointed out that it is important not to overinterpret the words "concept" or "category" (see also Lea 1984). They argued that even if animals can possess concept and categorize stimuli by using them, it is still not proven that they discriminated complex stimuli, for example color slides including people from those not containing them, by using the concept "people". Moreover, as we discussed previously it is doubtful that monkeys and pigeons recognize pictures as being fully equivalent to real objects. It is often impossible for us to establish the underlying semantic structure of the representation of objects that animals learned from experiences with categorization demands defined by experimenters, but we can examine how animals categorize different stimuli into the same classes depending on the task requirements. A comparative approach with animals may provide us opportunities for investigating categorization without language. Comparisons not only between human and nonhuman animals but also between closely or distantly related nonhuman species may contribute to our understanding of evolutionary and ecological significance of concept formation.

3.1 Generalization Within Classes and Discrimination Between Classes

Keller and Schoenfeld (1950) remarked that "concepts are not things possessed by organisms or held in their minds; instead, they constitute a particular pattern of behavior" (see Astley and Wasserman 1992, p. 193). Keller and Schoenfeld defined concepts as involving "generalization within classes and discrimination between classes," and thereafter concept learning in animals was often equated with animals' categorization performances. In most categorization experiments, animals were first trained to discriminate a certain number of different stimuli. Half the stimuli, defined as the positive set, depicted a particular natural object, for example, people. The other half of the stimuli, defined as the negative set, depicted other objects but not people. Animals were rewarded for responding to the positive stimuli and not rewarded for responding to the negative stimuli. After the animals learned to discriminate these stimuli according to the experimenters' own concept of "people," they were tested with new stimuli. If animals would emit more responses to novel pictures depicting people than the nonpeople pictures, then it was confirmed that the animals had learned to categorize pictures according to the concept held by the experimenters. Transfer to novel stimuli is thus assumed as indicating that categorization was not just a product of memorization of the positive and negative stimuli used for training.

Following the pioneering work by Herrnstein and Loveland (1964) that demonstrated pigeons could classify new instances of people/nonpeople slide pictures, it has been well documented that animals, particularly pigeons, can classify photographs of natural objects such as people, fish, trees, other pigeons, and bodies of water. In spite of a growing interest in abstract processes in nonhuman primates, little work on concept discrimination has been undertaken with monkeys. Schrier et al. (1984) reported that stump-tailed macaques discriminated categories of people and monkeys, but transfer to new instances was lower than had been reported for pigeons. Schrier and Brady (1987) presented new slides at every trial of every experimental session without repeating any individual slide, and they successfully trained rhesus monkeys to classify slides showing scenes with and without people. On the other hand, D'Amato and Sant (1988) trained capuchin monkeys with a limited number of exemplars and still obtained high levels of transfer to a wide variety of new instances. Although this finding seems to provide strong evidence for concept discrimination in monkeys, they argued that the significant transfer to new instances might not be governed by abstract processes. Instead, "the monkeys might have identified a limited number of features, alone or in combination, such as a pair of eyes contained within a closed oval, which, if present in a new slide to a reasonable degree of similarity, would elicit a person categorization" (D'Amato and Sant 1988, p. 54).

Roberts and Mazmanian (1988) showed that transfer to novel stimuli was largely determined by the degree of similarity among the items within a category. They trained rhesus monkeys and pigeons in categorization tasks that differed in degree of abstraction. At a low level of abstraction, the positive stimuli were pictures of a single bird species (kingfisher), and the negative stimuli were pictures of other bird species. At a more abstract level, the task was a discrimination between pictures of birds in general and those of any other kinds of animal. The most abstract task required subjects to discriminate pictures of animals in general from pictures of nonanimals. Although pigeons and monkeys successfully learned the tasks, reliable transfer to novel stimuli occurred only with the task at the lowest level of abstraction. The authors reported that there was far more similarity between the kingfisher pictures than between the bird or the animal pictures.

Wasserman and Astley (1994) correctly remarked that concept learning experiments reveal concepts in animals only in the sense that they show us how animals lump images together or perceive similarity between visual stimuli (see also Roberts 1996). Objects in the same natural categories are more perceptually similar to one another than objects from different categories. What has been shown by most of the findings of concept learning experiments in animals is that such categorical coherence is an essential property of natural objects and is exploited by the visual systems of diverse animal species, including humans. Note, however, that what animals tend to see in color slides or video pictures is another issue. It is quite likely that objects belonging to the same natural category and their corresponding pictures often possess some perceptual features in common, regardless of the absence or presence of the true picture-object equivalence.

Importantly, animals often show performances suggesting a categorical coherence even before they are explicitly trained in categorization tasks, as revealed in a chimpanzee by Fujita and Matsuzawa (1986) by using a sensory reinforcement procedure and in monkeys by Sands et al. (1982) by using a same/different conditional discrimination procedure. Thus, perceptual categorization takes place without any previous training of concept discrimination. It is true that human categorization behavior cannot be fully accounted for by perceptual similarity, but it is also true that similarity among the to-be-categorized stimuli may play an important role for humans and animals in learning object categories.

3.2 Polymorphous Categories

Categorization studies in animals have repeatedly shown that similarity within categories and dissimilarity between categories facilitate categorization learning that transfers to novel exemplars. Most findings on picture categorization by animals can be explained by conjunction of exemplar-based learning and similarity-based transfer to novel stimuli. Transfer by stimulus generalization based on perceptual similarity is thus well established as an empirical phenomenon, but it is poorly understood as a causal process. Herrnstein (1985) concluded that the mechanisms underlying the excellent discrimination of natural categories by animals still remain obscure.

Jitsumori and Yoshihara (1997) trained pigeons to discriminate the facial expressions of happiness and anger of 25 different human subjects. Two photographs of the same person with the different facial expressions (which were bound to be perceptually similar) had to be grouped into different classes, while faces of different persons with the same facial expression (which were not as likely to be perceptually similar) had to be grouped into the same class. Thus, the birds had to classify rather dissimilar individual faces into the same classes based on their common facial expressions. The birds learned the discrimination and then showed high levels of transfer to novel faces, a finding demonstrating that their discrimination was not restricted to the particular faces used for training and that pigeons are capable of classifying relatively dissimilar stimuli into the same classes. This finding confirmed previous results by Wasserman et al. (1989) on pigeons trained to discriminate four types of human facial expressions (happiness, anger, surprise, disgust). Jitsumori and Yoshihara (1997) further found that although the facial expressions could be readily discriminated on the basis of a single local feature, either "eyes-and-eyebrows" or "mouth", the response of the birds was under the control of multiple facial features. It was revealed, however, that the birds did not used the global configuration of facial features as a cue to discriminate the facial expressions, as it was demonstrated that they did not show the so-called Thatcher illusion. Instead, the birds treated the multiple facial features independently and additively integrated them to determine their responding.

One explanation of the ability to classify different items into the same group is that animals respond to constellations of features. We can list features of objects belonging to a particular natural category. For example, greenness, leafy, branchiness, verticality, woody, and many other features are parts of a "tree," but we cannot pick out a single feature or a particular combination of features that readily define the "tree" category versus the "non-tree" category. We make a "tree" response to a giant pine tree, a maple tree with scarlet leaves, a tree in winter with its branches covered by snow, and perhaps also even to a large bush or an ivy running up on a wall. No single feature, however, is likely to be a necessary or sufficient condition for deciding the category membership. Categories characterized by such constellations of features are referred to as *polymorphous* categories.

Most natural categories have a polymorphous nature. A giant pine tree and an ivy running up a wall may not be perceptually similar to one another, but the "tree" category as a whole involves members having feature-based similarities. The feature-based account assumes a process of multiple feature analysis and integration, which at the behavioral level involves responding coming under the control of a number of more or less independent features, each of which may only correlate weakly with category membership (Lea and Ryan 1983).

Positive Set   Negative Set   **Fig. 1.** An example of the 2-out-of-3 polymor-
                                      phous rule. Positive features; *uppercase*, "A", *bold*.
B   *A*   a   b   *B*   *a*   Negative features; *lowercase*, "B", *italic*

A   *b*

D   𝔸   *d*   𝔹

A simple polymorphous rule is the 2-out-of-3 rule. The stimuli differ along 3 two-valued (positive and negative) dimensions, and membership of the positive category depends on possession of two out of the three positive features. Figure 1 illustrates an example of the artificial 2-out-of-3 polymorphous rule. If this rule is learned with the stimuli involving only two of the three positive or negative features (top column), then transfer should occur to the novel stimuli having all three positive or all three negative features (middle column), and even to the stimuli having one of the three features replaced with a novel one (bottom column). If feature analysis is an adequate description of object categorization by animals, artificial categories constructed in a similar way should be discriminated easily and transfer should occur to novel stimuli. An advantage of using artificial categories is that it is possible to analyze the way in which well-defined features control performances.

Since the earlier work by Lea and Harrison (1978) in pigeons, it has been often shown that pigeons (Aydin and Pearce 1994; Huber and Lenz 1993, Jitsumori 1993; Lea et al. 1993; Lea and Ryan 1990; Fersen and Lea 1990), rhesus monkeys (Jitsumori 1994), and baboons (Dépy et al. 1997) could learn to discriminate artificial polymorphous categories. Jitsumori (1993) used arrays of symbols on colored backgrounds as stimuli. The relevant features were symbol color (black or white), symbol shape (circle or triangle), and background color (red or green). The symbols were presented in ten different arrays, with each pattern containing three, four, or five identical symbols. Pigeons learned to discriminate 60 stimulus patterns containing two of three positive or negative features. The pigeons demonstrated a high level of transfer with the novel stimuli containing all the three positive or negative features and even with the stimuli having one of the three features replaced with a novel one; symbol color was gray, symbol shape was star, or background color was blue, with the other two features both positive, one positive and one negative, or both negative. The novel stimuli with more positive features generally controlled higher rates of responding. This finding was explained by assuming that the relevant features were additively integrated to determine category membership. Its operations can be accounted for by a very simple artificial neural network with a hidden unit that receives convergent excitatory inputs from several feature units.

3.3 Prototype Learning

On the basis of feature additive learning, the most pronounced discrimination is predicted to occur between the stimuli containing the extreme positive or negative values in all feature dimensions, a "super-releaser" effect, as suggested by Lea and Harrison (1978). In the two-out-of-three polymorphous categories, the stimuli with all three positive and with all three negative features are denoted "super stimuli" or the best exemplars. Even when the super stimuli are not in use during acquisition phase, they will be best discriminated during testing phase if the feature structure have been learned. Aydin and Pearce (1994) obtained clear evidence for a super-releaser effect in pigeons and argued that the super stimuli could be regarded as prototypes because these stimuli contained all the positive or negative features that had appeared more frequently than any other in the set of training stimuli used during acquisition phase. This notion is in line with the frequency theories of feature abstraction and concept learning in the human literature (Goldman and Homa 1977; Neumann 1974, 1977).

Huber and Lenz (1996) used a prototype as defined by central tendency of category. They trained pigeons to discriminate positive and negative sets of Brunswik faces differing on four 9-valued features. Within the four-dimensional stimulus space, the positive and negative stimuli were located around a standard stimulus (prototype) on concentric hyperspheres with different radii. The positive set was located closer to the prototype than the negative set. Pigeons showed stronger responding to the prototype than to any other training stimuli, and this finding was accounted for by the formation of a prototype. The authors, however, pointed out that the prototype effect could have readily arisen through a traditional Spencian generalization peak-shift, because the prototype was located a greater distance away from the negative set than the positive set. A similar peak-shift explanation could also account for the super-releaser effect found with the polymorphous stimuli mentioned earlier.

Both Mackintosh (1995) and Jitsumori (1996) investigated this issue. Pigeons were trained to discriminate two artificial categories without being exposed to the prototypes of the categories. The prototype in each category was designed to represent the central tendency of the positive or negative training stimuli in the multidimensional stimulus space. After training, pigeons were tested with the positive and negative prototypes and other novel stimuli that were either closer to or further from the category boundary than the prototypes. Jitsumori (1996) obtained a category-centering prototype effect rather than a peak-shift effect, i.e., the best discrimination occurred between the prototypes rather than between the stimuli located further away from the category boundary. This prototype effect could not be explained by peak-shift, but feature frequencies were distributed with the peaks at the prototypes during acquisition phase. Thus, the prototypes contained the features that occurred most frequently in their respective training categories. The category features used by Mackintosh (1995), on the other hand, occurred equally frequently during the acquisition phase. Mackintosh (1995) obtained a prototype effect when the birds had received preliminary training to peck the positive stimuli before categorization training began. Conversely, a peak-shift effect was obtained when the birds received categorization training without previous training with the positive stimuli alone. The peak-shift effect was explained by the superpositive or supernegative stimulus being located far away from the category boundary and thus having fewer elements in common with members of the complementary category. Mackintosh suggested that the prototype and peak-shift effects are two possible solutions for the categorization problem.

Related to this issue, recent work by Herbranson et al. (1999) examined categorization of multidimensional stimuli by pigeons from the point of view of an optical decision model (Ashby and Maddox 1998). Further studies in this light are required to clarify the underlying behavioral processes of the prototype effect in animals. It is highly possible that the prototype learning shown by pigeons in Jitsumori (1996) was generated by the feature-frequency distributions with the peaks at prototypes. The features were thus highly correlated with one another in each category, as is often the case in natural categories; for example, greenness of a tree is correlated with leafy, and verticality or branchiness is correlated with woodiness (Herrnstein et al. 1976). Feature structure of a category is a critical factor that may determine how animals solve the categorization problem. The search for artificial features and feature structures that may correspond more closely to the way in which the animal perceptual system actually partitions the stimulus information will probably be necessary.

Polymorphous categories are often said to possess the so-called family resemblance (Wittgenstein 1953). Figure 2 shows a set of human faces used by Makino and Jitsumori (in press; experiment 2); A, B, C, and D are real faces, AB, AC, CD, and BD are 50% morphed images created from the paired real faces (AD and BC are not shown in the figure), and ABCD is a prototype created by averaging all four real faces. The stimulus set was constructed to mimic the family resemblance based on the finding in experiment 1 that pigeons discriminated a 50% morphed image from the real faces from which it was created, but that they still perceived the 50% morphed image as being similar to both the parents faces. In experiment 2, pigeons were trained to discriminate two sets of 50% morphed images created from the pairs of eight real faces that were randomly assigned into the two sets. Thus, there were six 50% morphed images in each of the two training sets. One set was positive and the other was negative in a go/no-go discrimination procedure. The birds were then tested in extinction with the stimuli including the 50% morphed images used for training, the real faces, and the prototypes. Facial characteristics of each of the 50% morphed images might have been strengthened in the real faces used as parents, so a super-releaser effect predicts the best discrimination with the real faces.

Figure 3 shows discrimination ratios during testing. The most pronounced discrimination, even slightly better than to the stimuli used for training, occurred to the prototypes rather than to the real faces. Thus, the birds showed a prototype effect rather than a super-releaser effect. It should be noted, however, that the birds still showed pronounced discrimination to the real faces that were not particularly similar to one another within each of the two sets. One may argue that this was

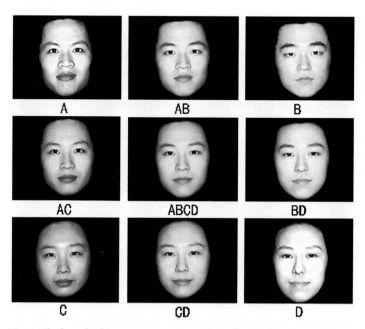

Fig. 2. Black-and-white representations of the faces belonging to the same category. *A, B, C,* and *D* are real faces; *AB, AC, CD,* and *BD* are their 50% morphed images (AD and BC are not shown) used for training; *ABCD* is defined as a prototype created by averaging the four real faces

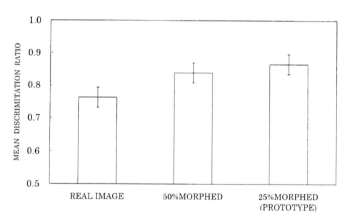

Fig. 3. Mean discrimination ratios for the real faces (REAL IMAGE), the 50% morphed images used for discrimination training (50% MORPHED), and the prototypes created by averaging the real faces in each of the two sets (25% MORPHED)

accomplished simply based on stimulus generalization from the 50% morphed images used for training. More importantly, we instead emphasize the fact that the real faces were joined together via the very existence of other members in each set. That is, perceptually dissimilar stimuli are treated more similarly to one another in the category as a whole, and this is the essential nature of the family resemblance of a natural category.

An abstraction of prototypes may simplify classification of a large number of stimuli because the category membership of each stimulus can be identified by generalization, or similarity to the prototypes. It should be noted, however, that the previously described behavioral studies do not provide any formal account as to the perceptual-cognitive processes by which animals distill a prototype. Theoretical work with human concept has demonstrated that prototype abstraction can be explained on the basis of distributed memory storage of exemplars or features. This position assumes that categories are formed automatically by the separate memory traces interacting in storage (cf. Anderson and Hinton 1981; Knapp and Anderson 1984; see also Schyns 1991). In this sense, prototypes may not be abstract entities, but rather a sum of previously learned exemplars or features held in memory.

3.4 Levels of Stimulus Control

We now proceed to distinguish different levels of categorization, by paraphrasing Herrnstein (1990). The most basic level (level 1) is the grouping of stimuli that are too physically similar for the organism to distinguish them perceptually. The next level (level 2) is categorization by rote that would involve stimuli that were in principle distinguishable to the organism but still subject to spontaneous interstimulus perceptual generalization. At a higher level of categorization (level 3), the process of perceptual generalization is broadened through experience to include more dissimilar stimuli and would form limitless open-ended categories. At a still higher level (level 4), the categorization is abstracted from coincidences over several different properties or nonsimilarity based functional qualities. Categorization at this level referred to as "concepts." At a further level (level 5), the categorization is based on common abstract relations between conceptual stimulus classes. We discuss categorization at level 4 and level 5 in the following sections. Note, however, that nothing prevents the processes responsible for the different levels of categorization to work in various degrees of conjunction, an issue that is of importance later.

3.5 Functional Equivalence Classes

Formation of functional equivalence classes originally reported by Vaughan (1988) with pigeons is often referred to as the categorization at level 4. Vaughan (1988) used 40 pictures of trees randomly divided into two sets of 20 pictures each for pigeons to learn functional equivalence classes. He trained pigeons to peck the pictures in one set and not to peck the pictures in the other set in a successive discrimination task. When pigeons learned to discriminate the two sets, the reinforcement allocations were reversed. The pigeons had to peck the pictures previously not rewarded, and conversely not to peck the pictures previously rewarded. After this reversal was learned, the reinforcement allocations were again reversed, and so several times more. After this treatment, pigeons only required experiences with a few stimuli after a reversal to correctly discriminate all the remaining stimuli of the sets. That is, the 20 pictures in each set had become functionally equivalent among each other. Importantly, because the stimuli were all tree pictures randomly divided into two sets, there were no specific perceptual features that could group the pictures into each set. It is thus clear that a common contingency of reinforcement among the members in each set enabled the pigeons to learn the functional equivalence classes.

In Jitsumori and Ohkubo's (1996) experiment pigeons were thoroughly trained to respond to upright-oriented slides depicting people or birds and not to respond to the same slides when presented in upside-down orientation. A subsequent reversal training conducted by using a subset of the people slides immediately transferred to the remainder of the slides of people used during original training but not used for the reversal training. The reversal transfer, however, did not spread at all to the bird slides. Conversely, an analogous partial reversal training employing a subset of the bird slides did not transfer at all to the slides of people, and this was so even after the pigeons had also undergone several repeated reversals (Jitsumori and Ohkubo, unpublished data). The dissimilarity between people and bird images clearly was an impediment for a shared classification of the "same" orientation stimuli that had been implemented by repeated synchronous reversal training. That is, although the pigeons, in the original training, successfully learned to categorize the people and birds pictures based on their orientations, the pigeons did not treat upright humans and upright birds as equivalent. Similarly, upside-down humans and upside-down birds were not equivalent. This finding suggests that, so far as concerns the pictorial stimuli we used in this study, pigeons may not be able to learn abstract categorical relations at level 5 as defined by Herrnstein (1990).

Note that Vaughan (1988) used pictures of trees that were just as similar to each other within classes and between classes. It is thus possible that similarity within classes favors equivalence formation even if there is no special dissimilarity between the classes. The advantage of the ability for animals to learn functional categories, as already mentioned, is expected to be linked to their necessity to identify diverse items, such as edible, mateable, threatening, water-offering, and so forth. The stimuli having such functional equivalence properties may share common responses and response outcomes but could also be perceptually similar to one another. Similarity or categorical coherence as a fundamental nature of natural objects may in practice help animals to learn about functional equivalence classes in their natural environments.

3.6 Nonsimilarity-based Associations

In a true concept rather than mere stimulus categorization, Herrnstein (1990) argued that "the effects of contingencies applied to members of the same set propagate to other members more than can be accounted for by the similarities among members of the set" (for a more detailed earlier argument, see Lea 1984). People often tag a verbal label to objects classified into the same category. Tagging a verbal label is a kind of common response that is symbolically associated to a set of different stimuli. In an attempt to study a higher-order process in animals, researchers have examined association among members of learned categories that are joined by common responses and/or response outcomes. Those studies employ physically different stimuli (colors, tilted lines, geometric shape, and so forth) to examine associations that are not based on perceptual similarity.

One group of studies in this line is those that examined the so-called mediated generalization or secondary generalization. Astley and Wasserman (1998; see also Wasserman et al. 1992) obtained evidence, albeit moderate, indicating that association with a common response joined perceptually different stimulus classes together into a superordinate category. Photographs of people plus chairs (categories C1 and C2) and of cars plus flowers (categories C3 and C4), for example, constituted two classes. The pigeons were required to classify pictures from different categories into the same class by responding to two different buttons, R1 and R2. R1 was the correct response when pictures from C1 and C2 were shown, whereas R2 was the correct response when pictures from C3 and C4 were shown. Reassignment training then linked new responses to one category from each pair, R3 being associated with C1 and R4 being associated with C3. In the test, transfer of the new responses to the nonreassigned members of the categories, C2 and C4, and to novel stimuli from all four categories was examined. Although performance accuracy with nonreassigned stimuli was far below the accuracy with novel reassigned stimuli, transfer occurred at levels that were significantly greater than chance. The result suggests that animals can treat stimuli from perceptually dissimilar categories in the same way via association with a common response.

Urcuioli, Zentall, and their colleagues have examined a phenomenon referred to as "common coding" that makes perceptually distinctive stimuli equivalent to one another (Urcuioli et al. 1989, 1995; Zentall et al. 1991, 1993, 1995). In these studies, pigeons were initially trained with a version of the symbolic MTS task in which one comparison stimulus is correct for two different samples and the other comparison stimulus is correct for the other two samples. The pigeons learned to match two (or more) sample stimuli to one comparison stimulus, thus the many-to-one matching-to-sample (MTO-MTS) designation of the task. Much as in the reassignment training used by Astley and Wasserman (1998; see earlier), a new comparison stimulus was then associated to one of the samples. The subsequent tests examined whether the new association would propagate to the other sample stimulus. It was shown that the untaught sample-comparison relations emerged.

This finding was explained by assuming that sample stimuli associated with the same comparison stimulus are commonly coded. In other words, as a result of original MTO-MTS training, the presentation of a sample gives rise to an anticipation, or memory recall, of the corresponding comparison (Urcuioli 1996). This idea is based on findings about the memorization strategy adopted by pigeons (Zentall et al. 1989), which suggests that the samples are coded prospectively (e.g., both red and vertical samples are coded as "circle comparison") rather than retrospectively (e.g., presentation of circle comparison gives rise to representation of the red or green sample) in MTO-MTS. It is said that the prospective code may act as an implicit mediator between the different samples being associated with a common comparison stimulus.

Another source of evidence for mediated generalization comes from transfer via a common outcome expectancy that can serve as a mediator between different samples (Edwards et al. 1982; Urcuioli 1990). As we reviewed earlier, much has been found out about mediated generalization in pigeons by using a variety of events as mediators, such as a common response, a common comparison stimulus, a common response outcome (see also Zentall 1998), and a common delay or probability of food reinforcement (Astley and Wasserman 1999). Urcuioli (1996) has remarked that mediated generalization is the primary process underlying acquired equivalences in nonhuman animals.

3.7 Stimulus Equivalence

A behavioral analysis of the stimulus equivalence underlying concept formation was originally proposed by Sidman and Tailby (1982). In the version commonly used with pigeons, a given arbitrary stimulus is shown on the middle key as a sample. When it is pecked, two comparison stimuli are presented on the left- and right-side keys. If the pigeon pecks the comparison stimulus determined by the experimenter as matching the sample stimulus, it is rewarded with food. If the pigeon pecks the other stimulus, defined as not matching the sample, it is penalized by time-out. The symbolic MTS training trials promote the animal learning to match the sample stimulus, say A1, with choosing the arbitrarily determined comparison stimulus B1 and to match A2 with choosing B2, and thus to establish "if A, then B" relation (A→B).

As evidence for the command of the equivalence class {A, B, C}, Sidman determined the emergence of untaught relations: reflexivity, symmetry, and transitivity (see Sidman 1994). That is, it is necessary to show that the organism which has learned A→B and B→C relations is able to cope with MTS trials that require the application of the additional A→A, B→B, C→C (reflexivity), B→A, C→B (symmetry), and A→C (transitivity) relations. These relationships should emerge as properties of the learned equivalence class {A, B, C}. Sidman considered that the symmetry property incorporated indirectly the reflexivity property. The demonstration of a symmetrical transitivity C→A relation thus represents a shortcut demonstration of the equivalence class {A, B, C}. The question asked in this paradigm is thus whether the equivalence class would be formed by learning the "if—then—" basic relations, A→B and B→C.

Language-competent humans usually pass all these tests without any difficulty, whereas language-deficient humans tend to fail at one or other tests and only master the equivalence after additional special training (Sidman et al. 1982). Children only begin to command equivalence classes when they reach the age of about 5 years and become linguistically fully capable. In animals, pigeons have been generally found not to exhibit symmetry and transitivity (D'Amato et al. 1985; Lipkens et al. 1988). Kuno et al. (1994) found that one pigeon of four exhibited transitivity but the issue of the symmetry was circumvented. Otherwise, positive findings were obtained only when pigeons were tested under the procedure in which the previously discussed mediating generalization was established via response outcomes or overt behavior (Steirn et al. 1991; Zentall et al. 1992; see also review in Zentall 1998).

Findings in nonhuman primates are controversial. Monkeys were observed to show clear evidence of transitivity by D'Amato et al. (1985) but not by Sidman et al. (1982). Response-mediated transitivity and symmetry were shown by McIntire et al. (1987), however. Chimpanzees would be a most likely species to demonstrate equivalence, in view of their similarity to humans and their excellent ability for object classification as reported in other chapters of this book. Yamamoto and Asano (1995) found that a chimpanzee demonstrated spontaneous transitivity but only when the chimpanzee had been previously and explicitly taught the symmetry relations. Tomonaga et al. (1991) demonstrated that one of three chimpanzees spontaneously showed evidence of symmetry, but the effect diminished rapidly across sessions, each involving eight unreinforced test trials. Under standard symbolic MTS procedures, there has been no strong evidence for emergence of symmetry in nonhuman primates.

Perhaps language-competent humans are biased to interpret the A→B relation as also implying its symmetrical relation B→A, to the extent that it often is a source of erroneous deductions (Rips 1994) or logical extensions. Young children, language-deficient humans, and animals seem not to be so disposed. As we discussed earlier, symmetry inevitably incorporates an overall reflexivity property. Reflexivity involves a kind of sameness recognition of a stimulus occurring in a variety of contexts. For example, the word "dog" for language-competent humans is the same even when it appears in different sentences. Thus, reflexivity is a part of basic language comprehension in humans. It is possible that the reflexivity property that is not directly taught in the symbolic MTS task in animal subjects might cause the difficulty with the emergence of the symmetry relation. During symmetry test trials, a given stimulus appears at a location and at a time that are different from those during training trials. Iversen et al. (1986) argued that pigeons do not treat, for example, a vertical line that on training trials appears as a sample on the center key as being identical to a vertical line that on symmetry test trials appears as comparison on a side key. This problem may be closely related to the issue of previously discussed invariance operations in object recognition by animals.

Another possibility is that the difficulty is the result of the characteristic of the symmetry tests that require subjects to show a not specifically trained behavior in a novel situation. For example, animals trained with red and green fields as samples and vertical and horizontal lines as comparisons have solely learned to compare and choose the line stimuli to make a choice response when a color sample is presented. The symmetry tests require animals to emit an untrained, spontaneous choice response with respect to the color stimuli now presented side by side under a new condition in which a line stimulus serves as a sample. This notion suggests that establishment of a special generic comparison response as they occur in the chimpanzee-language studies may possibly promote the emergence of symmetry (Savage-Rumbaugh et al. 1983; see also symmetry demonstration in a sea lion in Schusterman and Kastak 1993).

The emergence of symmetry and the emergence of untrained choice response may be the separate issues that are inevitably confounded in the standard symbolic MTS procedure. Moreover, the emergence of symmetry requires subjects to switch the functions of sample and comparison stimuli that have been taught thoroughly by extensive basic training. The symbolic MTS task may not be particularly suitable for animals to learn stimulus equivalences.

3.8 Equivalence Network Formation

Research on humans has shown that sequential exposure to individual associative linkages between stimuli facilitates the development of equivalence classes and that the associative strength between stimuli within an equivalence class is inversely related to the intervening nodal distance (Fields et al. 1995). Nodal distance is the number of nodes through which a stimulus has been linked by training to at least two other stimuli (Fields et al. 1984). The linkage network among category members is critical for the formation of any large class of stimuli. When a new stimulus is learned to be equivalent to a given class member, this relation would propagate spontaneously to other linked members of the category.

We have examined the formation of such a linkage network in pigeons using a synchronous reversal method (Jitsumori et al., submitted). Two groups of pigeons were trained to form linkages between members of two sets of dot patterns: A1, B1, C1, D1 in Set 1 and A2, B2, C2, D2 in Set 2. For one group of pigeons, the stimuli were designed to be physically dissimilar between the two sets yet physically similar within each of the sets (group Sim). For another group of pigeons, all stimuli were physically equally dissimilar from each other and arbitrarily assigned into the two sets (group Dis). Figure 4 shows the light-emitting diode patterns used as stimuli. The number of coincidentally lit diodes served as an index of physical similarity between given pairs of patterns.

We trained pigeons to learn two or more linkages step by step, and then tested the untrained relations that would possibly emerge from the trained relations. Figure 5 summarizes the overall structural design of the experiments, including stimuli X and Y used later for the Sim Group. Note that, because the two sets of stimuli were used in a fully complementary way to each other, we used the notation A, B, C, D in the figure. The thick lines with arrows represent the linkages trained and then tested, the thin lines with arrows represent the linkages only tested, and thick lines without arrows represent the linkages trained but not tested.

We first trained the pigeons to learn the associative linkage between A and B in each set. The pigeons were first trained with the stimulus pairs A1+A2-, A1+B2-, B1+A2-, B1+B2- (stimulus pairs with + and - signifying rewarded and penalized stimuli, respectively) in a forced-choice simultaneous discrimination task. When the pigeons reached a high discrimination criterion, the reinforcement allocations were reversed, thus, A1-A2+, A1-B2+, B1-A2+, B1-B2+. When the criterion was again reached, the reinforcement allocations were again reversed. Reversals were repeated until the birds reliably learned to switch their choices immediately after reversals. We expected that the pigeons would learn functional equivalence between A and B across several synchronously reversed sessions. In other words, it

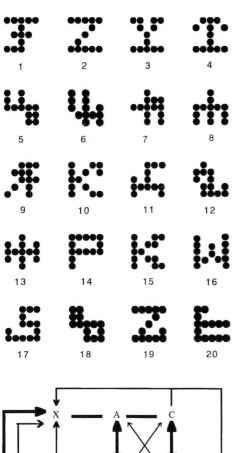

Fig. 4. Light-emitting-diode patterns used as stimuli. The patterns *1* to *8* were used with the similarity (Sim) group, with the patterns *1* to *4* in one set and *5* to *8* in the other set. The patterns *13* to *20* were used with the dissimilarity (Dis) group; the patterns *9* to *12* were novel dissimilar stimuli used with Sim group

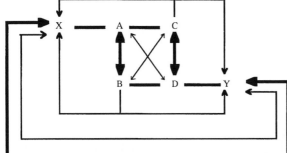

Fig. 5. Associative linkages trained and tested. The *thick lines with arrows* represent the linkages trained and then tested, the *thin lines with arrows* represent the linkages only tested, and the *thick lines without arrows* represent the linkages trained but not tested

would be learned that "A1 and B1 go together" and "A2 and B2 go together," regardless of whether they are positive or negative. Similarly, the birds were trained to form CD, AC, and BD linkages in sequentially introduced training stages. We then examined whether the pigeons would show untaught AD and BC associative linkages via transitivity.

The AD linkage was tested in the following manner. First, the A1+A2− training pair was presented until the birds discriminated it correctly, and then still within the same session, D1D2 was presented in unreinforced probe trials. In the next session, contingency of reinforcement for the training pair was reversed, i.e., the training and test pairs would be A1−A2+ and D1D2. If the birds formed the AD linkage, they would choose D1 during the first test session but D2 during the reversed session. For the remaining two test sessions, the training and test roles of the stimulus pairs were exchanged symmetrically. The training pair was D1+D2- in one session and D1-D2+ in the other; the test pair was A1A2. Only when pigeons showed a symmetrical bidirectional transfer ("A goes with D" and "D goes with A"), did we assume that the linkage was properly formed without direct training in each set ("A and D go together"). Similarly, the explicitly trained AB and CD linkages and the untaught BC linkage were tested. If the birds showed emergence of untaught AD and BC associative linkages, we may conclude that they learned the pairwise {A, B, C, D} equivalence classes.

The Dis group pigeons showed difficulties in learning the second-taught CD linkage. However, the untaught functional equivalence AD did emerge via transitivity (if "A and B go together" and "B and D go together", then "A and D would go together") in two of the four birds and thus they learned to form the {A, B, D} equivalence class. It was clearly demonstrated that a stimulus (B) functions as a node so as to connect the other stimuli (A and D) without direct training. It should be emphasized that there could be no other explicit or implicit mediators that might be able to bridge A and D. All the Sim group pigeons (five birds) learned to form the {A, B, C, D} equivalence class, suggesting that class-appropriate responding was clearly promoted by within-sets similarity in this group.

We then attempted to teach the Sim group additional associations using the stimuli labeled X1, Y1 and X2, Y2. These new stimuli were physically dissimilar to one another and also to the familiar stimuli A, B, C, and D. The pigeons failed to learn the AX linkage taught first, whereas they successfully learned the DY linkage taught second. Four of the five birds showed transfer of the newly learned DY association to the other familiar members, B and C, in each class. They then reasonably failed to spontaneously derive the XY association due to the disconnection between A and X (the AX nonequivalence). Thus, the birds learned to form the generalized equivalence class consisting of familiar similar and novel disparate stimuli {A, B, C, D, Y}. They were then given extensive XY training, and the training was not successful in teaching them this linkage. For these birds to learn the XY linkage, the previously learned equivalence network should be reorganized so as to include X into the network; otherwise, the network linkages would be inconsistent. This learning is a matter of behavioral plasticity, and we suggest that the pigeons were unable to reorganize the already learned AX nonequivalence relation. The remaining one bird did not learn the equivalences between the familiar and novel stimuli at all. This bird could, however, then learn the explicitly taught XY linkage. Thus, the equivalence classes {A, B, C, D} and {X, Y} were formed separately from one another. For this bird, the new disparate stimuli were disconnected from the familiar equivalence network, and so it might be not hindered from separately learning the XY association. All in all, the results obtained with the Sim group subjects indicated that pigeons are well capable of learning equivalences between stimuli that are physically similar but also just capable of learning equivalences between stimuli that are physically dissimilar when these equivalences are well trained and consistent with the already existing equivalence network. The equivalence network developed by our pigeons does not, of course, rely on language competence restricted to humans but must undoubtedly reflect processes of which the neural networks of advanced animals are capable.

Acknowledgment

This work was supported by grants from the Ministry of Education, Science, and Culture, Japan (no. 10610067) to M.J. and Deutsch Forchungsgemeinschaft (Bonn) to J.D.D.

References

Anderson JA, Hinton GE (1981) Models of information processing in the brain. In: Hinton GE, Anderson A (eds) Parallel models of associative memory. Erlbaum, Hillsdale, NJ, pp 9–48

Ashby FG, Maddox WT (1998) Stimulus categorization. In: Birnbaum MH (ed) Measurement, judgment, and decision making: handbook of perception and cognition. Academic Press, San Diego, pp 251–301

Astley SL, Wasserman EA (1992) Categorical discrimination and generalization in pigeons: all negative stimuli are not created equal. J Exp Psychol Anim Behav Process 18:193–207

Astley SL, Wasserman EA (1998) Novelty and functional equivalence in superordinate categorization by pigeons. Anim Learn Behav 26:125–138

Astley SL, Wasserman EA (1999) Superordinate category formation in pigeons: association with a common delay or probability of food reinforcement makes perceptually dissimilar stimuli functionally equivalent. J Exp Psychol Anim Behav Process 25:415–432

Aydin A, Pearce JM (1994) Prototype effects in categorization by pigeons. J Exp Psychol Anim Behav Process 20:264–277

Bovet D, Vauclair J (1998) Functional categorization of objects and of their pictures in baboons (*Papio anubis*). Learn Motiv 29:309–322

Bradshaw RH, Dawkins MS (1993) Slides of conspecifics as representatives of real animals in laying hens (*Gallus domesticus*). Behav Process 28:165–172

Cabe PA, Healey ML (1979) Figure-background color differences and transfer of discrimination from objects to line drawings with pigeons. Bull Psychon Soc 13:124–126

Cerella J (1977) Absence of perspective processing in the pigeon. Pattern Recogn 9:65–68

Cerella J (1990a) Pigeon pattern perception: limits on perspective invariance. Perception 19:141–159

Cerella J (1990b) Shape constancy in the pigeon: the perspective transformations decomposed. In: Commons ML, Herrnstein RJ, Kosslyn SM, Mumford DB (eds) Quantitative analysis of behavior: behavioral approaches to pattern recognition and concept formation, vol 8. Erlbaum, Hillsdale, NJ, pp 145–163

Cole PD, Honig WL (1994) Transfer of a discrimination by pigeons (*Columba livia*) between pictured locations and the represented environments. J Comp Psychol 108:189–198

Cook RG, Katz JS (1999) Dynamic object perception by pigeons. J Exp Psychol Anim Behav Process 25:194–210

D'Amato MR, Sant PV (1988) The person concept in monkeys (*Cebus apella*). J Exp Psychol Anim Behav Process 14:43–55

D'Amato MR, Salmon DP, Loukas E, Tomie A (1985) Symmetry and transitivity of conditional relations in monkeys (*Cebus apella*) and pigeons (*Columba livia*). J Exp Anal Behav 44:35–47

Delius JD (1992) Categorical discrimination of objects and pictures by pigeons. Anim Learn Behav 20:301–311

Delius JD, Hollard VD (1995) Orientation invariance in pattern recognition by pigeons and humans. J Comp Psychol 109:278–290

Delius JD, Emmerton J, Hörster W, Jäger R, Ostheim J (1999) Picture-object recognition in pigeons. Current Psychology of Cognition 18:621–656

Dépy D, Fagot J, Vauclair J (1997) Categorization of three-dimensional stimuli by humans and baboons: search for prototype effects. Behav Process 39:299–306

Edelman S, Bülthoff HH (1992) Orientation dependence in the recognition of familiar and novel views of three-dimensional objects. Vision Res 32:2385–2400

Edwards CA, Jagielo JA, Zentall TR, Hogan DE (1982) Acquired equivalence and distinctiveness in matching to sample by pigeons: mediation by reinforcer-specific expectancies. J Exp Psychol Anim Behav Process 8:244–259

Emmerton J (1983) Vision. In: Abs M (ed) Physiology and behavior of the pigeon. Academic Press, London, pp 245–266

Evans CS, Macedonia JM, Marler P (1993) Effects of apparent size and speed on the response of chickens, *Gallus gallus*, to computer-generated stimulations of aerial predators. Anim Behav 46:1–11

Fersen von L, Lea SEG (1990) Category discrimination by pigeons using five polymorphous features. J Exp Anal Behav 54:69–84

Fields L, Verhave T, Fath SJ (1984) Stimulus equivalence and transitive associations: a methodological analysis. J Exp Anal Behav 42:143–157

Fields L, Landon-Jimenez V, Buffington DM, Adams BJ (1995) Maintained nodal-distance effects in equivalence classes. J Exp Anal Behav 64:129–145

Fujita K, Matsuzawa T (1986) A new procedure to study perceptual world on animals with sensory reinforcement: recognition of humans by a chimpanzee. Primates 27:283–291

Gibson JJ (1986) The ecological approach to visual perception. Erlbaum, Hillsdale, NJ

Goldman D, Homa D (1997) Integrative and metric properties of abstracted information as a function of category discriminability, instance variability, and experience. J Exp Psychol Hum Learn Mem 3:375–385

Hendricks J (1966) Flicker threshold as determined by a modified conditioned suppression procedure. J Exp Anal Behav 9:501–506

Herbranson WT, Fremouw T, Shimp CP (1999) The randomization procedure in the study of categorization of multidimensional stimuli by pigeons. J Exp Psychol Anim Behav Process 25:113–135

Herman LM, Pack AA, Morris-Samuels P (1993) Representational and conceptual skills of dolphins. In: Roitblat HL, Herman LM, Nachtigall PE (eds) Language and communication: comparative perspectives. Erlbaum, Hillsdale, NJ, pp 403–422

Herrnstein RJ (1985) Riddles of natural categorization. Philos Trans R Soc B 308:129–144

Herrnstein RJ (1990) Levels of stimulus control: a functional approach. Cognition 37:133–166

Herrnstein RJ, Loveland DH (1964) Complex visual concept in the pigeon. Science 146:549–551

Herrnstein RJ, Loveland DH, Cable C (1976) Natural concepts in pigeons. J Exp Psychol Anim Behav Process 2:285–302

Hollard DV, Delius DJ (1982) Rotation invariance in visual pattern recognition by pigeons and humans. Science 218:804–806

Hopkins WD, Fagot J, Vauclair J (1993) Mirror-image matching and mental rotation problem solving in baboons (*Papio papio*): unilateral input enhance performance. J Exp Psychol Gen 122:61–72

Huber L, Lenz R (1993) A test of the linear feature model of polymorphous concept discrimination with pigeons. Q J Exp Psychol 46B:1–18

Huber L, Lenz R (1996) Categorization of prototypical stimulus classes by pigeons. Q J Exp Psychol 48:134–147

Iverson I, Sidman M, Carrigan P (1986) Stimulus definition in conditional discrimination. J Exp Anal Behav 45:297–304

Jitsumori M (1993) Category discrimination of artificial polymorphous stimuli based on feature learning. J Exp Psychol Anim Behav Process 19:244–254

Jitsumori M (1994) Discrimination of artificial polymorphous categories by rhesus monkeys (*Macaca mulatta*). Q J Exp Psychol 47:371–386

Jitsumori M (1996) A prototype effects and categorization of artificial polymorphous stimuli in pigeons. J Exp Psychol Anim Behav Process 22:405–419

Jitsumori M, Matsuzawa T (1991) Picture perception in monkeys and pigeons: transfer of rightside-up versus upside-down discrimination of photographic objects across conceptual categories. Primates 32:473–482

Jitsumori M, Ohkubo O (1996) Orientation discrimination and categorization of photographs of natural objects by pigeons. Behav Process 38:205–226

Jitsumori M, Yoshihara M (1997) Categorical discrimination of human facial expressions by pigeons: a test of the linear feature model. Q J Exp Psychol 50B:253–268

Jitsumori M, Natori M, Okuyama K (1999) Recognition of moving video images of conspecifics by pigeons: effects of individuals, static and dynamic motion cues, and movement. Anim Learn Behav 27:303–315

Keller FS, Schoenfeld WN (1950) Principles of psychology. Appleton-Century-Crofts, New York

Knapp AG, Anderson JA (1984) Theory of categorization based on distributed memory storage. J Exp Psychol Learn Mem Cogn 10(4):616–637

Kuno H, Kitadate H, Iwamoto T (1994) Formation of transitivity in conditional matching to sample by pigeons. J Exp Anal Behav 62:399–408

Lea SEG (1984) In what sense do pigeons learn concepts? In: Roitblat HL, Bever T, Terrace HS (eds) Animal cognition. Erlbaum, Hillsdale, NJ, pp 263–277

Lea SEG, Harrison SN (1978) Discrimination of polymorphous stimulus sets by pigeons. Q J Exp Psychol 30:521–537

Lea SEG, Ryan CME (1983) Feature analysis of pigeons' acquisition of concept discrimination. In: Commons ML, Herrnstein RJ, Wagner AR (eds) Quantitative analysis of behavior: discrimination processes, vol 4. Ballinger, Cambridge, pp 263–276

Lea SEG, Ryan CME (1990) Unnatural concepts and the theory of concept discrimination in birds. In: Commons ML, Herrnstein RJ, Kosslyn S, Mumford D (eds) Quantitative analysis of behavior: behavioral approaches to pattern recognition and concept formation, vol 8. Erlbaum, Hillsdale, NJ, pp 165–185

Lea SEG, Lohmann A, Ryan CME (1993) Discrimination of five-dimensional stimuli by pigeons: limitations of feature analysis. Q J Exp Psychol 46:19–42

Lipkens R, Kop PFM, Matthijs W (1988) A test of symmetry and transitivity in the conditional discrimination performances of pigeons. J Exp Anal Behav 49:395–409

Lohmann A, Delius JD, Hollard VD, Friesel M (1988) Discrimination of shape reflections and shape orientations by Columba livia. J Comp Psychol 102:3–13

Lumsden EA (1977) Generalization of an operant response to photographs and drawing/silhouettes of a three-dimensional object at various orientations. Bull Psychon Soc 10:405–407

Mackintosh NJ (1995) Categorization by people and pigeons: the twenty-second Bartlett memorial lectures. Q J Exp Psychol 48B(3):193–214

Makino H, Jitsumori M (in press) Category learning and prototype effect in pigeons: a study by using morphed images of human faces. Jpn J Psychol (in Japanese)

McIntire KD, Cleary J, Thompson T (1987) Conditional relations by monkeys: reflexivity, symmetry, and transitivity. J Exp Anal Behav 47:279–285

McQuoid LM, Galef BF Jr (1993) Social stimuli influencing feeding behavior of Burmese fowl: a video analysis. Anim Behav 46:13–22

Miller RJ (1973) Cross-cultural research in the perception of pictorial materials. Psychol Bull 80:135–150

Neumann PG (1974) An attribute frequency model for the abstraction of prototypes. Mem Cogn 2:241–248

Neumann PG (1977) Visual prototype information with discontinuous representation of dimensions of variability. Mem Cogn 5:187–197

Patterson-Kane E, Nicol CJ, Foster TM, Temple W (1997) Limited perception of video images by domestic hens. Anim Behav 53:951–963

Powell RW (1967) The pulse-to-cycle fraction as a determinant of critical flicker fusion in the pigeon. Psychol Rec 17:151–160

Rips L (1994) The psychology of proof: deductive reasoning in human thinking. MIT Press, Cambridge

Roberts WA (1996) Stimulus generalization and hierarchical structure in categorization by animals. In: Zentall TR, Smeets PM (eds) Stimulus class formation in humans and animals. Elsevier, Amsterdam, pp 35–54

Roberts WA, Mazmanian DS (1988) Concept learning at different levels of abstraction by pigeons, monkeys, and people. J Exp Psychol Anim Behav Process 14:247–260

Ryan CME, Lea SEG (1994) Images of conspecifics as categories to be discriminated by pigeons and chickens: slides, video tapes, stuffed birds and live birds. Behav Process 33:155–176

Sands SF, Lincoln CE, Wright AA (1982) Pictorial similarity judgments and the organization of visual memory in the rhesus monkey. J Exp Psychol Gen 3:369–389

Savage-Rumbaugh ES, Rumbaugh DM, Smith ST, Lawson J (1980) Reference: the linguistic essential. Science 210:922–925

Savage-Rumbaugh ES, Pate JL, Lawson J, Smith ST, Rosenbaum S (1983) Can a chimpanzee make a statement? J Exp Psychol Gen 112:457–492

Schrier AM, Brady PM (1987) Categorization of natural stimuli by monkeys (*Macaca mulatta*): effects of stimulus set size and modification of exemplars. J Exp Psychol Anim Behav Process 13:136–143

Schrier AM, Angarella R, Povar ML (1984) Studies of concept formation by stumptailed monkeys: concepts humans, monkeys, and letter A. J Exp Psychol Anim Behav Process 10:564–584

Schusterman RJ, Kastak D (1993) A California sea lion (*Zalophus californianus*) is capable of forming equivalence relations. Psychol Rec 43:823–839

Schyns PG (1991) A modular neural network model of concept acquisition. Cogn Sci 15:461–508

Shepard RN, Metzler J (1971) Mental rotation of three-dimensional objects. Science 171:701–703

Sidman M (1994) Equivalence relations and behavior: a research story. Authors Cooperative, Boston

Sidman M, Tailby W (1982) Conditional discrimination vs. matching to sample: An expansion of the testing paradigm. J Exp Anal Behav 37:5–22

Sidman M, Rauzin R, Lazar R, Cunningham S, Tailby W, Carrigan P (1982) A search for symmetry in the conditional discriminations of rhesus monkeys, baboons, and children. J Exp Anal Behav 37:23–44

Steirn JN, Jackson-Smith P, Zentall TR (1991) Mediational use of internal representations of food and no-food events by pigeons. Learn Motiv 22:353–365

Tolan JC, Rogers CM, Malone DR (1981) Cross-modal matching in monkeys: altered visual cues and delay. Neuropsychologia 19:289–300

Tomonaga M, Matsuzawa T, Fujita K, Yamamoto J (1991) Emergence of symmetry in visual conditional discrimination by chimpanzees (*Pan troglodytes*). Psychol Rep 68:51–60

Ullman S (1989) Aligning pictorial descriptions: an approach to object recognition. Cognition 32:193–254

Urcuioli PJ (1990) Some relationships between outcome expectancies and sample stimulus in pigeons' delayed matching. Anim Learn Behav 18:302–314

Urcuioli PJ (1996) Acquired equivalences and mediated generalization in pigeon's matching-to-sample. In: Zentall TR, Smeets PM (eds) Stimulus class formation in humans and animals. Elsevier, Amsterdam, pp 55–70

Urcuioli PJ, Zentall TR, Jackson-Smith P, Steirn JN (1989) Evidence for common coding in many-to-one matching: retention, intertrial interference, and transfer. J Exp Psychol Anim Behav Process 15:264–273

Urcuioli PJ, Zentall TR, DeMarse T (1995) Transfer to derived sample-comparison relations by pigeons following many-to-one versus one-to-many matching with identical training relations. QJ Exp Psychol 48B:158–178

Vauclair J, Fagot J, Hopkins WD (1993) Rotation of mental images in baboons when the visual input is directed to the left cerebral hemisphere. Psychol Sci 4:99–103

Vaughan W (1988) Formation of equivalence sets in pigeons. J Exp Psychol Anim Behav Process 14:36–42

Wasserman EA (1995) The conceptual abilities of pigeons. Am Sci 83:246–255

Wasserman EA, Astley SL (1994) A behavioral analysis of concepts: its application to pigeons and children. In: Medin DL (ed) The psychology of learning and motivation. Academic Press, New York

Wasserman EA, Tassinary LG, Bhatt RS, Sayasenh P (1989) Pigeons can discriminate emotional expression and individual identity from photographs of the human face. Presented at the annual meeting of the Psychonomic Society, Atlanta, GA

Wasserman EA, DeVolder CL, Coppage DJ (1992) Nonsimilarity-based conceptualization in pigeons via secondary or mediated generalization. Psychol Sci 3:374–379

Wasserman EA, Gagliardi JL, Astley SL, Cook BR, Kirkpatrick-Steger K, Biederman I (1996) The pigeon's recognition of drawings of depth-rotated stimuli. J Exp Psychol Anim Behav Process 22:205–221

Watanabe S (1993) Object-picture equivalence in the pigeon: an analysis with natural concept and pseudo-concept discriminations. Behav Process 30:225–232

Winner H, Ettlinger E (1978) Do chimpanzees recognize photographs as representations of objects? Neuropsychologia 17:413–420

Wittgenstein L (1953) Philosophical investigations. Macmillan, New York

Yamamoto J, Asano T (1995) Stimulus equivalence in a chimpanzee (*Pan troglodytes*). Psychol Rec 45:3–21

Zentall TR (1998) Symbolic representation in animals: emergent stimulus relations in conditional discrimination learning. Anim Learn Behav 26:363–377

Zentall TR, Urcuioli PJ, Jagielo JA, Jackson-Smith P (1989) Interaction of sample dimension and sample-comparison mapping on pigeons' performance of delayed conditional discriminations. Anim Learn Behav 17:172–178

Zentall TR, Steirn JN, Sherburne LM, Urcuioli PJ (1991) Common coding in pigeons assessed through partial versus total reversals of many-to-one conditional discrimination. J Exp Psychol Anim Behav Process 17:194–201

Zentall TR, Sherburne LM, Steirn JN (1992) Development of excitatory backward associations during the establishment of forward associations in a delayed conditional discrimination by pigeons. Anim Learn Behav 20:199–206

Zentall TR, Sherburne LM, Urcuioli PJ (1993) Common coding in a many-to-one delayed matching task as evidenced by facilitation and interference effect. Anim Learn Behav 21:233–237

Zentall TR, Sherburne LM, Urcuioli PJ (1995) Coding of hedonic and nonhedonic samples by pigeons in many-to-one delayed matching. Anim Learn Behav 23:189–196

Part 5
Recognition of Self, Others, and Species

14
Mirror Self-Recognition in Primates: An Ontogenetic and a Phylogenetic Approach

NORIKO INOUE-NAKAMURA

1 Mirror Self-Recognition

We often see a mother play with her baby while exposing it to a mirror. The baby tries to reach out its hand to its reflection in the mirror and laughs, or stares into its reflection with wonder. How does the baby recognize its reflection?

The present study focuses on self-recognition, which is one aspect of the notion of self. Mirrors have often been used in comparative studies of self-recognition to compare humans with other animals. An overview of research on mirror self-recognition follows.

There have been many studies on the development of mirror self-recognition in humans, with the subjects in most cases being infants (e.g., Gesell 1925; Shirley 1933; Wallon 1934). Wallon (1965) reported a three-stage developmental sequence of reactions of infants to their own image in a mirror. At the first stage (about 6 to 12 months of age) they showed surprise at their image in the mirror, and reacted by reaching out and laughing at their image as if it were real. At the second stage (about 12 to 24 months of age) they began to show interest in the relationship between their own movement and the reflection. At the third stage (after 24 months of age) they began to play with their image in the knowledge that it was not real.

The use of mirrors in these studies on self-recognition in human infants has long been considered to be a reliable method for studying the emergence of self-recognition. However, a problem with these early studies was that the experimenter in most cases assessed whether an infant recognized itself only by means of the verbal behavior of the infant. That is, an infant who had not yet acquired spoken language was likely to be regarded as not showing self-recognition. There were surely cases, however, in which the infant could recognize itself in the mirror but failed to say, "It's me."

Gallup (1970) resolved this problem by inventing an experimental procedure called the "mark test". He reported that the capacity for mirror self-recognition was not unique to humans. Gallup studied mirror self-recognition in non-human primates, species devoid of language. He reported that four young chimpanzees recognized themselves in a mirror. The chimpanzees initially showed social behav-

Primate Research Institute, Kyoto University, 41 Kanrin, Inuyama, Aichi 484-8506, Japan

iors to their reflected image, such as bobbing, vocalizing, and threatening. However, the social behaviors diminished over several days. The chimpanzees then began displaying self-directed behaviors, such as grooming parts of their own body which would otherwise be visually inaccessible without the mirror image: picking a bit of food from between the teeth while watching the mirror image, making faces at the mirror, blowing bubbles, manipulating food wads with the lips while watching the reflection, and so on.

After 10 days of exposure to the mirror, the chimpanzees were anesthetized and marked with a red alcohol-soluble dye above one eyebrow ridge and on the top half of the opposite ear. After the chimpanzees recovered from anesthesia, their behaviors were tested first without and then with a mirror. They touched the marks more frequently in the presence of the mirror than in the absence of the mirror. In some cases, mark-directed behaviors included direct visual inspection of the fingers that touched marked areas even though the dye had long since dried and was not transferable to the fingers. This marking technique provided a new, objective test to explore whether nonverbal human infants and non-human animals can recognize themselves in a mirror.

2 Ontogeny of Mirror Self-Recognition

2.1 Development of Mirror Self-Recognition in Infant Chimpanzees

Since Gallup (1970) conducted his landmark study, other researchers have reported that responses of mirror self-recognition in chimpanzees are very similar to those in human infants (Fig. 1). In most cases, however, mirror self-recognition among primates has been investigated in adults or adolescents (for exceptions see Lin et al. 1992; Miles 1994; Povinelli et al. 1993; Robert 1986). In other words, attention has been focused on the categorical question of whether chimpanzees could demonstrate mirror self-recognition or not. Developmental processes have rarely been explored. Inoue (1994) investigated the development of self-recognition in infant chimpanzees, adopting a complementary longitudinal and cross-sectional approach.

In the first longitudinal experiment, a human-raised female infant chimpanzee was exposed to a mirror (25 × 30 cm) for 10 min each day for 3 months at approximately 1.5 years of age (between 76 and 87 weeks). The mirror was hung from the ceiling in her home cage (230 × 120 × 70 cm), 20 cm away from one side and 30 cm from the floor. Mirror-related behaviors were video-recorded. Analysis of video-records revealed 50 different behavioral patterns toward the mirror. These were classified into five major categories: social, exploratory, contingent, self-directed, and complex behaviors. The definitions of the categories are given in Table 1. Over time, the subject's behavior changed from social, exploratory, and contingent behaviors, to self-directed and complex behaviors. Self-directed behaviors, a sign of self-recognition, were first observed at 82 weeks. Complex behaviors, which are not

Fig. 1. Self-directed behavior in a female chimpanzee (18 years old) toward her mirror image. She is grooming a blemish on the top of her head (**a**) and picking her teeth with fingers (**b**) while looking at her reflection

always considered to be a sign of self-recognition, were first observed at 85 weeks. Figure 2 shows the mean frequency of each behavioral category toward the mirror in each 10-min session, as a function of age. The subject that received extensive exposure showed self-recognition at the age of 1.5 years of age. This result is in clear contrast to previous studies reporting a later age of onset of self-recognition (Lin et al. 1992; Miles 1994; Povinelli et al. 1993).

In a second experiment, using a cross-sectional method, the author tested seventeen infant chimpanzees (six males, eleven females), ranging from 1 to 4.5 years of age (16 to 59 months). All of them were artificially raised and kept with conspecifics of the similar age. None of them had previous experience of mirrors. They were

Table 1. Behavioral categories and definitions of mirror-related behaviors

Categories	Definitions	Examples of behaviors
Social behavior	Subject directs social behaviors at the mirror.	Threatening, vocalizing, sexual presentations, lip smacking, or lip flips.
Exploratory behavior	Subject shows either active or passive exploration of the mirror.	Attempting to look over or behind the miror, or to reach the mirror.
Contingent behavior	Subject makes two or more repetitions of a bodily movement while looking into the mirror.	Waving a hand, licking the mesh, or moving upper body slowly while looking into the mirror.
Self-directed behavior	Subject uses fingers or hands to manipulate parts of the body otherwise not visible, while looking into the mirror.	Picking teech, or grooming forehead while looking into the mirror.
Complex behavior	Subject performs two or more mirror-related behaviors simultaneously while looking into the mirror.	Waving a hand and licking the other hand simultaneously while looking into the mirror.

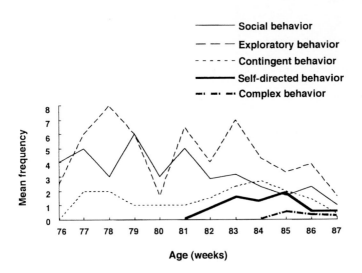

Fig. 2. The mean frequency of each behavioral category toward the mirror in each 10-min session, as a function of age

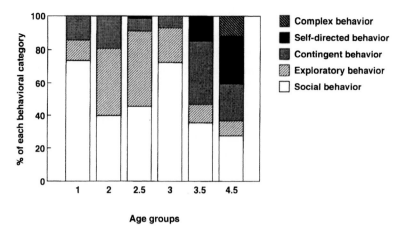

Fig. 3. Percentages of time spent in each behavioral category by each age group. 1-year-olds, n = 3; 2-year-olds, n = 2; 2.5-year-olds, n = 2; 3-year-olds, n = 4; 3.5-year-olds, n = 4; 4.5-year-olds, n = 2

grouped by age and given a 40-min session of mirror exposure. Mirror-related behaviors were video-recorded and classified as in the first experiment. Figure 3 shows the percentages of time spent in each behavioral category in each age group.

Social behaviors were particularly prominent in subjects up to 3 years of age, while self-directed and complex behaviors were more prevalent among subjects of 3 years of age. Within the 40-min session, behaviors changed especially in the two oldest groups. Social behaviors decreased in frequency in the first 10 min, to be replaced by self-directed and complex behaviors. This rapid, within-session behavioral change in the presence of the mirror recalls the developmental change seen over weeks or months. In sum, the emergence of mirror self-recognition was similar in the two experiments, i.e., longitudinally and cross-sectionally. However, self-recognition first occurred at about 3.5 years of age in the second experiment, which is about two years later than observed for the chimpanzee in the first experiment, with much more extensive mirror exposure.

These results clearly suggest that the onset of self-recognition depends on age, that is, the maturation of cognitive abilities. It also depends on the amount of learning experiences with a mirror. In conclusion, chimpanzees with no previous experience of a mirror may demonstrate mirror self-recognition in one 40-min session of mirror exposure at 3.5 years of age. Further, a 1.5-year-old chimpanzee may show similar behavior if it receives extensive exposure to mirrors.

2.2 Development of Mirror Self-Recognition in Human Infants

Amsterdam (1972) investigated the developmental process of mirror self-recognition in human infants, using much the same technique as that devised by Gallup

(1970). This objective and nonverbal technique involved placing a spot of rouge on the infant's nose (on one side close to the cheek). Behavior toward the spot served as a point of reference for evaluating self-recognition in the mirror.

The subjects were between 3 and 24 months of age. According to Amsterdam, the infants' behavior toward the mirror showed three developmental phases. In the first phase (from 6 to 12 months of age), infants reacted to their image as if it were a sociable "playmate". They smiled and vocalized toward the mirror, showing delight and enthusiastic playful approaches to the "other child". These behaviors decreased up to 18 months of age. In the second phase (from 13 to 20 months of age), infants no longer responded to the mirror with naive delight and enthusiasm, but became wary and often withdrew from it, although some infants still intermittently smiled or vocalized at the image. There are some other behaviors characterizing this period, such as searching for the image, displaying signs of embarrassment, and self-admiration. In the final phase (from 20 to 24 months of age), infants started to show recognition of their reflection. Some infants became aware of the spot of rouge on the side of their nose and touched it while looking at the mirror.

Many studies have confirmed the above-described results. Human infants at around the age of 2 years show self-directed behavior indicating that they recognize their mirror image (e.g., Amsterdam 1972; Bertenthal and Fisher 1978; Lewis and Brooks-Gunn 1979; Schulman and Kaplowitz 1977). However, most of these studies were conducted in advanced countries such as the USA, European countries, and Japan, where infants may be frequently exposed to mirrors from shortly after birth. The data from the study of chimpanzees described above suggest that learning experiences with a mirror can have an important influence on the development of mirror self-recognition. Thus, most studies with human infants probably lack control over the amount of experience the infant has with mirrors.

In view of the results of two experiments with infant chimpanzees it is open to question whether mirror self-recognition occurs at about 2 years of age in all human infants. How does mirror self-recognition develop in human infants who have had almost no previous experience with mirror exposure? The data from infant chimpanzees predict a delayed onset of mirror self-recognition in such infants. However, as far as I know, only one relevant study has been done (Priel and de Schonen 1986) reporting the normal onset of self-recognition in Bedouin-nomadic infants in spite of these infants' lack of previous experience with mirrors. This result contradicts my two-factor theory of maturation and experience jointly influencing the onset of mirror self-recognition. In order to control the factor of experience, Inoue-Nakamura (1997a) conducted an experiment with infants of the Manon community in West Africa. Manon people live in Bossou, a small village near where wild chimpanzees also live, on the southeastern edge of the Republic of Guinea (Inoue-Nakamura and Matsuzawa 1997). The Manon have no mirrors at all except for a tiny piece of broken mirror (approximately 5 x 8 cm) that is possessed by men for shaving and is seldom utilized for any other purpose. They do not have glass windows in their houses or other reflecting objects in their environment. There are some reflecting surfaces such as pools of rainwater. However, the muddy water cannot be considered as a substitute for a mirror. This means that the

Fig. 4. A Manon infant in front of a mirror

Manon infants I studied had very limited previous experience, if any, of exposure to mirrors.

The study used a cross-sectional method. Eight infants (four males, four females) ranging in age from 24 to 45 months were studied. There were two infants (one male and one female) at ages 2, 2.5, 3, and 3.5 years, respectively. In the test, all infants received mirror exposure with their mothers. The infant sat in front of the mirror (40 × 65 cm), which was just within arm's reach, while the mother sat close by on the left side of the infant (Fig. 4). The mother was instructed to allow her infant to behave freely, and not to talk or give any kind of verbal or gestural prompt to the infant. One experimenter sat to the side of the mother-infant pair and took responsibility for changing mirror position, giving verbal instructions, and so forth. Another experimenter, standing about 3 m away, video recorded the behavior of infants in front of the mirror. Each infant received a 30-min session of mirror exposure. The session consisted of three phases in an A-B-A design: 5-min exposure to the back of the mirror, 20 min of mirror exposure, and again 5-min exposure to the back of the mirror. A spot of rouge was put on the infant's forehead during the first 5-min period by the first experimenter, who pretended to touch the forehead of the infant innocently. Mirror-related behaviors were categorized into one of the following five behavioral categories according to a mutually exclusive and exhaustive coding system (see Table 1): looking into mirror, social behaviors, exploratory behaviors, contingent behaviors, and self-directed behaviors. When 15 min had elapsed from the start of the 20-min mirror exposure period, the first experimenter pointed at the infant's mirror image with his index finger and asked, "Who's that? What's the red spot? Please touch it", in the Manon language.

Figure 5 shows the amount of time spent in mirror-related behaviors for each

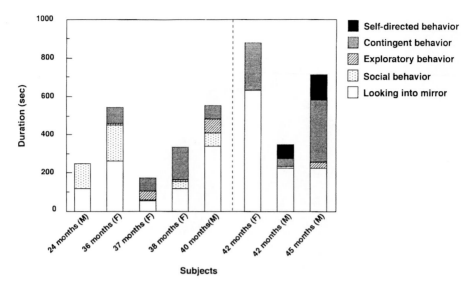

Fig. 5. The amount of time spent in each behavioral category by each infant

infant. The youngest male (24 months) showed only simple looking and social behaviors. The five infants between 36 and 42 months of age showed social behaviors, exploratory behaviors, and contingent behaviors. Only two infants, a 42-month-old male and a 45-month-old female, answered the question "Who's that?" by saying their own names. This study thus showed that two infants over 42 months (3.5 years) of age could demonstrate mirror self-recognition within a single 20-min session. Although the onset is very rapid, it is about 1.5 years later than subjects in previous studies who have probably had exposure to mirrors from birth in more advanced countries. In sum, it is not only in chimpanzees but also in human infants that the amount of experience with mirrors in conjunction with the maturation of cognitive abilities largely determines the age of onset of mirror self-recognition.

3 Phylogeny of Mirror Self-Recognition

3.1 Mirror Self-Recognition in Animals Other than Humans and Chimpanzees

As we have seen, the developmental process of mirror self-recognition is very similar in infant humans and infant chimpanzees. How do animals other than humans and chimpanzees behave toward their reflected image in a mirror? The following is a summary of previous studies on mirror self-recognition in various animals.

In general, the absence of self-directed behaviors has been reported in fish, birds, and non-primate mammals (Anderson 1994; Mitchell 1993; Povinelli 1989). However, there are some exceptions. Dolphins (*Tursiops truncatus*) have been reported to show self-directed behaviors (Marten and Psarakos 1994). Dolphins, which have big brains and impressive cognitive functions, including language-like competencies (Herman et al. 1984), may have acquired the capacity for self-recognition ability at some point in their evolution.

Mirror self-recognition has also been studied in a wide variety of primate species other than humans and chimpanzees (e.g., Parker et al. 1994; and review articles: Anderson 1984; Gallup 1982). However, these studies are characterized by variations in test procedures, and there has been controversy with regard to the criteria, validity, and reliability of self-recognition (Gallup et al. 1995; Heyes 1994, 1995). With the aid of a standardized procedure, Inoue-Nakamura (1997b) examined mirror-related behaviors in 12 species of nonhuman primates, including prosimians (ring-tailed lemur), New World monkeys (cottontop tamarin, squirrel monkey, and brown-capped capuchin), Old World monkeys (bonnet macaque, rhesus macaque, and Japanese macaque), gibbons (white-handed gibbon), and great apes (orangutan, gorilla, bonobo, and chimpanzee). The study employed a unified coding system to measure mirror self-recognition and the standardized test procedure with the same type of mirror, presented at the same distance from the subject and for the same duration of exposure, and so on. Three sizes of mirrors were prepared: large (60 x 90 cm), medium (45 x 67.5 cm), and small (30 x 45 cm). A mirror was presented depending on the body size of the subjects. The mirror was placed in front of the subjects' cage just out of reach. All subjects received a single 30-min session of mirror exposure consisting of three phases in an A-B-A design: 20 min of mirror exposure, 5-min exposure to the back of the mirror, and again 5 min of mirror exposure. Mirror-related behaviors were video-recorded and then categorized into one of five classes as in the other studies described above.

Figure 6 shows the relative amount of time that each species spent in each of the behavioral categories. There was a significant difference among the three groups of species in the amount of time spent in each behavioral category. Briefly, the first group (prosimian and New World monkeys) spent more time in looking into the mirror and in social behaviors. The second group (Old World monkeys and gibbons) spent more time in exploratory behaviors and contingent behaviors. The third group, containing great apes, was characterized by the occurrence of self-directed behaviors, which were never observed in the other two groups. It is important to note that the phylogenetic trends regarding mirror self-recognition parallel the ontogenetic developmental phases of behaviors toward the mirror found in human and chimpanzee infants.

Table 2 summarizes the results of the present study and other references on mirror self-recognition in primates. Due to limited space, the table is not exhaustive, but it clearly shows that prosimians show only social behaviors and exploratory behaviors toward their reflection, and do not show any other type of behavior. The New World monkeys and the Old World monkeys show some contingent behaviors, but do not show any evidence of self-directed behaviors; nor do gibbons. Only hominoids, the great apes and humans, demonstrate self-directed be-

Table 2. Summary of responses to mirror image in primates

	Social behavior	Exploratory behavior	Contingent behavior	Self-directed behavior	References
Prosimians					
Ring-tailed lemur (*Lemur catta*)	O	O	X	X	Inoue-Nakamura 1997b; Fornasieri et al. 1991
New World Monkeys					
Pygmy marmoset (*Cebuella pygmaea*)	O	O	O	X	Eglash and Snowdon 1983
Cottontop tamarin (*Saguinus oedipus*)	O O	O	O	X[b]	Inoue-Nakamura 1997b; Hauser et al. 1995
Squirrel monkey (*Saimiri sciureus*)	O	X	X	X	Inoue-Nakamura 1997b; Maclean 1964
Black-capped capuchin (*Cebus apella*)	O	O	O	X	Inoue-Nakamura 1997b; Anderson and Roeder 1989
Spider monkey (*Ateles spp.*)	O	—	—	X	Lethmate and Ducker 1973
Old World Monkeys					
Red-faced macaque (*Macaca arctoides*)[a]	O	O O O	—	X	Anderson 1983
Bonnet macaque (*Macaca radiata*)	O	—	O O	X	Inoue-Nakamura 1997b
Pig-tailed macaque (*Macaca nemestrina*)	O	—	—	X[b]	Boccia 1994; Anderson 1986
Lion-tailed macaque (*Macaca silenus*)	O	—	—	X	Lethmate and Ducker 1973
Tonkean macaque (*Macaca tonkeana*)	O	—	—	X	Anderson 1986
Crab-eating macaque (*Macaca fascicularis*)	O	—	O O	X	Gallup 1977; Anderson 1986
Rhesus macaque (*Macaca mulatta*)	O	O O	O	X	Inoue-Nakamura 1997b; Gallup et al. 1980
Japanese macaque (*Macaca fuscata*)	O	O O	O O	X[b]	Inoue-Nakamura 1997b; Platt and Thompson 1985; Itakura 1987
Patas monkey (*Erythrocebus patas*)	O	O	O	X	Hall 1962
Hamadryas baboon (*Papio hamadryas*)	O	—	—	X	Lethmate and Ducker 1973
Olive baboon (*Papio anubis*)[a]	O	—	—	X	Benhar et al. 1975
Mandrill (*Mandrillus sphinx*)	O	—	—	X	Lethmate and Ducker 1973

				References
Gibbons				
White-handed gibbon (Hylobates lar)	O	O	X[b]	Inoue-Nakamura 1997b; Lethmate and Ducker 1973; Hyatt 1998; Ujhelyi et al. 2000
Agile gibbon (Hylobates agilis)	O	—	X	Lethmate and Ducker 1973
Great apes				
Orangutan (Pongo pygmaeus)	O	O	O	Inoue-Nakamura 1997b; Suarez and Gallup 1981; Miles 1994
Gorilla (Gorilla gorilla)	O	O	O	Inoue-Nakamura 1997b; Patterson and Cohn 1994; Ledbetter and Basen 1982
Bonobo (Pan paniscus)	O	O	O	Inoue-Nakamura 1997b; Hyatt and Hopkins 1994; Savage-Rumbaugh 1986
Chimpanzee (Pan troglodytes)	O	O	O	Inoue-Nakamura 1997b; Gallup 1970; Povinelli et al. 1993
Human (Homo sapiens)[a]	O	O	O	Amsterdam 1972; Bertenthal and Fischer 1978

The subjects of the present study are represented in bold. Circles, Positive data; crosses, negative data; dashes, no available data.

[a] The subjects were infants.

[b] This behavior has been reported in a few individuals.

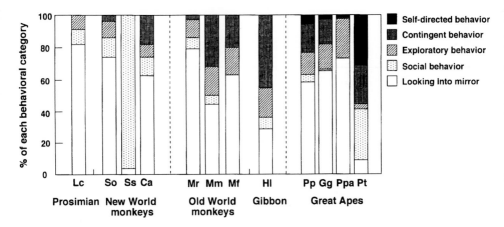

Fig. 6. Percentages of time spent in each behavioral category in 12 species of primates. Lc, *Lemur catta*; So, *Saguinus oedipus*; Ss, *Saimiri sciureus*; Ca, *Cebus apella*; Mr, *Macaca radiata*; Mf, *Macaca fuscata*; Mm, *Macaca mulatta*; Hl, *Hylobates lar*; Pp, *Pongo pygmaeus*; Gg, *Gorilla gorilla*; Ppa, *Pan paniscus*; Pt, *Pan troglodytes*

haviors that are thought to be a behavioral indicator of mirror self-recognition. However, there may be a few exceptions. A few studies have reported that gibbon (Ujhelyi et al. 2000), the Japanese macaque (Itakura 1987), pigtail macaque (Boccia 1994), and cotton-top tamarin (Hauser et al. 1995) showed evidence of self-recognition, although none of these so-called positive results were confirmed by follow-up studies. In sum, it is reasonable to conclude that, as shown in Table 2, there is a clear and qualitative gap in mirror self-recognition ability between Hominoids and the other primate species. There have been some other attempts to draw up a similar table (e.g., Mitchell 1993).

3.2 Phylogenetic Origin of Mirror Self-Recognition

Since Gallup (1970) reported his landmark study three decades ago, many studies have explored responses to mirror-image stimulation in various kinds of animals. The series of experiments described in the present study focused especially on primates including humans. Based on the available data, we can summarize the current position as follows.

1) Fish, birds, and non-primate mammals mainly show social behaviors toward their reflection. They apparently perceive their mirror image as "other".

2) Non-hominoid primates show exploratory and contingent behaviors, reacting to their mirror image as if it were a "strange other". However, as yet, they fail to display any evidence of self-recognition.

3) Only hominoids exhibit self-directed behaviors that are thought to be an indicator of self-recognition. There is still some controversy about the capacity for mirror self-recognition in gorillas. However, the data from the enculturated cap-

tive gorilla, Koko, studied by Patterson and Cohn (1994) clearly shows that these apes have the capacity. There should be no significant differences among the three genera of great apes in terms of cognitive ability. The difference between great apes and lesser apes (gibbons) may be much more marked. It is also important to note that the onset of mirror self-recognition depends on maturation interacting with experience of being exposed to mirrors.

How has the capacity for mirror self-recognition evolved? Based on the summarized results, especially the divergent data from the range of primate species described in this study let me speculate about the evolutionary emergence of mirror self-recognition. A plausible explanation is as follows: Suppose that mirror experiments had been possible in prehistoric times. The common ancestor of living primates would have had a strong tendency to show social behaviors to the mirror image. Exploration of the mirror image may also have been aroused in the common ancestor. However, the capacity for self-recognition emerged in the common ancestor of the Pongidae, around 12 to 22 million years ago. There is a marked difference in body size, which inevitably leads to differences in brain volume, between the hominoids and other primates. Humans, chimpanzees, gorillas, and orangutans are simply much larger than the other living primates. The human brain is almost three times as large as that of the great apes. However, the brains of great apes are in turn much larger than that of macaques. There should be a sort of quantum jump in the course of hominoid evolution from monkey-like creatures to apes, similar to the hominization process from ape-like creatures.

4 Summary

This paper reported a series of experiments on mirror self-recognition in primates, including humans. The present study is unique in the following three ways: First, it investigated the development of mirror self-recognition in infant chimpanzees, using both a longitudinal survey and a cross-sectional comparison. The developmental changes found in chimpanzees were very similar to those observed in human infants. Both species share common developmental stages of mirror-related behaviors; the change involves a progression from social behavior, to exploratory behavior, to contingent behavior, and finally to self-directed behavior. The reverse developmental sequence has never been observed.

Second, the study revealed the influence of experience with mirrors on the onset of mirror self-recognition. An infant chimpanzee with extensive exposure started to show self-directed behavior much earlier than mirror-naïve control chimpanzees. The same story can be applied to human infants: The onset of mirror self-recognition was delayed in human infants who had been reared in a cultural environment that provided only limited or no experience of mirrors. Thus, mirror self-recognition is clearly dependent on the interaction of the maturational factors such as age, with learning experiences with reflecting surfaces.

Third, the present study examined behaviors toward a mirror in 12 species of primates. The comparative data were obtained using a standardized test procedure and a unified coding system of behavioral categorization to measure mirror self-recognition. The results clearly showed that only hominoids displayed self-directed behaviors. By contrast, gibbon and Old World monkeys showed only social and exploratory behaviors, with some instances of contingent behaviors but no evidence of self-directed behaviors. Social behaviors were prominent in New World monkeys and the prosimian tested. There seems to be a qualitative gap in mirror self-recognition ability between hominoids and the other primate species. The results suggest that the capacity for mirror self-recognition emerged about 12 to 22 million years ago in the common ancestor of living hominoids, although self-recognition per se is a very recent psychological event.

In summary, the present study has shed light on the evolutionary sequence of mirror-related abilities by comparing primates, including humans. The results clarified the ontogenetic process of mirror-related behavior in two species, humans and chimpanzees; the process was highly similar and concurred with phylogenetic trends.

References

Amsterdam B (1972) Mirror self-image reactions before age two. Dev Psychobiol 5:297–305

Anderson JR (1983) Responses to mirror-image stimulation, and assessment of self-recognition in mirror- and peer-reared stumptail macaques. Q J Exp Psychol 35:201–222

Anderson JR (1984) The development of self-recognition: A review. Dev Psychobiol 17:35–49

Anderson JR (1986) Mirror-mediated finding of hidden food by monkeys (*Macaca tonkeana* and *M. fascicularis*). J Comp Psychol 100:237–242

Anderson JR (1994) The monkey in the mirror: A strange conspecific. In: Parker ST, Mitchell RW, Boccia ML (eds) Self-awareness in animals and humans. Cambridge University Press, pp 315–329

Anderson JR, Roeder J (1989) Responses of capuchin monkeys (*Cebus apella*) to different conditions of mirror-image stimulation. Primates 30:581–587

Benhar E, Carlton P, Samuel D (1975) A search for mirror-image reinforcement and self-recognition in baboons. In: Kondo S, Kawai M, Ehara A (eds) Contemporary primatology. Japan Science Press, Tokyo, pp 202–208

Bertenthal B, Fischer KW (1978) Development of self-recognition in the infant. Dev Psychol 14:44–50

Boccia ML (1994) Mirror behavior in macaques. In: Parker ST, Mitchell RW, Boccia ML (eds) Self-awareness in animals and humans. Cambridge University Press, Cambridge, pp 350–360

Eglash AR, Snowdon CT (1983) Mirror-image responses in pygmy marmosets (*Cebuella pygmaea*). Am J Primatol 5:211–219

Fornasieri I, Roeder J, Anderson JR (1991) Les reactions au miroir chez trois especes de lemuriens (*Lemur fulvus, L. macaco, L. catta*). Comptes Rendus de L'academie des Sciences 312:349–354

Gallup GG Jr (1970) Chimpanzees: Self-recognition. Science 167:86–87

Gallup GG Jr (1977) Absence of self-recognition in a monkey (*Macaca fascicularis*) following prolonged exposure to a mirror. Dev Psychobiol 10:281–284

Gallup GG Jr, Wallnau LB, Suarez SD (1980) Failure to find self-recognition in mother-infant and infant-infant rhesus monkey pairs. Folia Primatol 33:210–219

Gallup GG Jr (1982) Self-awareness and the emergence of mind in primates. Am J Primatol 2:237–248

Gallup GG Jr, Povinelli DJ, Suarez SD, Anderson JR, Lethmate J, Menzel EW Jr (1995) Further reflections on self-recognition in primates. Anim Behav 50:1525–1532

Hall KRL (1962) Behaviour of monkeys towards mirror images. Nature 196:1258–1261

Hauser MD, Kralik J, Botto-Mahan C, Garrett M, Oser J (1995) Self-recognition in primates: Phylogeny and the salience of species-typical features. Proc Natl Acad Sciences USA 92:10811–10814

Herman LM, Richards DG, Wolz JP (1984) Comprehension of sentences by bottlenosed dolphins. Cognition 16:129–219

Heyes CM (1994) Reflections on self-recognition in primates. Anim Behav 47:909–919

Heyes CM (1995) Self-recognition in primates: further reflections create a hall of mirrors. Anim Behav 50:1533–1542

Hyatt C (1998) Responses of gibbons (*Hylobates lar*) to their mirror images. Am J Primatol 45:307–311

Hyatt CW, Hopkins WH (1994) Self-awareness in bonobos and chimpanzees: A comparative perspective. In: Parker ST, Mitchell RW, Boccia ML (eds) Self-awareness in animals and humans. Cambridge University Press, Cambridge, pp 248–253

Inoue N (1994) Mirror self-recognition among infant chimpanzees: application of longitudinal and cross-sectional methods (in Japanese with English summary). Jpn J Dev Psychol 5:51–60

Inoue-Nakamura N (1997a) Development of mirror self-recognition in the infants at Bossou, Guinea. Paper presented at the 8th Conference of Japan Society of Developmental Psychology, Osaka

Inoue-Nakamura N (1997b) Mirror self-recognition in nonhuman primates: A phylogenetic approach. Jpn Psychol Res 39:266–275

Inoue-Nakamura N, Matsuzawa T (1997) Development of stone tool use by wild chimpanzees (*Pan troglodytes*). J Comp Psychol 111:159–173

Itakura S (1987) Use of a mirror to direct their responses in Japanese monkeys (*Macaca fuscata fuscata*). Primates 28:343–352

Ledbetter DH, Basen JA (1982) Failure to demonstrate self-recognition in gorillas. Am J of Primatol 2:307–310

Lethmate J, Ducker G (1973) Untersuchungen zum Selbsterkennen im Spiegel bei Orang-Utans und einigen anderen Affenarten. Z Tierpsychol 33:248–269

Lewis M, Brooks-Gunn J (1979) Social cognition and the acquisition of self. Plenum, New York

Lin AC, Bard KA, Anderson JR (1992) Development of self-recognition in chimpanzees (*Pan troglodytes*). J Comp Psychol 106:120–127

Maclean PD (1964) Mirror display in the squirrel monkey, *Saimiri sciureus*. Science 146:950–952

Marten K, Psarakos S (1994) Evidence of self-awareness in the bottlenose dolphin (*Tursiops tuncatus*). In: Parker ST, Mitchell RW, Boccia ML (eds) Self-awareness in animals and humans. Cambridge University Press, Cambridge, pp 361–379

Miles HL (1994) ME CHANTEK: The development of self-awareness in a signing orangutan. In: Parker ST, Mitchell RW, Boccia ML (eds) Self-awareness in animals and humans. Cambridge University Press, Cambridge, pp 254–272

Mitchell RW (1993) Mental models of mirror-self-recognition: Two theories. New Ideas Psychol 11:295–325

Parker ST, Mitchell RW, Boccia ML (1994) Self-awareness in animals and humans. Cambridge University Press, Cambridge

Patterson FG, Cohn RH (1994) Self-recognition and self-awareness in lowland gorillas. In: Parker ST, Mitchell RW, Boccia ML (eds) Self-awareness in animals and humans. Cambridge University Press, Cambridge, pp 273–290

Platt M, Thompson R (1985) Mirror responses in a Japanese macaque troop. Primates 26:300–314

Povinelli DJ (1989) Failure to find self-recognition in Asian elephants (*Elephas maximus*) in contrast to their use of mirror cues to discover hidden food. J Comp Psychol 103:122–131

Povinelli DJ, Rulf AB, Landau KR, Bierschwale DT (1993) Self-recognition in chimpanzees (*Pan troglodytes*): Distribution, ontogeny, and patterns of emergence. J Comp Psychol 107:347–372

Priel B, de Schonen S (1986) Self-recognition: A study of a population without mirrors. J Exp Child Psychol 41:237–250

Robert S (1986) Ontogeny of mirror behavior in two species of great apes. Am J Primatol 10:109–117

Savage-Rumbaugh ES (1986) Ape language: From conditioned response to symbol. Columbia University Press, New York

Schulman AH, Kaplowitz C (1977) Mirror-image response during the first two years of life. Dev Psychobiol 10:133–142

Suarez S, Gallup GG Jr (1981) Self-recognition in chimpanzees and orangutans, but not gorillas. J Hum Evol 10:175–188

Ujhelyi M, Merker B, Buck P, Geissmann T (2000) Observation on the behavior of gibbons (*Hylobates leucogenys*, *H. gabriellae*, and *H. lar*) in the presence of mirrors. J Comp Psychol 114:253–262

Wallon H (1934) Les Origines du caractère chez lénfant. P.U.F., Paris

15
The Level of Self-Knowledge in Nonhuman Primates: From the Perspective of Comparative Cognitive Science

SHOJI ITAKURA

1 Introduction

The origin of self-knowledge is arguably the most fundamental problem of developmental psychology. Recent progress in the study of infant behavior provides important and new insights regarding the origins of self-knowledge (Rochat 1995).

In the traditional view, infants were described as lacking the capacity for self-awareness and born into an adualistic state of fusion with the environment (Bahrick 1995). Piaget also described the "adualistic confusion" as characterizing much of the first year of life (cited by Bahrick 1995). According to Piaget, infants experience no distinction between the self and not-self, and they come to gradually differentiate themselves from other entities at the age of about 8 or 9 months (Bahrick 1995).

In contrast to such a view, another way to look at development is to view it as the expansion of fundamental fits or functional links between organism and environment. Rochat (1995) points out that the self-recognition or a mirror-based view on the origins of self-knowledge is inadequate in light of recent progress in infancy research. He insists that the ecological approach to perception and action, in particular the principle of cooperation and the theory of affordances formulated by Gibson (1979), provides a new, compelling way to look at the problem of the origins of self-knowledge. Rochat (in press) concludes, by reviewing a number of researches, that it is reasonable to postulate that from the onset of development, information specifying the self as a differentiated, agentive and situated entity in the environment does exist and is actively used by young infants. This information is at the origins of self-knowledge, announcing and preparing for the emergence of the conceptual self that has been extensively studied by developmental psychologists using mirror self-recognition tests.

Neisser (1995) also distinguished among five kinds of self-knowledge, each of which may intersect with rather differential periods of human development: 1) the ecological self, which is directly perceived with respect to the physical environment; 2) the interpersonal self, also directly perceived, which depends on emo-

Department of Psychology, Faculty of Letters, Kyoto University, Yoshida Honmachi, Sakyo-ku, Kyoto 606-8501, Japan

tional and other species-specific forms of communication; 3) the temporally extended self, which is based on memory and anticipation and implies a representation of self; 4) the private self, which reflects knowledge that our conscious experiences are exclusively of own, also dependent on representation; and 5) the conceptual self, defined as a theory of self based on sociocultural experience. To my knowledge, this is quite a new idea about the way to think of the self.

In the case of animals, a number of comparative psychologists and cognitive ethologists have explored the possibility of animal self-awareness. Spada et al. (1995) pointed out the stream of these kinds of studies as follows: Comparative psychologists have attempted to determine self-awareness in animals, particularly primates, by means of self-recognition tests with a mirror (e.g., Gallup 1970). Cognitive ethologists (e.g., Griffine 1976), on the other hand, have sought animal self-awareness in natural behavior, particularly behavior that appears to rest on a high level of intentionality or cognition. There have been differences between these two perspectives. According to the comparative psychologist, self awareness is limited to humans and some of the great apes (see Parker et al. 1994 for reviews), whereas cognitive ethologists believe self-awareness to be much more widespread (reviewed in Ristau 1991). Spada et al. (1995) hypothesized that one necessary condition is a certain sense of self as a reference point for social, affective, ecological, and cognitive aspects of the interaction between animals and their environment.

This chapter is an attempt to illustrate my own investigations to approach the "self" in nonhuman primates. First, mirror self-recognition in Japanese monkeys is discussed from the perspective of contingency. Second, acquisition of personal pronouns by a language-like trained chimpanzee is introduced. Third, possession of a symbolic level in a chimpanzee is demonstrated.

2 Mirror Self-Recognition in Nonhuman Primates

According to Gibson (1995), the first paper on the subject of "self-awareness" in an infant was published by Zazzo (1948). He observed his own child's responses to a mirror placed before him and to pictures of himself, through the child's first 3 years, and concluded: "By the way the child reacts to the image of his body, the mirror therefore reveals the origins of consciousness, the image of the body being essentially the consciousness of the self" (cited from Gibson 1995). Human children begin to recognize their contingent-independent image in photographs only after mirror self-recognition, with most children at 22 months of age correctly labeling a photograph of themselves among photographs of other infants (Lewis and Brooks-Gunn 1979).

Self-recognition in nonhuman primates is also often equated with mirror self-recognition (Anderson 1984; Gallup 1970; Povinelli 1987). Nevertheless, for humans, "it is the ability to recognize and respond to self independent of contingency which represents the important developmental milestone in self-recognition" (Lewis and Brooks-Gunn 1979, p.218).

Gallup (1970) was the first researcher who found that chimpanzees were capable of recognizing themselves in the mirror. He studied the reactions of chimpanzees to their mirror reflections. Initially, chimpanzees responded by engaging in a number of typical social responses, such as threatening, play-behavior, and sexual presentations. On about the third day of exposure, they began to show "self-directed" behavior which Gallup labeled. These behaviors suggested that the chimpanzees had discovered that real source of the images was themselves. For example, the chimpanzees explored parts of their bodies and removed something from their teeth while carefully monitoring the mirror images. After the exposure period, all subjects were anesthetized and bright red marks were placed on one eyebrow ridge and on the opposite side. The dye offered little or no olfactory or tactile cues. In another words, the chimpanzees would have had no way of knowing of its presence when they recovered from the anesthesia.

After recovery from the anesthesia, the chimpanzees were observed for 30 minutes in controlled conditions in which the mirror was not presented to the subjects. Any attempts to touch the marked regions were recorded. Next, the mirror was presented to the subjects, and again the number of mark-direct contacts was recorded. The chimpanzees made few if any touches to the dye spots during the controlled condition, but made a numbers of touches in the mirror condition. Gallup thought that the chimpanzee had discovered the source of the mirror image. Since then extensive work has been done on other nonhuman primates, such as macaques, lemurs, capuchins, and orangutans. However, the failure to find self-recognition in members of primate species outside the great ape-human clade has been widely replicated (Benhar et al. 1975; Gallup 1977; Gallup et al. 1980; Anderson 1984). Anderson (1984) and Gallup (1982) concluded that only chimpanzees and orangutans show evidence of self-recognition and therefore a concept of "self."

Again, what did Gallup's (1970) chimpanzees do in a mirror situation? What does "self-directed" behavior mean? In other words, we may be able to say that chimpanzees could use a mirror to inspect their bodies. During the exposure to the mirror chimpanzees learned the association between the movement of the mirror image and their own movement. Can monkeys not do such learning? How do the monkeys respond if they learn to use a mirror to guide their hands? Although there are various levels of self-recognition (see Neisser 1995), we start from this kind of sensory-perceptual level of self-recognition.

2.1 Training of Use of a Mirror in Japanese Macaques

Brown et al. (1965) reported that rhesus monkeys are capable of discriminating among objects by observing their mirrored images. They showed that monkeys could be trained to pull a cord connected to a blue box containing food rather than a cord attached to an empty red box when these items were seen in the mirror. In this experiment the mirror functioned as a reflector of the discriminative cues. Menzel et al. (1985) called this behavior "mirror-mediated object discrimination (MMOD)" to distinguish it from "mirror-guided behavior (MGB)". They suggested that MMOD is a prerequisite for MGB, behavior in which neither the goal object nor the hand can be viewed directly.

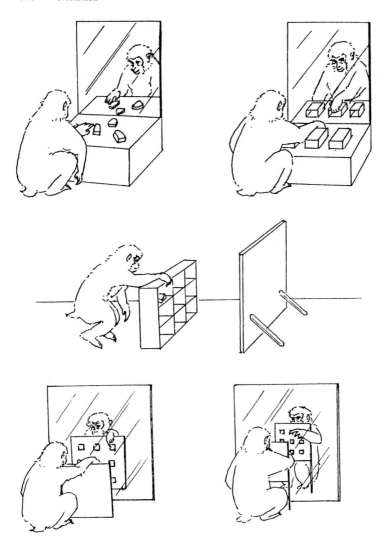

Fig. 1. Mirror-use training by Japanese monkeys

Itakura (1987a,b) trained two Japanese macaques to use a mirror to reach an object that could not be seen directly step by step (see Fig. 1), that is, monkeys exhibit MGB given appropriate training.

Training of mirror-using was carried out step by step as follows:

1) Task 1: Simple reaching for food. A piece of apple was placed on the surface. The subject had to reach directly through the cage bars to obtain food on the surface.

2) Task 2: Discriminatory reaching for food in a visible box. The task was to obtain a piece of apple that could be seen only in the mirror. A piece of apple was placed in one of three small red plastic boxes on the surface, placed in a row. They

were approximately 5 cm apart from one another with open sides facing the mirror. The subjects were required to choose the box containing the apple by observing it in the mirror through the cage.

3) Task 3: Mirror-guided indirect reaching for food in a nonvisible box. The behavior required in this task differed from that in Task 2 in that the subject could not look directly at the objects involved, or its own hand. That is, in this task the subject could see neither the box containing the apple nor his hand directly as he moved it towards the box. A panel was mounted on the center of the cage front to prevent visual access. Because it did not completely cover the front of the cage, the subjects had to use the 5 cm opening bordering the front of the cage to view the mirror and to reach for food hidden in boxes mounted on the side of the panel facing the mirror. There were eight such boxes. The subjects were required to reach around the side of the board to obtain a piece of apple placed in a box.

In pre-training, the subjects were taught to extend their hands through the 5 cm opening to take an apple piece from the box, because the experimental situation had been changed.

4) Task 4: Mirror-guided key pressing. In this task, the subjects were trained to press an illuminated key by observing its mirror image. The panel covering the front of the cage was removed and replaced by a gray wooden panel on which nine rectangular green keys could be mounted. The keys were illuminated from behind by a computer. The opening bordering the side of the panel was then 20 cm in width. When the monkey pressed an illuminated key, a piece of apple was delivered automatically through the tube connected to the feeder.

5) Task 5: Mirror-guided key tracking. The subjects were presented with a series of keys that were illuminated sequentially. The subjects were required to press the illuminated keys in the correct order. Twelve keys were arranged in four rows of three keys. The outer dimensions of the panel were 20 × 30 cm, with an outside opening available for reaching 20 cm. It was then possible for the subjects to reach all the keys from either side. The first key in the sequence was chosen randomly with the restriction that the programmed sequence could be executed.

With the exception of Task 1, in which there was no incorrect response, training during Task 2, 3, 4, and 5 continued to be administered until the subjects reached a criterion of 90% correct on two successive sessions.

6) Task 6: Test for self-directed behavior. A flower-test was used to investigate whether monkeys that had been given training to use a mirror for local objects and to guide their hands towards objects would use a mirror to inspect their own bodies (see Itakura 1987b who also tried a dye test). The flower-test is similar to Bertenthal and Fischer's hat-test (1978). A stick with a cloth flower at the end that weighed 0.05 g was attached behind the monkey's head by means of a collar (see Fig. 2). Prior to the test the monkeys were given four days to habituate to the collar. At the end of this time it was completely ignored. The stick and flower were attached under general anesthesia, to avoid giving them any cues other than those they might be obtain from the mirror. After recovery from anesthesia their behavior was observed without the mirror for 10 minutes and with the mirror for 10 minutes.

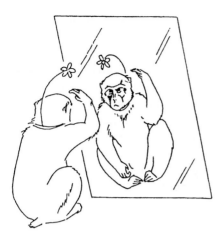

Fig. 2. Flower test

Two Japanese macaques, the subjects, reached the criterion in each task (Task 2 – Task 5), which meant that with proper training these Japanese macaques were capable of learning to use a mirror as a tool to guide their hands (Itakura 1987a). When presented with a flower attached behind their heads the monkeys used the mirror not only to observe it but also to grasp the flower (Task 6).

2.2 The Possibility for Another Test for Mirror Self-Recognition

As described before, self-recognition in nonhuman primates is often equated with mirror self-recognition (Anderson 1984; Gallup 1970; Itakura 1987a, b; Povinelli 1987). The indexes of mirror-self-recognition in nonhuman primates are the disappearance of social behavior and the appearance of self-directed behavior while watching images in a mirror. Finally, in general mirror-self-recognition studies the subjects receive a dye-test procedure to obtain evidence of objective self-directed behavior. However, this index is very controversial among psychologists and primatologists. For example, Gallup (1970) insisted that only chimpanzees and orangutans could recognize themselves in mirrors. In contrast, Swartz and Evas (1994) reported that not all chimpanzees showed evidence of self-recognition. Itakura (1987a, b) and Boccia (1994) reported that macaques (Japanese monkeys and pigtailed monkeys) could use mirrors to direct their responses to their bodies. Hauser reported that the cotton-top tamarin passed the dye-test (cited by Tomasello and Call 1997).

Platt et al. (1991) suggested that a physiological measurement that was correlated with data about the self would provide data on a second level. Boysen and Bernston (1986) reported that chimpanzee heart rates in response to pictures of humans and chimpanzees might be categorical or individual-specific. Platt et al. (1994) stated that the "Response to the animal's own picture (after familiarization with mirrors) should be measured as well as responses during mirror studies". (Platt et al. 1991).

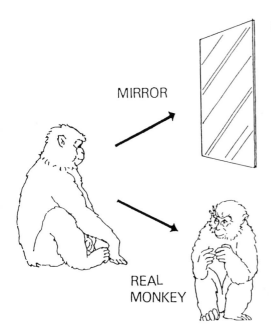

Fig. 3. Cardiac experiment

MIRROR

REAL
MONKEY

Itakura and Fukuda (1993) investigated the cardiac correlation of mirror-self-recognition in a Japanese macaque. The subject was a male Japanese macaque (*Macaca fuscata*), four years of age. The monkey had been trained long enough to habituate wearing a monkey-jacket which was used to settle the ECG-telemeter. The subject's heart rate was measured by disposable electrodes attached to a standard thoracic monitor site and secured by an elastic bandage. An ECG telemeter was settled on the monkey's back by a monkey-jacket. An ECG signal was received by the Telemeter-receiver and a heart rate counter simultaneously monitored the heart rate. A schematic representation of the experiment is shown in Fig. 3.

To explore the cardiac correlation of mirror-self-recognition in a Japanese macaque we measured heart rate response to presentation of a mirror or Japanese macaques (male or female) as the stimulus objects. The subject was tested for three days; two days were for real monkey stimuli (a male or a female) and the other day for mirror-self image (see Fig. 3). The test started with a baseline condition where there was no mirror or no monkeys. This condition lasted 3 min. The mirror or real monkey was then presented for 5 min. This procedure was repeated three times in a session. Heart rate measures were obtained throughout the test session. Momentary heart rate was displayed on the heart rate counter every 2 s.

The changes in heart rate from the baseline in response to each stimulus for the first 12 s (3 blocks) were depicted, and the differential heart rate responses shown across the stimulus condition. The differential heart rate changes of real monkeys were similar to each other. Heart rate increased at first sight of the other individual and then decreased immediately during the second time block. In contrast, the self-mirror-image heart rate decreased during the first and second time blocks and then increased slightly during the third time block. These results suggest that changes in the heart rate pattern of the Japanese macaques were different for the mirror self-image and for other macaques. If so, it may be possible that heart rate provides a new measure of self-recognition. However, we cannot draw conclusions from the single subject used in our experiment. In the future more data should be collected when employing such a paradigm and procedure.

In conclusion, macaques with proper training are capable of learning to use a mirror as a tool to guide their hands (Itakura 1987a). This skill can be applied to inspect an object attached to their bodies, which is considered to be a similar behavior to that used by Gallup (1970) in his demonstration of "self-recognition". Although there are many self-recognition levels in psychology, mirror image self-recognition is a sensory-perceptual self-recognition that represents some aspects of "self." Usually, we observe a change of subject's behavior as evidence for mirror self-recognition. However, we demonstrated that the heart rate pattern changed with the change of the stimulus, real monkey or mirror-imaged monkey, and this may indicate a possibility for using another measure of mirror self-recognition.

3 Understanding and Use of Personal Pronouns

The personal pronoun "I" represents the self strongly at a language level in humans. When we use personal pronouns "I" or "me" they refer directly to ourselves, and we know our awareness of the self. Can a chimpanzee, then, understand the function of personal pronouns and use of personal pronouns? There are several reports on apes trained to use sign-language using the personal pronouns "I" or "you" by signing (Paterson and Linden 1981). However, it is not clear that these apes understood the relationship or shift of reference depending on the pronoun user, which is the most important character of personal pronouns.

3.1 Understanding of Function of Personal Pronouns by a Chimpanzee

Itakura (1992a) demonstrated that a chimpanzee can be trained to use the personal pronouns "me," "you," "him," and "her" in a way similar to that used by humans, even when the referent individuals are shifted with respect to the "speaker." The use of pronouns would appear to require the chimpanzee to understand a relative point of view. For example, in a conversation, a given person is "me" while speaking, but "you" while listening.

The subject was a female chimpanzee named Ai. She had extensive experience in tasks involving the use of visual symbols, "lexigrams", that represented objects and object attributes. Before this training, Ai had demonstrated that she was able to acquire individual names and to describe the behavior of others by using these names (Itakura and Matsuzawa 1993).

Training to use personal pronouns proceeded step-by-step as follows.

Fig. 4. Ai's console in the room where experiments were conducted

3.1.1 Labeling of Personal Pronouns

In this training, Ai learned to use lexigrams representing the personal pronouns "me," "you," "he," and "her" to refer to the nine individuals. She was required to use "you" to refer to the experimenter, who could be seen by the chimpanzee through the wall separating the subject and experimenter (see Fig. 4). The chimpanzee was to refer to herself as "me". Other individuals were to be given the pronouns "him" or "her", according to their respective sex. Details of the procedure have been published elsewhere (Itakura 1992a).

3.1.2 Acquistion of Role-Reversal Usage of Personal Pronouns

In this training, the subject was shown that the individuals designated by the first- and second-person pronouns could be reversed if speech roles were changed, while the individuals designated by the third-person pronouns would remain the same.

3.1.3 Transfer Training in Role-Reversal Usage of Personal Pronouns

In this training, the role of companion was assigned to a human female whose picture had previously been the correct choice only following "her" pronouns. The experimenter, whose picture had previously been the correct choice following either "me", when he presented it, or "you" when Ai presented it, remained visible throughout this experiment, but was available as a correct choice following the personal pronouns "him". This experiment required the chimpanzee to generalize the referential function of the personal pronouns to familiar individuals in novel roles.

3.1.4 Generalization Test for Role-Reversal Usage with a New Companion

In this test session, the chimpanzee was tested to determine whether she could use personal pronouns correctly in referring to another new individual. This new companion was a human Ai knew well, but who had previously been designated only for "him".

Ai reached the criterion of learning at each training session, from labeling of personal pronouns to transfer training in role-reversal usage of personal pronouns. She showed a high percentage of correct responses in the test session with a new companion.

The results suggest that this chimpanzee was able to use the pronouns correctly in the presence of yet another new person. In order to respond correctly from the beginning of the generalization test, the chimpanzee must have learned the following: (a) the pictures of individuals represent real individuals. (b) The person who is interacting with the chimpanzee is the listener to be called "you" by the chimpanzee. (c) The referents of personal pronouns are determined by the relation between a speaker and the personal pronouns. The chimpanzee clearly reached the solution to those problems. This study suggests that a chimpanzee can understand the deictic function of personal pronouns and, in particular, that the referent individual depends on the speaker and user of the personal pronouns.

3.2 Individual Recognition of Line Drawings

Tyrell et al. (1987) reported that human infants are able to recognize individuals from line drawings if the line drawings accurately represent the individual's features. In their experiments, infants were presented with two trials of a novelty preference task. The infants recognized familiar faces and could discriminate novel faces of the same sex. In the case of nonhuman primates, Hayes and Hayes (1953) reported that a home-raised chimpanzee recognized a large number of pictured objects (toys, foods, etc.), even when they were presented as simple black-and-white line drawings. However, individual line drawings were not included in the stimuli.

Can a chimpanzee, then, recognize line drawing representations of individuals, be they conspecific or other species, including humans? Itakura (1994a) evaluated the ability of a chimpanzee to recognize individuals portrayed in line drawings, including the subject herself. In this study, a female chimpanzee with extensive prior experience in the use of visual symbols matched the line drawings of chimpanzees, humans, and an orangutan with a specific letter of the alphabet. When a line drawing of a familiar individual was presented on the computer screen, the chimpanzee responded by touching a key with the letter of the alphabet that corresponded to the individual's name. We demonstrated the chimpanzee's ability to discriminate line drawings representing humans, chimpanzees (including the subject herself), and an orangutan. The results suggest the possibility that the chimpanzee can abstract individuality from line drawing representations, a fact that might indicate a higher level of cognitive ability than has been reported previously.

In conclusion, a chimpanzee can understand the deictic function of personal pronouns and, in particular, that the referent individuals depend on the speaker and user of the personal pronouns. A chimpanzee is also able to categorize individuals, including herself, from novel line drawing representations.

4 Possession at a Symbolic Level

In the case of humans, possession is one of the most important factors of self-awareness. When a child points to an object and names its owner (e.g., pointing to the father's shoe and saying, "Daddy"), this utterance is generally interpreted as an indication of the child's awareness of the relation between the object and its owner (cited from Oshima-Takane 1995). Rodgon and Rashman (1976) claimed that such utterances indicate that child's preliminary awareness of the possessor-possession relationships and of the conventional linguistic expression of possession.

4.1 Symbolic Representation of Possession in a Chimpanzee

Although possession in nonhuman primates has been studied in the context of control over objects (Kummer 1973; Thierry et al. 1989; Torii 1975), there is no evidence that possession in primates is mediated by a self-concept or a concept of other, such that they view these objects as symbolically representing themselves or another. Itakura (1992b) explored whether a symbol-trained chimpanzee can symbolically associate objects and individuals (including herself) after observing an arbitrarily created relationship between them. In particular, I showed that a chimpanzee can not only recognize and name photographs of herself and others, but also can recognize the symbolic relation between these individuals and objects distinctly of their own.

4.2 Feeding Training with Fixed-Color Bowls

Chimpanzee Ai, described above, received feeding training together with a male chimpanzee, Akira, and a male orangutan named Doudou (see Fig. 5).

Ai was the target subject and the two other individuals were stimulus subjects. In the feeding procedure, Ai was always fed with a green bowl, Akira was fed with a red bowl, and Doudou was fed with a yellow bowl. The trial ended after the three individuals were fed with these bowls, and each session consisted of 12 trials according to the combination of feeding order. Each trial was followed by a 60-s inter-trial interval. This training was conducted for five sessions. Before-feeding time, Ai's food-demanding behavior was coded by a one-zero type sampling method. Before-feeding time was defined as the time from when the experimenter picked up one of three bowls until the experimenter moved to the individual. Food-demanding behavior was defined as reaching a hand to, opening the mouth to, or following the experimenter.

At the first, Ai showed food-demanding behavior toward bowls of all three colors, and showed some aggressive behavior to Akira when he fed. Food-demanding behavior decreased, however, when the experimenter held red and yellow bowls, used for Akira and Doudou, respectively. Eventually, Ai came to ignore the experimenter when he picked up the red and yellow bowls, and came to show food-demanding behavior only when the green bowl was picked up.

Fig. 5. Feeding training situation in the ape's cage

4.3 Feeding with Different Bowls

In this test, each individual was fed with a different bowl; for example, Ai was fed with the red bowl, Doudou with the green bowl, and Akira with the yellow bowl. When Ai was fed with the red or yellow bowl, she accepted them with no hesitation; however, when the green bowl was used for Doudou, Ai followed the experimenter, then gave up quickly. When the green bowl was used for Akira, Ai showed violent, aggressive behavior to Akira as if she asserted her privilege to be fed with the green bowl.

4.4 Preference Test of Bowls

In a second test, Ai's preference for the green bowl was tested by pairing two of three bowls in counterbalanced right-left position. Two bowls, each containing a peanut, were presented to Ai. We then coded which bowl's peanut was taken first by Ai.

Although when presented with two bowls Ai could take both peanuts at once, initially she took the peanuts only one at a time. When the green bowl was paired with another bowl, she took the peanut from the green bowl first and then the one from the other bowl. When the red and yellow bowls were paired, Ai preferred to take the one in the left position first, then took the peanut from the other bowl. In the case of green bowl, color preference was dominant to position preference.

4.5 Symbolic Association Between Individuals and Bowls

In the final test, Ai sat in from of a TV monitor and a 5 × 6 matrix keyboard on a console in the experimental room. The TV monitor displayed still-photograph stimuli via a laser disc player. Letters were mounted on the keyboard. There was a tray by the TV monitor. A food reward, such as a peanut, was automatically delivered after a correct response. Ai would begin a trial by pressing a start key, which presented photographs of each bowl (green, red, and yellow), on the TV monitor. After a 2 s delay, one of the two rows of keys on the console was illuminated. Each row contained the name of Ai, Akira, and Doudou. Ai's task was to choose the correct name among the three letters in the illuminated row that corresponded to the photograph presented on the TV monitor. In this task, Ai showed high accuracy in associating individuals and bowls. She chose the correct letter from the first trial, when each bowl was presented on the TV monitor, making only one error, which occurred after 30 trials.

In conclusion, these results clearly indicate that the chimpanzee Ai developed the representation of a symbolic relationship between objects and individuals, including herself, solely by participating in the feeding protocol described.

5 General Discussion

In this chapter, I reviewed several studies of my own concerning the self in nonhuman primates. The important issue is what level and kind of self appears in nonhuman primates. We may be able to postulate an "ecological self" for them, because they also have biological system perceived functional goal of environment as well as humans. Then what kinds of higher-order self can we hypothesize in them?

I think mirror self-recognition itself can be reduced to the problem of understanding contingency between the mirror image and their inner sense of movement, and this may be a stage of the ecological self. However, the difference in the response pattern to a mirror between chimpanzees and monkeys may cause the difference in their social relationship. Chimpanzees may pay close attention to the existence of other chimpanzees, their movement, interactions, and their effects on others. On the other hand, although macaques have an established and sophisticated social system and social interaction with each other, their responses seem to be just stereotypic aggressive behavior or threatening behavior.

In the situation of mirror exploration, the chimpanzee looks at the mirror and it finds a chimpanzee in the mirror. When the chimpanzee tries to interact with the mirrored chimpanzee, that chimpanzee never reacts like the subject chimpanzee expects. The mirrored chimpanzee also never initiates interaction with the subject chimpanzee by him- or herself. The subject chimpanzee is very curious about the mirrored chimpanzee. With time, the chimpanzees come to understand the synchronization of the movement between the subject chimpanzees and mirrored chimpanzee. I call this "contingency". In other words, the chimpanzee understands the causality of their inner sense of movement and the movement of the mirrored chimpanzee at a body sense level. In the case of macaques, the first part of the responses may be the same as those of chimpanzees. However, their responses do not progress after they are aware of the curious macaque in the mirror. After that they do not tend to have any relationship with the mirrored macaque. That means they do not have a chance to touch upon the contingency or causality. I think these are the reasons for the difference of mirror behavior between chimpanzees and macaques. Recently, Inoue-Nakamura (1997) attempted to explore the phylogenetic origin of mirror self-recognition. She examined behavior towards a mirror in 12 species of nonhuman primates. Only the great apes exhibited self-directed behaviors that were thought to be the critical evidence of mirror self-recognition. She concluded that there is a large difference between great apes and the other nonhuman primates in terms of the capacity for mirror self-recognition.

The chimpanzee used here was capable of understanding the symbols of personal pronouns that represent the deictic relationship among some individuals, including the subject herself. The chimpanzee could also connect her own bowl and her name, and another's bowl and name. This fact suggests that the chimpanzee was able to link the objects and its owner at a symbolic level. There have been several studies on the chimpanzee ability for joint visual attention, social referencing, and empathic behavior (Itakura 1994b, 1995; Itakura and Tanaka 1998; Russel et al. 1997). However, there have been few such studies on such abilities in macaques.

Chimpanzee behavior may be based on such a strong and deep social relationship, whereas in macaques such social relationships may be weaker than that of chimpanzees. I think that such social relationships or social bonds are strongly related to the level of "self". From the perspective of the level of relationships among individuals, chimpanzees seem to have higher-order and more complex relationships than macaques.

In conclusion, the "self" seems to be like a net stretched in a society. Only when we have sensitivity to our own contingency, and the level of relationship with others becomes stronger, can higher-order self appear from time to time. The appearance of self changes dynamically at intervals along the relationship between physical environment and social environment. From Neisser's classification of self, both chimpanzees (apes) and macaques have "ecological self" and "interpersonal self", because they interact with physical and social environments. Besides, there are some studies which show that chimpanzees have a very basic ability for joint attention or understanding of the knowledge of others. If so we may be able to postulate a higher-order self in chimpanzees than that of macaques. However, from the perspective of "theory of mind" as in humans, we do not have established evidence yet even for chimpanzees. That is, there is not enough evidence that chimpanzees can understand or infer the thoughts, beliefs, and intentions of others at the same level as humans. Therefore we cannot postulate higher-order social self even in chimpanzees. However, this does not mean chimpanzees never have a higher order self, it just means we have no evidence about that, so we cannot make a conclusion yet. The important thing is that research in this area is under way and additional data should be collected in the future.

References

Anderson JR (1984). Monkeys with mirrors: Some questions for primates psychology. Int J Primatol 5:81-98

Bahrick LE (1995) Intermodal origins of self-perception. In: Rochat P (ed) The self in infancy: Theory and research. Elsevier, Amsterdam, pp 349-374

Benhar EE, Carlton PL, Samuel D (1975). A search for mirror-image reinforcement and self-recogntion in the baboon. In: Kondo S, Kawai M, Ehara, A (eds) Contemporary primatology: Proceedings of the 5th International Congress of Primatology. Karger, New York, pp 202-208

Bertenthal BI, Fischer KW (1978) Development of self-recognition in the infants. Dev Psychol 14:44-50

Boccia ML (1994) Mirror behavior in macaques. In: Parker ST, Mitchell BW, Boccia ML (eds) Self-awareness in animals and humans: developmental perspectives. Cambridge University Press, Cambridge, pp 350-360

Boysen ST, Bernston GC (1986) Cardiac correlate of individual recognition in the chimpanzee (*Pan troglodytes*). J Comp Psychol 100:321-324

Brown WL, McDowell AA, Lobinson EM (1965) Discrimination learning of mirrored cues by rhesus monkeys. J Gen Psychol 106:123-128

Gallup GG Jr (1970) Chimpanzees: self-recognition. Science 167:86-87

Gallup GG Jr (1977) Self-recognition in primates: a comparative approach to the bidirectional properties of consciousness. Am Psychol 32:329-338

Gallup GG Jr (1982) Self-awareness and the emergence of mind in primates. Am J Primatol 2:237-248

Gallup GG Jr, Wallnau LB, Suarez SD (1980) Failure to find self-recognition in mother-infant and infant-infant rhesus monkeys pairs. Folia Primatol 33:210–219

Gibson JJ (1979) The ecological approach to visual perception. Houghton Mifflin, Boston

Gibson EJ (1995) Are we automata? In: Rochat P (ed) The self in infancy: theory and research. Elsevier, Amsterdam, pp 3–14

Griffine DR (1976) The question of animal awareness: evolutionary continuity of mental experience. Rockfeller University Press, New York

Hayes KJ, Hayes C (1953) Picture perception in a home-raised chimpanzee. J Com Physiol Psychol 46:470–474

Inoue-Nakamura N (1997) Mirror self-recognition: a phylogenetic approach. J Psychol Res 39:266–275

Itakura S (1987a) Mirror guided behavior in Japanese monkeys (*Macaca fuscata fuscata*). Primates 28:149–161

Itakura S (1987b) Use of a mirror to direct their responses in Japanese monkeys (*Macaca fuscata fuscata*). Primates 28:343–352

Itakura S (1992a) A chimpanzee with the ability to learn the use of personal pronouns. Psychol Rec 42:157–172

Itakura S (1992b) Symbolic association between individuals and objects as an initiation of ownership in a chimpanzee. Psychol Rep 70:539–544

Itakura S (1994a) Recognition of line-drawing representation by a chimpanzee. J Gen Psychol 121:189–197

Itakura S (1994b) Differential responses to different human conditions by a chimpanzee. Percept Mot Skills 79:1288–1290

Itakura S (1995) An exploratory study of social referencing in chimpanzees. Folia Primatol 64:44–48

Itakura S, Fukuda S (1993) Cardiac responses in a Japanese monkey to self-mirror-image. J Yokohama Nat Univ 33:221–226

Itakura S, Matsuzawa T (1993) Acquisition of personal pronouns by a chimpanzee. In: Ritoblat HL, Herman LM, Nachtigall PE (eds) Language and communication: comparative perspective. Lawrence Earlbaum Associates, Hillsdale, NJ, pp 347–363

Itakura S, Tanaka M (1998) Use of experimenter-given cues during object choice tasks by chimpanzees (*Pan troglodytes*), an orangutan (*Pongo pygmaeus*), and human infants (*Homo sapiens*). J Com Psychol 120:119–126

Kummer H (1973) Dominance versus possession: an experiment on hamadryas baboons. In: Menzel EW (ed) Perceptual primate behavior. Karger, Basel, pp 226–231

Lewis M, Brooks-Gunn J (1979) Social cognition and the acquisition of self. Plenum, New York

Menzel EW Jr, Savage-Rumbaugh ES, Lawson J (1985) Chimpanzee (*Pan troglodytes*) spatial problem solving with the use of mirrors and televised equivalents of mirrors. J Com Psychol 99:211–217

Neisser U (1995) Criterion for an ecological self. In: Rochat P (ed) The self in infancy: theory and research. Elsevier, Amsterdam , pp 17–34

Oshima-Takane Y (1995) Development of possessive forms in English-speaking children: Functional approach. Jpn Psychol Res 37:59–69

Parker ST, Mitchell RW, Boccia ML (1994) Self-awareness in animals and humans: developmental perspectives. Cambridge University Press, Cambridge

Paterson F, Linden E (1981) The education of Koko. Owl Books, New York

Platt MM, Thompson RL, Boatright SL (1991) The question of mirror-mediated self-recognition in apes and monkeys: some new results and reservations. In: Parker ST, Mitchell BW, Boccia ML (eds) Self-awareness in animals and humans: developmental perspectives. Cambridge University Press, Cambridge, pp 330–349

Povinelli DJ (1987) Monkeys, apes, mirrors, and mind: the evolution of self-awareness in primates. Human Evol 2:493–509

Ristau CC (1991) Cognitive ethology: The minds of other animals. Lawrence Erlbaum Associates, Hillsdale, NJ

Rochat P (1995) Early objectification of the self. In: Rochat P (ed) The self in infancy: theory and research. Elsevier, Amsterdam, pp 53–72

Rochat P (in press) Early development of the ecological self. In: Dent-Read C, Zukow-Golding P (eds) Evolving explanation of development. American Psychological Association, Washington, DC, pp 91–122

Rodgon MM, Rashman SE (1976) Single-word usage, cognitive development and the beginnings of combinatorial speech: A study of ten English-speaking children. Cambridge University Press, New York

Russel CL, Bard KA, Adamson LB (1997) Social referencing by young chimpanzees (*Pan troglodytes*). J Comp Psychol 111:185–193

Spada EC, Aureli P, Verbeek P, deWaal FBM (1995) The self as reference point: can animals do without it? In: Rochat P (ed) The self in infancy: theory and research. Elsevier, Amsterdam, pp 193–215

Swartz K, Evas S (1994) Social and cognitive factors in chimpanzees and gorilla mirror behavior and self-recognition. In: Parker ST, Mitchell BW, Boccia ML (eds) Self-awareness in animals and humans: developmental perspectives. Cambridge University Press, Cambridge, pp 350–360

Thierry B, Wundderlich D, Gueth G (1989) Possession and transfer of object in a group of brown capuchins (*Cebus apella*). Behaviour 110:294–305

Tomasello M, Call J (1997) Primate cognition. Oxford University Press, New York

Torii M (1975) Possession by non-human primates. In: Kondo S, Kawai M, Ehara A (eds) Contemporary primatology. Karger, Basel, pp 310–314

Tyrell DJ, Anderson JT, Clubb M, Bladbury A (1987) Infants' recognition of the correspondence between photographs and caricatures of human faces. Bull Psychonomic Soc 25:41–43

Zazzo R (1948) Images du corps et conscience de soi. Enfance 1:29–43

16
Self- and Other-Control in Squirrel Monkeys

James R. Anderson

1 Introduction: Extending "Self"-Related Research in Nonhuman Primates

The last three decades of the 20th century have seen keen interest in the study of self-awareness in nonhuman primates. This area has been dominated by experiments using mirror-image stimulation. Typically, monkeys or apes are observed in the presence of their reflection and monitored for any signs that they correctly interpret the image as a representation of their own body. Self-recognition is admitted for a primate that spontaneously uses the image to check its visual appearance, more specifically to explore parts of its body that cannot be seen without the aid of a mirror. Confirmatory evidence may take the form of the individual using the reflection to visually and tactually explore a mark surreptitiously placed on an normally unseen body part (usually the head) which is ignored in the absence of the mirror (for reviews see Anderson 1984a; Gallup 1994; Parker et al. 1994).

Self-recognition research has given rise to a wealth of information on how prosimians, monkeys, and apes respond to their own mirror-images, with corresponding theoretical developments regarding the ontogeny and phylogeny of self-awareness and its underlying mechanisms. A spin-off of this approach has been some insights into socioperceptual mechanisms in those species which respond to their reflection as if they were in the presence of a conspecific (Gallup 1982, 1991; Anderson 1984a, 1994; Mitchell 1993; Anderson and Gallup 1999).

The heavy reliance on visual self-recognition as a window into nonhuman primates' sense of self contrasts with the wide range of approaches used to study the development of sense of self in human infants. Although experiments using mirror and video feedback also figure prominently in the literature on human infants (Lewis and Brooks-Gunn 1979; Anderson 1984b), this broad approach is complemented by a multitude of other approaches which focus to varying degrees on the sensory, social, emotional, linguistic, and autobiographical aspects of self, among others (Damon and Hart 1982; Lewis 1994; Rochat 1995). The sense of self in humans is widely recognized as being multifaceted, and it can be argued that this broader view of self-awareness should be adopted when it comes to considering

Department of Psychology, University of Stirling, Stirling FK9 4LA, Scotland, UK

Fig. 1. Squirrel monkeys (*Saimiri sciureus*), the species studied in the experiments

nonhuman primates. Correspondingly, research efforts should be extended to include the application of a range of methods adapted from research on humans (e.g., Gallup 1991; Mitchell 1994).

In this chapter I describe two studies which I will argue are relevant to the attempt to understand self-awareness in other species. The primate under study is a New World monkey, the squirrel monkey (*Saimiri sciureus*) (Fig. 1). The first study looks at the question of self-control in a situation in which squirrel monkeys are required to inhibit their natural tendency to reach for the greater of two quantities of food. The second assesses the development of the ability to control another individual's behavior, in the context of a task involving deception. At the time of writing (late in 1999), work on both these topics is continuing, but the early findings clearly extend current knowledge about self-control and tactical deception in nonhuman primates. The experiments to be described have been conducted in the laboratory of Professor Kazuo Fujita of Kyoto University Department of Psychology, with the assistance of several graduate students (see Acknowledgments).

2 On Self-Control by Squirrel Monkeys

2.1 Background

As the human infant's sense of self develops, there is an increasing ability to control many aspects of behavior. This is most readily observed in basic postural and locomotor functions. However, these improving abilities for self-regulation are also in the mental domain, as the infant learns to delay immediate gratification, i.e., postponing immediate rewards until later. The early development of self-control is strongly influenced by the family and then the preschool and school environments, as children are taught to resist immediate gratification in order to obtain benefit at a subsequent time across a range of contexts. As such self-control is considered central to planning and foresight (Mischel et al. 1989), the degree to which individuals show self-control as opposed to impulsiveness may be fundamental to the developing individual's notion of self. In typical experimental investigations of self-control, subjects are given the choice between a smaller, immediate reward and a larger, delayed reward. In preschool children, the likelihood of showing self-control in experimental situations is related to other cognitive competencies (Mischel et al. 1989).

Self-control and impulsiveness have been studied in comparative perspective through experiments with several species of birds, rats, and nonhuman primates. In general, when rats and pigeons are given a choice between stimuli leading to a smaller, immediate reward or a larger, delayed reward, they tend to select the former, thus failing to maximize rewards in the long term (Logue 1988). The picture regarding nonhuman primates and self-control is less clear. In one study, two macaques (*M. fascicularis*) were found to show self-control in experimental conditions in which rats and pigeons show impulsiveness (Tobin et al. 1996). The authors related self-control by the macaques to their relatively advanced general cognitive abilities, a function of encephalization, and to their evolutionary history in an environment in which food was relatively abundant and stable. Three chimpanzees (*Pan troglodytes*) were tested in another paradigm based on child research: they either rang a bell during a trial to receive an immediate, less preferred food reward or they refrained from ringing the bell in order to receive another, more preferred food reward. The chimpanzees learned to delay gratification in this situation, but marked individual differences emerged when photographs or lexigram symbols replaced the food during the delay (Beran et al. 1999).

An original study relevant to self-control was reported by Boysen and Berntson (1995). One chimpanzee (the selector) was shown two different quantities of food (1 versus 2, 4, or 6) in the presence of a second chimpanzee. When the selector reached or pointed toward one of the quantities of food (candies), the experimenter gave this quantity to the second chimpanzee and the remaining candies to the selector. Thus, the selector received a "reverse reward" contingency, i.e., getting to eat only the quantity of food other than the one it selected. Remarkably, neither chimpanzee was reliably able to select the smaller of the two arrays in order to receive the larger one, even after several hundred trials. However, when Arabic numerals representing the quantities replaced the candies as stimuli, the chimpanzee who was experienced in the use of these symbols overcame the tendency to reach toward the larger array. The authors concluded that symbolic stimuli (the numerals) served as a representational aid for overcoming a perceptual–motor bias for the larger array. Both findings, that of an inability to inhibit reaching toward the larger quantity of food and that of symbol-mediated control over this bias, were confirmed in a subsequent study of five chimpanzees which were experienced in the use of symbols (Boysen et al. 1996).

In a learning-theory-based approach to the interference caused by the "intrinsic incentive properties" of the larger food array in the chimpanzee studies, Silberberg and Fujita (1996) pointed out that the chimpanzees were rewarded regardless of their selections. In other words, even if the selector did not get the four candies because it pointed to them, it nevertheless received a reward in the form of the remaining candy. These authors suggested that this reinforcement might be strong enough to maintain the chimpanzees' natural tendency to "always reach for the higher value of two alternatives" (Silberberg and Fujita 1996, p. 144). To test this hypothesis, an experimenter showed Japanese macaques (*M. fuscata*) one piece of food held in one hand and four pieces of food in the other hand. A strong bias to reach toward the larger array was confirmed, i.e., the monkeys consistently reached toward the hand holding the larger array although this resulted in their getting only the smaller one as a reward. In the second phase of the study, a "large or none" reward contingency was instigated, in which the experimenter gave the monkey the four pieces of food if it reached toward the hand holding the single piece, but the four pieces were immediately dropped and the hand withdrawn if the monkey reached to the hand with four pieces. Under this condition, the monkeys soon learned to reach toward the hand holding the smaller quantity of food in preference to the hand holding the larger quantity.

We decided to look at how squirrel monkeys would perform in a self-control study similar to those described above for chimpanzees and Japanese macaques (Anderson et al. 2000). Although considerably smaller than macaques, squirrel monkeys also have a larger brain than their body size would predict (Gibson 1986). They have shown relatively good performance in some studies of transitivity and numeracy (Thomas et al.1980; McGonigle and Chalmers 1992; Olthof et al. 1997), and like macaques, they achieve a positive transfer index on discrimination reversal learning (Rumbaugh 1997). On the other hand, compared with macaques and great apes they cope less well with delayed-response tasks (French 1959). We were therefore interested to see whether the squirrel monkeys' mental abilities and limits would be compatible with self-control when the monkeys were faced with two quantitatively different food arrays.

2.2 The Initial Preference: Methods and Results

The subjects were eight socially housed adult squirrel monkeys. They were tested individually in a standard transport box which they had previously been trained to enter for positive reinforcement, or in the case of one reluctant individual, in an isolated compartment of the home cage. The monkeys were not food-deprived. Small pieces of apple (for six subjects) or almond (for two subjects) were used as the food stimuli.

An experimenter presented two food arrays on a hand-held wooden tray (36 cm × 22 cm) which had one line bisecting it into left and right halves and another line perpendicular to this one, 2 cm from the front edge. Pairs of spots approximately 10, 15, and 20 cm from the front edge indicated potential locations for the food stimuli. The experimenter prepared a trial by arranging the food on the tray while his back was turned to the subject. To start a trial, he turned around to face the subject, held the tray out of the subject's reach for 3–5 s, and then extended his arms to bring the front edge of the tray into contact with the front of the transport box or cage, thus allowing the subject to reach through the bars toward the food. In every trial, one piece of food was on one of the spots on the tray and either two or four

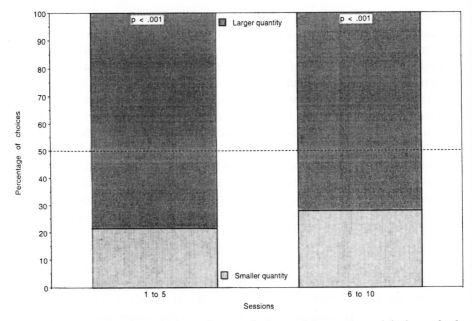

Fig. 2. Percentage of trials in which monkeys reached toward the smaller and the larger food arrays in phase 1 of the self-control study

pieces were at the corresponding location on the other side of the bisecting line (the left and right locations of the different quantities were quasirandom).

In phase 1, the pairings always consisted of one and four pieces of food. The two lines on the tray facilitated detection and orientation of reaches. As soon as the subject reached toward one of the arrays, the experimenter turned the tray 90° so as to bring only the other array within reach of the subject. Thus, a subject that reached toward the four pieces of food was allowed only to retrieve and eat the single piece, and vice versa. A trial ended when the subject took the food offered (which it always did), and the experimenter turned around to prepare the following trial. Ten 20-trial sessions were run with each subject.

Figure 2 shows the mean frequency of reaches toward four pieces as opposed to one piece of food for sessions 1 to 5 and 6 to 10. A one-sample t-test against a 50% chance level confirmed a highly significant bias toward reaching for the larger array in the first five sessions ($P < 0.001$). Individual subjects' selections were analyzed using binomial tests, and every subject showed a significant bias toward the larger array at $P < 0.001$. Across sessions 6 to 10, the group-level bias toward reaching for the larger array remained highly significant ($P < 0.001$), with seven of the eight subjects showing an individually significant effect at $P < 0.001$, and the remaining subject developing an exclusive side preference instead. Thus, like chimpanzees and macaques, squirrel monkeys consistently reached toward an array of four pieces of food in preference to one piece, even though this response resulted in them receiving only the smaller array every time. All monkeys did sometimes select the smaller array (between 31 and 62 times each) and thus obtained the larger array as a reward on those occasions, but this never led to any of them developing self-control, i.e., reaching toward the single piece of food instead of the larger quantity.

2.3 The Large-or-None Intervention: Methods and Results

In order to determine whether squirrel monkeys could learn to overcome their persistent tendency to reach for the larger array, the next phase employed the procedure of Silberberg and Fujita (1996) of only reinforcing selections of the smaller array. In other words, if a monkey reached toward the single piece of food, the larger array was given as a reward, whereas if it reached toward the larger array, it received no reward. This "large-or-none" procedure was used with half of the monkeys. To assess whether further experience of the original reverse-reward contingency might eventually lead to a change in response strategy, the other half of the monkeys continued under the same general conditions as in phase 1, but with the larger array of food reduced to two pieces. In view of the finding by Boysen and Berntson (1995) that chimpanzees' bias for the larger array was more marked as the disparity between the array sizes increased, we wondered whether squirrel monkeys might learn to choose the smaller array if the size disparity was reduced.

Thus, the same eight squirrel monkeys were tested, four under the same reverse-reward condition as in phase 1 but with array sizes of one and two, and four under the large-or-none condition using arrays of one and four. For the latter four mon-

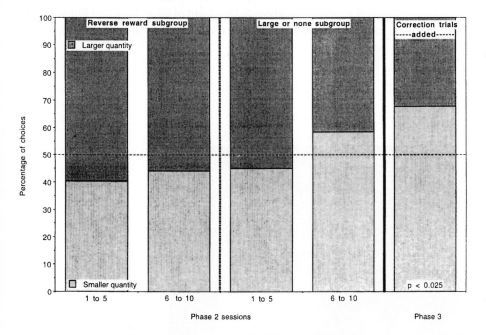

Fig. 3. Percentage of trials in which monkeys reached toward the smaller and the larger food arrays in phases 2 and 3 of the self-control study

keys, a reach toward the larger array caused the experimenter to withdraw the stimulus tray and turn around to prepare the next trial, which commenced after approximately 20 s. Ten 20-trial sessions were run for each subject.

The four left-hand columns of Fig. 3 show the results for the two subgroups of monkeys across the first five and then the second five sessions. In this phase, neither subgroup showed an overall preference for either array. However, the group-level analyses hide important individual differences in response patterns. Of the four "reverse-reward" monkeys, two reached preferentially toward the larger array on the first five sessions and one of these continued to do so in the second five sessions; all showed highly significant preferences for one side rather than one array (at $P < 0.001$). Of the four "large-or-none" monkeys, two also reached preferentially for the larger array across sessions 1 to 5, but in the next five sessions these two and one other monkey selected the arrays equally frequently due to significant side preferences (all at $P < 0.001$). More importantly, however, one of the four large-or-none monkeys selected the smaller array significantly more frequently than the larger one in three of the last five sessions, producing a significant overall preference.

Thus, at a group level, the strong tendency to reach impulsively toward the larger of two food arrays was confirmed in this phase. Neither the imposition of a large-or-none contingency nor continued experience of the reverse-reward contingency produced significant group-wide changes in the monkeys' response strategies other than the development of side preferences. However, some important individual differences in response patterns emerged in terms of the persistence of the initial bias for the larger array, with one of the monkeys in the large-or-none test developing a bias for the smaller array, and thereby showing self-control (Fig. 4). Thus, some limited support was given to the claim of Silberberg and Fujita (1996) that monkeys can learn to overcome their initial tendency always to reach for the higher-value incentive.

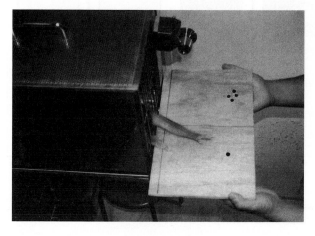

Fig. 4. Monkey Karin reaches toward the smaller food array. (Photo: S. Awazu)

2.4 Addition of Correction Trials: Methods and Results

The development of self-control in one monkey as a result of the large-or-none procedure in the previous phase prompted us to ask whether further training might lead to success with other subjects. One potentially important difference between the procedure applied in phase 2 and that of Silberberg and Fujita (1996) was that the latter study used correction trials in the event of nonmaximizing responses. In other words, if a macaque selected the larger array, this array was dropped and the smaller array withdrawn, and then the same array configuration was re-presented as necessary until the macaque switched its choice. Therefore, the seven squirrel monkeys who had not yet learned to reach toward the smaller array were tested using the large-or-none contingency associated with correction trials. For these monkeys, the array sizes used in the previous phase were maintained. In the event of an "error" (i.e., selection of the larger array), a correction trial started after approximately 20 s, during which the experimenter stood with his back to the monkey ("time out"). The one squirrel monkey to show self-control in the previous phase was again tested without correction trials but with the larger array size reduced from four to two. This phase consisted of five sessions.

As a group, the monkeys now showed a significant bias for reaching toward the smaller array when compared with a hypothetical mean of 50% ($P < 0.025$) (Fig. 3, right-hand panel). At the individual level, two monkeys showed a significant bias toward the smaller array in all five sessions, three did so only in later sessions, one switched from a nonsignificant bias toward the larger array to a similar bias for the smaller array, while one subject continued to show a significant preference for the larger array.

Thus, by the end of phase 3, six of the eight monkeys had overcome their initial tendency to reach toward the larger food array, and were now reliably selecting the smaller array. The next step was to assess whether such self-control would be maintained if the correction procedure was withdrawn.

2.5 Return to Original Reverse Reward: Methods and Results

Because of the variability in individual subjects' performances and experiences in the previous phases, in phase 4 the procedure was varied for individual subjects. Two monkeys were tested on reverse-reward as in phase 1, one with arrays of one and two pieces of food, and one with array sizes one and four. The large-or-none procedure was used with four of the remaining six monkeys, but without correction trials. For three of these four monkeys, array sizes were one and four, while for the fourth they were one and two. Finally, in a last effort to get the two remaining nonlearners to develop the self-control strategy, they were run again using the large-or-none contingency combined with correction trials. This phase consisted of five sessions per subject.

Table 1 shows the performance of the eight individual subjects. Seven of the eight subjects reached toward the smaller of the two food arrays in order to receive

Table 1. Percentage of trials in which individual subjects reached for the larger array in phase 4 of the self-control study

Subjects	Kar	Pan	Cob	Gye	Dau	Boo	Hom	Yay
Array sizes	1v2	1v4	1v4	1v4	1v2	1v2	1v2	1v4
Condition[a]	RR	LN	RR	LN	LN	LN	LN	LN
						Cor		Cor
Percentage	3***	14***	8***	18***	48ns	38*	9***	4***

[a]RR, "reverse reward" contingency; LN, "large-or-none" contingency.
*$P < 0.05$; ***$P < 0.001$.

the larger one as a reward a significant number of times; the group-level analysis was highly significant ($P < 0.001$). There was no clear effect of the size of the disparity between the arrays (one versus two or four). One of the two subjects which received correction trials now showed a bias for the smaller array, while the eighth subject never showed any preference other than an extremely persistent side preference.

To summarize, by the end of phase 4, seven of the eight squirrel monkeys were reliably selecting the smaller array of pairs consisting of either one versus two or one versus four pieces of food. The two monkeys tested in the original reverse-reward condition showed a high level of self-control. Even the occasional error, i.e., selection of the larger array resulting in the delivery of the smaller reward, did not undermine the monkeys' performance.

2.6 Summary of Results

The main finding of the study is that squirrel monkeys are capable of showing self-control when confronted with two food arrays, i.e., they can overcome an initial, strong tendency to reach for the larger of the two arrays in order to maximize gain. When tested with a reverse-reward contingency and visible food, chimpanzees find it extremely difficult to inhibit their natural tendency to reach for the larger array, but they can overcome this tendency if the food is replaced by symbols (Boysen and Berntson 1995; Boysen et al. 1996). The results of our study, along with those of Silberberg and Fujita (1996), show that symbol-training is not the only way to achieve self-control responses in nonhuman primates in this type of situation. Instead, appropriate manipulation of environmental contingencies can lead to the learning of maximizing responses. It should also be recalled that in other situations chimpanzees can also show delay of gratification in the presence of foods (Beran et al. 1999); it remains to be determined whether squirrel monkeys can do likewise.

3 On Deception by Squirrel Monkeys

3.1 Background

Gallup (e.g., 1991) has suggested that only species (and by extension, individuals) with self-awareness should be capable of intentional deception. Deceptive acts are widespread among animals, but interest here is restricted to deception as an intentional act performed to alter another individual's knowledge state. In order to engage in higher-order deceptive acts (see Mitchell 1986; Byrne and Whiten 1992), the individual must generate a plan in which some notion of self takes a central role. The plan may be rehearsed and worked through with the aid of mental images of the environment, other individuals in the plot, and self. In support of Gallup's view that deception should be limited to species with self-awareness is the fact that chimpanzees, the species most likely to show mirror self-recognition, are also by far the most accomplished deceivers, particularly in advanced forms of deception involving "mind reading" skills (Byrne and Whiten 1992).

However, there is continuing debate over the quality of the data relevant to deception in nonhuman primates (Heyes 1994, 1998, Kummer et al., 1990). Most descriptions of deceptive acts are anecdotal, with all the associated problems of incompleteness of observations, observer bias, and over-elaborate interpretation. While these faults can also exist in experimental research, the latter affords repeatability, control, and the elimination of a good many extraneous variables. Surprisingly, no one has yet improved upon the single experimental demonstration of deception in chimpanzees, the classic study by Woodruff and Premack (1979). Chimpanzees could obtain food by pointing to one of two containers, one of which concealed food. Pointing to the correct, i.e., baited, container in the presence of a cooperative trainer resulted in the trainer giving the food to the ape. Once communicative pointing was established, a competitive trainer was introduced, who kept the food if the chimpanzee's behavior revealed which container was baited. If the competitive trainer selected the unbaited container, however, the chimpanzee was allowed to eat the food. Under these conditions, over many months of testing, two chimpanzees eventually systematically misled the competitive trainer by directing him to the unbaited container.

Although these findings by Woodruff and Premack (1979) are generally considered to be the strongest experimental data on deception by apes, they can be interpreted in terms of simple conditional discrimination learning rather than mental-state manipulation. Mitchell and Anderson (1997) tested capuchin monkeys in an analogue of the Woodruff and Premack procedure, and found evidence of withholding of information and active deception in these monkeys, in spite of the fact that there is virtually no anecdotal or observational evidence of higher-order deception by capuchin monkeys (Byrne and Whiten 1992). Furthermore, capuchins show no signs of self-recognition (Anderson 1996; Anderson & Marchal, 1994). The learning curves for the capuchin monkeys' deceptive acts strongly suggested the establishment of conditioned response through trial and error (Mitchell and Anderson 1997). In that study, there were no follow-up experiments or observations to examine the flexibility of the monkeys' deceptive behavior, or to probe for the possible involvement of mental-state manipulation. We therefore sought to assess whether conditioning procedures might lead squirrel monkeys to show deception similar to that established in the above studies (Anderson et al., unpublished work), and to follow up with additional analyses to clarify the mental processes involved in any deceptive acts. The results of the first part of this study are presented below.

3.2 Establishment of "Honest" Reaching: Methods and Results

The subjects were one group-living adult female ("Gye") and two pair-housed (later group-housed) adult males ("Boo" and "Coboo"). These three monkeys were also used in the study on self-control described above. They were not food-deprived, but received part of their daily rations as rewards during testing and the remainder in their home cages after testing each day. For testing, each monkey was transported to the test room and placed in a test cage (46 cm × 46 cm × 52 cm). The four sides and top of the test cage were made of transparent Perspex and the floor was wire mesh. The bottom edge of the door of the cage was secured 4 cm from floor level so that the monkey could extend its arms out from the cage toward two containers on a tabletop (40 cm × 80 cm) which was level with the floor of the cage.

Following preliminary training in which the monkeys learned to respond to the approach of a cooperative trainer by reaching toward one of two containers (Anderson et al. unpublished work), testing consisted of the two phases described below.

For sessions with the cooperative trainer, the experimenter approached and sat down at the table behind the containers (one white, one brown), opposite the monkey. The experimenter placed the bait (a piece of banana or apple, or a peanut) on one side of the table, and then slowly and deliberately covered the bait with one of the containers (held in one hand) while simultaneously manipulating the other container, held in the other hand, in the same way on the other side of the table. The experimenter then lifted and lowered both containers once or twice until quite satisfied that the monkey had seen the location of the bait. The distance between the nearest side of the containers and the cage was 21 cm, and the containers were 23 cm apart. Once baiting was completed, the experimenter stood up, turned around and left the test area, immediately following which a cooperative trainer ("CoopT") entered. Throughout the study, the CoopT wore a distinctive yellow cardigan. The CoopT, unaware of which container was baited, approached the table, sat down, and silently placed a hand at the midpoint between the two containers. As soon as the monkey reached clearly toward one of the two containers the CoopT lifted that container up, and gave the bait by hand to the monkey if it was the correct (i.e., baited) container. In the event of the monkey reaching toward, and the CoopT therefore selecting, the unbaited container, the CoopT simply left the selected container upturned and left the test area. After leaving the test area, the CoopT informed the experimenter of the outcome of the trial, and then sat down facing away from the test area. The experimenter then prepared the next trial (or correc-

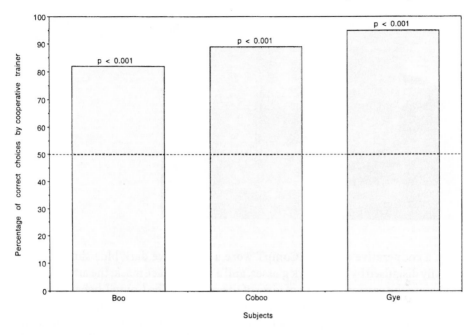

Fig. 5. Percentage of correct choices by the cooperative trainer in the last five sessions of phase 1 of the deception study

tion trial, see below); the intertrial interval was approximately 20 s. Each session consisted of 20 trials and correction trials as necessary. After 15 sessions a correction procedure was introduced as follows: if the CoopT selected the wrong (i.e., unbaited) container, the trial was repeated as often as necessary until the correct selection was made. Note that in correction trials the trainer knew which container covered the food, but the instructions remained the same, i.e., selection was to be based on the monkeys' first reach toward a container. Additional procedures, some individually tailored according to the progress of the three subjects, were used to facilitate learning, but these need not be described here (Anderson et al. unpublished work).

Figure 5 shows the percentage of correct responses for all three monkeys in the last five sessions (i.e., 100 trials) with CoopT (sessions 46–50). All three subjects were reliably gesturing toward the correct cup in the presence of CoopT; the performance of all three was highly significant, as indicated by binomial tests. Indeed, in the last five sessions performance never fell below 17/20 correct (Gye), 16/20 correct (Coboo), or 15/20 correct (Boo).

3.3 Introduction of the Competitive Trainer: Methods and Results

Following the establishment of reliable "honest" responding to the CoopT, another type of session was introduced, called "CompT" sessions. These were identical to CoopT sessions except that a competitive trainer ("CompT") participated rather

Fig. 6. A monkey responding in a trial with the competitive trainer

than a cooperative one. The CompT wore a distinctive dark blue shirt and was heavily disguised by a hat, dark glasses, and a surgical face mask; the attire was put on before the session and out of view of the monkey. The CompT behaved identically to the CoopT throughout a trial until a container was selected. If it was the correct one (i.e., covering the bait), instead of giving the bait to the monkey the CompT pretended to eat it in full view of the monkey and then left the test area. If the wrong (unbaited) container was selected, the CompT said "Wrong!", left the wrong container upturned, and left the test area empty-handed, upon which the experimenter came forward quickly to praise the monkey and uncover and give it the bait. One CoopT or one CompT session was run each day, with no more than three consecutive sessions of one type allowed in the series. Each monkey took part in 25 sessions for a total of 500 trials with each type of trainer. Figure 6 shows a monkey responding in a CompT trial.

Figure 7 shows the performance of the three subjects in the last five sessions with each type of trainer. There was at best only weak evidence of deception of the CompT in these sessions, and then only for one subject (Boo). Indeed in the last five sessions neither Gye nor Coboo sent the CompT to the wrong container on more than 11 of 20 trials, whereas Boo's gesturing did deceive the CompT 16 times in one session; however, in the next CompT session, Boo indicated the correct container to the CompT in 12 trials. More striking than the monkeys' failure to deceive the CompT in this phase, however, was the disruption to their previously "honest" reaching during CoopT sessions. Only Coboo continued to send the CoopT to the correct container significantly frequently, whereas Boo did so *less* frequently than would be expected by chance. The success of all three monkeys in these sessions with the CoopT was markedly lower than in the earlier phase, as if the introduction of the CompT had disrupted the monkeys' understanding of their own behavior, that of the CoopT, or the link between the two.

What the monkeys often did in this phase was to stretch out an arm and then rapidly alternate between the two containers, making it difficult for the trainer to select one as being the monkey's choice. In order to eliminate such ambiguity, 30 further sessions were run during which a piece of card was stuck to the cage door

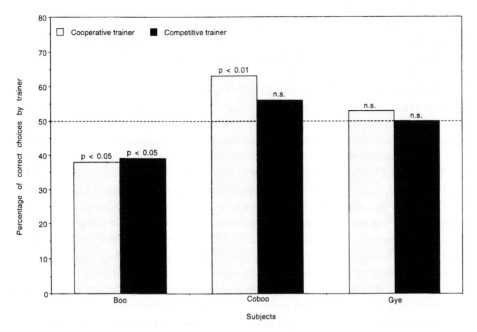

Fig. 7. Percentage of correct choices by the trainers in phase 2 of the deception study

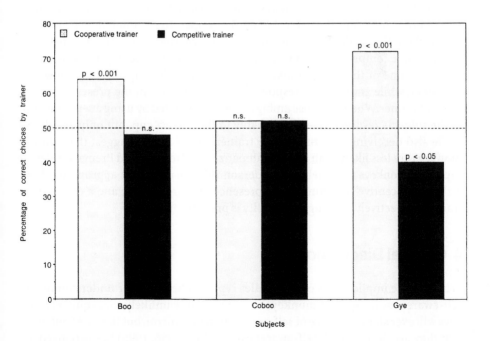

Fig. 8. Percentage of correct choices by the trainers in phase 3 of the deception study

to prevent the monkey from sweeping its arm to and fro between the containers. In other words, the initial response of the monkey was rendered much clearer for the trainer, who now chose the container nearest the gap through which the monkey first reached on any given trial.

The first ten sessions with the divider in place were with the CoopT only, in order to reestablish reliable gesturing toward the correct container in the presence of the cooperative trainer. This was achieved for all three monkeys, and then CoopT and CompT sessions were run quasirandomly as before for a total of ten sessions each (constituting phase 3). Performances in the last five sessions are shown in Fig. 8. There are a number of differences between the results of this phase and the previous one. Whereas previously monkey Boo scored significantly below chance with both types of trainer, he was again reaching toward the correct container significantly more than expected by chance in the presence of the CoopT. However, in CompT sessions, Boo's reaching was random. Coboo's reaching was at chance levels regardless of the type of trainer present, while Gye showed the clearest differentiation, reaching highly significantly toward the baited container in the presence of the CoopT, and significantly less frequently than expected by chance during the CompT sessions. Thus, individual differences precluded the emergence of any general tendency to engage in deceptive reaching toward the competitive trainer.

3.4 Brief Discussion

Although all three squirrel monkeys initially learned to reach toward the baited cup in order to obtain the reward from the cooperative trainer, they showed no ready adaptation of their response for use as a deceptive device. Indeed, the introduction of a competitive trainer revealed the "honest" reaching response to be fragile, as one of the three monkeys began sending *both* trainers to the wrong container, while another one responded at chance levels in the presence of both types of trainer. When response ambiguity was eliminated by using a separator, two of the subjects resumed reaching reliably to the correct cup, but only one did so while also deceiving the competitive trainer. These results suggest that squirrel monkeys are less likely than either chimpanzees (Woodruff and Premack 1979) or capuchin monkeys (Mitchell and Anderson 1997) to engage in appropriate "honest" and "deceptive" reaching in the presence of a cooperative and a competitive trainer, respectively, although the ability is present.

4 General Discussion

What are the implications of the studies reported here for our understanding of self-awareness in squirrel monkeys? It seems highly unlikely that squirrel monkeys will ever show evidence of self-recognition in a mirror, but this does not mean that they are devoid of all self-awareness. Neisser (1988, 1995) has proposed the existence of five kinds of self-knowledge which can be considered in investigations of self in both humans and animals. These five "selves" are the ecological self, the interpersonal self, the conceptual self, the temporally extended self, and the private self. Looking at the studies reported here from this perspective, the self-control required in study 1 can be seen as drawing upon the monkeys' "ecological self". At first the monkeys expressed what is likely to be a natural tendency, i.e., to reach for the larger of two quantities of a desirable food. This perception–action bias appears to be widespread among animals. However, all of the monkeys changed their strategies in response to changing environmental contingencies. Most of them developed a side preference (which resulted in them obtaining 50% of the available rewards) before eventually overcoming their initial tendency to reach for the larger array, and reaching systematically for the smaller array instead. In showing such self-control, the monkeys revealed an ability to monitor their own muscular activity, indeed to anticipate it, and to decide to inhibit a natural act and replace it with an alternative. It remains to be seen whether the ability for self-control in squirrel monkeys extends to situations in which the benefit of gratification accrues at a later stage rather than immediately, as in the study conducted here. The latter type of self-control requires the ability to retain information about contingencies and the continuity of self over extended time (i.e., Neisser's "temporally extended self"); it is exercised by humans (Mischel 1974) and by chimpanzees, at least over a shorter time-scale (Beran et al. 1999).

Self-control is also necessary for deception of the type shown by chimpanzees in the study by Woodruff and Premack (1979), by capuchin monkeys in the study by Mitchell and Anderson (1997), and sought in study 2. In order to deceive the competitor, the subject must inhibit the tendency to reach toward the container which hides the food, and instead reach toward the container which is empty. This appears similar to what the squirrel monkeys were doing by the end of study 1 when they were presented with two different sized food arrays. However, whereas all but one of the squirrel monkeys eventually showed self-control in study 1, only one of the three monkeys in the deception study reliably deceived the competitor while continuing to respond "honestly" in the presence of the cooperative trainer.

Because of the explicit social overtones in the deception study—the monkey's reaching can be seen as a signal to the trainer who responds by selecting a container and either giving the monkey the food or keeping it—it is reasonable to think that in the deception scenario the monkey's "interpersonal self" (perhaps better rephrased as "social self") is involved. According to Neisser (1995), interpersonal self-knowledge is available to humans from earliest infancy, and it may be a valid conceptual tool for examining animal behavior (Cenami Spada et al. 1995). Indeed, the latter authors have explicitly argued for the differentiation of one's own knowledge states and those of others in the context of deceptive acts in primates.

Squirrel monkeys are notably absent from the deception database compiled by Byrne and Whiten (1990, 1992), in spite of the many studies of these monkeys in the field and in captivity. The results of study 2 show that deceptive acts are not beyond the abilities of squirrel monkeys, but in contexts such as the one we employed the likelihood of deception appears quite low, and variable among individuals. As we have shown, squirrel monkeys are capable of the basic form of self-control necessary for the deceptive act, and they are also capable of conditional discriminations in studies using nonsocial stimuli (Thomas and Kerr 1976). So why were the results of the deception study so modest? One possibility is that the monkeys' "social self" was not implicated as much as one might at first think. Although all three monkeys learned to act appropriately in sessions with the cooperative trainer, this response was disrupted when the competitive trainer was introduced, suggesting that the monkeys had at best only a limited understanding of the communicative function of their reaching.

It is entirely conceivable that learning to switch from "honest" to "deceptive" reaching in the presence of a cooperative and a competitive trainer reflects conditional discrimination learning without any attempt to influence the knowledge state of the trainer (Heyes 1994, 1998; Mitchell and Anderson 1997). This is true for chimpanzees and capuchin monkeys, and also for squirrel monkeys. Additional experiments with the three squirrel monkeys after study 2, which elaborated upon the conditional discrimination aspects of the procedure, have led to the establishment of much stronger "deceptive" responses in all three monkeys (Anderson et al. unpublished work). Future reports will describe results of experiments aimed at clarifying the psychological mechanisms involved in the "honest" and "deceptive" reaching of these monkeys.

One striking aspect of the data in both studies concerned individual differences in response patterns. In study 1, for example, one monkey started to show self-control responses much earlier than the others, and she never required correction trials, while at the other extreme another monkey never learned to reach for the smaller array, but either persevered in reaching for the larger array or expressed a strong side preference. In study 2, the introduction of the competitive trainer initially caused one subject to respond at below chance levels to both kinds of trainer. By the end of training, two subjects were again responding effectively in the presence of the cooperative trainer, and one of these subjects (the only one to do so) reliably sent the competitive trainer to the wrong container. Individual differences among squirrel monkeys in other areas of learning have been reported (Rumbaugh 1984). It would be of interest to assess the stability and the correlates of individual differences across several situations requiring some form of self-control. There are well-documented differences in self-awareness in humans, and similar variability exists for at least one expression of self-awareness in great apes, namely self-recognition (Swartz and Evans 1991; Povinelli et al. 1993), which would appear to be founded upon a "conceptual self" (in the terms used by Neisser 1995). A further assessment of abilities and limitations in a range of situations drawing upon different forms of self-knowledge will provide further insights into the evolution of "self" or "selves" among the primates.

Acknowledgments

I am grateful to Prof. Kazuo Fujita for inviting me to work in his laboratory, and to the following students for their help at various stages of the studies: Shunji Awazu, Satoru Ishikawa, Hika Kuroshima, Hiro Kuwahata, Aika Tokuhisa, and Sarah Vick.

References

Anderson JR (1984a) Monkeys with mirrors: some questions for primate psychology. Int J Primatol 5:81–98

Anderson JR (1984b) The development of self-recognition: a review. Dev Psychobiol 17:35–49

Anderson JR (1994) The monkey in the mirror: A strange conspecific. In: Parker ST, Mitchell RW, Boccia ML (eds) Self-awareness in animals and humans. Cambridge University Press, New York, pp 315–329

Anderson JR (1996) Chimpanzees and capuchin monkeys: comparative cognition. In: Russon A, Bard K, Parker ST (eds) Reaching into thought. Cambridge University Press, Cambridge, pp 23–56

Anderson JR, Gallup GG Jr (1999) Self-recognition in nonhuman primates: past and future challenges. In: Haug M, Whalen RE (eds) Animal models of human emotion and cognition. American Psychological Association, Washington, pp 175–194

Anderson JR, Marchal P (1994) Capuchin monkeys and confrontations with mirrors. In: Roeder JJ, Thierry B, Anderson JR, Herrenschmidt N (eds) Current primatology, vol 2. Social development, learning and behaviour. Universit Louis Pasteur, Strasbourg, pp 371–380

Anderson JR, Awazu S, Fujita K (2000) Can squirrel monkeys (Saimiri sciureus) learn self-control? A study using food array selection tests and reverse reward contingency. J Exp Psychol: Anim Behav Process 26:87–97

Beran MJ, Savage-Rumbaugh ES, Pate JL, Rumbaugh DM (1999) Delay of gratification in chimpanzees (Pan troglodytes). Dev Psychobiol 34:119–127

Boysen ST, Berntson GG (1995) Responses to quantity: perceptual versus cognitive mechanisms in chimpanzees (Pan troglodytes). J Exp Psychol: Anim Behav Process 21:83–86

Boysen ST, Berntson GG, Hannan MB, Cacioppo JT (1996) Quantity-based interference and symbolic representations in chimpanzees (Pan troglodytes). J Exp Psychol: Anim Behav Process 22:76–86

Byrne RW, Whiten A (1990) Tactical deception in primates: the 1990 database. Primate Rep 27:1–101

Byrne RW, Whiten A (1992) Cognitive evolution in primates: evidence from tactical deception. Man 27:609–627

Cenami Spada E, Aureli F, Verbeek P, de Waal FBM (1995) The self as reference point: can animals do without it? In: Rochat P (ed) The self in infancy: theory and research. Elsevier, Amsterdam, pp 193–215

Damon W, Hart D (1982) The development of self-understanding from infancy through adolescence. Child Dev 53:841–864

French GM (1959) Performance of squirrel monkeys on variants of delayed response. J Comp Physiol Psychol 52:741–745

Gallup GG Jr (1982) Self-awareness and the emergence of mind in primates. Am J Primatol 2:237–248

Gallup GG Jr (1991) Toward a comparative psychology of self-awareness: species limitations and cognitive consequences. In: Goethals GR, Strauss J (eds) The self: an interdisciplinary approach. Springer, New York, pp 121–135

Gallup GG Jr (1994) Self-recognition: research strategies and experimental design. In: Parker ST, Mitchell RW, Boccia ML (eds) Self-awareness in animals and humans: developmental perspectives. Cambridge University Press, New York, pp 35–50

Gibson KR (1986) Cognition, brain size and the extraction of embedded food resources. In: Else JG, Lee PC (eds) Primate ontogeny, cognition and social behaviour. Cambridge University Press, Cambridge, pp 93–103

Heyes C (1994) Social cognition in primates. In: Mackintosh NJ (ed) Animal learning and cognition. Academic Press, London, pp 281–305

Heyes C (1998) Theory of mind in nonhuman primates. Behav Brain Sci 21:101–134

Kummer H, Dasser V, Hoyningen-Huene P (1990)Exploring primate social cognition: some critical remarks. Behaviour 112:84–98

Lewis M (1994) Myself and me. In: Parker ST, Mitchell RW, Boccia ML (eds) Self-awareness in animals and humans: developmental perspectives. Cambridge University Press, New York, pp 20–34

Lewis M, Brooks-Gunn J (1979) Social cognition and the acquisition of self. Plenum, New York

Logue AW (1988) Research on self-control: an integrating framework. Behav Brain Sci 11:665–709

McGonigle BO, Chalmers M (1992) Monkeys are rational! Q J Exp Psychol 45B:189–228

Mischel W (1974) Processes in delay of gratification. In: Berkowitz L (ed) Advances in experimental social psychology, vol. 7. Academic Press, New York, pp 249–292

Mischel W, Shoda Y, Rodriguez ML (1989) Delay of gratification in children. Science 244:933–938

Mitchell RW (1986) A framework for discussing deception. In: Mitchell RW, Thompson NS (eds) Deception: perspectives on human and nonhuman deceit. State University of New York Press, Albany, pp 3–40

Mitchell RW (1993) Mental models of mirror-self-recognition: two theories. New Ideas Psychol 11:295–325

Mitchell RW (1994) Multiplicities of self. In: Parker ST, Mitchell RW, Boccia ML (eds) Self-awareness in animals and humans: developmental perspectives. Cambridge University Press, New York, pp 81–107

Mitchell RW, Anderson JR (1997) Pointing, withholding information, and deception in capuchin monkeys (Cebus apella). J Comp Psychol 111:351–361

Neisser U (1988) Five kinds of self-knowledge. Philos Psychol 1:35–59

Neisser U (1995) Criteria for an ecological self. In: Rochat P (ed) The self in infancy: theory and research. Elsevier, Amsterdam, pp 17–34

Olthof A, Iden CM, Roberts WA (1997) Judgements of ordinality and summation of number symbols by squirrel monkeys (Saimiri sciureus). J Exp Psychol: Anim Behav Process 23:325–339

Parker ST, Mitchell RW, Boccia ML (eds) (1994) Self-awareness in animals and humans. Cambridge University Press, New York

Povinelli DJ, Rulf AB, Landau KR, Bierschwale DT (1993) Self-recognition in chimpanzees (Pan troglodytes): distribution, ontogeny, and patterns of emergence. J Comp Psychol 107:347–372

Rochat P (1995) The self in infancy. Elsevier Science B.V., New York

Rumbaugh DM (1984) Primates' learning by levels. In: Greenberg G, Tobach E (eds) Behavioral evolution and integrative levels. Lawrence Erlbaum, Hillsdale, pp 221–240

Rumbaugh DM (1997) Conpetence, cortex, and primate models: a comparative primate perspective. In: Krasnegor NA, Lyon GR, Goldman-Rakic PS (eds) Development of the prefrontal cortex. Paul H. Brookes, Baltimore, pp 117–139

Silberberg A, Fujita K (1996) Pointing at smaller food amounts in an analogue of Boysen and Berntson's (1995) procedure. J Exp Anal Behav 66:143–147

Swartz KB, Evans S (1991) Not all chimpanzees (Pan troglodytes) show self-recognition. Primates 32:483–496

Thomas RK, Kerr RS (1976) Conceptual discrimination in Saimiri sciureus. Anim Learn Behav 4:333–336

Thomas RK, Fowlkes D, Vickery JD (1980) Conceptual numerousness judgements by squirrel monkeys. Am J Psychol 93:247–257

Tobin H, Logue AW, Chelonis JJ, Ackerman KT (1996) Self-control in the monkey Macaca fascicularis. Anim Learn Behav 24:168–174

Woodruff G, Premack D (1979) Intentional communication in the chimpanzee: the development of deception. Cognition 7:333–362

17
Evolutionary Foundation and Development of Imitation

MASAKO MYOWA-YAMAKOSHI

1 Introduction

Although imitating actions is a familiar daily human experience, the mechanisms underlying imitation involve complex cognitive operations. We transform visual information into matching motor movements. Why have humans evolved this critical ability?

In this chapter, I compare the imitative abilities of humans (*Homo sapiens*) and chimpanzees (*Pan troglodytes*), our closest evolutionary relatives. This approach may be described as Comparative Cognitive Science (CCS), and explores human cognition from an evolutionary perspective. The research method is characterized by a comparison of the performance of different species based on a unified objective scale (Matsuzawa 1996). To this end, I used the same test apparatus and following the same test procedure to compare the performance of humans and chimpanzees. I propose a model that explains the mechanism of imitation that incorporates both the shared and different characteristics of the two species.

I focus on the basic visual–motor or visual–kinesthetic cross-modal level to reveal the differences in imitation. Matsuzawa (1996) suggests that the difference between the two species is the depth of the hierarchical self-embedding structure, i.e., the "relationship of the relationship of the relationship." I suggest that basic differences at the perceptual–motor level reflect the main differences between humans and chimpanzees with respect to higher cognitive–social abilities.

2 Why is Imitation Important?

How is imitation defined? When an individual learns a nongenetic behavior similar to that performed by another individual, one may call this imitation. Recently, there has been much controversy over the evidence for imitation in animals. For example, Byrne (1994) argues that "the acquisition of novel behavior can be safely used in the diagnosis of imitation." On the other hand, Whiten and Ham (1992)

Department of Behavioral and Brain Sciences, Primate Research Institute, Kyoto University, 41 Kanrin, Inuyama, Aichi 484-8506, Japan

define imitation as when one individual learns some aspect(s) of the form of an act from another. However, it is very difficult to identify whether an action is totally novel to the observers in each individual's lifetime (Whiten and Custance 1996). In this study, I define imitation as "performing similar behavior after observing the behavior of another individual, irrespective of whether the behavior is already in the animal's repertoire."

Many developmental psychologists have emphasized two aspects of function in imitation. One is social learning, which contributes to adaptive skills in the human environment. In the second year of life, human infants are capable of acquiring a wide variety of novel actions by imitation (e.g., Abravanel and Gingold 1985; Meltzoff 1988). The other is from the perspective of communication. Many psychological reports on human infants have suggested that imitation plays an important role in developing social–cognitive abilities. For example, the ability to imitate others is considered to be the foundation for developing a normal "theory of mind (Premack and Woodruff 1978)" (Barresi and Moore 1996; Meltzoff and Gopnik 1993; Rogers and Pennington 1991), the development of self-awareness, and the awareness of others (Meltzoff 1990), and is a precursor to the capacity to represent symbols (Piaget 1962; Werner and Kaplan 1963).

3 The Development of Imitation in Human Infants

The most comprehensive theory on the development of imitation is that of Piaget (1962). Piaget postulated six stages of action imitation by infants. These may be divided into three main levels. At the first level (from birth to 8 months), human imitation is restricted to imitating simple hand-opening. This type of imitation can be accomplished through an intramodal matching process. Human infants can directly compare the demonstrator's hand movements with those of their own visible hand. The second level appears as early as 9–12 months of age. Human infants begin to imitate facial gestures without intramodal guidance. Infants cannot see their own facial gestures, such as opening their mouth. Instead, facial imitation depends on a cross-modal matching process. At the third level, the important development is deferred imitation. Deferred imitation is not performed in the presence of the demonstration. Piaget postulated that deferred imitation involves the infants' representational capacities and is a precursor to representing symbols.

In opposition to the hypothetical framework of Piaget, Meltzoff and Moore (1977) showed experimentally that human neonates can imitate some demonstrators' facial gestures (tongue protrusion, mouth protrusion, and mouth opening). There have been many studies of neonatal imitation, including the imitation of other facial expressions (Abravanel and Sigafoos 1984; Heimann 1989; Field et al. 1982), eye blinking, head and finger movement, and cheek movements (Fontaine 1984; Meltzoff and Moore 1989, 1992, 1994; Reissland 1988; Vinter 1986). Meltzoff and Moore (1977, 1983) speculated that the human infant's capacity to imitate motor acts performed by others is accomplished through active intermodal matching (AIM) mediated by an innate representational system. According to the AIM hypothesis, human neonates can cross-modally process visual and motor information and detect the equivalent motor response.

4 Does Imitative Ability Arise from Neonatal Imitation?

The most important topic in the development of imitation in humans is whether neonatal imitation should be interpreted as imitation. Many researchers have been preoccupied with this question. Meltzoff (1996) insists that evidence of neonatal imitation has been shown in independent laboratories in more than 24 experiments.

However, an alternative view of how to interpret this phenomenon has been proposed. In this hypothesis, neonatal imitation is mediated by an "innate releasing mechanism" based on simple reflexes such as the Moro reaction. Jacobson (1979) found that moving a pen toward and away from a 6-week-old infant's mouth elicited as much tongue protrusion as did tongue protrusion by a demonstrator. In a review of neonatal imitation studies, Anisfeld (1996) suggested that tongue protrusion was the only modeled gesture that clearly produced matching behavior. Furthermore, a powerful argument supporting the second view is that neonatal imitation disappears or declines at approximately 2–3 months of age and later reappears (Abravanel and Sigafoos 1984; Fontaine 1984; Maratos 1982).

Meltzoff and Moore (1977, 1983, 1992) have provided two arguments to counter the second hypothesis. First, it seems implausible that human infants would have evolved release mechanisms for apparently meaningless gestures like tongue protrusion. Second, facial imitation did not disappear at 2–3 months of age. In fact, the opposite occurs. In some imitation studies, human infants make more elaborate imitations and more accurately match a demonstrator as an experiment progresses.

Meltzoff and Moore (1992) emphasized that neonatal imitation serves social-communicative functions, and pointed out that early cognitive ability makes infants understand "person" as opposed to "thing" and helps to identify people. Engaging in face-to-face social interactions with adults, infants may use imitation as a means of verifying identification. In addition, Meltzoff and Moore argued that, contrary to the second view, the "apparent" disappearance of imitative responses in the 2–3 month period is due to the fact that older infants respond to people by engaging in social interacting games more vigorously than neonates.

At approximately 2–3 months of age, the performance of human infants changes remarkably. They begin to vocalize and smile at others spontaneously in a face-to-face social–communicative context (Butterworth and Harris 1994; Field et al. 1986). Johnson (1990) proposed that a subcortical visual pathway involving the superior colliculus controls tracking moving face stimuli in the first month of life. Around 2 months of age, there is a shift in processing from the subcortical visual pathway to a second system that appears in plastic cortical visual pathways. This second system is thought to allow the recognition of characteristic facial expressions (Johnson and Morton 1991).

Is neonatal imitation an imitative ability? To date, this question is still open.

However, both views are very similar with respect to the role of neonatal imitation. Researchers espousing either view emphasize that neonatal imitation may serve a social–communicative function. For survival, it is important for infants to orient towards human-like stimuli. Neonatal imitation might play a crucial role in attracting adults' attention and increasing opportunities to interact.

5 The Origin of Imitative Ability from an Evolutionary Perspective

Chimpanzees have much in common with humans, and some studies suggest that there are similarities in the early abilities of human and chimpanzee neonates (Bard et al. 1990a, b, 1992; Hallock et al. 1989; Mathieu and Bergeron 1981). If infant chimpanzees imitate facial expression, we can determine the characteristics of neonatal imitation shared by the two species. These may have a common evolutionary origin. Myowa (1996) investigated the imitation of facial gestures by an infant chimpanzee following exactly the same procedure as Meltzoff and Moore (1977).

5.1 Method

5.1.1 Subject

The subject was an infant female chimpanzee. Humans raised her in a nursery, starting within 24 h of birth, because her biological mother provided inadequate maternal care. She was placed in an incubator and given nursery care (Fig. 1). The experiment was conducted once a week from 5 to 15 weeks of age, except at 9 and 13 weeks of age.

Fig. 1. An infant chimpanzee in an incubator

5.1.2 General Procedure

Each session began with a 90-s period in which the experimenter (author) sat in front of the subject and presented an unreactive "passive face" (lips closed, neutral facial expression) to the subject. The infant was then shown one of the following four gestures in a random order: tongue protrusion, lip protrusion, mouth opening, and a sequential finger opening movement. Each gesture was demonstrated four times in a 15-s stimulus-presentation period. A 20-s response period was allowed immediately after the stimulus-presentation period. In the response period, the experimenter stopped performing the gesture and displayed the passive face.

Videotape recordings of the response periods were edited and scored. Separately, six researchers who are familiar with chimpanzees scored the infant's responses in 36 segments (4 gestures × 9 weeks) of the edited videotape. They were informed that the infant in each segment was shown one of the four gestures (tongue protrusion, mouth opening, lip protrusion, or passive face). In this experiment, the passive face was the response when sequential finger movement was demonstrated. The scorers were instructed to order the four gestures by four ranks from the one they thought it most likely that the infant in each segment was imitating to the one they thought was least likely.

For the purpose of analysis, the two highest ranks and the two lowest ranks were collapsed. This procedure yields dichotomous judgments of whether it was likely or unlikely (hereafter referred to as "yes" or "no") that the infant was imitating a particular gesture.

5.2 Results

5.2.1 The Change in the Total Number of "Yes" Judgments

The results showed that the infant could imitate human facial gestures (Fig. 2). The total number of "yes" judgments of six scorers was shown on each facial gesture at each week of age. The infant's performance underwent a remarkable shift. From 5 to 10–11 weeks of age, the infant tended to exhibit tongue protrusion and mouth opening when the corresponding gesture was demonstrated by the experimenter. After 11–12 weeks of age, however, the infant no longer exhibited the corresponding gesture. On the other hand, when lip protrusion was demonstrated, this kind of developmental change of gesture was not found (Fig. 3).

5.2.2 Imitative Response for each Gesture

All response periods were divided into two phases: the periods when the infant showed the corresponding gesture demonstrated by the experimenter, and the periods when she did not. Figure 4 shows the distribution of "yes" judgments as a function of the gestures shown to the infant. The results revealed that from 5 to 10 weeks of age, the infant produced tongue protrusion significantly more frequently than any other gesture when the experimenter demonstrated the gesture of tongue protrusion. From 5 to 11 weeks of age, the infant produced mouth opening signifi-

354   M. Myowa-Yamakoshi

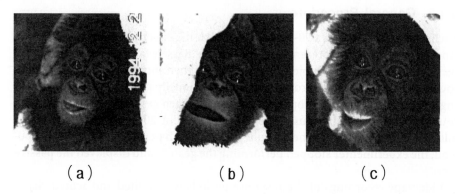

(a)   (b)   (c)

Fig. 2. Sample photographs from videotape recordings of an infant chimpanzee: **a** tongue protrusion; **b** mouth opening; **c** lip protrusion. These actions had been demonstrated by a human demonstrator

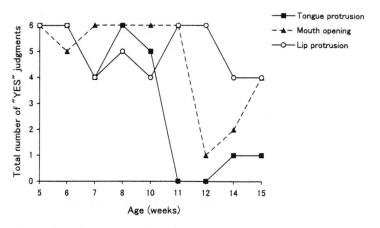

Fig. 3. Developmental changes in the total number of "yes" judgments

(a)   (b)   (c)

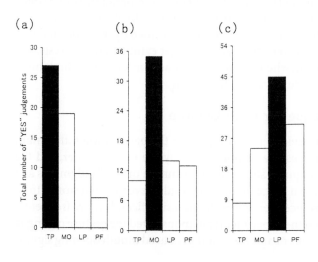

Fig. 4. Distribution of "yes" judgments as a function of the gesture shown to the infant. *Shaded bars* indicate the imitative response. **a** Number of judgments that the infant responded with tongue protrusion (*TP*) from 5 to 10 weeks of age; **b** mouth opening (*MO*) judgments from 5 to 11 weeks; **c** lip protrusion (*LP*) judgments from 5 to 15 weeks

cantly more frequently than any other gesture when the experimenter demonstrated the gesture of mouth opening. As for lip protrusion, in all response periods, the infant produced lip protrusion more frequently than any other gesture when the experimenter demonstrated the corresponding gesture.

5.2.3 Summary of Results

These results suggest that, for a limited period, infant chimpanzees respond to facial gestures like human infants. It is reasonable to assume that human infants are not unique among primates in their ability to imitate adult facial gestures. Recently, Bard and Russell (1999) also confirmed these results.

6 Can Chimpanzees Imitate?

If imitative ability arises from neonatal imitation, does the imitative ability of chimpanzees develop in the same way as that of humans? In a review of imitation studies from a variety of species, Whiten and Ham (1992) concluded that there is little convincing evidence of imitation in nonhuman animals, except in great apes and dolphins.

However, recent experimental studies have shown that chimpanzees do not imitate a broad range of actions to the degree exhibited by humans (Custance et al. 1995; Hayes and Hayes 1952; Nagell et al. 1993; Myowa-Yamakoshi and Matsuzawa 1999; Tomasello et al. 1987; Whiten et al. 1996). In the first empirical study, Hayes and Hayes (1952) trained a human-raised chimpanzee, Viki, to reproduce various actions on the command "Do this!" She had been trained for more than 17 months before the "Do this!" test began in her daily life. After training, some novel actions were introduced. Hayes and Hayes reported that Viki could imitate at least ten completely novel acts. More recently, Custance et al. (1995) conducted a "Do this!" experiment in a similar manner, but in a more rigorous fashion. Of 48 novel gestures demonstrated by a human, two chimpanzees as young as 4 years old reproduced 13 and 20 arbitrary gestures, respectively. On the other hand, most 4-year-old children imitated them all perfectly (Whiten 1996).

7 Factors Influencing Imitation in Chimpanzees

It seems to be reasonable to say that the level of fidelity of chimpanzee imitations does not reach that of humans. Myowa-Yamakoshi and Matsuzawa (1999) systematically investigated the factors that determine the degree of difficulty for chimpanzees when they imitate human actions in a face-to-face situation.

7.1 Method

7.1.1 Subjects

The subjects were five female nursery-reared chimpanzees, 12–19 years old, at the Kyoto University Primate Research Institute.

7.1.2 General Procedure

Four pairs of objects were used as test stimuli. Each pair consisted of two objects that differed from each other and had no explicit relationship.

Each subject participated in a total of 16 experimental sessions. Four sessions were conducted with each pair of objects. Each session consisted of three conditions: (a) one object, (b) one object to self, and (c) one object to another. In the one object (O) condition, the subjects watched the demonstrator manipulate one object (e.g., hitting the bottom of a bowl). In the one object to self (O to S) condition, the demonstrator manipulated one object against his body (e.g., putting a bowl on his/her head). In the one object to the other object (O to O) condition, the demonstrator manipulated one object towards another (e.g., putting a ball into a bowl). These three conditions involved many different motor patterns (e.g., hitting, pulling, etc.). Table 1 shows the list actions demonstrated on each set of objects.

During the sessions, the human demonstrator and the subject sat face to face. Before the test started, a pair of objects was presented to the subject for approximately 3 min free play. During this time, the subject interacted in some way with each of the objects. The demonstrator then retrieved the objects and began to demonstrate an action. Each action was demonstrated two or three times to ensure that the subject paid attention to the action. After an action was demonstrated, the subject was then given the objects and told to "Do this!" (Fig. 5).

We conducted the test in two phases, depending on the subject's responses. In the first phase, we observed the subject's responses in the first attempt to determine whether the chimpanzee could reproduce the action (imitation phase). If both experimenters judged that the subject was able to perform the demonstrated action, the next action was demonstrated. If the subject did not perform the demonstrated action, we proceeded to the teaching phase, in which the demonstrator trained the subject to perform the action using verbal guidance, gestures, molding, and shaping with verbal praise and food reinforcements. The demonstrator then repeated the model trial to show the action again and handed the objects to the subject. When both experimenters judged that the subject performed the action three times in succession, we proceeded to the next action. A trial began with the initial response of the subject, and ended either when the subject successfully performed the demonstrated action or after the demonstrator had taught the action. Any one action was repeated a maximum of 20 times.

The motor patterns involved in the demonstrated actions and the subject's responses were recorded on videotape and the subject's responses in each trial were identified as one of 23 mutually exclusive types. These motor patterns were classified into two main categories: (a) general motor patterns that had been observed in free-play manipulation, and (b) nongeneral motor patterns that were not observed.

Table 1. List of demonstrated actions on each set of objects

Set of objects	O	O to S	O to O
		Condition	
Bowl and ball	Hit bottom of bowl	Put bowl on one's head	Put ball into bowl
	Rub bottom of bowl	Push bowl against one's chest	Cover ball with bowl
	Shake bowl	Push bowl against one's arm	Put ball on bottom of upside-down bowl
	Slide bowl on the floor	Push bowl against one's foot	Hit ball with bowl
Lid and towel	Throw lid	Push lid against one's hip	Put lid on towel
	Hit lid	Push lid against one's mouth	Hit lid with towel
	Rotate lid on the floor	Put lid on one's palm	Cover lid with towel
	Poke at lid	Put lid on one's thigh	Wipe lid with towel
Hose and stool	Stretch hose	Hang hose around one's neck	Put hose into a hole in stool
	Hit hose	Hang hose around one's arm	Hang hose around back of stool
	Roll hose	Bind hose around one's waist	Hang hose around leg of stool
	Drop hose	Hang hose around one's foot	Push hose against stool
Scoop and can	Slide scoop on the floor	Push scoop against one's hand	Hit can with scoop
	Hit scoop	Hit one's shoulder with scoop	Insert scoop into a hole in can
	Drop scoop	Push scoop against one's mouth	Insert scoop into space between can and the floor
	Slide scoop on the floor	Hit one's palm with scoop	Push scoop against the slide of can

O, The one object condition; O to S, the one object to self condition; O to O, the one object to the other object condition.

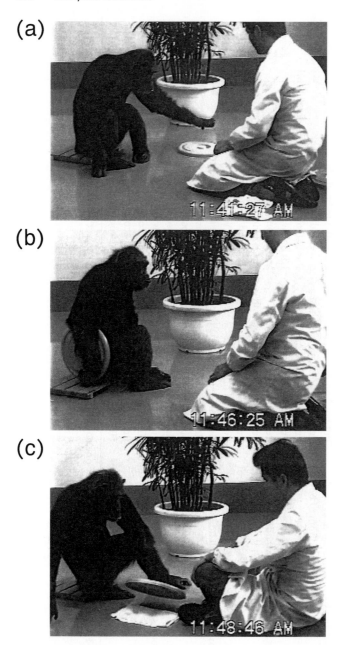

Fig. 5. A chimpanzee sitting face to face with a human demonstrator and performing the demonstrated actions in the three conditions. **a** Throwing the lid (the one object condition); **b** pushing the lid against her lip (the one object to self condition); **c** putting the lid on the towel (the one object to the other object condition)

To assess how difficult an action was to reproduce, we counted the total number of trials that each subject needed in order to perform each demonstrated action successfully, and compared the mean number of trials across the three conditions and two categories of motor patterns.

7.2 Results

7.2.1 Success at the First Attempt

Table 2 shows the total number of subjects performing each action successfully in the first trial. The chimpanzees rarely reproduced a demonstrated action at the first attempt (less than 6% of the overall actions). Moreover, this occurred only in the O to O condition.

Table 2. Total number of subjects in each action reproduced in the first trial

Action	Total number of subjects
Put ball into bowl	4 (Ai, Chloé, Popo, Pan)
Put ball on the bottom of bowl	2 (Ai, Chloé)
Wipe lid with towel	1 (Ai)
Put hose into a hole in stool	1(Ai)
Hang hose around back of stool	2 (Ai, Chloé)
Push hose against stool	1 (Ai)
Insert scoop into a hole in can	2 (Pendesa, Popo)

7.2.2 Factors Influencing Imitation

We made three important findings. First, actions involving novel motor patterns were more difficult for chimpanzees to perform than actions involving familiar motor patterns (Fig. 6). It was noted that the chimpanzees seldom reproduced demonstrated actions at their first attempt, even when these actions involved motor patterns that they had already acquired.

Second, when an object was directed to another object, the actions involving motor patterns that subjects already possessed were easy for chimpanzees to perform (Fig. 7). It seems likely that the chimpanzees focused on the direction in which objects were manipulated for visual cues to reproduce the demonstrator's actions.

Third, we found some very specific types of error in the imitative tasks. Table 3 shows the proportions of four different types of error responses in the first trial for each subject. The chimpanzees persistently repeated actions that had been taught in a previous session, and also continued to manipulate each object in familiar ways. This indicates that their responses were highly restricted; their performance

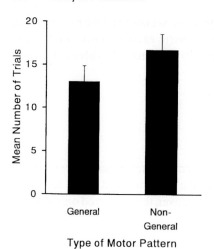

Fig. 6. Mean number (plus standard error) of trials needed to perform the demonstrated actions involving general and nongeneral motor patterns

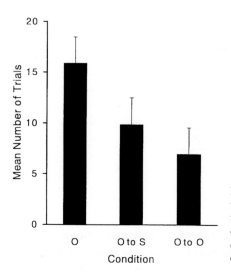

Fig. 7 Mean number (plus standard error) of trials needed to perform the demonstrated actions in each of three conditions. *O*, The one object condition; *O to S*, the one object to self condition; *O to O*, the one object to the other object condition

Table 3. Proportions of four different types of error response in the first trial for each subject

	Subject					
Type of error (n = 227)	Ai (n = 42)	Pendesa (n = 47)	Chloé (n = 46)	Popo (n= 47)	Pan (n = 47)	Average
Taught action	.51	.50	.81	.47	.72	.60
Familiar action	.15	.04	.04	.17	.07	.10
Other responses	.02	.13	.07	.17	.14	.11
No response	.32	.33	.05	.19	.07	.19

was stereotyped and perseverate for each object. These phenomena, characterized as failure to account for context, suggests that their manipulatory actions are somehow stimulus-bound.

7.2.3 Summary of the Results

This study suggested that it is easier for chimpanzees to perform an action in which an object is directed towards some external location than to manipulate a single object alone. In addition, chimpanzees were less likely to focus on the details of a demonstrator's body movements involved in a manipulation; they paid more attention to where the manipulated object was being directed than to the body movements of the demonstrator performing the manipulation. There were some constraints in the cognitive processes required to transform visual information into matching motor acts when chimpanzees imitated human actions.

8 A Model of Imitation Focusing on Objects and Body Movements

Based on the findings of Myowa (1996) and Myowa-Yamakoshi and Matsuzawa (1999), I have tried to postulate the mechanism of imitation in chimpanzees. There follows a model of imitation illustrating that humans and chimpanzees differ in the way they process visual–motor information (Fig. 8). The model is broken down into several steps.

(P-o, P-b): The model assumes that there are no fundamental differences between the two species at the visual and perceptual levels (Matsuzawa 1985, 1990; Tomonaga and Matsuzawa 1992).

(C-o, C-b): Visual information is transmitted to higher cognitive processing steps. With respect to "information processing on objects (C-o)", humans and chimpanzees identify the manipulated objects from their experience. They recognize an object's physical characteristics (e.g., color, shape, and size) and function by retrieving the represented information (Gillan et al. 1981). In this regard, chimpanzees can process information on object directionality appropriately (Myowa-Yamakoshi and Matsuzawa, 1999).

However, the two species may be different in the way that they process "information on body movements (C-b)." Myowa-Yamakoshi and Matsuzawa (1999) found that actions involving novel motor patterns were more difficult for chimpanzees to perform than those involving familiar motor patterns. This shows that chimpanzees also retrieve represented information on body movements. Of equal importance, however, is the finding that chimpanzees seldom reproduced demonstrated actions on their first attempt, even though these actions involved motor patterns that they had already acquired. In this respect, there is a remarkable species difference in information processing of body movements.

(R): The difference in how information on body movements is processed may influence the next processing steps (dotted line in Fig. 8). It seems to be more difficult for chimpanzees to process information on the demonstrator's body move-

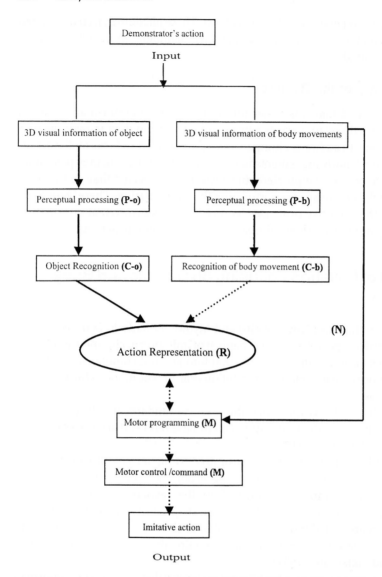

Fig. 8. Information processing model of imitation. The dotted lines indicate differences between humans and chimpanzees. *P,* Perceptual process; *C,* cognitive process; *M,* motor output processes; *N,* the pathway of neonatal imitation; *o,* information on manipulated objects; *b,* information on body movements

ments than it is for humans. Therefore, chimpanzees rely mainly on information about the manipulated object to comprehend observed actions (Myowa-Yamakoshi and Matsuzawa 2000). When the manipulated object is familiar to the chimpanzee, knowledge relating to the object is recalled to facilitate understanding of the cause and effect of actions. That is, the information on manipulated objects is processed independently and the chimpanzees recognize how to manipulate the objects from past experience. For example, when presented with a can and a can opener that are familiar, a chimpanzee understands the cause–effect relation between the two objects without necessarily processing the information on the body movements of the demonstrator who is performing the manipulation (Gillan et al. 1981; Premack and Premack 1983).

However, when familiar objects (e.g., the can and can opener) are manipulated in an unfamiliar way (e.g., hitting the can with the can opener), it might be difficult for chimpanzees to integrate the information on the familiar objects and the unfamiliar movements. As I have already mentioned, there are constraints on a chimpanzee's ability to cognitively process information on body movements. This view is reflected in the fact that the chimpanzees made specific types of errors that were stereotyped and perseverate for each object in an arbitrary action imitation task (Myowa-Yamakoshi and Matsuzawa 1999). These errors suggest that chimpanzees recognize the demonstrated action on the basis of familiar objects (e.g., a screwdriver) related to specific body movements stored in their memory (e.g., twisting), irrespective of the body movements demonstrated (e.g., hitting, throwing, etc.).

(M): Constraints on the input process result in different performances in humans and chimpanzees in a process involving programmed and controlled motor output. The chimpanzee's performance is primarily based on information about familiar objects. For actions manipulating familiar objects, actions involving familiar motor patterns are easier for chimpanzees to reproduce than actions involving unfamiliar motor patterns.

(N): I reported previously that an infant chimpanzee can reproduce several facial gestures (Myowa 1996). Hence, it seems reasonable to suppose that a visual–motor processing pathway for neonatal imitation may exist in both humans and chimpanzees.

The proposed model holds for previous studies on imitation in chimpanzees. Interestingly, most of the cases that provide evidence for imitation in chimpanzees are those involving imitation of actions on objects (see Custance et al. 1995 for an exception). When observing an action on an object, chimpanzees could reproduce the action using information about the object (pathway (C-o) → (R) → (M) in Fig. 8). In this case, a chimpanzee would not depend on information about the demonstrator's body movements (pathway (C-b) → (R) in Fig. 8) to reproduce the actions. This implies that performances with an object have often been interpreted as evidence of imitation in chimpanzees.

The capacity for neonatal imitation is likely a characteristic which is common to humans and chimpanzees, and which has resulted from natural selection. I postulate that neonatal imitation should be interpreted as an innate adaptive capacity that enables newborns to orient preferentially to the faces of conspecifics in both species. On the other hand, the ability to imitate a broad range of whole-body actions, especially actions not involving objects, seems to be an ability that evolved after the human lineage separated from that of chimpanzees.

The next question is how this basic difference in visual–motor information processing is reflected in differences in higher social–cognitive aspects of humans and chimpanzees. In the following section, I explore the characteristics of social behavior in wild chimpanzees.

9 Imitation in Cultural Differences in Wild Chimpanzees

What nongenetic social behaviors exist in wild chimpanzees? I divide chimpanzee social behaviors into two types: (a) behavior directed toward objects, and (b) behavior directed toward other individuals.

The former includes using tools to obtain food, for personal hygiene, and so forth. In nature, chimpanzees display a wide variety of population-specific tool-using and tool-making behavioral traditions (Matsuzawa and Yamakoshi 1996; McGrew 1992; Whiten et al. 1999). The latter includes gestural signals. Wild chimpanzees use gestural signals for communication in a variety of contexts, such as greeting, grooming, and presenting (Goodall 1986). At several study sites, however, pairs of chimpanzees use most of the gestural signals reported. Furthermore, individuals use the same gestures in different contexts (Tomasello and Call 1997). With regard to learning processes, this could be explained by a stimulus–response contingency-shaped relationship between specific pairs. To date, there is little evidence that all members of a given population use population-specific gestures as sign language as humans do.

In contrast, the existence of a few population differences in gestural signals using objects has been reported. There are two well-known cases in wild chimpanzees. The first example is "leaf-clipping" observed in the Mahale K group (Nishida 1980), Bossou (Sugiyama 1981), and Tai (Boesch 1995) chimpanzees. In this behavior, a chimpanzee pulls one or more dried leaves through its mouth and strips them, making a distinctive loud noise. Its function is interpreted as attention getting. The second is "leaf-grooming" observed in Gombe (Goodall 1986). In this behavior, a chimpanzee directs typical grooming motor patterns, such as peering, mouthing, and lip smacking, towards randomly picked leaves. It occurs when a lone chimpanzee seems bored and its function is not clear (Goodall 1986).

It is important to note that both of these behaviors are inconspicuous or rare. In addition, these behaviors are used in different contexts, and there are distinct individual differences within the groups in which these behaviors occur (McGrew 1992; Tomasello et al. 1994; Tomasello and Camaioni 1997). It is controversial whether these two behaviors are really used for communication or simply to manipulate objects. It is likely that wild chimpanzees only rarely use body movements as population-specific gestural signals to communicate with another individual.

Why are nongenetic population-specific gestures in chimpanzees rare? I propose that the chimpanzees' limited capacity for whole-body imitation may limit the characteristics of social behaviors as mentioned. For a behavior to function as a signal, that signal must be recognized and performed by both partners. For signaling reciprocity to occur, the receiver must be able to understand the intentions of the signal. A human can anticipate and imitate another human's actions by processing information on body movements. This may enable humans to understand and perform signals involving body movement. Moreover, humans can share a common signal between multiple partners. On the other hand, chimpanzees have difficulty imitating body actions due to cognitive constraints on processing information about body movements. Hence, it would be difficult for a chimpanzee to understand another chimpanzee's intentions and imitate gestural signals not involving objects. For this reason, wild chimpanzees may naturally rarely display population-specific gestural signals.

In this chapter, I have explored the evolutionary roots of imitation, focusing on visual–motor information processing in humans and chimpanzees. Basic differences in information processing may be reflected in the different social–cognitive abilities of the two species.

Acknowledgments

This study was financed by Grant 07102010 from the Ministry of Education, Science, Sports, and Culture, Japan. The preparation of this article was supported by Research Fellowship 2867 from the Japan Society for the Promotion of Science for Young Scientists, and by the Cooperation Research Program of the Primate Research Institute, Kyoto University. I gratefully acknowledge the help of T. Matsuzawa, J.R. Anderson, J. Call, M. Koyasu, S. Kojima, and G. Yamakoshi in their comments on a draft of this article and generous guidance throughout this study.

References

Abravanel E, Gingold H (1985) Learning via observation during the second year of life. Dev Psychol 21:614-623
Abravanel E, Sigafoos AD (1984) Exploring the presence of imitation during early infancy. Child Dev 55:381-392
Anisfeld M (1996) Only tongue protrusion modeling is matched by neonates. Dev Rev 16:149-161
Bard KA, Russell CL (1999) Evolutionary foundations of imitation: social cognitive and developmental aspects of imitative processes in non-human primates. In: Nadel J, Butterworth G (eds) Imitation in infancy. Cambridge University Press, Cambridge, pp 89-123
Bard KA, Hopkins WD, Fort CL (1990a) Lateral bias in infant chimpanzees (Pan troglodytes). J Comp Psychol 104:309-321
Bard KA, Platzman KA, Lester BM (1990b) Comparisons of neurobehavioral integrity in chimpanzees (Pan troglodytes). Am J Primatol 20:171-172
Bard KA, Platzman KA, Lester BM, Suomi SJ (1992) Orientation to social and nonsocial stimuli in neonatal chimpanzees and humans. Infant Behav Dev 15:43-56
Barresi J, Moore C (1996) Intentional relations and social understanding. Behav Brain Sci 19:107-122
Boesch C (1995) Innovation in wild chimpanzees (Pan troglodytes). Int J Primatol 16:1-16
Butterworth G, Harris M (1994) Principles of developmental psychology, vol 1. In: Eysenck MW, Green S, Hayes N (eds) Principles of psychology. Lawrence Erlbaum, London
Byrne R (1994) The thinking ape: evolutionary origins of intelligence. Oxford University Press, New York
Custance DM, Whiten A, Bard KA (1995) Can young chimpanzees (Pan troglodytes) imitate arbitrary actions? Hayes and Hayes (1952) revisited. Behaviour 132:839-858

Field TM, Woodson R, Greenberg R, Cohen D (1982) Discrimination and limitation of facial expressions by neonates. Science 218:179–181

Field TM, Goldstein S, Vega-Lahr N, Porter K (1986) Changes in imitative behavior during early infancy. Infant Behav Dev 9:415–421

Fontaine R (1984) Imitative skills between birth and six months. Infant Behav Dev 7:323–333

Gillan DJ, Premack D, Woodruff G (1981) Reasoning in the chimpanzee: analogical reasoning. J Exp Psychol: Anim Behav Process 7:1–17

Goodall J (1986) The chimpanzees of Gombe. Harvard University Press, Cambridge

Hallock MB, Worobey J, Self PS (1989) Behavioral development in chimpanzee (Pan troglodytes) and human newborns across the first month of life. Int J Behav Dev 12:527–540

Hayes KJ, Hayes C (1952) Imitation in a home-raised chimpanzee. J Comp Physiol Psychol 45:450–459

Heimann M (1989) Neonatal imitation, gaze aversion, and mother–infant interaction. Infant Behav Dev 12:495–505

Jacobson SW (1979) Matching behavior in the young infant. Child Dev 50:425–430

Johnson MJ (1990) Cortical maturation and the development of visual attention in early infancy. J Cogn Neurosci 2:81–95

Johnson MJ, Morton J (1991) Biology and cognitive development: the case of face recognition. Blackwell, Oxford

Maratos O (1982) Trends in the development of imitation in early infancy. In: Bever TG (ed) Regressions in mental development: basic phenomena and theories. Erlbaum, Hillsdale, pp 81–101

Mathieu M, Bergeron G (1981) Piagetian assessment on cognitive development in chimpanzee (Pan troglodytes). In: Chiarelli AB, Corruccini RS (eds) Primate behavior and sociobiology. Springer, Berlin, pp 142–147

Matsuzawa T (1985) Color naming and classification in a chimpanzee (Pan troglodytes). J Hum Evol 14:283–291

Matsuzawa T (1990) From perception and visual acuity in a chimpanzee. Folia Primatol 55:24–32

Matsuzawa T (1996) Chimpanzee intelligence in nature and in captivity: isomorphism of symbol use and tool use. In: McGrew WC, Marchant LF, Nishida T (eds) Great ape societies. Cambridge University Press, New York, pp 196–209

Matsuzawa T, Yamakoshi G (1996) Comparison of chimpanzee material culture between Bossou and Nimba, West Africa. In: Russon AE, Parker ST, Bard KA (eds) Reaching into thought: the minds of the great apes. Cambridge University Press, Cambridge, pp 211–232

McGrew WC (1992) Chimpanzee material culture: implications for human evolution. Cambridge University Press, Cambridge

Meltzoff AN (1988) Infant imitation after a 1-week delay: Long-term memory for novel acts and multiple stimuli. Dev Psychol 24:470–476

Meltzoff AN (1990) Foundations for developing a concept of self: the role of imitation in relating self to other and the value of social mirroring, social modeling, and self practice in infancy. In: Cicchetti D, Beeghly M (eds) The self in transition: infancy to childhood. University of Chicago Press, Chicago, pp 139–164

Meltzoff AN (1996) The human infant as imitative generalist: a 20-year progress report on infant imitation with implications for comparative psychology. In: Galef BG Jr, Heyes CM (eds) Social learning in animals: the roots of culture. Academic Press, New York, pp 347–370

Meltzoff AN, Gopnik A (1993) The role of imitation in understanding persons and developing a theory of mind. In: Baron-Cohen S,. Tager-Flusberg H, Cohen D (eds) Understanding other minds: perspectives from autism. Oxford University Press, New York, pp 335–366

Meltzoff AN, Moore MK (1977) Imitation of facial and manual gestures by newborn infants. Science 198:75–78

Meltzoff AN, Moore MK (1983) Newborn infants imitate adult facial gestures. Child Dev 54:702–709

Meltzoff AN, Moore MK (1989) Imitation in newborn infants: Exploring the range of gestures imitated and the underlying mechanisms. Dev Psychol 25:954–962

Meltzoff AN, Moore MK (1992) Early imitation within a functional framework: the importance of person identity, movement, and development. Infant Behav Dev 15:479-505

Meltzoff AN, Moore MK (1994) Imitation, memory, and the representation of persons. Infant Behav Dev 17:83-99

Myowa M (1996) Imitation of facial gestures by an infant chimpanzee. Primates 37:207-213

Myowa-Yamakoshi M, Matsuzawa T (1999) Factors influencing imitation of manipulatory actions in chimpanzees (Pan troglodytes). J Comp Psychol 113:128-136

Myowa-Yamakoshi M, Matsuzawa T (2000) Imitation of intentional manipulatory actions in chimpanzees (Pan troglodytes). J Comp Psychol 114:381-391

Nagell K, Olguin R, Tomasello M (1993) Processes of social learning in the imitative learning of chimpanzees and human children. J Comp Psychol 107:174-186

Nishida T (1980) The leaf-clipping display: A newly discovered expressive gesture in wild chimpanzees. J Hum Evol 9:117-128

Piaget J (1962) Play, dreams and imitation in childhood. Norton, New York

Premack D, Premack AJ (1983) The mind of an ape. Norton, New York

Premack D, Woodruff G (1978) Does the chimpanzee have a theory of mind? Behav Brain Sci 4:515-526

Reissland N (1988) Neonatal imitation in the first hour of life: observations in rural Nepal. Dev Psychol 24:464-469

Rogers S, Pennington B (1991) A theoretical approach to the deficit in infantile autism. Dev Psychopathol, 3:137-162

Sugiyama Y (1981) Observations on the population dynamics and behavior of wild chimpanzees at Bossou, Guinea, 1979-1980. Primates 22:432-444

Tomasello M, Call J (1997) Primate cognition. Oxford University Press, New York

Tomasello M Camaioni L (1997) A comparison of gestural communication of apes and human infants. Hum Dev 40:7-24

Tomasello M, Davis-Dasilva M, Camak L, Bard K (1987) Observational learning of tool-use by young chimpanzees. Hum Evol 2:175-183

Tomasello M, Call J, Nagell K, Olguin R, Carpenter M (1994) The learning and use of gestural signals by young chimpanzees: a trans-generational study. Primates 35:137-154

Tomonaga M Matsuzawa T (1992) Perception of complex geometric figures in chimpanzees (Pan troglodytes) and humans (Homo sapiens): analysis of visual similarity on the basis of choice reaction time. J Comp Psychol 106:43-52

Vinter A (1986) The role of movement in eliciting early imitations. Child Dev 57:66-71

Werner H, Kaplan B (1963) Symbol formation: an organismic-developmental approach to language and the expression of thought. Wiley, New York

Whiten A (1996) Imitation, pretense, and mindreading: secondary representation in comparative primatology and developmental psychology? In:Russon AE, Parker ST, Bard KA (eds) Reaching into thought: the minds of the great apes. Cambridge University Press, Cambridge, pp 278-299

Whiten A, Custance DM (1996) Studies of imitation in chimpanzees and children. In: Galef BG Jr, Heyes CM (eds) Social learning in animals: the roots of culture. Academic Press, New York, pp 300-324

Whiten A, Ham R (1992) On the nature and evolution of imitation in the animal kingdom: reappraisal of a century of research. In: Slater P, Rosenblatt J (eds) Advances in the study of behavior. Academic Press, New York, pp 239-283

Whiten A, Custance DM, Gómez J-C, Teixidor P, Bard KA (1996) Imitative learning of artificial fruit processing in children (Homo sapiens) and chimpanzees (Pan troglodytes). J Comp Psychol 110:3-14

Whiten A, Goodall J, McGrew WC, Nishida T, Reynolds V, Sugiyama Y, Tutin CEG, Wrangham RW, Boesch C (1999) Culture in chimpanzees. Nature 399:682-685

18
Species Recognition by Macaques Measured by Sensory Reinforcement

KAZUO FUJITA

Recognizing social objects is essential to many animals. Identifying their own species is clearly most fundamental of all, not only for animals living social lives, but also for those living alone as long as they have to mate with conspecifics to reproduce their genes. Hybridization is often prevented by differences in sexual organs, in time and place of mating, in number of chromosomes, etc. Hybridization usually results in the mortality or sterility of the hybrids even it succeeds in making a zygote. However, mating usually consumes much time and energy. Particularly for animals who mate only once in their lifetime, such as insects, copulation with a different species leading to failure of reproduction is a complete waste of their lives. As a result, many such animals have evolved a variety of behavioral mechanisms to recognize species. One of the most impressive examples is the species-specific flashing-light communication in many fireflies (Lloyd 1966). Although the cost of mating a different species might be a little less serious in animals who reproduce more than once in their lifetime, it is still maladaptive in terms of time and energy. The fascinating variety of animals in the world is a result of such reproductive isolation.

The situation is different in nonhuman primates. Although they differ greatly from each other in appearance, nonhuman primates often make fertile hybrids when different species are kept together experimentally (Chiarelli 1973; Bernstein and Gordon 1980). In some cases hybrid individuals even make wild groups, e.g., baboons (Nagel 1973; Sugawara 1979) and macaques (Groves 1980; Watanabe and Matsumura 1991). However, wild hybrids are still rare and found only in places where artificial disturbance of the environment is suspected. Primate species in the wild seem to have established reproductive isolation in some manner.

Yoshikubo (1985, 1987) proposed that primate individuals may actively choose to mate conspecific individuals. This is a type of behavioral reproductive isolation, but no ritualized sequence of behavior is hypothesized; a mere psychological preference is enough for this mechanism to work. Yoshikubo called it a "psychological reproductive isolation mechanism."

In this chapter it will be shown that at least some nonhuman primate species actually do have such a preference for their own species. After describing the meth-

Department of Psychology, Graduate School of Letters, Kyoto University, Yoshida Honmachi, Sakyo-ku, Kyoto 606-8501, Japan

odology used in the first section, some studies with macaque monkeys will be presented.

1 Sensory Reinforcement: A New Method to Study Recognition of Social Objects

A transportable apparatus was attached to the home cage of a language-trained chimpanzee, Ai. This contained a touch-sensitive key, a screen, and a slide projector. Ai was verbally encouraged to touch the key. When she touched the key, the slide projector presented a colored picture on the screen. No other reinforcers, such as food or water, were contingent upon the response; slides were the sole reinforcer of this response. A variety of mammals have been shown to respond to such sensory stimuli (Kish 1966; Matsuzawa 1981), which is called sensory reinforcement. Primates have been shown to be particularly sensitive to subtle differences in visual stimuli (e.g., Humphrey 1971, 1974; Kish 1966; Matsuzawa and Fujita 1981). In other words, primates show differential interest in different visual stimuli. An important implication of this is that we may be able to discover how the primates recognize objects in the environment from an analysis of such differential interests.

The procedure used to show the slides was as shown in Fig. 1 (Fujita and Matsuzawa 1986). When the subject touched the key, the slide in the current slot of the rotary magazine of the projector was shown on the screen. The picture remained there as long as the subject kept touching the key. The picture disappeared when the subject released the key. The same slide was repeatedly presented if the subject touched the key again within 10 s after releasing it. The rotary magazine

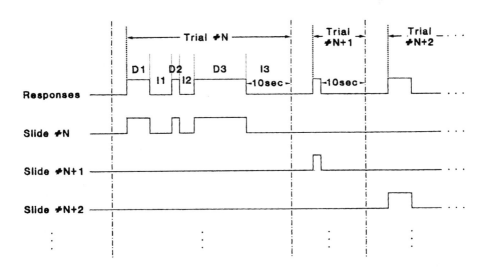

Fig. 1. Diagram of the experimental procedure used. The responses were reinforced by slide pictures. (From Fujita and Matsuzawa 1986, with permission)

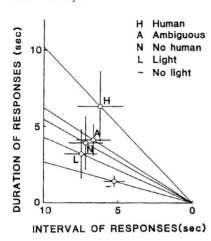

H Human
A Ambiguous
N No human
L Light
− No light

Fig. 2. Chimpanzee responses to a variety of slide pictures. The horizontal axis is the response interval. Note that the value is smaller on the right-hand side. The vertical axis is the duration of the responses. *Circles* denote the center of distribution of the responses for each of the following five groups of slides: slides with humans (*H*), slides with very small images of humans (*A*), slides with no humans (*N*), simple light (*L*), and no light (*-*). The *vertical* and *horizontal bars* denote the standard deviation of the duration and the interval of responses. (From Fujita and Matsuzawa 1986, with permission)

advanced by one slide when 10 s had passed without response. This new slide was presented the next time the subject touched the key. The sequence of events between the first touch for one slide and the change of slide is called a trial. With this arrangement, the subject was allowed to see each picture for as long as she wanted during each trial, as many times as she wanted, and with any interval shorter than 10 s that she wanted. If I were to instruct the chimpanzee verbally, I would tell her: "Touch the key for the slide. You may see the slide for as long and for as many times as you want. Refrain from touching for 10 s if you are bored with the same slide and want to see the next one."

The subject was first shown some preliminary slides. Then we presented 100 commercial slides of a variety of scenes of Japan. After the subject had completed 1000 trials, i.e., 10 trials per slide, the mean duration (*D*) and the mean interval (*I*) of her key-touching responses were calculated for each slide. Based on these two values, each slide was plotted on a two-dimensional plane with axes *D* and *I*.

The slides were classified into the following five groups: human, no human, ambiguous, light, and no light. Human slides had at least one image of a human in the photograph, no-human slides had no image of humans, ambiguous slides had tiny human images which were very difficult to detect, light slides had no film on the slide, and no-light slides had thick paper on the slide. Figure 2 shows the center of distribution of the slides in each stimulus group. The horizontal axis is *I* and the vertical axis is *D*. Note that *I* has the shorter value to the right along the abscissa. Each group of slides had a different location on this plane. The no-light slides had short *D* and *I*, and occupied the bottom–right area. This extreme placing of the no-light slides is probably a consequence of a temporary extinction of the touch responses. The most important finding was that the human slides tended to have a shorter *I* and a longer *D* than the others, and thus occupied the top–right area on this plane. This suggests that humans are a readily distinguishable category for the chimpanzee. Note that this is not a result of training to discriminate slides with humans from slides without humans, as in traditional studies on concept formation (e.g., Herrnstein et al. 1976; Schrier et al. 1984; Yoshikubo 1985). Training to form a concept may not be needed when investigating what kinds of concepts the animals already have.

The inclination of the lines connecting each datum point and the origin where $I = D = 0$ shows the ratio of D to I. This D/I ratio can be regarded as a score of preference for the stimulus, because D is expected to be large (long) and I is expected to be small (short) for preferred slides. In terms of this score, the human slides were clearly the most preferred, ambiguous slides were next, and no-human, light, and no-light slides were least preferred.

In summary, this simple procedure using a sensory reinforcement provides a useful tool to analyze the discrimination and categorization of social objects as well as preferences for them.

2 Species Recognition in Five Macaque Species Measured by Sensory Reinforcement

Four Japanese monkeys (*Macaca fuscata*), three rhesus monkeys (*M. mulatta*), two pigtail monkeys (*M. nemestrina*), two bonnet monkeys (*M. radiata*), and three stump-tailed monkeys (*M. arctoides*) were tested in the same procedure as described above for their recognition of closely related species (Fujita 1987, 1989). All were adults. Some of the monkeys had experience of pressing a lever or key to obtain food.

After the monkeys were trained to press and hold a lever for 2 or 3 s to obtain food, they were given several sessions in which the same lever-pressing responses were reinforced by a variety of pictures of animals and plants as preliminary stimuli. The experimental sessions started after these preliminary sessions.

In experimental sessions, 24 different colored slides of each of the seven macaque species were presented, including the five species listed above, plus crab-eating monkeys (*M. fascicularis*) and Taiwan monkeys (*M. cyclopis*). An additional 16 slides of light and 16 slides of no light were used as the controls. In total there were 200 slides. The procedure was the same as the one used to test the chimpanzee except that the maximum duration of the lever-pressing response was limited to 10 s.

After the subjects had completed 2000 trials, i.e., 10 trials per slide, the mean duration (D) and the mean interval (I) of lever-pressing responses were calculated for each slide. For all subject species except stump-tailed monkeys, slides of their own species tended to have relatively shorter Is and longer Ds. The D/I ratio was then taken. This score is referred to as the preference index (PI). Figure 3 shows relative PIs for each of the seven stimulus species averaged across individuals of each subject species. The values plotted are the deviations of the PI from the average for each stimulus species:

$$(PI_i / PI_t - 1/7) \times 100$$

where PI_i is the PI for stimulus species i, and PI_t is the sum of all PI_is. If the subjects press the lever randomly, all these deviation scores are 0. Positive values mean that

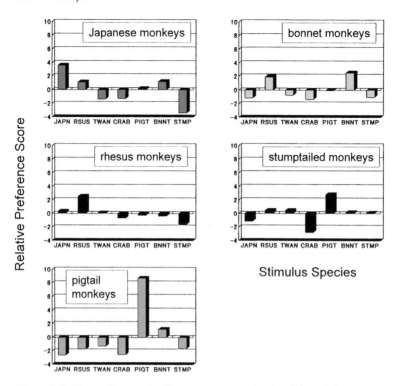

Fig. 3. Relative preferences by five macaque species for slides of the seven macaque species shown on the horizontal axis (*JAPN*, Japanese monkeys; *RSUS*, rhesus monkeys; *TWAN*, Taiwan monkeys; *CRAB*, crab-eating monkeys; *PIGT*, pigtail monkeys; *BNNT*, bonnet monkeys; *STMP*, stump-tailed monkeys). The data are averages of two–four monkeys. *Upward bars* indicate relatively preferred slides and *downward bars* indicate relatively avoided slides. See text for the values shown on the vertical axis. All the species except the stump-tailed monkeys preferred their own species. (Based on data from Fujita 1987, with permission)

the subjects relatively prefer the stimuli and negative values mean that they relatively avoid the stimuli.

As the graph shows, Japanese, rhesus, pigtail, and bonnet monkeys preferred the images of their own species. This was the case for all the individual subjects of these species. These results were consistent with those of previous studies (Sackett 1970; Swartz and Rosenblum 1980). The only exception was stump-tailed monkeys. No subjects of this species showed a preference for their own species. This was inconsistent with the findings of DeMaria and Thierry (1988), who reported that stump-tailed monkeys showed the strongest interest in pictures of their own species rather than those of other primates, including macaques. However, they did not include pictures of pigtail monkeys, for which the subjects of the present study showed the strongest preference. The experiment of DeMaria and Thierry (1988) also failed to separate effects of unidentified bias in the pictures used because they did not show the same slides to other species.

The visual preference for their own species shown here by most of the macaque species seems consistent with the hypothesis of Yoshikubo (1985, 1987) that the monkeys may have a tendency to choose conspecifics as a mate rather than closely related species. A potential reason why stump-tailed monkeys failed to show the same preference could be that they may not need to discriminate species very strictly because they have unique external sexual organs. It has been suggested that these are useful to prevent copulation with other macaque species (Fooden 1967), and stump-tailed monkeys are actually classified as a distinctive subgroup of the genus *Macaca* (Fooden 1980).

3 Development and Determinants of Visual Preference in Macaques

The next series of experiments used the same procedure to test the visual preferences of Japanese monkeys and rhesus monkeys who had had their social experience controlled (Fujita 1989, 1990, 1993).

3.1 Hand-Reared Monkeys

The subject monkeys were separated from their mothers within a week after birth and were reared by humans thereafter. They were paired, or in one conspecific case put in a group of three, either with a conspecific peer or a heterospecific peer. This resulted in monkeys having social experiences with only one or two individuals. Some of them had experience of social interactions only with heterospecifics.

The visual preferences of these infant monkeys were repeatedly tested from the age of 3 months until they were 27 months old, using the same procedure as described above. The monkey pictures used were only those of Japanese monkeys or rhesus monkeys. Figure 4 shows the relative *PI* for pictures of rhesus monkeys. This time the plotted deviations of *PI* are

$$(PI_r / PI_t - 1/2) \times 100$$

where PI_r is the *PI* for the pictures of rhesus monkeys and PI_t is the total *PIs* for the pictures of Japanese and rhesus monkeys. Hence, positive values mean a relative preference for rhesus monkeys over Japanese monkeys. As is clear from Fig. 4, all the monkeys, regardless of rearing experience and testing age, showed relative preference scores larger than 0. That is, they always preferred pressing the lever for pictures of rhesus monkeys in terms of the *PI*. This tendency became stronger with age.

This might be a consequence of a bias in the pictures used. To test this possibility, one adult Japanese monkey who had had normal social experiences was tested with the same set of pictures. This subject showed a clear preference for pictures of Japanese monkeys. Therefore the observed preference for pictures of rhesus monkeys is unlikely to be due to biased pictures.

The same infant monkeys were also shown the same slides of the seven macaques

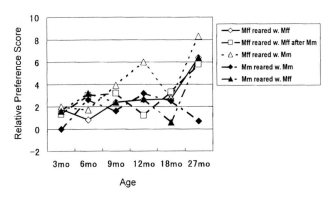

Fig. 4. Developmental change of relative preferences for pictures of rhesus monkeys over those of Japanese monkeys by hand-reared infant Japanese and rhesus monkeys who had had a variety of restricted social experiences (see *key*). The horizontal axis is the age in months. (Based on data from Fujita 1989, 1990, with permission)

that were used in the experiment to test adults of five macaque species. This test was done at either 9, 14, or 16 months of age, and resulted once again in a preference of all subjects for rhesus over Japanese monkeys.

3.2 Mother-Reared Japanese Monkeys

The results above suggest that both Japanese and rhesus monkeys have a native tendency to prefer seeing pictures of rhesus monkeys rather than Japanese monkeys. Then how do adult Japanese monkeys come to prefer their own pictures?

One possible factor is the quality of early social interaction. The hand-reared subjects described above had interacted only with cage mates of the same age. Experience with conspecific peers may not be enough for Japanese monkeys to learn to prefer their own species.

To learn the potential effects of this quality of social experience, two Japanese monkey infants reared by their biological mothers were tested with the pictures of the seven macaques in the same procedure. The test was done immediately after

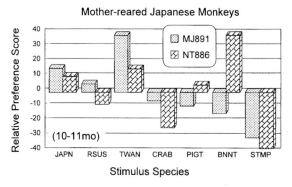

Fig. 5. Relative preferences of two mother-reared infant Japanese monkeys for pictures of seven macaque monkeys. Other details as in Fig. 3. (Based on data from Fujita 1987, with permission)

they were separated from their mothers when they were 10 or 11 months old. These mother-reared monkeys relatively preferred seeing pictures of Japanese monkeys rather than rhesus monkeys, although they actually preferred some of the other species best (Fig. 5).

3.3 Cross-Fostered Monkeys

If the quality of social interaction during the first year of life is critical, then monkeys raised by mothers of different species should show a preference for the foster-mother species. Two cases of interspecific cross-fostering between Japanese and rhesus monkeys were conducted. We anesthetized the mothers and then exchanged the babies. These attempts were not perfectly successful. One rhesus monkey mother rejected the Japanese monkey baby, but the attempt resulted in two rhesus monkey infants being successfully reared by Japanese monkey mothers, and one Japanese monkey infant being reared by a rhesus monkey mother. These monkeys were separated from their foster mothers when they were 13 or 14 months old, and were then tested immediately.

The test was conducted with the same procedure using only pictures of Japanese and rhesus monkeys. The Japanese monkey clearly preferred seeing pictures of rhesus monkeys. Interestingly, neither of the two rhesus monkeys showed a clear differential preference, although they still showed a very slight preference for rhesus pictures. These results suggest an interaction of native tendency and experience. That is, the visual preferences of monkeys may change with social experience, but the extent of this change may be restricted by their genes. In fact, this constraint seems so powerful in rhesus monkeys that good-quality social experience with a Japanese monkey mother for 1 year cannot reverse the preference in rhesus infants.

3.4 Long-Term Change

It is possible that Japanese monkeys do not show a preference for own species until they are much older. To investigate this possibility, most of the hand-reared infant subjects and cross-fostered subjects described above received one or two additional tests with the pictures of seven macaque species when they were either 3, 4, or 5 years old (Fujita 1993). At the time of this test, the hand-reared monkeys had been reared in the same pairs as long as the partner was healthy. The cross-fostered monkeys had basically been reared in individual cages. However, the cross-fostered Japanese monkey had been paired with one of the cross-fostered rhesus monkeys for alternating periods between the ages of 18 months and 2 years.

Throughout these follow-up tests, not only hand-reared but also cross-fostered rhesus monkeys showed a preference for rhesus over all the other macaque monkeys presented as stimuli. That is, preference by the cross-fostered rhesus monkeys for their own species, which seems to have been suppressed at the time of separation from their foster mother, came back later (4 years old). The effects of early social experience with a heterospecific mother were not permanent. The patterns

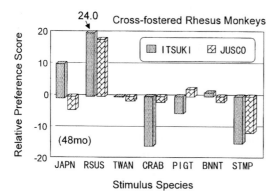

Fig. 6. Relative preferences of two cross-fostered rhesus monkeys reared by Japanese monkey mothers for pictures of seven macaque monkeys tested at 4 years old. Other details as in Fig. 3. (From Fujita 1993, with permission)

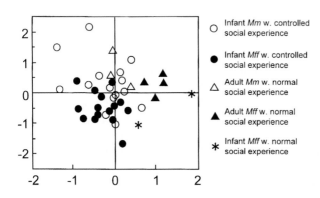

Fig. 7. Result of the multidimensional scaling (MDS) of preference patterns of infant and adult Japanese and rhesus monkeys who had had a variety of social experiences (see *key*). (From Fujita 1993, with permission)

of the visual preferences for the pictures of all the seven species were strikingly similar to those of adult rhesus monkeys from the first experiment (Fig. 6).

On the other hand, Japanese monkeys, regardless of their social experience, came to prefer pictures of pigtail monkeys or bonnet monkeys, although they still relatively preferred rhesus over Japanese monkeys. These patterns of preference were completely different from those of normally raised adult Japanese monkeys in the first experiment of this series. Instead, they were more like the preference patterns of rhesus monkeys.

The patterns of visual preference for the pictures of all the seven species were entered into a multidimensional scaling procedure (MDS). Figure 7 gives the results of this analysis and shows how similar the patterns of preference in each test are to one another. It is clear that the positions of the datum points for rhesus infants (open circles) and rhesus adults (filled circles) overlap to a large extent, but the points for Japanese monkey infants (open triangles) are separated from those for adult Japanese monkeys (filled triangles). This suggests that the preference patterns of adult Japanese monkeys may be a consequence of learning through abundant social interactions with conspecific individuals, but that the same patterns for adult rhesus monkeys may be innately determined.

Such dissociation of the determinants of this most fundamental social preference may be partly understandable in terms of the distribution and history of wild populations of these species. Rhesus monkeys share their distribution area with several macaque species who live on the Asian continent (Fooden 1980). Innately determined social preference might have worked to prevent interbreeding of this species. On the other hand, Japanese monkeys are the only species who live in the Japanese islands, and there is no risk of interbreeding. Japanese monkeys are supposed to be the descendants of prototype rhesus monkeys who migrated to the Japanese islands in the late Middle Pleistocene (Delson 1980). If this prototype population had no selection pressure to prefer their own species genetically, their preference patterns may have remained almost unchanged, or may possibly have changed randomly.

4 Visual Preference for Sulawesi Macaques by Sulawesi Macaques

Seven closely related species of the genus *Macaca* (*M. nigra, M. nigrescens, M. hecki, M. tonkeana, M. maurus, M. ochreata,* and *M. brunnescens*) live in Sulawesi and the surrounding islands of Indonesia. Their distribution does not overlap but, except in one case, there seem to be no strict geographic barriers which could work to prevent interbreeding. Although there have been reports of suspected hybrid groups in restricted areas (Groves 1980; Watanabe and Matsumura 1991), in general these species seem to be isolated reproductively. Therefore this group of macaques seems a suitable subject on which to test Yoshikubo's psychological isolation hypothesis.

The visual preferences of Sulawesi macaques for Sulawesi macaques (Fujita and Watanabe 1995; Fujita et al. 1997) were tested in a series of experiments. The subjects were monkeys kept as pets by local people. The experiments were conducted at five places in Sulawesi and on one neighboring island. A simple experimental setup, as shown in Fig. 8, was built at each place. There was a screen on which stimulus pictures were presented by a slide projector, and the subject was restrained by a rope or chain attached to a pole near the screen.

In this series of experiment we did not use operant responses in order to save the time needed to train each monkey. We simply presented stimulus slides one by one for 50 or 60 s each in sequence, and the experimenter recorded the visual fixation of the subject monkeys on the stimulus. At least two individuals of each sex of each species were tested. The estimated ages of the subjects ranged from 1 year to full adult, but most of them were younger than 5 years.

The stimulus slides were eight different photographs of each of the seven Sulawesi macaques plus Japanese monkeys and pigtail monkeys. These photographs were put into the rotary magazine of a slide projector in a quasirandom order. Before

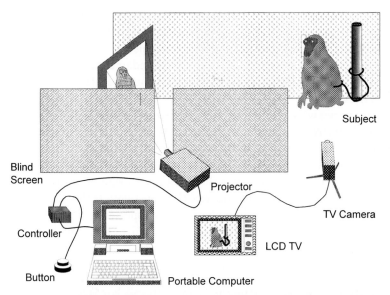

Fig. 8. Sketch of the experimental setting to test the visual preferences of Sulawesi macaques for Sulawesi macaques. (From Fujita and Watanabe 1995, with permission)

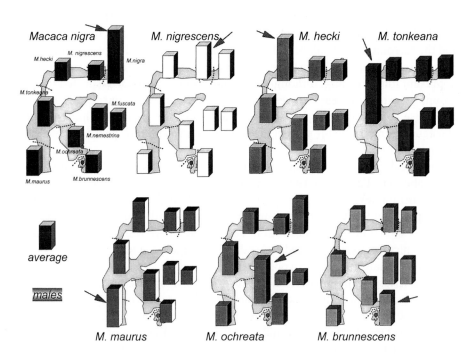

Fig. 9. Relative preference scores for pictures of Sulawesi macaques from male Sulawesi macaques. The *bars* showing the relative preferences are located on the distribution areas (separated by *broken lines*) of each stimulus species in Sulawesi and surrounding islands. *Arrows* denote a subject's own species. *Bars located off the map* are for pictures of Japanese and pigtail monkeys. The overall average is shown at the left of the figure. All the species showed the greatest visual preferences for their own species. (From Fujita et al. 1997, with permission)

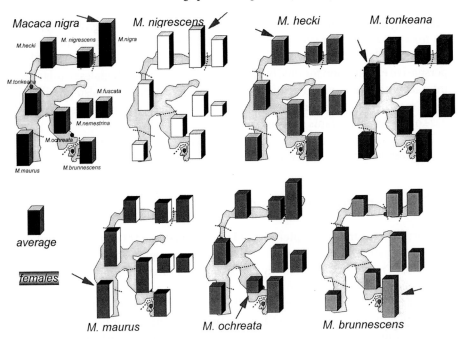

Fig. 10. Relative preference scores for pictures of Sulawesi macaques from female Sulawesi macaques. Females' preference patterns are somewhat variable compared with those of males. Other details as in Fig. 9. (From Fujita et al. 1997, with permission)

testing, 20 preliminary pictures of a variety of animals and plants were presented to habituate the subjects to the situation.

Figures 9 and 10 show the relative total duration of visual fixation on slides of each category superimposed on a map of Sulawesi showing the distribution areas of each species. The isolated bar at the bottom–left of each figure denotes the average viewing duration for all the monkey pictures. Figure 9 is for male subjects and Fig. 10 is for females. Viewing durations for the subject's own species are marked by arrows. The data for the pictures of Japanese and pigtail monkeys are shown to the right of the island map.

As is clear from Fig. 9, male subjects of all species looked at pictures of their own species longest. This result is consistent with Yoshikubo's psychological isolation hypothesis. An interesting finding was that subjects did not necessarily like to watch species living in neighboring areas. For example, *M. nigra* avoided looking at their neighbor *M. nigrescens*. In Sulawesi species, the external morphology of the macaques is closest between neighboring species, and patterns of preference cannot be explained by the simple generalization of a preference for their own species.

Figure 10 shows that female subjects had a similar tendency, but in general it was weaker than that of the males. This is at least partly because female monkeys tended to show weaker social reactions, such as lip-smacking, grimaces, etc., to the slides, and it was more difficult for the experimenter to recognize a visual fixation. One surprising finding, however, was that although most species (*M. nigra, M. nigrescens, M. tonkeana,* and *M. brunnescens*) preferred their own species, *M. ochreata* females actually tended to avoid watching their own species. There seems to be no persuasive explanation for this, but it is possible that the way females of this species interact with others might be different from other species. It is also possible that there may be something in the stimulus pictures used which works to avert the gaze of the females of this species.

This last, different, series of experiments showed that Sulawesi macaques share a visual preference for their own species.

5 Summary and Discussion

The first section in this chapter showed that simple operant responses reinforced by sensory stimuli may be used to analyze the recognition of social objects in primates. The chimpanzee readily distinguished humans from nonhumans without explicit training in this concept.

The next series of experiment used this procedure to demonstrate that a visual preference for their own species is widespread among macaques. In the first experiment, mother-reared adult macaques pressed a lever most frequently to see conspecifics. The only exception was stump-tailed monkeys. In the experiments using infant macaques who had had a variety of controlled early social experiences, rhesus monkeys were found to develop preference patterns which were similar to those of adults regardless of their social experience. On the other hand, Japanese monkeys, who did not have rich social experiences with conspecifics, developed preference patterns which were different from those of adults who had had normal social experiences. The preference patterns seem to be innately determined in rhesus monkeys, but they are probably learned in Japanese monkeys.

In the third series of experiments, males of all seven species of Sulawesi macaques were demonstrated to have a visual preference for their own species in a procedure which simply measured differential interests in monkey pictures, although this tendency was more variable in females.

All of these findings seem to support the "psychological isolation" theory suggested by Yoshikubo (1985, 1987). One may argue that a visual preference does not necessarily mean a mating preference. This is true in theory, but a visual preference is more likely to be positively correlated with a mate preference. It is actually difficult to imagine a situation in which visual preference is not correlated, or even is negatively correlated, with mate preference. An animal has to approach a potential mate first before courtship. It is likely that primates complete this first step using visual clues, although the animals may rely on other clues such as odors or subtle facial movements as they come closer. Previous reports suggest that primates also show stronger interest in urine odors of their own species than those of other species (DeMaria and Roeder 1989; Ueno 1994). How the animals use clues from different sensory modalities is an interesting question for future research. Also, the correlation of visual preferences and mating preferences should be studied empirically.

If psychological isolation is really working, it will produce a strong selective pressure from inside the genetic population. In primates, the variation in external morphology is surprisingly large in comparison with that of genetic traits. For example, variations in blood proteins between primate species are as small as those found between subspecies of other taxonomic groups (Nozawa et al. 1977). If selection pressure by external morphology from inside the population is strong, the morphology of the genetic population may change quickly in a relatively short time, but not enough to separate the reproductive mechanisms from the mother population and modify them. This might explain why primates are so variable in their appearance.

Finally, as briefly noted in the introduction, olive and hamadryas baboons make permanent hybrid troops in which the appearances of the animals gradually change from one species to the other along the Awash Valleys in Ethiopia (e.g., Nagel 1973; Sugawara 1979). They may have met after a period of separation which was not long enough to differentiate the two species in reproductive physiology. As in the case of the Japanese monkeys, one gene population may not develop a native preferences for its own species if it is separated from other populations by strict geographic barriers. In such cases, two populations may easily mix when the barrier is no longer present. It may be interesting to test the visual preferences of these baboon species.

References

Bernstein IS, Gordon TP (1980) Mixed taxa introductions, hybrids, and macaque systematics. In: Lindburg DG (ed) The macaques: studies in ecology, behavior, and evolution. Van Nostrand Reinhold, New York, pp 125–147

Chiarelli B (1973) Check-list of Catarrhina primate hybrids. J Hum Genet 2:301-305

Delson E (1980) Fossil macaques, phyletic relationships and a scenario of deployment. In: Lindburg DG (ed) The macaques: studies in ecology, behavior, and evolution. Van Nostrand Reinhold, New York, pp 10–30

DeMaria C, Roeder JJ (1989) Responses to urinary stimuli in pigtailed (Macaca nemestrina) and stump-tailed (Macaca arctoides) macaques. Primates 30:111–115

DeMaria C, Thierry B (1988) Responses to animal stimulus photographs in stump-tailed macaques (Macaca arctoides). Primates 29:237–244

Fooden J (1967) Complementary specialization of male and female reproductive structures in the bear macaque, Macaca arctoides. Nature 214:939–941

Fooden J (1980) Classification and distribution of living macaques (Macaca lacepede, 1799). In: Lindburg DG (ed) The macaques: studies in ecology, behavior, and evolution. Van Nostrand Reinhold, New York, pp 1–9

Fujita K (1987) Species recognition by five macaque monkeys. Primates 28:353–366

Fujita K (1989) Species recognition in nonhuman primates (in Japanese with English summary). Jpn Psychol Rev 32:66–89

Fujita K (1990) Species preference by infant macaques with controlled social experience. Int J Primatol 11:553–573

Fujita K (1993) Development of visual preference for closely related species by infant and juvenile macaques with restricted social experience. Primates 34:141–150

Fujita K, Matsuzawa T (1986) A new procedure to study the perceptual world of animals with sensory reinforcement: recognition of humans by a chimpanzee. Primates 27:283–291

Fujita K, Watanabe K (1995) Visual preference for closely related species by Sulawesi macaques. Am J Primatol 37:253–261

Fujita K, Watanabe K, Widarto TH, Suryobroto B (1997) Discrimination of macaques by macaques: the case of Sulawesi species. Primates 38:233–245

Groves CP (1980) Speciation in Macaca: the view from Sulawesi. In: Lindburg DG (ed) The macaques: studies in ecology, behavior, and evolution. Van Nostrand Reinhold, New York, pp 84–124

Herrnstein RJ, Loveland DH, Cable C (1976) Natural concepts in pigeons. J Exp Psychol: Anim Behav Process 2:285–302

Humphrey NK (1971) Colour and brightness preferences in monkeys. Nature 229:615–617

Humphrey NK (1974) Species and individuals in the perceptual world of monkeys. Perception 3:105–114

Kish GB (1966) Studies of sensory reinforcement. In: Honig WK (ed) Operant behavior: the design and conduct of sensory experiments. Appleton–Century–Crofts, New York, pp 109–159

Lloyd JE (1966) Studies on the flash communication system in Photinus fireflies. Miscellaneous Publications of the Museum of Zoology, University of Michigan, No.130

Matsuzawa T (1981) Sensory reinforcement: the variety of reinforcers (in Japanese with English summary). Jpn Psychol Rev 24:220–251

Matsuzawa T, Fujita K (1981) Behavioral analysis of sensory reinforcement in infant Japanese monkeys in terms of time allocation (in Japanese with English summary). Jpn J Psychol 51:351–355

Nagel U (1973) A comparison of anubis baboons, hamadryas baboons and their hybrids at a species border in Ethiopia. Folia Primatol 19:104–165

Nozawa K, Shotake T, Ohkura Y, Tanabe Y (1977) Genetic variations within and between species of Asian macaques. Jpn J Genet 52:15–30

Sackett GP (1970) Unlearned responses, differential rearing experiences, and the development of social attachments by rhesus monkeys. In: Rosenblum LA (ed) Primate behavior: Developments in field and laboratory research, Vol. 1, Academic Press, New York, pp 111–140

Schrier AM, Angarella R, Povar ML (1984) Studies of concept formation by stump-tailed monkeys: concepts of humans, monkeys, and letter A. J Exp Psychol: Anim Behav Process 10:564–584

Sugawara K (1979) Sociological study of a wild group of hybrid baboons between Papio anubis and P. hamadryas in the Awash Valley, Ethiopia. Primates 20:21–56

Swartz KB, Rosenblum LA (1980) Operant responding by bonnet macaques for color videotape recordings of social stimuli. Anim Learn Behav 8:311–321

Ueno Y (1994) Responses to urine odor in the tufted capuchin (Cebus apella). J Ethol 12:81–87

Watanabe K, Matsumura S (1991) The borderlands and possible hybrids between three species of macaques, M. nigra, M. nigrescens, and M. hecki, in the northern peninsula of Sulawesi. Primates 32:365–369

Yoshikubo S (1985) Species discrimination and concept formation by rhesus monkeys (Macaca mulatta). Primates 26:285–299

Yoshikubo S (1987) A possible reproductive isolation through a species discrimination leaning in genus Macaca (in Japanese with English summary). Primate Res 3:43–47

19
Evolution of the Human Eye as a Device for Communication

Hiromi Kobayashi and Shiro Kohshima

1 Introduction

Recognizing the gaze-directions of others is one of the important cognitive bases for communication in humans (Gibson and Pick 1963; Kendon 1967). To clarify the biological basis of this ability, especially in relation to the evolution of social intelligence, researchers have experimentally examined the cognitive ability to detect the gaze direction of others in nonhuman primates (Gomez 1991; Itakura and Anderson 1996; Tomasello et al. 1998). However, little attention has been given to the external morphology of the eye although this ability of humans might be supported by a unique morphology of the human eye. For example, in humans, the widely exposed white sclera (the white of the eye) surrounding the darker colored iris makes it easy for others to discern the gaze direction and has been said to be a characteristic of humans not found in other primate species (Morris 1985). However, this has not been examined in detail, partly because of the difficulty in measuring the soft parts of living animals.

In this study, we measured the external eye morphologies of nearly half of all extant primate species with video camera and computer-aided image analyzing techniques to clarify the morphological uniqueness of the human eye and to understand the adaptive meanings of external eye-morphology in primates. The results clearly showed exceptional features of the human eye in both shape and coloration.

In order to understand the adaptive meaning of these exceptional features of the human eye, we postulated some hypotheses and examined them. To explain the close correlation of eye-shape parameters (width/height-ratio of the eye-outline and the proportion of exposed sclera in the eye-outline) with habitat type or body size of the species examined, we postulated a hypothesis that these features are adaptations for extending the visual field by eyeball movement, especially in the horizontal direction. This hypothesis was examined and supported by analyzing the eye-movements of video-recorded primates and the developmental change of eye morphology in humans and olive baboons. To explain the unique coloration of

Basic Biology, Faculty of Bioscience and Biotechnology, Tokyo Institute of Technology, 2-12-1 Ookayama, Meguro-ku, Tokyo 152-8551, Japan

the human eye with its exposed white sclera void of any pigmentation, we postulated a hypothesis that only coloration of the human eye is adapted to enhance the gaze signal while eye-coloration of other primates is adapted to camouflage the gaze direction against other individuals and/or predators. This hypothesis was examined and supported by analyzing the relationships among iris coloration, sclera coloration and facial coloration around eye.

Our results suggested that unique features of the human eye started to evolve as adaptations to large body size and terrestrial life and were completed as a device for communication using gaze signal.

2 Methods

2.1 Eye-Shape Measurements

A total of 874 adult animals (88 species: Prosimii; 10, Ceboidea; 26, Cercopithecoidea; 43, Hominoidea; 9) were studied. Facial images of 80 species were recorded by video camera at the Japan Monkey Centre. Facial images of 8 species (*Microcebus, Loris tardigradus, Perodicticus potto, Tarsius, Saguinus imperator, Pithecia monachus, Cacajao rubicundus, Cercopithecus hamlyni*) were collected from books (Itani and Uehara 1986; Yoshino 1994). For humans, facial images of 244 Japanese, 347 Caucasian and 68 Afro-Caribbean adults that were video recorded or collected from books (Ohara 1970; Gomi 1994) were studied. Frontal full-face images without obvious facial expression by subjects were recorded by video camera. These images were processed and analyzed on a Macintosh Quadra 840AV computer using the public domain NIH Image program. For each image, (a) the distance between the corners of the eye, (b) the longest perpendicular line between the upper and lower eyelid, (c) width of the exposed eyeball, and (d) diameter of the iris were measured (Fig. 1).

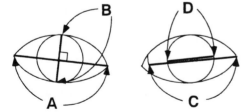

Fig. 1. The following parameters were measured: (*A*) the distance between the corners of the eye; (*B*) the longest perpendicular line to (*A*) between the upper eyelid and lower eyelid; (*C*) the width of the exposed eyeball, and (*D*) the diameter of the iris. Width/height ratio (WHR) is (*A*)/(*B*) and ratio of exposed sclera in the eye-outline (SSI) is (*C*)/(*D*)

2.2 Eye-Shape Parameters

To analyze eye-shape, the following two parameters were measured for each species: the width/height-ratio of the eye-outline (WHR) and the ratio of exposed sclera in the eye-outline (SSI). WHR is (a)/(b) and SSI is (c)/(d).

2.3 Body-Size Parameters

Data for weight, crown-rump length, sitting-height, and habitat type of primates were collected from reference books (Itani and Uehara 1986; Napier and Napier 1985), since we did not have permission for physical contact with primates. Walking-height and sitting-height of primates were measured in the Japan Monkey Centre using marks on the wall of the cage.

2.4 Eye-Coloration Measurements

Coloration of the exposed sclera (including the conjunctiva to be precise), iris and face around the eye was recorded for 91 species by direct observation of living animals (82 species) and eyeball specimens (55 species, 124 animals) kept in the Japan Monkey Centre.

2.5 Observation of Section Samples

Eyeball specimens of the Japanese macaque and crab-eating macaque were supplied from a cooperative program of the Primate Research Institute, Kyoto University, Inuyama, Aichi, Japan. Eyeballs were fixed with 4% paraformaldehyde in 0.1 M phosphate buffer (pH 7.2) at 4 °C overnight. The tissue including the conjunctiva and cornea separated from eyeballs were washed several times in cold phosphate-buffered saline (PBS), dehydrated in an ethanolic series finishing xylene, and embedded in paraffin. Serial sections with a 4 mmm thickness were cut with disposable blades, floated onto water, and placed on slides. These sections were deparaffinized in xylene, washed in ethanol and PBS, and studied by light microscopy.

2.6 Analysis of Eyeball-Movement for Visual Field Extension

To analyze the relationship between eye-shape and visual field extension by eyeball movement, we video-recorded various primates eating food by hand in cages and analyzed the videotape. Movements of the eyeball and the head in horizontal and vertical directions, respectively, were counted (total observation time: 10037 s) for 29 individuals of 18 species: *Lemur catta, Cebus apella, Ateles paniscus, A. geoffroyi, Presbytis vetulus, Nasalis larvatus, Colobus angolensis, Cercocebus galeritus, C. torquatus, Presbytis cristata, P. entellus, Cercopithecus ascanius, Erythrocebus patas, Papio hamadryas, Hylobates lar, H. pileatus, Pan troglodytes,*

Homo sapiens (Japanese). Frequency and duration of horizontal and vertical scanning was measured (total observation time = 12579 s)s for 40 individuals of 26 species: Arboreal species; *Lemur catta, Cebus apella, C. albifrons*, Pithecia pithecia*, Ateles belzebuth*, A. geoffroyi, A. paniscus, Cercocebus galeritus, Cercopithecus cephus*, Colobus angolensis, Presbytis cristata, P. vetulus, P. francoisi*, Nasalis larvatus, Hylobates lar* and *H. pileatus,* Semi-arboreal species; *Macaca fuscata*, Cercocebus torquatus, Mandrillus sphinx*, M. leucophaeus*, Cercopithecus ascanius, Presbytiss entelllus* and *Pan troglodytes,* Terrestrial species; *Papio hamadryas, Erythrocebus patas* and *Homo sapiens* (Japanese) (*: duration only).

3 Results and Discussion

3.1 Unique Shape of the Human Eye

Figure 2 shows that among primates human eyes have the largest exposed scleral area and show extraordinary horizontal elongation of the eye-outline. Both WHR and SSI increased in the following order: Prosimii (primitive type) < Ceboidea < Cercopithecoidea < Hominoidea. Even among the Hominid species, the SSI and WHR of humans were exceptionally high (Fig. 2b). We measured both sexes of three human races: Mongoloids, Caucasians and Afro-Caribbeans, and sexual and racial difference were slight in these parameters relative to interspecies difference. This result seems to reflect an evolutionary trend of visual function and/or adaptation of each phylogenic group to their common habitat and body size. We then investigated the relationship between eye-shape parameters and these factors.

3.1.1 Relationship Between Eye-Shape and Visual Function

Primates are mammals with a well-developed visual function. Many primate species have (1) cone cells for color vision, (2) the fovea (a dense concentration of cones in the retina focused on the center of gaze) for a high resolution visual image, (3) forward-facing eyes for wide stereoscopic vision by both eyes, and (4) a well developed postorbital plate behind the eyes. It is possible that the differences of eye-shape parameters among phylogenic groups reflects a difference in visual function related to these anatomical structures. However, eye-shape parameters can not be explained by the difference in these visual functions because most primate groups except the prosimians have all these anatomical structures. For example, some nocturnal prosimian species are the only primates which lack both cone cells and a fovea in their eyes (Castenholtz 1984; Alfieri et al. 1976; Debruyn et al. 1980; Webb and Kaas 1976; Wolin and Massopust 1970). Figure 3 shows the relationship of eye-shape parameters with orbital axis angle measured from the cranium (Shigehara 1996) which reflects the ability of stereoscopic vision. In this figure, no significant correlation is observed between orbital axis angle and eye-shape parameters. These facts suggest that the variation in the eye-shape parameters of each phylogenic group does not reflect an evolutionary trend in these visual functions.

a

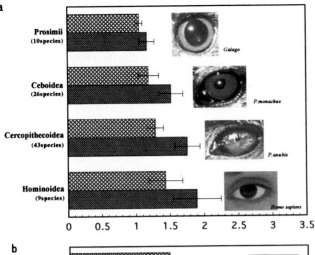

Prosimii
(10 species)

Ceboidea
(26 species)

Cercopithecoidea
(43 species)

Hominoidea
(9 species)

Galago

P. monachus

P. anubis

Homo sapiens

0   0.5   1   1.5   2   2.5   3   3.5

b

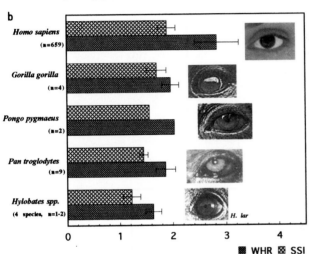

Homo sapiens
(n=659)

Gorilla gorilla
(n=4)

Pongo pygmaeus
(n=2)

Pan troglodytes
(n=9)

Hylobates spp.
(4 species, n=1-2)

H. lar

0   1   2   3   4

■ WHR ▨ SSI

Fig. 2. Variation of WHR and SSI among the phylogenic groups of primates (**a**), and in Hominoidea (**b**). In the former, the difference between phylogenic groups was significant as tested by ANOVA (SSI: $F_{(3, 90)} = 14.68$, $P < 0.01$, WHR: $F_{(3, 90)} = 32.77$, $P < 0.01$). With multiple comparisons (LSD), differences between phylogenic groups were significant, except for WHR between Cercopithecoidea and Hominoidea (SSI; MSe = 0.02, $P < 0.01$, WHR; MSe = 0.04, $P < 0.01$)

3.1.2 Relationship Between Body Size and SSI

SSI correlated well with various body size parameters (weight; r = 0.59, P < 0.001, crown-rump length; r = 0.59, P < 0.001, sitting-height; r = 0.65, P < 0.001, walking-height; r = 0.72, P < 0.001). The best correlation was observed with walking-height (Fig. 4). This means that species with larger body size had a larger exposed scleral area. A larger SSI means a smaller iris relative to the eye-outline and probably a greater ability for visual field extension by eyeball movement; in eyes with a large SSI the small iris has a wider space to move within the open eye-outline. If we suppose that a larger exposed sclera is an adaptation for extending the visual field by eyeball movement, the correlation between SSI and body size can be explained by the theory of scaling. This is because, as body size becomes larger, visual field extension by eyeball movement becomes more effective than that by head or body movement. This is so since as body height becomes greater, the weight of the head and body increases proportionally to the cube of body height. In contrast, the force

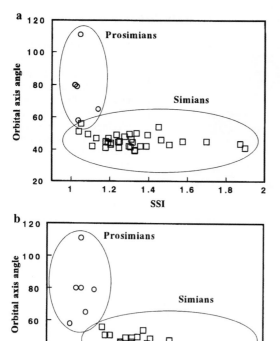

Fig. 3. Orbital axis angle and WHR / SSI. Orbital axis angle is the angle formed when the right and left orbital axes meet. Orbital axis is the line between the lowest point of the optic canal and the center of the orbital width. *Circles*, Prosimians; *squares*, simians

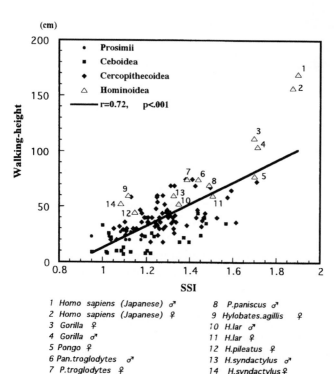

1 Homo sapiens (Japanese) ♂	*8 P.paniscus* ♂
2 Homo sapiens (Japanese) ♀	*9 Hylobates.agillis* ♀
3 Gorilla ♀	*10 H.lar* ♂
4 Gorilla ♂	*11 H.lar* ♀
5 Pongo ♀	*12 H.pileatus* ♀
6 Pan.troglodytes ♂	*13 H.syndactylus* ♂
7 P.troglodytes ♀	*14 H.syndactylus* ♀

Fig. 4. SSI and walking-height

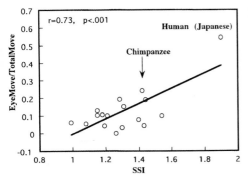

Fig. 5. SSI and eye-movement. *EyeMove/TotalMove*, Frequency of gaze-direction change only by eye-movement/frequency of all gaze-direction change

required for moving them only increases with the square of body height as it depends on the size of the muscle cross section. Moreover, since the relative growth of the eyes to body height is smaller than that of head and body size, comparative eyeball size becomes smaller in larger animals (Schultz 1940). Thus, to save energy when changing the direction of gaze, a large-sized species would move the eyeball more often than a small-sized species, and have a larger exposed scleral area. Besides, in small species with comparatively large eyeballs in a small skull, space for muscles moving the eyeballs may be seriously limited.

We then video-recorded various primates (18 species, 29 individuals) eating food by hand in cages and counted the movements of head and eyeball when they changed the direction in which they were looking. The results showed that the proportion of scanning performed only by eyeball movement was correlated with SSI (Fig. 5; $r = 0.73$, $P < 0.001$). The proportion was exceptionally high in humans ($61 \pm 28\%$ of horizontal scans, $n = 5$) compared to other primates (4.3%–24.4%, mean = 10.6%). The highest proportion in nonhuman primates was observed in chimpanzees (20%–35%, $n = 3$), the largest nonhuman species examined.

3.1.3 Relationship Between SSI and the Sclera-Ratio in the Eyeball

We also examined the correlation between SSI and the ratio of iris-diameter to eyeball-diameter by measuring 70 eyeball specimens of 26 primate species. No correlation ($r = 0.46$) was found between them. This result indicates that the relative size of exposed sclera in the eye-outline (SSI) never simply reflects the proportion of sclera in the eyeball.

3.1.4 Relationship Between Habitat Type and WHR

The mean value of WHR is greatest in terrestrial species, moderate in semi-arboreal species and lowest in arboreal species (Fig. 6). This result suggests that a horizontally elongated eye-outline is adaptive to terrestrial life in some way. We speculated that horizontal elongation of the eye-outline is adaptive in extending the visual field horizontally by eyeball movement, and terrestrial life needs more horizontal scanning than vertical scanning. We then observed various primates eating food by

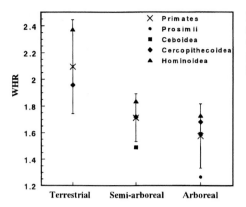

Fig. 6. WHR and habitat type. Difference among habitat types was significant as tested with ANOVA and by multiple comparison (LSD) ($F_{(2, 127)} = 24.63$, $P < 0.01$, LSD; MSe = 0.052, $P < 0.01$)

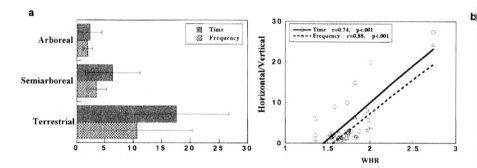

Fig. 7. a Habitat type and the ratio of horizontal scanning to vertical scanning. Differences among habitat types were significant as tested by ANOVA ($F_{(2, 23)} = 14.5$, $P < 0.01$ for time, $F_{(2, 15)} = 4.53$, $P < 0.05$ for frequency). The difference between terrestrial and arboreal species was significant using multiple comparison (LSD) (MSe = 20.76, $P < 0.01$ for time, MSe = 20.05, $P < 0.01$ for frequency). **b** Ratio of horizontal scanning to vertical scanning and WHR

hand in cages, and measured the time and frequency of horizontal scanning and vertical scanning. The result shows that the ratio of horizontal scanning to vertical scanning is higher in terrestrial species than in arboreal species, tested with ANOVA (Fig. 7a). These ratios were also correlated with WHR (Fig. 7b: time; $r = 0.74$, $P < 0.001$, frequency; $r = 0.88$, $P < 0.001$). The results support our hypothesis. These investigations suggest that the shape of the eye-outline and relative size of the exposed scleral area are the result of an adaptation for visual field extension by eyeball movement.

3.2 Unique Coloration of the Human Eye

3.2.1 The Color of the Exposed Sclera

Morris (1985) pointed out that among primates only human eyes have an exposed white sclera. However, this has not been examined in detail. It is probably because almost no one cares about the adaptive significance of external eye color. In this

nose   **temporal**   Fig. 8. Three types of scleral coloration in
nonhuman primates

a

Guinea baboon

b

Pig-tailed macaque

c

Red-handed tamarin

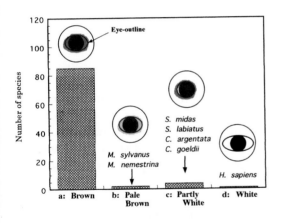

Fig. 9. Variation of scleral color. The *shaded portion* of the nonhuman primate eyeballs shows the general area where color was noted. The eye-outline is represented by the *solid line* surrounding the cornea

section, we would like to clarify whether only human eyes have an exposed white sclera, and if it is so, its adaptive significance.

The following four colorations of exposed sclera were observed (Figs. 8, 9); (a) in almost all nonhuman primates (85 species out of 92 species, or 92%) the exposed part of the sclera is uniformly brown or dark brown, (b) *Macaca sylvanus* and *M. nemestrina* with a pale brown body color had sclera colored pale brown, (c) *Saguinus midas, S. labiatus, Callithrix argentata* and *Callimico goeldii* had brown sclera with a white part in the corner of the eye, (d) humans were the only primates having white sclera without any pigmentation. Microscopic analysis of the eyeball section specimens from the Japanese macaque and crab-eating macaque revealed

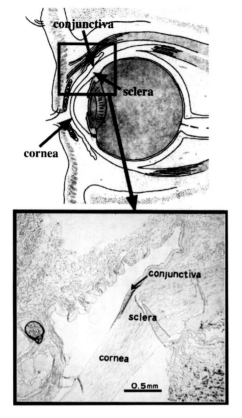

Fig. 10. Eyeball section of crab-eating macaque

that the brown coloration of the exposed sclera was due to pigmentation in the epithelium cornea, conjunctiva and sclera (Fig. 10). External observations of other nonhuman primate eyes suggested that their dark coloration is also due to similar pigmentation. Humans have transparent conjunctiva and white sclera without pigmentation. The inner part of the sclera in nonhuman primates was also white like that of humans.

3.2.2 Adaptive Meaning of Dark-Colored Sclera

Nonhuman primates have sclera colored brown. As pigmentation costs some energy, the dark coloration of the exposed sclera probably has some adaptive function. Brown coloration of the exposed sclera was observed in many other mammal species too, and the following two hypotheses have been proposed on the adaptive meanings of sclera color. (1) Anti-glare theory: it was pointed out that the pigmentation may be an anti-glare device because it seemed to be absent in nocturnal or crepuscular species (Duke-Elder 1985). However, our results on primate species were contrary to this expectation; nocturnal species (*Galago senegalensis, Tarsius syrichta, Perodicticus potto, Nycticebus coucang* and *Aotus trivirgatus*) also had colored sclera and diurnal humans had no pigmentation. Therefore, our results cannot be explained by the "anti-glare theory". (2) Gaze-camouflage theory: in many nonhuman primates, gaze direction is important in intraspecific communication. For example, direct eye contact is associated with gestures predominantly showing a tendency to attack in many monkeys. In macaques, it is reported that sclera pigmentation obscures gaze direction and may be adaptive for escaping the attacks of other individuals (Perrett and Mistlin 1990). Colored sclera obscuring gaze direction may serve to deceive natural predators too, by making it difficult for predators to know if the prey has them in their gaze. If prey animals can make it known to the predator that they already know of their presence, its chances of survival may increase (Sherman 1977). If the dark coloration of exposed sclera is adaptation for gaze camouflage (Gaze camouflage theory), the color of exposed sclera should be similar to the color of iris and/or face around eyes, to make it difficult to detect the position of iris in the eye-outline and/or the eye-outline in the face.

3.2.3 Relationship Between Sclera, Iris and Face Color

Figure 11 shows the relationship between sclera color, iris color and face color around the eye. The 82 primate species were classified by direct observation of living animals into the following 4 types by the differences in color and contrast among sclera, iris and face.

Type 1) face ≈ sclera ≈ iris (43 species): sclera color is similar to that of face and iris; both eye-outline in the face and iris position in the eye-outline are unclear.

Type 2) face < sclera ≈ iris (37 species): sclera color is darker than face color but

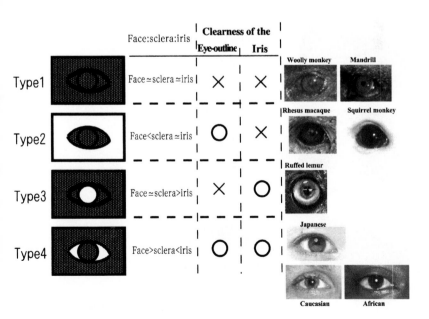

Fig. 11. Color patterns of sclera, iris, and face skin around the eye

sclera color is similar to iris color; eye-outline in the face is clear but iris position in the eye-outline is unclear.

Type 3) face ≈ sclera > iris (one species: Ruffed lemur): sclera color is darker than iris color but sclera color is similar to face color; eye-outline in the face is unclear but iris position in the eye-outline is clear.

Type 4) face > sclera < iris (one species, *Homo sapiens*): sclera color is paler than face color and iris color; both eye-outline in the face and iris position in the eye-outline are clear.

Almost all nonhuman primate species observed (80 out of 81 species) belonged to Type 1 or Type 2 coloration. In these coloration types, the position of the iris in the eye-outline was unclear because of similarity between sclera color and iris color ("Gaze camouflage type"). In Type 1 coloration (43 species), the position of the eye-outline in the face was also unclear because of similarity between sclera color and face color around the eyes. In addition, the Ruffed lemur (*Varecia variegata*), the only species that had Type 3 coloration, has a very small exposed sclera area (SSI = 1.08) and almost all the area of its eye-outline is occupied by the iris. Thus, the ruffed lemurs eye also can be seen as a "Gaze camouflage type". The results thus support the "Gaze camouflage theory".

In contrast, humans were the only species that had Type 4 coloration, in which both eye-outline in the face and iris position in the eye-outline were very clear because the color of the exposed sclera is paler than that of the facial skin and iris ("Gaze signalling type"; showing gaze direction to others). The human was the

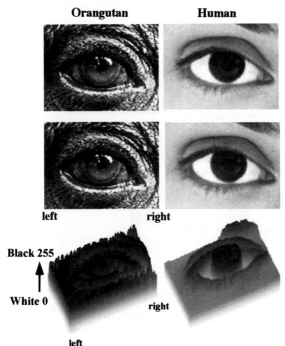

Fig. 12. Difference of gaze stimulus between human and orangutan

only species with sclera much paler than the facial skin. Because of this coloration, it is very easy to discern the gaze direction in humans, in contrast to the gaze-camouflaging eyes of the other primates. Figure 2b shows the contrast between sclera color and face/iris color of human and orangutans (*Pongo pygmaeus*). In this figure, the darkness of color was shown by the 256 gradations in grey-scale number (white = 0, black = 255, see Fig. 12). Even Great apes that have SSI and WHR near that of humans (Fig. 2b), had "Gaze camouflage eye" type with brown sclera, brown facial skin and brown iris, in which the eye position and iris position is unclear. In contrast to the gaze-camouflaging eyes of other primates, the human sclera is remarkably paler than the facial skin and the iris, and it is very easy to discern the gaze direction.

3.3 Sex Differences and Developmental Change of the Human Eye

3.3.1 Sex Differences and Developmental Change of External Eye Morphology

In previous sections, we made an interspecies comparison of the external morphology of primate eyes and presented the following hypotheses (Kobayashi and Kohshima 1997, 2001):

1. Relative size of exposed sclera in the eye outline (SSI) increases as body size increases, because it is an adaptation for visual-field extension by eyeball movement, an energy-saving way of visual-field extension for large-sized animals.
2. Width/height ratio of eye outline (WHR) increased to extend visual field in the horizontal direction.
3. Color of exposed sclera is an adaptation to camouflage or enhance gaze direction.

In this section, we examine these hypotheses on primate eye-morphology by studying sex differences and developmental change of external eye morphology in two species (*Papio anubis* and *Homo sapiens*).

Subjects

Olive baboon (*Papio anubis*): 60 animals of ages 0–23 years belonging to one group at the Japan Monkey Centre were observed in 1996. Body weight was estimated by data collected at the Japan Monkey Centre (Hamada and Tsuji 1985 unpublished data).
Human: 307 Japanese of ages 0–42 years living in Japan were observed.

3.3.2 Sex Differences and the Developmental Change of SSI

Figure 13 shows the relationship between age and body weight in the olive baboon. Females grew to a weight of about 20 kg at the age of about 8 years and males to about 35 kg at the age of about 10 years. Sexual maturation in the olive baboon occurs at the age of about 8 years in females and about 10 years in males. The figure

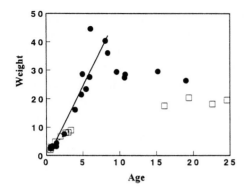

Fig. 13. Age and weight in olive baboon. *Circles*, Male, $r = 0.95$, $P < 0.001$; *squares*, female, $r = 0.98$, $P < 0.001$

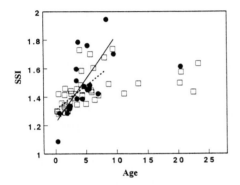

Fig. 14. Age and SSI in olive baboon. *Circles*, Male, $r = 0.71$, $P < 0.001$; *squares*, female, $r = 0.62$, $P < 0.005$

shows clear sex differences in body weight in this species: adult males are about twice as heavy as adult females. The weight of males increased more rapidly than that of females during the growth period.

SSI of the olive baboon is plotted against age in Fig.14. SSI increased during the growth period (to the age of 8 years in females and to the age of 10 years in males). SSI of humans also increased during the growth period (Fig.15); SSI of both males and females increased to an age of about 20 years, when the increase in body size almost stops in both sexes. These results suggest that SSI increases as the body size of the individual increases during the growth period, both in the olive baboon and human.

In the olive baboon (Fig.14), the SSI of adult males (1.75) was significantly higher than that of adult females (1.55) (t-test: $t_{11d.f} = 1.91$, $P < 0.01$), and the SSI of males increased more rapidly than that of females.

In contrast to the olive baboon, such a significant sex difference of SSI was not observed in humans (Fig.15). This is probably because sexual differences of body size in humans are much smaller than that for the olive baboon.

These results support the correlation between SSI and body-size suggested by our interspecies comparison and the hypothesis that relative size of exposed sclera in the eye outline (SSI) increased to extend the visual field by eyeball movement. Though we still have no quantitative data for this, infants with a small body-size

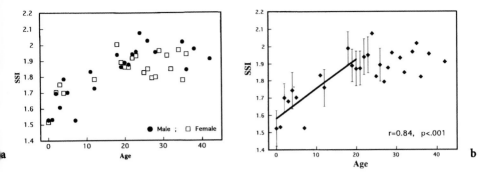

Fig. 15. Age and SSI in Japanese human. **a** *Circles,* Male; *squares,* female. **b** *Diamonds,* Average

seemed to move their head and/or body more frequently than adults to change gaze direction.

3.3.3 Sex Difference and the Developmental Change of WHR

The WHR of olive baboons is plotted against age in Fig.16. In both females and males, WHR increased up to the age of 8 years. In males of olive baboons, WHR stops increasing before their sexual and body-size maturation (at about 10 years old). There were no significant differences in WHR between adult males (2.02) and adult females (1.92).

WHR of humans is plotted against age in Fig.17. In both males and females, WHR increased up to the age of 10 years. In humans, the WHR of both sexes stopped increasing before their sexual and body-size maturation. No significant difference in WHR was observed between adult males and adult females.

These results suggest that WHR changes depending on some factor other than body size. Then, why does WHR in humans and olive baboons increase in their infancy? Though it is still unclear, we speculate as follows. Our studies by interspecies-comparison suggested that terrestrial species, such as olive baboons and humans, have horizontally elongated eye-outline because they live in a visual space where they need more horizontal scanning than vertical scanning. However, since even in terrestrial species infants spend a long period clinging to their moth-

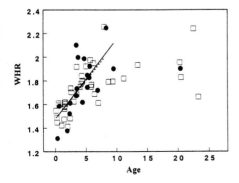

Fig. 16. Age and WHR in olive baboon. *Circles,* Male, $r = 0.67$, $P < 0.005$; *squares,* female, $r = 0.77$, $P < 0.001$

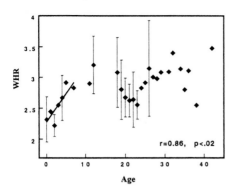

Fig. 17. Age and WHR in Japanese human

ers body, their visual space seems similar to that of arboreal species. They seem to need more vertical scanning than adults need, and have an eye-outline less horizontally elongated.

3.3.4 Developmental Change of Sclera Color

In order to clarify the sclera color of new-born babies of nonhuman primates, seven living animals (from 2 days old to 2 weeks old) of four species were directly

Table 1. Color of exposed sclera in newborn babies

Species		Age (days)	Color	Color of adult
Lemur catta	S	14	Pale brown	Brown
Lemur catta	S	4	Pale brown	Brown
Varecia variegata	S	2	Pale brown	Brown
Varecia variegata	S	3	Pale brown	Brown
Galago senegalensis	S	13	Pale brown	Brown
Callithrix jacchus	S	4	White	Brown
Saguinus oedipus	S	2	Pale brown	Brown
Callimico goeldii	S	4	White	Brown
Aotus trivirgatus	S	3	Pale brown	Brown
Macaca cyclopis	S	8	Pale brown	Brown
Macaca fascicularis	L	7	Pale brown	Brown
Macaca fuscata	L	7	Pale brown	Brown
Macaca fuscata	L	12	Pale brown	Brown
Macaca fuscata	L	10	Pale brown	Brown
Macaca fuscata	L	5	Pale brown	Brown
Mandrillus leucophaeus	S	2	White	Brown
Cercopithecus neglectus	S	8	Pale brown	Brown
Hylobates agilis	L	14	Pale brown	Brown
Pan troglodytes	L	14	Pale brown	Brown
Pan troglodytes	S	2	Brown	Brown

Data from living animals (*L*) and fixed specimens (*S*).

observed and 14 formalin fixed specimens of 12 species kept in Japan Monkey Centre were analyzed (Table 1). Most of these newborn babies had white sclera color or obviously paler sclera color than that of adults. In contrast to the sclera color, iris color of babies was almost same as that of adults. The result suggests that sclera coloration of primates gets darker during their development. Though the meaning of the paler coloration of infant sclera is still unclear, it is possible that they have paler sclera because they don't need to camouflage gaze direction as adults do.

3.4 Evolution of the Unique Morphology of the Human Eye

Why have only humans discarded scleral pigmentation? It may because the necessity for gaze camouflage decreased and that of gaze-signal enhancement increased in human evolution. The risk of predation might also decrease because of the enlarged body size and the use of tools and fire. Gaze-signal enhancement might aid the conspecific communication required for increased co-operative and mutualistic behaviours to allow group hunting and scavenging. Co-operative and mutualistic behaviours might need refined communication systems such as language, to inform one's intention to other members of the group. The human eye, with a large scleral area surrounding the iris and a great of eyeball movement ability, would have provided a chance for a drastic change from "gaze-camouflaging eyes" into "gaze-signalling eyes" through a small change in scleral coloration. The SSI and WHR of human eyes are even greater than those of gorillas, the largest primate, which suggests adaptation for gaze-signal enhancement. The uniqueness of the human eye morphology among primates illustrates the difference between human and other primates in the ability to communicate using the gaze signals.

3.4.1 Eye Morphology and Gaze-Signal Communication

Baron-Cohen (1995) postulated a neural mechanism (eye-direction detector) in the human brain specialized to detect others' eye direction, and discussed the possibility that such a mechanism might be related to the evolution of a "theory of mind." In recent years, many studies have been carried out to examine the cognitive ability to detect others' gaze-direction in various nonhuman primates. The studies have suggested that such ability is limited (Itakura and Anderson 1996; Tomasello et al. 1998). However, there seems to be some confusion in defining "gaze-direction" in these studies. Because gaze-direction can be changed by eyeball movement, by head movement, and by body movement, it should be defined by considering all those factors: eyeball direction, head direction, and body direction. Most of the studies, however, defined "gaze-direction" mainly by eyeball direction. Our results (Fig. 5) suggest that the contribution of eyeball movement to the change in gaze-direction is extremely high in humans compared with other primate species. The gaze-signaling eye coloration of humans also suggests that the contribution of eyeball direction to the gaze signal is exceptionally high in humans. Therefore, for nonhuman primates, head direction and body direction might be more important than eyeball direction in detecting others' gaze direction. We should pay more attention to head direction and body direction in future analyses of gaze signals in nonhuman primates.

Acknowledgements

We would like to thank Shigetaka Kodera for his interest in our study and for allowing us to observe the animals of the Japan Monkey Centre. It is a pleasure to acknowledge the hospitality and encouragement of the members of JMC. We wish to express our gratitude to Tetsuro Matsuzawa, Takashi Kageyama and Nobuo Shigehara for stimulating discussions under the cooperative research program in Primate Research Institute, Kyoto University, and Manabu Ogiso, Nobuyuki Saitoh and Teruhiko Hamanaka for providing eyeball samples. We are indebted to a number of our colleagues at the Tokyo Institute of Technology and the Primate Research Institute, Kyoto University, especially to Michael A. Huffman, Sou Kanazawa, Masami Yamaguchi and Kazuhide Hashiya, for their constructive criticisms of this paper.

References

Alfieri R, Pariente G, Sole P (1976) Dynamic electroretinography in monochromatic lights and fluorescence electroretinography in lemurs. Doc Ophthal Proc Ser 10:169–178

Baron-Cohen S (1995) Mindblindness: An essay on autism and theory of mind. MIT press, Cambridge

Castenholtz A (1984) The eye of *Tarsius*. In: Niemitz C (ed) Biology of tarsiers. Gustav Fischer, Stuttgart, pp 303–318

Debruyn EJ, Wise VL, Casagrande VA (1980) The size and topographic arrangement of retinal ganglion cells in the *Galago*. Vision Res 20:315–327

Duke-Elder SS (1985) The eye in evolution. In: Duke-Elder SS (ed) System of ophthalmology. Henry Kimpton, London, p 453

Gibson JJ, Pick AD (1963) Perception of another person's looking behavior. Am J Psychol 76:386–394

Gomez JC (1991) Visual behaviour as a window for reading the mind of others in primates. In: Whiten A (ed) Natural theories of mind. Blackwell, Oxford, pp 195–207

Gomi A (1994) Americans 1.0 1994 Los Angeles. Fuga Shobo, Tokyo

Itakura S, Anderson LR (1996) Learning to use experimenter-given cues during an object-choice task by a capuchin monkey. Current Psychology of Cognition 15:103–112

Itani J (1986) Vol. 3, Primates. In: Macdonald DW (ed) The encyclopaedia of animals. Heibonsha, Tokyo

Kendon A (1967) Some function of gaze-direction in social interaction. Acta Psychologica 26:22–63

Kobayashi H, Kohshima S (1997) Unique morphology of the human eye. Nature 387:767–768

Kobayashi H, Kohshima S (2001) Unique morphology of the human eye and its adaptive meaning: Comparative studies on external morphology of the primate eye. J Hum Evol (in press)

Morris D (1985) Body watching. Equinox, Oxford

Napier JR, Napier PH (1985) The natural history of the primates. MIT, Cambridge, Mass

Ohara K (1970) One. Tsukui Shokan, Tokyo

Perrett DI, Mistlin AJ (1990) Perception of facial characteristics by monkeys. In: Stebbins WC, Berkley MA (eds) Comparative perception—complex signals. John Wiley & Sons, New York, pp 187–215

Schultz AH (1940) The size of the orbit and of the eye in primates. AJPA Old Series 26:389–408

Sherman PW (1977) Nepotism and the evolution of alarm call. Science 197:1246–1253

Shigehara N (1996) Metrical study of the direction of the orbits in primates. Primates Res 12:165–178

Tomasello M, Call J, Hare B (1998) Five primate species follow the visual gaze of conspecifics. Anim Behav 55:1063–1069

Webb SV, Kaas JH (1976) The size and distribution of ganglion cells in the retina of the owl monkey, *Aotus trivirgatus*. Vision Res 16:1247–1254

Wolin LR, Massopust LC Jr (1970) Morphology of the primate retina. In: Noback CR, Montagna W (eds) The primate brain. Appleton Century Crofts, New York, pp 1–27

Yoshino S (1994) Animal face. Nikkei Saiensu, Tokyo

Part 6
Society and Social Interaction

20
A Review of 50 Years of Research on the Japanese Monkeys of Koshima: Status and Dominance

Kunio Watanabe

1 Introduction

Since the first scientific observations were made of wild Japanese macaques (*Macaca fuscata fuscata*) on the island of Koshima (32 ha), half a century has passed. The Koshima group was first provisioned in 1952, at which point all of 22 individuals were identified. Since then, observations on all of the individually identified monkeys have been carried out. Since the average life span of a Japanese macaque is around 20 years and as most females bear their first infant between the ages of 5 and 10, several generations of monkeys have been observed on the island, all descendants of the original members. By 1999 the population on the island had risen to 450 individuals (Fig. 1), a little over half of which (57%) had reached puberty and were over 5 years of age. The Koshima group of Japanese macaques is unique in that nearly 40% of the offspring die within their first year. Such a high first year mortality may be the result of poor nutritional conditions prevailing on the island, in turn resulting from there being a large population confined to such a small island (Mori 1979; Watanabe et al. 1992).

When the early studies revealed that the monkeys on Koshima lived in a well-established and organized society comprised of unique individuals with recognizable characteristics, it came as a surprise to the scientific community, since at the time it was impossible to imagine such a highly-organized society in any species other than human beings. The discovery of an organized society in another primate led scientists to expect that by close examination of these primate societies, it would be possible to deduce not only the origin of human society and organization, and the basic organizing rules and principles involved, but also the line of evolution towards human beings (e.g., Imanishi 1957). Kawai (1964) was the first to point out three important principles intrinsic to Japanese macaque society; i.e., leadership, dominance hierarchy, and matrilineal relationships.

The initial aim of the long-term study of the Koshima monkeys was, therefore, to clarify the social principles or rules governing their society by following its history on a continuous basis (Imanishi 1957). The resulting observations on social interaction in the Koshima group have generated an enormous amount of data. The

Field Research Center, Primate Research Institute, Kyoto University, 41 Kanrin, Inuyama, Aichi 484-8506, Japan

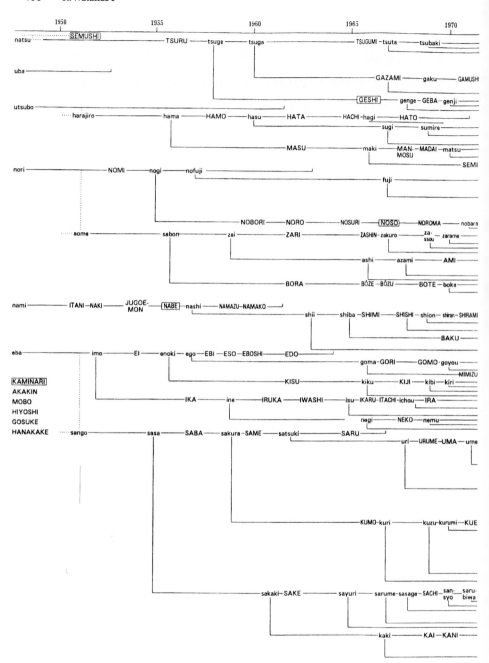

Fig. 1. The genealogy of Koshima monkeys. Males are indicated by *capital letters* and females by *lowercase letters*. Individual names are arranged by birth year and *horizontal lines* indicate female' life span. The *arrow* attached to the *horizontal lines* indicate females who left the island when it was connected to the main land by a sandbar. Names enclosed in *open squares* are those of six adult males who reached alpha male status in the main group. Only individuals living for more than five years are shown

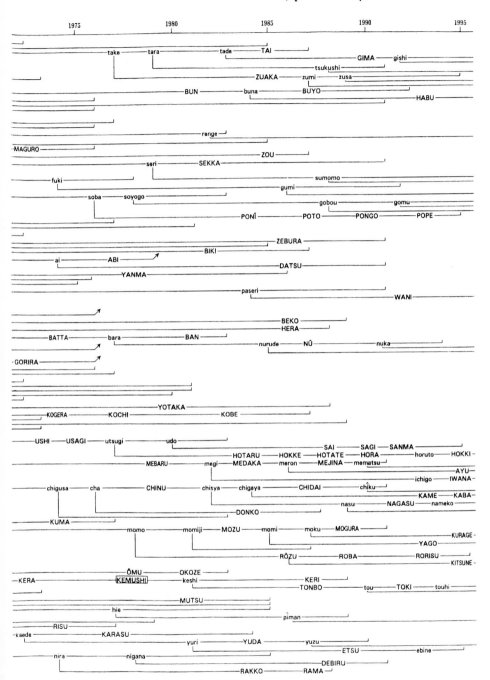

Fig. 1. *Continued*

most famous finding during the early period of the study was the propagation process of "pre-cultural behaviors" exemplified by the potato- or wheat-washing behavior (e.g., Kawai 1965; Watanabe 1994). In addition, the relations between population parameters and nutritional conditions (e.g., Mori 1979; Watanabe et al. 1992), seasonal and situational differences of daily food intake in relation to the social status of monkeys (e.g., Iwamoto 1974, 1982), and many other studies have been conducted. Although the number of reports concerning the Koshima monkeys now exceeds 400 (Board of Education, Kushima City 1997), the initial aim of the study (seeking to delineate social principles by analyzing social history) has not been fully realized. There are serious methodological difficulties in that, for such an analysis, so many variables should be recorded systematically during much longer periods than are available for any ordinary scientific study.

Nevertheless, as a result of the enormous accumulation of data, certain specific aspects of macaque society have been revealed. Here, I describe the long history of Koshima monkeys and discuss the social implications of the principles proposed by Kawai (1964), especially on leadership and dominance hierarchy.

2 Alpha Male Status in the Koshima Group

"Leadership" described by Kawai (1964) refers to the social role that high-ranked male(s) have in controlling conflicts between group members, in defending group members from predators and/or outside enemies and in deciding the time and direction of movements. Two to four of the adult males in a group were referred to as "leader males" on the basis that they always were in the central part of the group and took such a social role. The concept was accepted widely, especially during the first 20 years of the study until the 1970s, and was used to explain the different social structures resulting from the individual differences or "personalities" of those alpha males (e.g., Yamada 1966).

It has been said that the behavior of the high-ranking males changed gradually as the years passed and artificial provisioning was continued. Koshima is not an exception. Mito, who has continued her observations on the Koshima monkeys since her first visit with researchers in 1949, has also testified that those high-ranking males gradually reduce their interference in intra-group conflicts, and sometimes even remain at the feeding ground alone apart from the group members (Mito, personal communication). For such reasons, many researchers have complained and stressed that those males with the special social roles, described by Kawai (1964), are not recognized within groups of Japanese macaques (e.g., Izawa 1982). Their arguments imply that the existence of those higher-ranking males or "leader males" may have been rooted in the anthropomorphic sympathy projected by former researchers rather than based on reality.

The issue that requires clarification is whether there is a clear-cut difference in the social relationships between adult males and certain high-ranking males engaged in the previously defined specific social roles or not. The term "leader" undoubtedly involves an anthropomorphic element; however, even putting this aside, the concept of such specialized social roles within a Japanese macaque group is still in dispute among Japanese researchers.

In this chapter, I use the term "alpha male" to refer to the highest-ranking male, "main male" to refer to second, third or fourth ranking adult males, and "ordinary male" to refer to the other males. These distinctions are practical since such differences in position and social interaction within the group are most easily recognized (Itani 1954).

The concept and reality of the alpha male are so stable, that a change in alpha male has only been recorded five times during the 47 years from 1952 to 1999. The current alpha male is the sixth in line (see Fig. 2). It is obvious that high-ranking males retain their position for fairly long periods, usually several years, especially if they reach alpha male status. Kaminari, a large adult male, who was present when all individuals were first identified in 1952, was the first alpha male. He had, presumably, already been in that position for some years before. He retained his position for the following 18 years. The second alpha male, Semushi, kept his rank for six years, the third, Nabe, for four years, the fourth, Geshi, for 12 years, and the fifth, Noso, for 6 years.

Once an individual had reached alpha male status, that rank was retained until the time of death. During the reign of all these alpha males, other large and well-built adult males were also in the group and sometimes threatened their alpha rank. Despite these challenges, there exists only one example of an alpha male being ousted by an adult male, and this event occurred just one month before the deposed alpha male, Semushi, died of old age. At that time Semushi was already 25 years old and was infirm (Mito 1980).

It appears usual in Koshima that the alpha male rank is occupied by the oldest male in the group. Given that Kaminari was a full adult male and already in the alpha position when the Koshima group was first provisioned in 1952, it is possible that he was more than 30 years old when he died. Semushi achieved alpha status when he was 19 years old and Nabe when he was 21. Geshi became alpha male at 14 years of age and retained that rank until he was 27. Noso attained alpha status at 25 and kept the position until he was 31. The current, sixth, alpha male, Kemushi, was 20 when he took over the position. It is clear that the alpha male position has been occupied by old adult males who had already passed their prime and were, therefore, no longer really involved in fathering offspring. Interestingly, the results of DNA analysis of the members of the Koshima group clearly indicate that neither the alpha nor the main males actually inseminate most of the females of the group (Inoue personal communication).

It has been argued that adult males gather females as a resource, or that they behave ultimately in order to increase their own reproductive potentials (e.g., Wrangam 1979). The case of the alpha males of Koshima does not, however, support this argument. Thus, what is the advantage of being such a male? What is clear is that the alpha male lived longer than other males (see Fig. 3), in other words, retention of their rank as alpha males was linked to longer life spans rather than with greater production of offspring.

It should be noted that in 1990, DNA was collected from all individuals living on

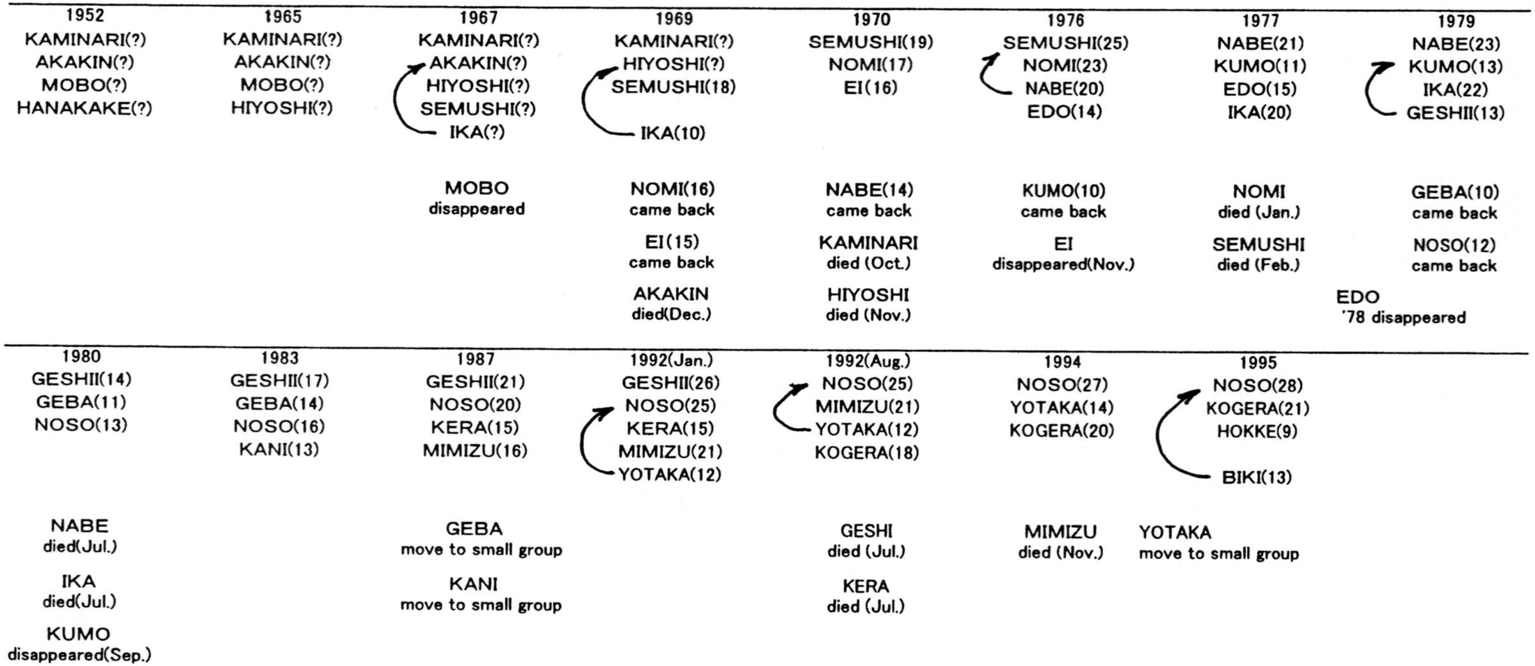

1952	1965	1967	1969	1970	1976	1977	1979
KAMINARI(?)	KAMINARI(?)	KAMINARI(?)	KAMINARI(?)	SEMUSHI(19)	SEMUSHI(25)	NABE(21)	NABE(23)
AKAKIN(?)	AKAKIN(?)	AKAKIN(?)	HIYOSHI(?)	NOMI(17)	NOMI(23)	KUMO(11)	KUMO(13)
MOBO(?)	MOBO(?)	HIYOSHI(?)	SEMUSHI(18)	EI(16)	NABE(20)	EDO(15)	IKA(22)
HANAKAKE(?)	HIYOSHI(?)	SEMUSHI(?)			EDO(14)	IKA(20)	GESHII(13)
		IKA(?)	IKA(10)				
		MOBO disappeared	NOMI(16) came back	NABE(14) come back	KUMO(10) come back	NOMI died (Jan.)	GEBA(10) come back
			EI(15) come back	KAMINARI died (Oct.)	EI disappeared(Nov.)	SEMUSHI died (Feb.)	NOSO(12) come back
			AKAKIN died(Dec.)	HIYOSHI died (Nov.)			
							EDO '78 disappeared

1980	1983	1987	1992(Jan.)	1992(Aug.)	1994	1995
GESHII(14)	GESHII(17)	GESHII(21)	GESHII(26)	NOSO(25)	NOSO(27)	NOSO(28)
GEBA(11)	GEBA(14)	NOSO(20)	NOSO(25)	MIMIZU(21)	YOTAKA(14)	KOGERA(21)
NOSO(13)	NOSO(16)	KERA(15)	KERA(15)	YOTAKA(12)	KOGERA(20)	HOKKE(9)
	KANI(13)	MIMIZU(16)	MIMIZU(21)	KOGERA(18)		
			YOTAKA(12)			BIKI(13)
NABE died(Jul.)		GEBA move to small group		GESHI died (Jul.)	MIMIZU died (Nov.)	MIMIZU YOTAKA move to small group
IKA died(Jul.)		KANI move to small group		KERA died (Jul.)		
KUMO disappeared(Sep.)						

Fig. 2. Changes in alpha and main males in the Koshima group. Alpha males are indicated at the *top* and arranged by descending rank order. The age of each male is shown in the parentheses. "Main males" usually include second- and third-ranking males, but also occasionally a fourth-ranked male. Classification was possible on the basis of their position in the group and on certain behavioral characteristics. In some cases, however, it is difficult to discriminate since social position changed regularly. The *arrow* indicates the frequent challenge of relevant males observed

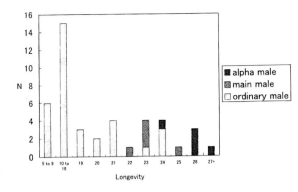

Fig. 3. The longevity of adult males that presumably died on the island. Males that died when younger than 5 years of age are excluded

Koshima and the possible fathers of each offspring born in the preceeding years were examined. At that time, the alpha male, Geshi, was 24 years old and other main males, Geba and Noso, were 21 and 23 years of age respectively.

3 Who Became the Alpha Male?

Why is it that young powerful males do not attain high ranks in their group, and why do they not try to take over the alpha male role? It has been reported that certain younger and well-built ambitious males have challenged old alpha males and/or main males (e.g., Mito 1971, 1980; Kawai 1973). However, all these challenges have so far failed. In all cases, second ranked males attained alpha male status after the death of the former alpha male, and after having usually been main males for several years. Inevitably then, alpha males will be older individuals. The only exception to this pattern was Geshi who achieved alpha status at 14 by chasing the second male, Kumo, off the island just before Nabe, the former alpha male, died. Even then, Geshi did not outrank Nabe and retained his second position until Nabe died.

Japanese macaques more than 20 years old usually have worn teeth and look infirm. As the older males age, younger adult males no longer keep a distance from them at the feeding ground. Even though the younger males are still subordinate to them, they may be aware of the senility of the older monkeys. Only rarely have such younger males fought with the old alpha males and supplanted them during feeding sessions.

It is still not known why younger powerful males do not take over the alpha status from the old, infirm alpha males. However, two possible explanations exist. First, monkeys of the group may prevent this cooperatively, and second, some restraint or deference on behalf of the younger monkeys may prevent them from doing so.

It is interesting that the change from one alpha male to the next often coincided with changes among the main males who have kept their positions for several years. Within the year preceding Kaminari's death, for example, both second-

Table 1. Alliance behavior among the males of the Koshima group

	GESHI	GEBA	NOSO	KANI	GAZAMI	Others
Winner-support	17	11	2	5	3	7
Loser-support	23	17	4	4	2	4

Data collected during the summers of 1983-84 indicated that higher-ranking males supported more losers while other males supported more winners (see Figs. 1, 2 for the age and status of the males). Furthermore, Geshi and Geba rarely supported adult males who attacked or were attacked by other group members (only once did Geba support Geshi who attacked a young male), yet 13 times they supported group members who attacked adult males (winner-support) and 16 times those who were attacked by adult males (loser support). The trend becomes more obscure as the rank of those adult males declined.

(Akakin) and third-ranking (Hiyoshi) males died. This situation was by no means unique. Within three months of Nabe succeeding Semushi as alpha male, second ranking Nomi died and third ranking Ei disappeared. Both Nomi and Ei had lived as main males throughout Semushi's reign (Fig. 2). When Geshi took over the alpha rank from Nabe, the second male, Kumo, left the island, and the third male, Ika, died almost at the same time as Nabe. Then when Noso took over the alpha role from Geshi, the old third male Noso became very weak. However, Kera, the new second male, died suddenly, perhaps as a result of illness. Noso then exerted himself to maintain his status with Mimizu who became the second male after Kera against Yotaka who just achieved a higher rank and actively attacked many other group members during the period.

Changes among the higher ranking males occurred over very short periods of time, whereas the intervening stable periods were very much longer (see Fig. 2). During these prolonged periods, rank order remained quite stable, particularly that of the alpha and second males.

When outlining the history of challenges to and acquisitions of the alpha or main male status, it seems that those males able to establish special social relationships with the alpha and main males retained stable positions for fairly long periods. It is thought that the social role of such higher-ranking males may be somewhat different from other males (Kawai 1964). When higher-ranking males intervene in conflicts within their group, the direction of attacks can be different from that of other males. Watanabe (1979) showed that when an intra-group conflict occurs, alpha males support losers rather than winners, and they also support adult females or younger monkeys. In contrast, ordinary males usually support winners. In 1984, the same trend was confirmed for alpha male Geshi, the main males, and especially the second ranked male Geba (Table 1). Old alpha males, and main males, sometimes cooperated when confronting dogs or human beings that disturbed the group (e.g., Toei Film Company 1969). Such close ties between adult males may be particularly important in the effective organization of larger groups, which often exceed more than one hundred monkeys.

4 Dominant Families and Why Their Sons Do Not Achieve Alpha Male Status

When the Koshima group was first provisioned in 1952, six females were present and all their descendants can be traced within their family group. It is apparent that not all families produced descendants equally (Fig. 1). The dominant Eba family has produced many descendants, whereas the other families have not been so successful. At present, about sixty percent of Koshima monkeys are descendants from the Eba matriline. The offspring of the most dominant females were able to obtain more food (Iwamoto 1974) and therefore grew faster and larger than those of subordinate mothers (Mori 1979; Watanabe unpublished data).

Although the Eba matriline has produced many offspring, it did not produce an alpha male until Kemushi achieved that rank in 1999, a first for an Eba family male. At present, many of the monkeys in the group belong to the Eba family, which has become differentiated into several sub-families. Kemushi belongs to a small subordinate subfamily within the Eba family, since his mother died many years ago (Fig. 1).

It seems that the sons of higher-ranking females are not normally able to achieve alpha male status even if they are large-bodied. Instead, however, individuals such as Ei, Kumo, and Yotaka during Semushi's, Nabe's, and Geshi's reign, respectively, became second-ranked males, although they left the group without ever reaching alpha male status. Even in a small group which budded out of the main group during the mid 1970s (Kudo 1984) such males have not reached alpha status. It seems, therefore, that such adult males either lack some necessary quality or something restrains them from achieving alpha rank.

Research has shown that adult male Japanese macaques rarely copulate with their mother and sisters (e.g., Itani 1972). Perhaps, therefore, when a male's mother and/or sisters live in close relationship to the alpha male, or main males, in the central part of the group, there may be some behavioral inhibition. Moreover, staff who have observed the Koshima monkeys closely every day for 32 years consider that males born to higher ranked females are unpretentious and relatively quiet (Yamaguchi and Kanchi personal communication), perhaps making it difficult for them to exert leadership qualities.

As previously mentioned, among the Koshima monkeys, even when males were clearly stronger and more powerful than the alpha male, they rarely took over the alpha male's position. Koshima is a small island with little population inter-change. For example, in the 18 years since 1982 only three adult males have arrived from outside Koshima, and even so they were presumed to have originated from the island and to have returned after several years of absence on the mainland of Kyushu (Watanabe unpublished data).

The monkeys on the island have all lived together since birth and thus know each other very well. Among humans, sons may respect their father throughout their lives. Could a similar psychological link function among the males on Koshima, particularly when the alpha male is old?

Year 1986	1987 Oct	1988 Jan	1988 Feb	1988 Apr	1989 Jul	1990 Mar	1991 Sep	1993 Mar	1995 May	1996 Aug	1997 Mar	1997 Jul
NOSURI	GEBA	GEBA	GEBA	GEBA	GEBA	GEBA	GEBA	RISU	YOTAKA	YOTAKA	YOTAKA	YOTAKA
KOGERA	KARASU	RISU	RISU	KOTI	RISU	RISU	MEDAKA	MEBARU	BAN	TONBO	TONBO	TONBO
KOTI	KOTI	KEMUSHI	KOTI	BAN	KEMUSHI	MEBARU	YUDA	BAN	YUDA	RAKKO	RAKKO	RAKKO
KARASU	YANMA	KOTI	KEMUSHI	BUN	MEBARU	KEMUSHI	KOBE	YUDA	BONII	DEBIRU	ZEBURA	BUYO
	BAN	BAN	KARASU	RISU				MEDAKA			BUYO	DEBIRU
	BUN	BUN									DEBIRU	KERI
											KERI	

Fig. 4. The change of dominance rank order within a subsidiary break-away group that was formed after the mid 1970s

5 Changes Within the Dominance Hierarchy

It is obvious that whereas the alpha male's status remains stable for a prolonged period of time, as one descends the rank order, the positions change more often (Fig. 2). Ordinary males change rank so frequently, that it is difficult to record all the changes over long periods (but see Fig. 4). These changes occur less frequently among dominant males and more frequently among the subordinate males, which often leave and then re-join the group. Once they have left the group they may be chased away by the remaining group members. If they return, they must start again from the bottom of the dominance hierarchy.

Females rarely leave their natal group, and while they are still subordinate they experience many changes in their rank (Fig. 5). Dominant females are able to obtain food even when subordinate group members are present, whereas for subordinate females, food access is more difficult. Thus, their rank order in the dominance hierarchy within the group influences their daily lives directly by affecting their access to food. In the Koshima group, the highest-ranking individuals were able to take more food at the feeding ground (Iwamoto 1974) and the top five females produced more young than the other females (Watanabe et al. 1992). For this reason, the dominant monkeys strongly resist any challenge, while low ranking monkeys do not since for them minor differences in the hierarchy have a negligible impact on their daily lives. Furthermore, if several dominant monkeys resist any change in their status then their rank order cannot change. The important questions raised by these observations are: why, how and how long can the rank order remain stable among group members, and under what kind of situations do changes occur.

Acknowledgements

I would like to give my sincere thanks to Satsue Mito, who has continued her observations and also assisted scientific studies of the Koshima monkeys since 1948. I also extend my thanks to Naotsugu Yamaguchi and Fujio Kanchi of the Koshima Field Station, Primate Research Institute, Kyoto University and to the many other researchers who have visited and who kept valuable notes on monkey behavior. Prof. Tetsuro Matsuzawa of the same institute gave me an opportunity to compile long-term data on the Koshima monkeys and to write this chapter. Dr. Mark Brazil of Rakuno Gakuen University helped me to correct the English.

1957	1962	1969	1970	1973	1974	1975	1977	1978	1980	1983	1984	1985	1986

Fig. 5. The change of dominance rank order within the females of Koshima group (adapted from Mori et al. 1989). Successive changes are indicated by the *lines* that follow the position of each individual

References

Board of Education, Kushima City, Japan (1997) Management plan of the national monument, "Habitat for Koshima monkeys" (in Japanese). Kushima City (Japan) Board of Education, p 115

Imanishi K (1957) Identification: a process of socialization in the subhuman society of *Macaca fuscata* (in Japanese). Primates 1:1–29

Itani J (1954) Takasakiyama no saru (Japanese monkeys in Takasakiyama) (in Japanese). Kobunsha, Tokyo

Itani J (1972) A preliminary essay on the relationship between social organization and incest aviodance in nonhuman primates. In: Poirier FE (ed) "Primate socialization." Random House, New York, pp165–171

Iwamoto T (1974) A bioeconomical study on a provisioned troop of Japanese monkeys (*Macaca fuscata fuscata*) at Koshima Islet, Miyazaki. Primates 15:241–262

Iwamoto T (1982) Food and nutritional condition of free ranging Japanese monkeys on Koshima Islet during winter. Primates 23:153–170

Izawa K (1982) Nihonzaru no seitai: gohsetsu no hakusan ni yasei wo tou (Ecology of Japanese macaques: searching for wild natures on snowy Mt. Hakusan) (in Japanese). Dobutsusha, Tokyo

Kawai M (1964) Nihonzaru no seitai (Ecology of Japanese macaques) (in Japanese). Kawade Shobo, Tokyo

Kawai M (1965) Newly acquired precultural behavior of the natural troop of Japanese monkeys on Koshima Island. Primates 6:1–30

Kawai M (1973) A life of Hiyoshimaru, fourth-ranked male, in the Koshima group of Japanese macaques (in Japanese). Anima 1(1):50–55, (2):21–25

Kudo H (1984) Dynamics of social relationships in the process of troop formation of Japanese monkeys in Koshima—"working upon" and "impressing behavior" (in Japanese). J Anthrop Soc Nippon 92:253–272

Mito S (1971) Koshima no saru: 25nen no kiroku (Monkeys on Koshima island:25 years of history) (in Japanese). Popura-sha, Tokyo

Mito S (1980) Bosuzaru heno michi (The way of becoming an alpha male) (in Japanese). Popura-sha, Tokyo

Mori A (1979) Analysis of population changes by measurement of body weight in the Koshima troop of Japanese monkeys. Primates 20:371–397

Mori A, Watanabe K, Yamaguchi N (1989) Longitudinal changes of dominance rank among the females of the Koshima group of Japanese monkeys. Primates 30:147–173

Toei Film Co (1969) The society and life of Japanese monkeys (16mm film, color, Japanese soundtrack, 30 min) Toei, Tokyo

Watanabe K (1979) Alliance formation in a free-ranging troop of Japanese macaques. Primates 20:459–474

Watanabe K (1994) Precultural behavior of Japanese macaques: longitudinal studies of the Koshima troops. In: Gardner RA, Gardner BT, Chiarelli B, Plooij FX (eds) The ethological roots of culture. Kluwer Academic, Dordrecht, pp 81–94

Watanabe K, Mori A, Kawai M (1992) Characteristic features of the reproduction of Koshima monkeys, *Macaca fuscata fuscata*: a summary of thirty-four years of observation. Primates 33:1–32

Wrangam RW (1979) On the evolution of ape social systems. Soc Sci Information 18:334–368

Yamada M (1966) Five natural troops of Japanese monkeys in Shodoshima Island: distribution and social organization. Primates 7:315–362

21
Mother–Offspring Relationship in Macaques

Masayuki Nakamichi

1 Introduction

In the 1950s, research on free-ranging Japanese macaques (*Macaca fuscata*) started in several places, and included the provisioning and identification of monkeys. Researchers succeeded in describing the social structure and dominance rank system (Kawamura 1958; Kawai 1958), and the transmission of precultural behavior (Itani 1958; Kawamura 1959; Kawai 1965). However, systematic behavioral studies on mother–infant relationships or infant development were not conducted in free-ranging situations until the first half of the 1960s.

Itoigawa (1973), who began studying a free-ranging, provisioned group of Japanese macaques living in the north of Katsuyama, Okayama Prefecture, in 1958, tried to describe the social organization of the group. From 1963, he also started to collect quantitative data on mother–infant relationships in this group by using the focal animal sampling method, and reported sex differences in mother–infant relationships during the first 8 months of infant life (Itoigawa 1973). Struhsaker (1971) observed wild vervet monkey (*Cercopithecus aethiops*) mothers and their infants during the 12-month period from 1963 to 1964. Thus, these two studies may be the first quantitative ones on mother–infant relationships in nonhuman primates conducted in the wild. Itoigawa's research (1973) also included quantitative data on mother–infant relationships in captive Japanese macaques. Importantly, Itoigawa (1973) stressed that behavioral studies in the wild and in captivity should be complementary to each other.

Since these studies, several Japanese primatologists have examined mother–offspring relationships in Japanese and other macaques. This chapter is a review of several aspects of the relationship between mothers and their offspring, focusing on research conducted by Japanese primatologists. I first briefly review the literature for differences in mother–infant relationships in macaques, and discuss the mother's role in infant independence. Next, I describe the continuity in mother–immature offspring relationship based on long-term studies. The flexibility of mother–infant relationships are also pointed out, based on some case studies. Since

Laboratory of Ethological Studies, Faculty of Human Sciences, Osaka University, 1-2 Yamada-oka, Suita, Osaka 565-0871, Japan

females stay among their natal groups throughout their lives, they tend to interact with their mothers after maturation. A few adult males also interact with their mothers, albeit infrequently. Therefore, the relationship between mothers and their adult offspring is also discussed. Lastly, laboratory studies on mother–infant relationships conducted in a big breeding colony of macaques are reviewed.

2 Factors Influencing Mother–Infant Relationships

2.1 Traditional Factors

While an infant's social environment is mainly limited to its mother during the early stages of development, infants rapidly become independent from their mothers and extend their social interactions with other group members. Japanese macaque infants become able to distinguish their mothers from other adult females by 8–12 weeks of age (Nakamichi and Yoshida 1986), and then they become primarily responsible for the maintenance of proximity to their mothers and increasingly expand their social environment. It is well known that some factors tend to influence not only the mother–infant relationship but also the social relationships of infants with group members other than their mothers, such as their peers. Mother's parity, dominance, and infant's sex are "traditional" factors (Berman 1984) for which a great number of studies have been conducted, both in the wild and in captivity.

Primiparous macaque mothers generally tend to be more protective than multiparous mothers (Kuyk et al. 1977; Berman 1984), although White and Hinde (1975) found no clear differences in mother–infant relationships between primiparous and multiparous rhesus (*Macaca mulatta*) mothers living in captive social groups. Hiraiwa (1981) reported that the method used by primiparous Japanese macaque mothers to handle their infants in a free-ranging group seemed awkward. Suzuki et al. (1984) observed eight infants in a free-ranging, provisioned group of Japanese macaques at Arashiyama once a week throughout the daytime during the first 6 months of infant life. They reported that primiparous mothers spent more time in contact with their infants and retrieved them more frequently than multiparous mothers, but the former also rejected the infants more frequently than the latter. The fact that primiparous mothers were more protective while they acted negatively and aggressively toward their infants might indicate instability of maternal behavior in primiparous mothers. Negayama et al. (1986) analyzed parturitional behavior in wild-born, caged Japanese macaques and found that primiparous mothers looked at their neonates more frequently but spent less time holding them than multiparous mothers for the first hour after parturition. This observation suggests that primiparous mothers tend to show unstable maternal behavior directly after giving birth.

Itoigawa (1973) first reported quantitative differences between male and female infants in the Katsuyama group of Japanese macaques: male infants spent less time with their mothers, and received grooming, restraining, and retrieving from their mothers less frequently than female infants. He also confirmed similar differences in captive Japanese macaques (Itoigawa 1973). Japanese macaque mothers in a large enclosure show differential treatment of their infants based on sex (Eaton et al. 1985), while infant sex hardly influenced associations with social partners other than their mothers during the first year of life in the same group (Glick et al. 1986). Nakamichi et al. (1990) documented clear sex differences in mother–infant interactions in wild-born, individually caged cynomolgus macaques (*M. fascicularis*) during the first 14 weeks of infant life. Regardless of the infant's sex, these mothers did not restrain their infants as frequently as mothers living in social groups, and male infants, who were heavier than female infants, moved about more actively than female infants. It is possible that differences in the activity level of infants might produce differences in mother–infant relationships between male and female infants as early as during the first 14 weeks of life.

Some studies of captive macaque groups have reported that low-ranking mothers tend to spend more time in contact with their infants, and restrain their infant more frequently, than high-ranking mothers (White and Hinde 1975; Tartabini et al. 1980). On the other hand, Tanaka (1989) did not find rank-related differences in mother–infant relationships in a free-ranging group of Japanese macaques. Thus, the mother's dominance might not have as great an influence on mother–infant relationships as previously thought.

With respect to the direction of individual differences in mother–infant relationships, there were no apparent inconsistencies among the studies reviewed above. In other words, we could not find any cases in which some studies demonstrated a clear direction for one traditional factor while others showed the opposite direction for the same factor. Therefore, it is clear that the three traditional factors actually control the variations in mother–infant relationships. However, one should not think that these three factors control mother–infant relationships to the same degree. Rather, it is probable that the strong influence of one factor tends to mask or weaken the other factors. According to Tanaka (1989), who observed 28 Japanese macaque mother–infant dyads in a free-ranging group, among the three traditional factors, maternal parity was the strongest factor affecting the variability in mother–infant relationships throughout the first 8 months of life.

2.2 Nontraditional Factors

As Berman (1984) pointed out, certain factors other than the three traditional factors described above should also be analyzed. She found that in a free-ranging group of rhesus macaques, each traditional factor was associated only weakly with variations in mother–infant relationships, and that maternal age and total experience strongly correlated with the variations. This correlation was not due to age or experience per se, but to the implications of age and experience. Infants of older and more experienced mothers tended to have more immature siblings within the group, and the presence of siblings controlled the variations in mother–infant relationships (Berman 1984, 1992). Hooly and Simpson (1983) reported that the presence of a sibling 1 year older encouraged an infant to be independent of its mother. Silk (1991) also pointed out that family composition, such as the number of siblings, was an important predictor of variation in proximity among captive bonnet macaque (*Macaca radiata*) mothers and their infants. Moreover, the birth of an infant influences the relationships between its older siblings and their mother: brothers, but not sisters, who were 1 year older tended to become independent of their mothers after a new sibling was born (Holman and Goy 1988). Matriline size as well as the presence of a sibling could be an important factor affecting the quality of mother–infant relationships (Schino et al. 1995).

The rate of infant mortality among orphaned primiparous Japanese macaque mothers living in a free-ranging group is higher than that of ordinary mothers, and some orphaned primiparous mothers showed incompetence in maternal care (Hasegawa and Hiraiwa 1980). Moreover, regardless of parity, orphaned mothers displayed aggressive behavior toward their infants more frequently than nonorphaned ones (Hiraiwa 1981). These findings indicate that it may be more important than has usually been thought for juvenile females to maintain interactions with their mothers after weaning to acquire the ordinal maternal ability.

Some rhesus and pig-tailed macaque (*Macaca nemestrina*) mothers living in captive social groups show violent behavior toward their infants as early as the first week of infant life. This behavior includes crushing, dragging, stepping on, and throwing. Such abusive mothers alternate short bouts of abuse with long periods of appropriate care-giving behavior (Maestripieri et al. 1997). Abusive mothers maltreated successive infants (Maestripieri et al. 1999), and they were concentrated in some kin-groups (Maestripieri et al. 1997). To our knowledge, there are no systematic studies of infant abuse for Japanese macaques to date, although some demographic data from free-ranging groups indicate the presence of some females who lost their infants repeatedly. For example, one Japanese macaque female of a free-ranging group lost all her seven infants within a few weeks after birth. She did not show any of the abusive behavior described above, but she sometimes rejected her infant's nipple contact or pulled the infant clinging to her ventrum. However, she carried her dead infants for several days after death (H. Nobuhara 1995, personal communication). As the rate of abusive mothers appears to be higher than was usually believed (Maestripieri et al. 1997), more attention should be paid to child abuse among macaques. Behavioral information on abusive mothers could be useful for understanding some aspects of the nature of macaque mothers which might be concealed in ordinary mothers.

Itoigawa (1973) compared quantitative data on Japanese macaque mother–infant relationships in the wild with those on individually caged mother–infant dyads and found no clear differences between these groups. Berman (1980) also pointed out that general aspects of rhesus macaque mother–infant relationships in the wild were similar to those in captive social groups. Johnson and Southwick (1984) compared the development of rhesus macaque mother–infant relationships across three habitats in India and Nepal but found no clear differences, although the physical environments of these groups were very different. In contrast to these studies on macaques, Hauser and Fairbanks (1988) found that, based on the rate at which vervet monkey mothers rejected their infants, the level of mother–

infant conflict was responsive to differences in the environment. Field studies on Japanese macaques have been conducted in a variety of habitats for almost 50 years. Some of these studies have provided useful data that allow comparison of various aspects among groups, such as their ecology, social organization, and social behavior (Koyama et al. 1981; Yamagiwa and Hill 1998), but not mother–infant relationships. Several Japanese macaque groups have been observed continuously over many years, making it possible to analyze intraspecific variations in mother–infant relationships with respect to environmental factors.

Macaques species are characterized by despotic dominance styles and strong bonds among related individuals, whereas interspecies variation in dominance style has been reported (Aureli et al. 1997). Such interspecies variation may be expected to influence mother–infant relationships. In fact, Maestripieri (1994a, b) has shown differences in mother–infant relationships during the first 3 months of infant life among three macaque species living in captive social groups. However, comparable data from different species living in similar physical and social conditions are limited.

3 Maternal Aggression and Infant Independence

The positive maternal aspects of satisfying the physical and psychological needs of their infants have been emphasized in many studies. Since Trivers (1974) provided an evolutionary explanation for mother–infant conflict, however, the negative aspects of maternal behavior have also been emphasized. Negayama (1981) recorded aggressive behavior by mothers toward their infants in individually caged Japanese macaque mother–infant pairs, and pointed out that maternal aggression served to promote the infant's independence from its mother. Japanese macaque mothers in social groups have also been reported to increase their acts of maternal aggression or rejection when they were estrous or during the weeks in which they were being mounted in the mating season (Collinge 1987; Worlein et al. 1988). However, few studies have reported a significant correlation between the frequency of maternal aggression and that of behavior associated with an infant's independence from its mother. To evaluate the importance of maternal aggression or rejection on the infant's independence from its mother, correlation analysis might not be the most appropriate method. Instead, we should pay attention to behavioral sequences. Acts of maternal aggression were elicited mainly by the infant's behavior directed toward its mother, such as approaching and making playful contact with her, and the infant tended to react to maternal aggression by stopping this behavior or by retreating from her (Negayama 1981; Nakamichi et al. 1990). However, it is important to know how the mother or the infant behaves following such behavior performed by the infant directly after maternal aggression.

Opponents in agonistic conflicts are likely to seek one another for friendly interactions soon after fights. This phenomenon is referred to as reconciliation: a postconflict affiliative reunion between former opponents. Reconciliation, not only between adults but also between juveniles, has been demonstrated in many nonhuman species, including at least ten species of macaques (for a review, see Kappeler and van Schaik 1992; Aureli et al. 1997), whereas little is known about reconciliation between a mother and her infant. If, after retreating from the mother in response to her aggression, an infant is permitted to make affiliative contact with her, or she actively retrieves her infant, this social interaction can be regarded as reconciliation between the mother and her infant.

Reconciliation between adults reduces social tension after conflicts and restores disturbed relationships (Aureli et al. 1989; Aureli and van Schaik 1991). Therefore, we can expect that reconciliation between a mother and her infant is probably helpful in reducing the infant's tension or anxiety, which could be produced by maternal aggression. Maternal aggression may make the distance between the mother and her infant extend temporarily due to the infant's retreat, but cannot directly lead to an infant's independence from its mother. Instead, the infant is more likely to seek affiliative contact with the mother soon after maternal aggression. If the infant is permitted to reconcile itself with its mother smoothly, she could be considered to be a good secure base for her infant. Moreover, the infant who has a good secure base could explore the physical and social environment voluntarily and more frequently.

As reviewed above, there are great individual differences among mother–infant pairs. Such differences could, in part, be dependent on how smoothly reconciliation could be performed between the mother and her infant. The following hypothesis can be offered. The independence of infants from their mothers is not directly related to the frequency of maternal aggression per se, but to the method used by mothers and infants to be reconciled after maternal aggression. That is, we can expect that the more smoothly the infant is reconciled with the mother after receiving maternal aggression or rejection, the more frequently it becomes able to leave her voluntarily to extend its social environment.

4 Long-Term Stability in Mother–Offspring Relationships

Many developmental studies on macaques as well as other nonhuman primates have shown variations in mother–infant relationships. Most of these studies, however, focused on monkeys in their first year of life, during which their absolute dependence on their mothers decreased markedly (see previous section). How and to what extent do young monkeys interact with their mothers when they are more independent? How long do individual differences in mother–offspring relationships which emerge in the early stages of development continue? In order to answer questions on long-term changes in mother–offspring relationships, we have to conduct longitudinal studies extending from infancy to adolescence. Only a few studies can meet this demand.

Hinde and Spencer-Booth (1967) observed eight rhesus macaque infants living with their mothers in small social groups during the first 2.5 years of life and found a considerable stability of individual differences in mother–offspring relationships. For example, some infants spent more time with their mothers than other infants throughout the 2.5 years. Since their data on the eight infants were not divided according to the sex of the infants to assess stability, it is probable that such stability of individual mother–offspring pairs could be caused by the well-known fact that female infants spend more time with their mothers than male infants.

Since the birth season of Japanese macaques is limited to about 3 months per year, from spring to summer, the body size and locomotor skills of young animals during the first few years of life are different from those of animals who are born in different years. Thus, it may be reasonable to follow all members of an age cohort from infancy to adolescence to describe long-term relationships between mother–offspring pairs.

Nakamichi (1989) observed Japanese macaques of an age cohort in the Awajishima free-ranging group during the first 4 years of life, focusing on maturational changes in the relationships of immature macaques of both sexes with their mothers and with group members other than their mothers. He found clear sex differences in mother–offspring relationships during the 3rd and 4th years of life, but not during the first 2 years. The main sex difference in the social development of macaques is that females stay with their natal group throughout their lives, while males change their spatial positions from the center to the periphery, and then leave the natal group. Thus, it is not surprising that males tended to interact with their mothers less frequently than did females, particularly during the later stages of development.

More interestingly, Nakamichi (1989) found that certain individuals interacted with their mothers more frequently than with others of the same sex throughout the first 4 years of life, which indicates the existence of stability (or consistency) of individuality among same-sex immature macaques for mother–offspring relationships. A young monkey is not allowed to take the mother's nipple in its mouth or to cling to her ventrum after the birth of a new sibling, and the mother tends to concentrate on the new baby. Thus, the birth of a new sibling influences the relationship of the older sibling with the mother. Some immature macaques in Nakamichi's study had a younger sibling while others did not, and his immature macaques had different numbers of older siblings. Regardless of these factors, which could have short-term influences on mother–offspring relationships, Nakamichi (1989) confirmed the stability of individuality for mother–offspring relationships. In other words, young macaques tend to interact with their mothers less and less frequently with increasing age, thus experiencing short-term changes in their mother–offspring relationship which may be caused by many factors, including the birth of new siblings. However, some mother–offspring pairs tend to maintain relatively stronger bonding than other pairs throughout the first 4 years. Nakamichi (1989) did not find any associations between such behavioral consistency and the mother's parity and dominance.

Japanese macaque females give birth several times throughout their lives after they experience their first parturition at 5 or 6 years of age (Itoigawa et al. 1992). Therefore, it is very interesting to know whether mothers maintain consistent maternal styles from one infant to another (i.e., consistency in maternal style within mothers). Moreover, it may be expected that adult daughters adopt maternal styles which are similar to those of their own mothers (i.e., cross-generational consistency in maternal style). Unfortunately, no quantitative behavioral data are available at present to answer these two questions for Japanese macaques.

Based on data of 92 infants born to 27 different mothers in naturally composed captive groups of vervet monkeys and those of 110 infants born to 37 different mothers in a free-ranging group of rhesus macaques, Fairbanks (1989) and Berman (1990), respectively, clearly answered these two questions. According to Fairbanks (1989), there are strong and consistent individual differences among mothers for the average amount of mother–infant contact during the first 6 months of life. Such differences are not related to the mother's rank. Moreover, Fairbanks (1989) found a significant correlation between the amount of mother–infant contact a female experienced as an infant and that she experienced as a mother. Thus, it seems that individual differences in mothering behavior in one generation can be passed on to the next generation. Berman (1990) showed that rejection rates for individual mothers were both consistent from infant to infant and similar to those of their own mothers.

As reviewed above, it seems that maternal attitudes or behavior toward offspring seem to be very robust or stable regardless of the development of the offspring, birth history, and generation. We can say that each mother is very likely to maintain her own maternal style, and that this maternal style is transferred over generations. However, these tendencies must be greatly dependent on the long-term stability of the physical and social environment in which the females live. Nakamichi (1989) and Berman (1990) observed free-ranging, but artificially provisioned, macaque groups, and Hinde and Spencer-Booth (1967) and Fairbanks (1989) observed captive groups. The interesting findings of these studies might be dependent on the artificially maintained physical environment. However, these studies can make us recognize anew the value of longitudinal studies conducted on the same individual or on the same group over many years. Short-term observational study could never have detected the existence of the consistency of individual differences in behavior.

Longitudinal studies are also useful in understanding the social development of young monkeys. Social play is very important in developing social skills and social bonds. Thus, social play has been featured as a target behavioral measure in studies concerning social relationships among monkeys. For example, immature male Japanese macaques tended to play with animals of the same sex, the same or a similar age, and a similar rank, whereas females tended to play with younger animals of both sexes (Mori 1974; Koyama 1985; Imakawa 1990). However, young monkeys interact with each other in ways other than social play (Hayaki 1983; Imakawa 1988). Thus, it would be more appropriate to adopt a more inclusive behavioral measure than to focus on a single behavior measure. By using the spatial proximity measure, Nakamichi (1989) described the sex differences in social development throughout the first 4 years after birth. Males maintained close proximity relationships with males of the same or similar age both before and after they moved from the center to the periphery: the male's preference for association with other males of similar age preceded peripheralization. Females preferred relationships with animals of the central part of the group: female group members of various ages, infants of both sexes, and fully adult males. Based on these data, Nakamichi (1989) concluded that Japanese macaques do not change their social relationships with their mothers and other group members suddenly and markedly when they become adults, but they begin to prepare for adult male/female roles when they are young and immature.

In another study, Nakamichi (1996) also analyzed data from the same immature Japanese macaques of age cohorts to characterize the proximity relationships among them throughout their immature years, and to assess the degree to which the proximity relationships among them were affected by those among their mothers. Like the long-term stability of mother–offspring relationships, most male and female immature macaques showed a consistent, long-term preference for proximity within their cohort to certain same-sex individuals whose dominance ranks were immediately adjacent to their own. Moreover, such prolonged proximity relationships between peers of the same sex were largely a reflection of those between their mothers. During the early stages of development, infants spend more time near their mothers, and mothers with infants also tend to stay near one another. Thus, it is probable that infants whose mothers have affiliative relationships can develop close relationships among them mainly through social play. Therefore, the proximity relationships between peers seem to be formed under the influence of social relationships between mothers during the first year of the infant's life. In other words, mothers seem to determine the complexity of the social environment in which their young offspring grow in the early stages of development. Thereafter, such close relationships between peers can still be maintained after young macaques grow beyond infancy through social play or other behavior such as foraging or traveling together.

5 Flexibility in Mother–Offspring Relationships

The fact that each macaque mother tends to maintain a consistent maternal style of her own should not be interpreted as a lack of flexibility or plasticity in the mother's behavior toward her infant. Instead, some studies have shown that macaque mothers can respond to their infants in a flexible manner. Congenital malformations of limbs in Japanese macaques have been observed over a wide range of their habitat since the first confirmed case in 1955 (Yoshihiro et al. 1979; Nakamichi et al. 1997). The birth rates of malformed infants ranged from 4% to 30% in several provisioned groups in the 1960s and 1970s (Yoshihiro et al. 1979). In and after the 1980s, the number of provisioned groups into which malformed infants were born seemed to decrease, although correct data on the births of malformed infants have been reported in only one provisioned group. In the Awajishima group, malformed infants have been born every year for the last three decades. The birth rate of malformed infants in the Awajishima group from 1978 to 1995 was 14.2% of the total births (Nakamichi et al. 1997). Unfortunately, the causes of malformations of limbs remain unknown. Japanese macaques in whom both hands and/or both feet are

Fig. 1. Two 1-year-old severely malformed male Japanese macaques in a free-ranging group at Awajishima. The infant on the *right-hand side* has no hands or feet. The infant on the *left-hand side* has distorted feet but no hands

missing are classified as severely malformed (Fig. 1), while animals with only one hand or whose hands and feet are disfigured but have some fingers are considered slightly malformed.

Like infants of other simian species, infant Japanese macaques are usually able to cling to their mother's ventrum with their limbs immediately after birth. Therefore, the mothers do not need to support the infants while they move on the ground or climb up and down trees, or while they nurse them. Moreover, the act of the infant clinging to the mother may be important for eliciting maternal care (Hansen 1966). This means that the clinging ability is essential for the survival of macaque infants. Regardless of the degree of limb malformations, however, almost all malformed infants are unable to cling to their mothers. It is known that mothers of some nonhuman primate species can offer adequate care to abnormal or immobile infants. For example, macaque mothers cradled and carried their anesthetized infants (Rosenblum and Youngstein 1974; Negayama 1988). However, it is possible that maternal care given to infants who were temporarily unable to cling was merely a continuation of behavior patterns established in response to previous clinging. It is also possible that maternal care would have stopped if the infant's inability to cling had continued. In fact, most mothers stopped carrying their dead infants within a few days after death (Nakamichi et al. 1996a).

On the other hand, congenital malformations of limbs are not temporal deficits. It is very interesting to know how mothers respond to malformed infants. Through studies on such handicapped monkeys, we can provide additional insight into mother–infant relationships that cannot be derived from observations of only normal monkeys.

Nakamichi (1986) observed Japanese macaque infants with severe and slight malformations, in addition to a normal infant, during their first 3 months of life in the Awajishima free-ranging group. A severely malformed infant, who had no feet and hands consisting of only two digits, showed more severe growth retardation, particularly in the development of posture and locomotion, compared with the infant with slight malformation and the normal infant. However, all infants spent approximately the same amount of time in contact with their mothers. This is because the mothers of severely malformed and slightly malformed infants provided maternal behavior appropriate for the deficits of their infants, such as holding and carrying.

The same observation was also true for another severely malformed male infant who was watched during the first year of life by the same group (Nakamichi et al. 1983a). Since the infant had no hands and the hind feet were malformed, he was unable to cling. During the first few weeks of life, the handicapped infant was unable to move by himself and was almost completely dependent on his mother's help. Although he walked bipedally on his distorted feet after the 6th month of life, locomotion was essentially inferior to that of normal infants. The mother was observed to carry the infant with one arm while she moved on the other three limbs even after he acquired a relatively stable locomotor pattern of bipedal walking. Therefore, it was obvious that the mother's compensatory care allowed the infant to survive in the free-ranging situation. While normal infants developed social relationships with same-aged infants or older juveniles through active behavior such as play, restricted locomotor ability did not permit the severely malformed male infant to move around with other young monkeys. By way of compensation, he developed relations particularly with adults other than his mother through behavior such as just sitting close to them, making physical contact with them, and social grooming. He was never excluded from the group. These findings indicate that not only compensatory care by the mother of the malformed infant, but also tolerant responses of group members other than the mother are necessary for malformed infants to survive in the field.

Nakamichi et al. (1983b) also reported that a malformed male juvenile lacking both hands developed bipedal walking on his normal hind legs. The juvenile monkey maintained a stronger bond with his mother than did normal males of the same age throughout the first 4 years of life, and acquired the same high dominance rank as his mother, while he did not show peripheralization as did normal juvenile males.

These three studies showed that four mothers provided maternal behavior appropriate for their infant's defects. However, the number of mother–infant dyads observed is too small to conclude that female Japanese macaques can generally provide appropriate care for their malformed infants. The birth and mortality rates of malformed infants in the Awajishima group have been reported previously (Nakamichi et al. 1997). Of the 606 infants born between 1978 and 1995, 86 (14.2%) showed limb malformations. The mortality rates within the first year after birth of severely and slightly malformed infants were 22.7% and 24.4%, respectively. The corresponding value for normal infants was 10.0%. This difference in the mortality rate between malformed and normal infants appears to be caused by the fact that most malformed infants had great difficulties in clinging to their mothers and in developing their locomotor patterns. More importantly, however, approximately 70% of such malformed infants were able to survive in the field during the first year of life. No mothers have been recorded as abandoning their malformed infants. We have previously shown that even a severely malformed mother lacking both hands was able to bring up her infant whose hands and feet were completely

Fig. 2. A 9-year-old severely malformed female who has no hands walks bipedally on her normal hind legs while carrying her severely malformed infant, which lacks the ability to cling, on her thigh, with her distorted forelimb covering the infant. (From Nakamichi et al. 1997, with permission)

missing (Fig. 2; Nakamichi et al. 1997). Lindburg (1969) conducted behavioral studies of caged rhesus macaques with thalidomide-induced malformations of the forelimbs and reported that the mothers compensated for their infants' inability to cling by supporting them. It can therefore be concluded that the maternal care-taking ability in macaques is flexible enough to cope with infants who are disabled and have motor deficits, and that the infant's ability to cling is not necessary for the display of maternal care.

Maternal response to twin infants is another example which shows the flexibility of a macaque mother's behavior. In simian species, with a litter size usually of one, the birth of twins is rare. More than 900 infants were born into the Katsuyama free-ranging group of Japanese macaques and no twin births have ever been recorded. In the Awajishima group, only two pairs of twin infants have been recorded in over 600 births. One of the two pairs was observed during the first year of the infants lives (Nakamichi 1983). There were no clear differences in behavioral measures such as tactile contact with the mother, nursing, grooming, and maternal rejection and aggression between the twin infants. That is, the mother treated each twin in the same manner. Another Japanese macaque mother in a different group (Matsui 1979) and a rhesus macaque mother in captivity (Spencer-Booth 1968) have been reported to accept and nurse their twins for 1 year. Moreover, a Japanese macaque mother in the Katsuyama group adopted one infant within a few days of the birth of her own and succeeded in rearing both of them (Kurokawa 1974). Another mother in the same group adopted or kidnapped a 1-week-old infant approximately 1 month after she gave birth and nursed the two infants for at least several months (M. Nakamichi and E. Kato 1994, personal observation).

Mothers of simian species in the wild carry their dead infants about for some time (see Nakamichi et al. 1996a for a review about maternal behavior toward dead infants). Like macaque mothers carrying their anesthetized infants, mothers may hold and carry their dead infants as a continuation of behavior patterns established in response to live infant's movements, including clinging and suckling. Thus, it is interesting to know how a mother responds to a stillborn baby. A Japanese macaque female living in a social captive group was observed during the birth of her stillborn baby (Nakamichi 1999). Since the dead baby was hairless, it was estimated that its age was around 100 days after conception. As soon as the dead baby was delivered with the placenta, the mother ate the placenta and licked the dead baby. Thereafter, she carried the dead baby for a while. Her behavior at parturition was to a large extent similar to that of mothers who give birth to live babies (Negayama et al. 1986; Nakamichi et al. 1992). A multiparous Japanese macaque mother in the Awajishima group was also observed to carry a hairless, dead baby with the placenta (H. Nobuhara 1992, personal communication). Consequently, we can conclude that a macaque mother's ability to care for her infants is essentially flexible enough to help infants with deficiencies or multiple infants survive, or to display maternal behavior even toward stillborn babies.

To further examine the flexibility of a mother's behavior toward her infant, several studies were reviewed. Each of these studies described only one or a few animals with particular characteristics, although quantitative data on the behavior of such animals were offered for comparison with those of normal animals. Thus, they were all so-called case studies. Generally speaking, we can generalize certain behavioral traits more strongly if a larger number of samples are available. However, some behavioral patterns that might seem very improbable or nonexistent tend to occur under certain conditions, and it is impossible to obtain many samples of such patterns. We should not leave such behavioral patterns unreported for no other reason than the small sample size or the rare occurrence. Instead, we should try to describe more precisely any behavioral patterns that tend to occur very infrequently, so as to enhance our understanding of the complexity and flexibility of these animals.

6 Relationships Between Mothers and Their Adult Daughters

As female macaques stay with their natal group throughout their lives, they associate with their mothers even after they grow to adulthood. Koyama (1991) observed the Arashiyama free-ranging group of Japanese macaques and found that mothers groomed their daughters more frequently than vice versa when their daughters were young juveniles, whereas the mothers received grooming from their daughters more often than they groomed them after the daughters were over 4 years of age. However, whether or not grooming interactions occur between all possible mother–adult daughter dyads was not dealt with in Koyama's study (1991).

Nakamichi (1984, 1988, 1991) conducted a series of studies to clarify the behavioral characteristics of older females both in the Arashiyama group and in a transplanted, confined group in Texas, USA, which was originally a branch group of the Arashiyama group. He found that older females, particularly those over 20 years of age, showed a marked decrease in social interactions with group members. This is because most older females in the period of senile sterility had few nonadult offspring, and these were primarily responsible for the high frequency of social interactions of younger females. Kato (1999) also found that older mothers spent more time resting and alone than their middle-aged adult daughters in the Katsuyama group of Japanese macaques.

It may be expected that if we restrict social interactions to only adult group members, older mothers could maintain more frequent social interactions than younger ones, since older mothers tend to have a larger number of adult daughters than younger ones. The reverse was the case. Mothers did not interact with all of their adult daughters at similar levels, but they were likely to have exceedingly close relations with their youngest adult daughters through social grooming, maintaining close proximity, etc. Since adult daughters tend to concentrate around their offspring (i.e., the grand-offspring of older mothers) after giving birth, older mothers tend to become socially less active (Nakamichi 1991).

Working with a free-ranging Japanese macaque group consisting of 11 adult females, Kawamura (1958) was the first to describe two principles about dominance relations among adult females: females acquire the rank just below that of their mothers and adult sisters in reverse order of age. Koyama (1967) confirmed Kawamura's principles while working with the Arashiyama group, and pointed out that almost all members of one kin-group were collectively ranked above or below individuals of other kin-groups. Since then, it has been confirmed that dominance relations between mothers and their adult daughters almost adhere to Kawamura's principles not only in other groups of Japanese macaques (Koyama 1967; Takahata 1991; Nakamichi et al.1995), but also in other macaques (Gouzoules and Gouzoules 1987). However, according to Takahata (1991), who observed the Arashiyama group, and Nakamichi (1984), who observed the same group 7 years later, some mothers aged over 25 years, who had lost their reproductive ability due to advanced age, were subordinate to some of their adult daughters. This was also true in the transplanted Arashiyama group in Texas (Nakamichi 1991). Therefore, it is probable that physical decline in older mothers could result in rank reversals. Generally speaking, adult daughters tend to develop their interactions with their own offspring along with an increase in the number of offspring, whereas the affiliative ties between older mothers and mature daughters do not remain as strong as before. Such weakening of the psychological ties may help mature daughters to out-rank their mothers.

On the other hand, according to Nakamichi et al. (1995), who observed the Katsuyama group of 27 mother–daughter dyads, daughters dominated their mothers in three cases. These mothers were 25, 23, and 18 years old and belonged to middle- or low-ranking kin-groups. In the Madingley colony of rhesus macaques, most mothers from low-ranking kin-groups were out-ranked by at least one daughter, but most mothers from high-ranking kin-group were not (Datta 1992). These findings indicate that relationships between mothers and their mature daughters are different according to their dominance ranks. Social cohesiveness among members of a kin-group is weaker for middle- or low-ranking kin-groups than it is for high-ranking kin-groups (Nakamichi et al. 1995). Therefore, differences in cohesiveness are expected to produce differences in dominance relations between older mothers and their mature daughters.

Takahata (1991) reported that the first daughters of the first-ranking kin-group in the Arashiyama group superseded their mothers (i.e., the alpha female) and became the alpha females for three generations. He concluded that these rank reversals did not result from senility but from social factors such as competition for the alpha-female position (see also Gouzoules 1980). On the other hand, an older female of the Katsuyama group maintained her alpha position until her death at 32 years of age without being out-ranked by her mature daughters (N. Itoigawa 2000, personal communication). The youngest daughter of this female became the alpha female at 16 years of age, after her mother's death. Since then, she has maintained her alpha position for more than 13 years, without being out-ranked by her four adult daughters (M. Nakamichi, unpublished data).

The fact that older mothers who have lost their reproductive ability can still remain within the group whose primary function is to produce the next generation is of great importance. Thus, older mothers can expect to have some social role in a well-organized group. They are likely to maintain social interactions with their mature daughters, albeit infrequently, even after rank reversals between them occur (Nakamichi 1988, 1991). Thus, it may be expected that such social interactions of older mothers with their mature daughters could increase the frequency of social interactions among mature daughters. In other words, the presence of the mother might give opportunities for her daughters and other relatives to maintain or develop social networks among these individuals. Conversely, the death of the mother might result in a decline in cohesiveness among her relatives. However, no data to support this speculation have been reported to date.

The presence of vervet grandmothers in captive social groups significantly influenced the reproduction life of their adult daughters: young adult females with mothers produced significantly more surviving offspring (Fairbanks and McGuire 1986) and were more relaxed and less restrictive than those without mothers (Fairbanks 1988). Unfortunately, there are no comparable data in Japanese macaques, although older females have been reported to take care of young orphans (Itoigawa 1982).

In a long-lived species with overlapping generations and kin-correlated complex social relationships, we should not place all mothers in a single category, irrespective of their ages, but try to describe life-span behavioral changes for females. More longitudinal studies are necessary to develop a more inclusive and complete theory for mother–daughter relationships.

7 Relationships Between Mothers and Adult Sons

Unlike young female macaques, most male macaques leave their natal group upon maturation (Norikoshi and Koyama 1975), and thus associations of adult sons with their mothers appear to be rare. However, some young or full adult males that stayed with their natal groups even after growing up have been reported to interact with their mothers.

When male Japanese macaques were 4 years of age they were on the periphery of the group, and thus were rarely recorded in close proximity to their mothers (Nakamichi 1989). However, one peripheral male at 4.5 years of age lost his left foot, probably after being caught in a wire trap. As soon as he was severely injured, he was noticed to stay near his mother in the central area of the group and was groomed by her at least once a day. The frequent proximity to and grooming interactions with the mother continued for more than 1 month until the wound healed up. After recovery, the male monkey again moved to the periphery of the group. Such close relationships between other ordinary males of the same age and their mothers were not observed (Nakamichi 1999). According to Itoigawa (1982), when a 5-year old male Japanese macaque returned to his natal free-ranging group after a 3-month disappearance, his mother tried to rescue him from repeated acts of aggression from central adult males or high-ranking females, and he continued to remain close to his mother. Consequently, his mother's support helped him to stay within the group.

Sons of alpha females or of the first-ranking kin-group reached the alpha-male position with support from their mothers (Chapais 1983; Itoigawa 1993). However, there is little information about the degree to which such mothers from the first-ranking kin-group contribute to the maintenance of their sons' alpha-male position. Generally, few interactions occur between central adult males and their mothers in the transplanted group of Japanese macaques in Texas (Nakamichi 1991). In the Katsuyama group, there were several high-ranking central adult males who were from middle-ranking kin-groups, but the aquisition of high-ranking positions by these males did not lead to a rise in the dominance ranks of their female relatives (Itoigawa 1997).

Itoigawa (1993) reported a group division which occurred in the Katsuyama group. The alpha male from the first-ranking kin-group associated with females of middle-ranking kin-groups after mating with some of them, and thereafter distinctive social conflicts occurred between these females and females of the high-ranking kin-groups, including the alpha male's relatives. The latter was chased off from the main group by the former after repeated agonistic interactions between two subgroups, although the alpha male did not actively support either of the two subgroups. Similar social conflict between females with whom the alpha male had no blood relationship and those with whom he had a blood relationship resulted in a group fission in the Arashiyama group (Koyama 1970). These findings indicate the difficulties of maintaining affiliative relationships between macaque mothers and their adult sons over several years, even though they were in the same group.

8 Studies on Mother–Infant Relationships in a Breeding Colony

Tsukuba Primate Center for Medical Science (TPC), Japan, was set up in 1978 to breed successive generations of nonhuman primates of good quality to be used as laboratory animals. Since that time, TPC has maintained a large colony of cynomolgus macaques (*Macaca fascicularis*) (Honjo 1985). TPC has some significant problems that hinder the successful breeding of nonhuman primates. A proportion of laboratory-born individuals are defective in mating and maternal behavior (Honjo et al. 1984; Cho et al. 1986). Young individuals tend to develop stereotyped behaviors, although they are housed with same-aged individuals of both sexes after weaning (Minami 1996). Therefore, behavioral studies are necessary to develop strategies for the production of successive generations of cynomolgus macaques.

Nakamichi et al. (1990) observed interactions between wild-born, individually caged cynomolgus mothers and their infants during the first 14 weeks of life. The wild-born mothers nurtured their infants in almost the same manner as macaque mothers living in social groups. On the other hand, the wild-born mothers tended to allow their infants to move about freely without maternal restriction, unlike macaque mothers in social groups. The lack of such maternal restriction could derive from the experience of living in individual cages for at least 5 years, without receiving social restriction. At TPC, 97% of wild-born females took care of their babies, whereas only 71% of laboratory-born mothers did so (Cho 1991). Moreover, the wild-born mothers nursed orphan babies as foster mothers (Cho et al. 1986). These findings indicate that the maternal behavior of wild-born cynomolgus macaque mothers appears to be relatively robust in terms of the influence of prolonged individual caging.

Are there any differences in maternal behavior between wild-born mothers and laboratory-born mothers? Are any behavioral differences found among laboratory-born animals of different generations? These are interesting questions. Nakamichi et al. (1996b) found that wild-born mothers held their infants more frequently than F1 mothers (laboratory-born progeny of wild-born females), and that the former displayed maternal acts of aggression more often than F1 mothers. However, infants of F1 mothers, as well as infants of wild-born mothers, interacted with their mothers through approaching and making playful contact with them, although this behavior by infants sometimes triggered acts of maternal aggression. These behavioral differences between mothers of the two different generations indicate that wild-born mothers tended to interact with their infants more actively than do F1 mothers. In other words, the shift from an active to a passive maternal style appears to occur in successive generations from wild-born mothers to F1 laboratory-born mothers at TPC. This trend might be strengthened along with the production of successive generations of laboratory-born individuals. This is an important and interesting question to be answered in future studies.

A great number of studies have shown that maternal separation produces not only behavioral changes, but also physiological and immunological changes in nonhuman primates (Reite and Capitanio 1985; Koyama and Terao 1992). At TPC, many infants tended to exhibit depressive behavior, diarrhea, and body weight loss during the period in which infants at 4–5 months of age were separated from their mothers and were then housed with several unfamiliar age-mates of both sexes. To counteract these adverse affects, infants who were separated from their mothers were housed with one unfamiliar adult female (nurse) for 4 weeks after maternal separation. This treatment reduced the incidence of diarrhea from 48% without a nurse to 16% with a nurse (Hanari et al. 1987). Koyama et al. (1991) also confirmed the effect of the nurse, using both behavioral and physiological measures. However, separation from the nurse produced adverse stress responses in infants, and marked between-infant variability was noted. Thus, checks for clinical symptoms, and physiological and behavioral assessments are necessary for valid characterization of the individual weaning procedures adopted at TPC. Most maternal separation studies tended to analyze behavioral and/or physiological measures after such separation. The quality of mother–infant relationships must vary from pair to pair. Thus, it is also important to analyze the relations between the quality of mother–infant relationships before maternal separation and infant response to maternal separation.

Negayama and Honjo (1986) tried to elucidate the developmental process of discrimination between mothers and infants by exchanging mother–infant combinations at TPC. Mothers and their infants were separated from each other and then housed with alien infants and females, respectively. The mothers and infants did not discriminate between their partners behaviorally during the first half of the first month after birth, but did so during the period from the second half of the first month to the third month after birth. Based on these data, Negayama and Honjo (1986) concluded that maternal behavioral discrimination between infants is gradually established from the second half of the first month onwards. Their conclusion agrees with the fact that Japanese macaque infants living in a captive social group can visually identify their mothers by 8–10 weeks of age (Nakamichi and Yoshida 1986). Moreover, Negayama and Honjo (1986) pointed out the importance of olfactory cues in discrimination of the infants by their mothers. Negayama (1988) also reported data indicating the active role of a young infants' odor in arousing the mothers' positive attention. In a series of studies, Negayama used new experimental methods such as switching between mothers and infants and olfactory hypoesthesia, and was successful in demonstrating how and when the mother and infant living in individually caged conditions identified each other. Such information from experimental research should be combined with that from observational research both in captivity in various conditions and in the wild. The results of such studies should enhance our understanding of various aspects of mother–infant relationships, and at the same time we could contribute to the maintenance and development of breeding colonies of nonhuman primates.

9 Summary

In this chapter, I first described some factors that explain individual differences in macaque mother–infant relationships. Factors other than the mother's parity and dominance, and the infant's sex (three well-known traditional factors) have a significant influence on mother–infant relationships. More attention should be paid to the facts that orphaned mothers are likely to fail to nurse their infants, and there are some abusive mothers, although all such mothers grew up in social groups. I also discussed maternal aggression and infant's independence from the mother, and pointed out that how smoothly the infant could reconcile itself with the mother may be responsible for the future independence of the young monkey from the mother.

Mothers tend to maintain consistent maternal styles throughout the immature years of their offspring, from one infant to another and throughout successive generations, showing the existence of stability in mother–offspring relationships. These findings are completely dependent on longitudinal observations on the same individuals or on the same group over many years. While macaque mothers tend to maintain a consistent maternal style, flexibility in the mother's behavior has also been demonstrated. Most Japanese macaque mothers are able to take care of congenitally malformed infants who are unable to cling to the mother or other objects.

Since female macaques stay with their natal group throughout their lives, they associate with their mothers even after maturation. However, their interactions with their mothers tend to decrease with increasing numbers of offspring. Some older mothers are out-ranked by their mature daughters, which is contrary to the well-known principle of dominance relations among macaques. Since older mothers remain within the group even after losing their reproductive ability, we should try to examine the social roles of such females. Although uncommon, a few young adult sons are dependent on their mothers, whereas mature sons do not show any tendency to support their relatives, including their mothers, who interact agonistically with females of other kin groups.

Lastly, I reviewed studies on mother–infant relationships conducted on a large breeding colony of long-tailed macaques. I then pointed out the importance of combining experimental research with observational studies from both captivity and the wild for a greater understanding of various aspects of mother–offspring relationships and to contribute to the maintenance of breeding colonies of nonhuman primates.

References

Aureli F, van Schaik CP (1991) Post-conflict behaviour in long-tailed macaques (*Macaca fascicularis*). II. Coping with the uncertainty. Ethology 89:101–114

Aureli F, van Schaik CP, van Hooff JARAM (1989) Functional aspects of reconciliation among captive long-tailed macaques (*Macaca fascicularis*). Am J Primatol 19:39–51

Aureli F, Das M, Veenema HC (1997) Differential kinship effect on reconciliation in three species of macaques (*Macaca fascicularis, M. fuscata*, and *M. sylvanus*). J Comp Psychol 111:91–99

Berman CM (1980) Mother–infant relationships among free-ranging rhesus monkeys on Cayo Santiago: a comparison with captive pairs. Anim Behav 28:860–873

Berman CM (1984) Variation in mother–infant relationships: traditional and nontraditional factors. In: Small MF (ed) Female primates: studies by women primatologists. Alan R Liss, New York, pp 17–36

Berman CM (1990) Intergenerational transmission of maternal rejection rates among free-ranging rhesus monkeys. Anim Behav 39:329–337

Berman CM (1992) Immature siblings and mother–infant relationships among free-ranging rhesus monkeys on Cayo Santiago. Anim Behav 44:247–258

Chapais B (1983) Matriline membership and male rhesus reaching high ranks in their natal troop. In: Hinde RA (ed) Primate social relationships: an integrated approach. Blackwell Scientific, Oxford, pp 171–175

Cho F (1991) Pregnancy rate and successful nursing rate. TPC News 10:6–7

Cho F, Suzuki M, Honjo S (1986) Adoption success under single-cage conditions by cynomolgus macaque mothers (Macaca fascicularis). Am J Primatol 10:119–124

Collinge NE (1987) Weaning variability in semi-free-ranging Japanese macaques (Macaca fuscata). Folia Primatol 48:137–150

Datta SB (1992) Effects of availability of allies on female dominance structure. In: Harcourt AH, de Waal FBM (eds) Coalitions and alliances in humans and other animals. Oxford University Press, New York, pp 61–82

Eaton GG, Johnson DF, Glick BB, Worlein JM (1985) Development in Japanese macaques (Macaca fuscata): sexually dimorphic behavior during the first year of life. Primates 26:238–248

Fairbanks LA (1988) Vervet monkey grandmothers: effects on mother–infant relationships. Behaviour 104:176–188

Fairbanks LA (1989) Early experience and cross-generational continuity of mother–infant contact in vervet monkeys. Dev Psychobiol 22:669–681

Fairbanks LA, McGuire MT (1986) Age, reproductive value, and dominance-related behaviour in vervet monkey females: cross-generational influences on social relationships and reproduction. Anim Behav 34:1710–1721

Glick BB, Eaton GG, Johnson DF, Worlein J (1986) Social behavior of infant and mother Japanese macaques (Macaca fuscata): effects of kinship, partner sex, and infant sex. Int J Primatol 7:139–155

Gouzoules H (1980) A description of genealogical rank changes in a troop of Japanese monkeys (Macaca fuscata). Primates 21:262–267

Gouzoules S, Gouzoules H (1987) Kinship. In: Smuts BB, Cheney DL, Seyfarth RM, Wrangham RW, Struhsaker TT (eds) Primate societies. University of Chicago Press, Chicago, pp 299–305

Hanari K, Komatsuzaki K, Ogawa H, Cho F, Honjo S (1987) Management of cynomolgus monkey infants after separation from their mothers (in Japanese with English summary). Exp Anim 36:281–284

Hansen EW (1966) The development of maternal and infant behaviour in the rhesus monkey. Behaviour 27:107–149

Hasegawa T, Hiraiwa M (1980) Social interactions of orphans observed in a free-ranging troop of Japanese monkeys. Folia Primatol 33:129–158

Hauser MD, Fairbanks LA (1988) Mother–offspring conflict in vervet monkeys: variation in response to ecological conditions. Anim Behav 36:802–813

Hayaki H (1983) The social interactions of juvenile Japanese monkeys on Koshima islet. Primates 24:139–153

Hinde RA, Spencer-Booth Y (1967) The behaviour of socially living rhesus monkeys in their first two and a half years. Anim Behav 15:169–196

Hiraiwa M (1981) Maternal and alloparental care in a troop of free-ranging Japanese monkeys. Primates 22:309–329

Holman SD, Goy RW (1988) Sexually dimorphic transition revealed in the relationships of yearling rhesus monkeys following the birth of siblings. Int J Primatol 9:113–133

Honjo S (1985) The Japanese Tsukuba Primate Center for medical science (TPC): an outline. J Med Primatol 14:75–89

Honjo S, Cho F, Terao K (1984) Establishing the cynomolgus monkeys as a laboratory animal. In: Hendrickx AG (ed) Advances in veterinary science and comparative medicine, vol 28. Research on nonhuman primates. Academic Press, New York, pp 51–80

Hooley JM, Simpson MJA (1983) Influence of siblings on the infant's relationships with mother and others. In: Hinde RA (ed) Primate social relationships: an integrated approach. Blackwell, London, pp 139–142

Imakawa S (1988) Development of co-feeding relationships in immature free-ranging Japanese monkeys (Macaca fuscata fuscata). Primates 29:493–504

Imakawa S (1990) Playmate relationships of immature free-ranging Japanese monkeys at Katsuyama. Primates 31:509–521

Itani J (1958) On the acquisition and propagation of a new food habit in the natural group of the Japanese monkey at Takasaki-Yama (in Japanese with English summary). Primates 1:84–98

Itoigawa N (1973) Group organization of a natural troop of Japanese monkeys and mother–infant interaction. In: Carpenter CR (ed) Behavioral regulators of behavior in primates. Bucknell University Press, Lewisburg, pp 229–250

Itoigawa N (1982) Aging and behavior in Japanese monkeys (in Japanese with English summary). Jpn J Ethnol 46:376–389

Itoigawa N (1993) Social conflict in adult male relationships in a free-ranging group of Japanese monkeys. In: Mason WA, Mendoza SP (eds) Primate social conflict. State University of New York Press, New York, pp 145–169

Itoigawa N (1997) A history of a monkey group: thirty-six-year record of Katsuyama group in Okayama Prefecture (in Japanese). Doubutsu-sha, Tokyo

Itoigawa N, Tanaka T, Ukai N, Fujii H, Kurokawa T, Koyama T, Ando A, Watanabe Y, Imakawa S (1992) Demography and reproductive parameters of a free-ranging group of Japanese macaques (*Macaca fuscata*) at Katsuyama. Primates 33:49–68

Johnson RL, Southwick CH (1984) Structural diversity and mother–infant relations among rhesus monkeys in India and Nepal. Folia Primatol 43:198–215

Kappeler PM, van Schaik CP (1992) Methodological and evolutionary aspects of reconciliation among primates. Ethology 92:51–69

Kato E (1999) Effects of age, dominance, and seasonal changes on proximity relationships in female Japanese macaques (*Macaca fuscata*) in a free-ranging group at Katsuyama. Primtes 40:291–300

Kawai M (1958) On the rank system in a natural group of Japanese monkey. I. The basic and dependent rank (in Japanese with English summary). Primates 1:111–130

Kawai M (1965) Newly acquired pre-cultural behavior of the natural troop of Japanese monkeys on Koshima islet. Primates 6:1–30

Kawamura S (1958) The matriarchal social order in the Minoo-B group (in Japanese with English summary). Primates 1:149–156

Kawamura S (1959) The process of sub-culture propagation among Japanese macaques. Primates 2:43–60

Koyama N (1967) On dominance and kinship of a wild Japanese monkey troop in Arashiyama. Primates 8:189–215

Koyama N (1970) Changes in dominance rank and division of a wild Japanese monkey troop in Arashiyama. Primates 11:335–390

Koyama N (1985) Playmate relationships among individuals of the Japanese monkey troop in Arashiyama. Primates 26:390–406

Koyama N (1991) Grooming relationships in the Arashiyama group of Japanese monkeys. In: Fedigan M, Asquith PJ (eds) The monkeys of Arashiyama: thirty-five years of research in Japan and the West. State University of New York Press, New York, pp 211–226

Koyama T, Terao K (1992) Psychological stress of maternal separation in cynomolgus monkeys. I. The effect of housing with a nurse female. In: Itoigawa N, Sugiyama Y, Sackett GP, Thompson RKR (eds) Topics in primatology, vol 2. Behavior, ecology and conservation. University of Tokyo Press, New York, pp 101–113

Koyama T, Fujii H, Yonekawa F (1981) Comparative studies of gregariousness and social structure among seven feral *Macaca fuscata* groups. In: Chiarelli AB, Corruccini RS (eds) Primate behavior and sociobiology. Springer, Berlin, pp 64–71

Koyama T, Terao K, Sackett GP (1991) Depressive behavior and serum cortisol of *Macaca fascicularis* after maternal separation and housing with a "nurse." Primate Res 7:1–11

Kurokawa T (1974) Development of behavior in wild Japanese monkeys (in Japanese). Anima 15:31–35

Kuyk K, Dazey J, Erwin J (1977) Primiparous and multiparous pigtail monkey mothers (*Macaca nemestrina*): restraint and retrieval of female infants. J Biol Psychol 18:16–19

Lindburg DG (1969) Behavior of infant rhesus monkeys with thalidomide-induced forelimb malformations: a pilot study. Psychon Sci 15:55–56

Maestripieri D (1994a) Mother–inafnt relationships in three species of macaques (*Macaca mulatta, M. nemestrina, M. arctoides*). I. Development of the mother–infant relationship in the first three months. Behaviour 131:75–96

Maestripieri D (1994b) Mother–infant relationships in three species of macaques (*Macaca mulatta, M. nemestrina, M. arctoides*). II. The social environment. Behaviour 131:97–113

Maestripieri D, Wallen K, Carroll KA (1997) Infant abuse runs in families of group-living pigtail macaques. Child Abuse Neglect 21:465–471

Maestripieri D, Tomaszycki M, Carrol KA (1999) Consistency and changes in the behavior of rhesus macaque abusive mothers with successive infants. Dev Psychobiol 34:29–35

Matsui T (1979) The twin infant Japanese monkeys survived (in Japanese). Monkey 23:6–15

Minami T (1996) Locomotive stereotyped behavior in cynomolgus macaques, *Macaca fascicularis*. Percept Mot Skills 83:935–938

Mori U (1974) The inter-individual relationships observed in social play of the young Japanese monkeys of the natural troop in Koshima islet (in Japanese with English summary). J Anthropol Soc Nippon 82:303–318

Nakamichi M (1983) Development of infant twin Japanese monkeys (*Macaca fuscata*) in a free-ranging group. Primates 24:576–583

Nakamichi M (1984) Behavioral characteristics of old female Japanese monkeys in a free-ranging group. Primates 25:192–203

Nakamichi M (1986) Behavior of infant Japanese monkeys (*Macaca fuscata*) with congenital limb malformations during the first three months. Dev Psychobiol 19:335–341

Nakamichi M (1988) Aging and behavioral changes of female Japanese macaques. In: Laboratory of Ethological Studies, Faculty of Human Science, Osaka University (ed) Research reports of the Arashiyama West and East groups of Japanese monkeys. Osaka University, Osaka, pp 87–97

Nakamichi M (1989) Sex differences in social development during the first 4 years in a free-ranging group of Japanese monkeys, *Macaca fuscata*. Anim Behav 38:737–748

Nakamichi M (1991) Behavior of old females: comparisons of Japanese monkeys in the Arashiyama East and West groups. In: Fedigan M, Asquith PJ (eds) The monkeys of Arashiyama: thirty-five years of research in Japan and the West. State University of New York Press, New York, pp 175–193

Nakamichi M (1996) Proximity relationships within a birth cohort of immature Japanese monkeys (*Macaca fuscata*) in a free-ranging group during the first four years of life. Am J Primatol 40:315–325

Nakamichi M (1999) Japanese monkey mothers and offspring (in Japanese). Fukumura, Tokyo

Nakamichi M, Yoshida A (1986) Discrimination of mother by infant among Japanese macaques (*Macaca fuscata*). Int J Primatol 7:481–489

Nakamichi M, Fuji H, Koyama T (1983a) Behavioral development of a malformed infant in a free-ranging group of Japanese monkeys. Primates 24:52–66

Nakamichi M, Fuji H, Koyama T (1983b) Development of a congenitally malformed Japanese monkey in a free-ranging group during the first four years of life. Am J Primatol 7:479–487

Nakamichi M, Cho F, Minami T (1990) Mother–infant interactions of wild-born, individually caged cynomolgus monkeys (*Macaca fascicularis*) during the first 14 weeks of infant life. Primates 31:213–224

Nakamichi M, Imakawa S, Kojima Y, Natsume A (1992) Parturition in a free-ranging Japanese monkey (*Macaca fuscata*). Primates 33:413–418

Nakamichi M, Itoigawa N, Imakawa S, Machida S (1995) Dominance relations among adult females in a free-ranging group of Japanese monkeys at Katsuyama. Am J Primatol 37:241–251

Nakamichi K, Koyama N, Jolly A (1996a) Maternal responses to dead and dying infants in wild troops of ring-tailed lemurs at the Berenty Reserve, Madagascar. Int J Primatol 17:505–523

Nakamichi M, Minami T, Cho F (1996b) Comparison between wild-born mother–female infant interactions and laboratory-born mother–female infant interactions during the first 14 weeks after birth in individually caged cynomolgus macaques. Primates 37:155–166

Nakamichi M, Nobuhara H, Nobuhara T, Nakahashi M, Nigi H (1997) Birth rate and mortality rate of infants with congenital malformations of the limbs in the Awajishima free-ranging group of Japanese monkeys (*Macaca fuscata*). Am J Primatol 42:225–234

Negayama K (1981) Maternal aggression to its offspring in Japanese monkeys. J Hum Evol 10:523–527

Negayama K (1988) Effect of infants' behavior on mothers' responsiveness toward them in *Macaca fascicularis*: an experimental analysis by anesthesia (in Japanese with English summary). Primate Res 4:1–10

Negayama K, Honjo S (1986) An experimental study on developmental changes of maternal discrimination of infants in crab-eating monkeys (*Macaca fascicularis*). Dev Psychobiol 19:49–56

Negayama K, Negayama T, Kondo K (1986) Behavior of Japanese monkey (*Macaca fuscata*) mothers and neonates at parturition. Int J Primatol 7:365–378

Norikoshi K, Koyama N (1975) Group shifting and social organization among Japanese monkeys. In: Kondo S, Kawai M, Ehara A, Kawamura S (eds) Proceedings of the Symposium of the 5th Congress of the Primatological Society. Japanese Science Press, Tokyo, pp 43–61

Reite M, Capitanio JP (1985) On the nature of social separation and social attachment. In: Reite M, Field T (eds) The psychobiology of attachment and separation. Academic Press, New York, pp 223–255

Rosenblum LA, Youngstein KP (1974) Developmental changes in compensatory dyadic responses in mother and infant monkeys. In: Lewis M, Rosenblum LA (eds) The effects of the infant on its caregiver. Wiley, New York, pp 141–161

Schino G, D'Amato FR, Troisi A (1995) Mother–infant relationships in Japanese macaques: sources of inter-individual variation. Anim Behav 49:151–158

Silk JB (1991) Mother–infant relationships in bonnet macaques: source of variation in proximity. Int J Primatol 12:21–38

Spencer-Booth Y (1968) The behaviour of twin rhesus monkeys and comparisons with the behaviour of single infants. Primates 9:75–84

Struhsaker TT (1971) Social behaviour of mother and infant vervet monkeys (*Cercopithecus aethiops*). Anim Behav 19:233–250

Suzuki H, Matsuura N, Kawamichi T (1984) Effects of mothers' kinship and parity on behavioral development of the Japanese macaque (in Japanese). Mamm Sci 48:19–30

Takahata Y (1991) Diachronic changes in the dominance relations of adult female Japanese monkeys of the Arashiyama B group. In: Fedigan M, Asquith PJ (eds) The monkeys of Arashiyama: thirty-five years of research in Japan and the West. State University of New York Press, New York, pp 123–139

Tanaka I (1989) Variability in the development of mother–infant relationships among free-ranging Japanese macaques. Primates 30:477–491

Tartabini A, Genta ML, Bertacchini PA (1980) Mother–infant interaction and rank order in rhesus monkeys (*Macaca mulatta*). J Hum Evol 9:139–146

Trivers RL (1974) Parent–offspring conflict. Am Zool 14:249–264

White LE, Hinde RA (1975) Some factors affecting mother–infant relations in rhesus monkeys. Anim Behav 23:527–542

Worlein JM, Eaton GG, Johnson DF, Glick BB (1988) Mating season effects on mother–infant conflict in Japanese macaques, *Macaca fuscata*. Anim Behav 36:1472–1481

Yamagiwa J, Hill DA (1998) Intraspecific variation in the social organization of Japanese macaques: past and present scope of field studies in natural habitats. Primates 39:257–273

Yoshihiro S, Goto S, Minezawa M, Muramatsu M, Saito Y, Sugita H, Nigi H (1979) Frequency of occurrences, morphology, and causes of congenital malformation of limbs in the Japanese monkey. Ecotoxicol Env Saf 3:458–470

22
The Myth of Despotism and Nepotism: Dominance and Kinship in Matrilineal Societies of Macaques

SHUICHI MATSUMURA

1 Introduction

The genus *Macaca,* or macaques, has been considered as typical species that form "female-bonded" groups, which are characterized by female philopatry and cooperation on the basis of matrilineal kinship (Wrangham 1980). It has been believed that social interactions among group members, in particular, among females, are strongly influenced by dominance and kinship. That belief seemed to have its origin in intensive studies on provisioned groups of Japanese and rhesus macaques. The emergence of sociobiology (Wilson 1975) appeared to accelerate this trend. Recently, studies of Japanese macaques in their natural habitat have thrown some doubt on this belief (Maruhashi et al. 1986; Sugiyama 1992; Hill 1999). Furthermore, comparative studies among macaques have revealed a considerable variation between species in the effects of dominance and kinship on social interactions (Thierry 1985a; de Waal and Luttrell 1989; Thierry et al. 1994). In egalitarian macaques, differences in behavior of dominant and subordinate individuals are relatively small. Nepotistic tendencies in cooperative or affiliative interactions are not recognized clearly.

Egalitarian macaques throw doubt on a simplistic view of dominance and kinship. Why don't dominant individuals behave arbitrarily? Why don't they support relatives? What kind of selective force has formed these tendencies? How do developmental processes influence the differences? Macaques have now become one of the most interesting subjects in evolutionary ecology in general. Studies on macaque behavior enable us to know how evolutionary and ontogenetic processes work together under certain ecological conditions. In this chapter, the history of studies on dominance and kinship in macaque societies is reviewed after a brief description of this genus. Then, differences in dominance styles are described, with special reference to studies on moor macaques in their natural habitats. Finally, possibilities for future studies are discussed.

Primate Research Institute, Kyoto University, 41 Kanrin, Inuyama, Aichi 484-8506, Japan

2 The Genus *Macaca*

The genus *Macaca,* or macaques, consists of 19 species (Fooden 1976; Fa 1989; Table 1). This genus has many advantages as the subject for comparative studies from evolutionary perspectives. First, they have the widest distribution of any primate genera other than *Homo.* The radiation of macaques is considered to have taken place relatively recently, around 5 million years ago. They show a great diversity in their habitats—from tropical rain forests to semi-arid, mangrove, riverine, warm and cold temperate forests. Some species inhabit the Himalayan Highland or the snowy areas of Japan. Second, since the pioneering stage of primatology, some species of macaques such as Japanese macaques and rhesus macaques have been studied intensively in their natural habitats, at artificial provisioning sites, and in captivity. In comparison with other genera, much information about their ecology, behavior, and social organization is available. In addition, the phylogenetic relationships among species within the genus have also been well studied from the viewpoint of morphology (e.g., Delson 1980) or genetics (e.g., Fooden and Lanyon 1989; Hoelzer and Melnick 1996). Knowledge about phylogenetic relationships is indispensable when comparison between species is carried out from an evolutionary viewpoint (Harvey and Pagel 1991).

2.1 Evolution and Dispersal of Macaques

Macaques distribute in South, Southeast, or East Asia with the exception of barbary macaques, which are found in North Africa. A gap of more than 9 000 km separates Asian macaques and barbary macaques. Differentiation of the Papionini into *Theropithecus, Papio, Cercocebus,* and *Macaca* is postulated to have occurred in Africa at the end of Miocene (reviewed by Fa 1989). Fossils of ancestor macaques are found in the circum-Mediterranean area. They are difficult to distinguish morphologically from the living barbary macaques. Fossil evidence shows that macaques became the most common cercopithecid in Europe during the Villafranchian and Pleistocene. Later, European and Mediterranean macaques became extinct, and the present populations of barbary macaques are recognized as relicts. There is not much fossil evidence that enables us to know about the evolution and dispersal of macaques in Asia.

Fooden (1976) divided living macaque species into four species groups (Table 1) based primarily on structure of male external genitalia. The *silenus-sylvanus* group includes barbary macaques, lion-tailed macaques, pig-tailed macaques, and seven species that inhabit Sulawesi Island. The *sinica* group consists of toque macaques, bonnet macaques, Assamese macaques, and Tibetan macaques. The *arctoides* group includes the stumptail macaques. The *fascicularis* group includes long-tailed macaques, rhesus macaques, Formosan macaques, and Japanese macaques. This classification has received general support (Fa 1989; Hoelzer and Melnick 1996). Phylogenetic trees constructed on the basis of morphology or protein polymorphism do not show a great discrepancy within the species groups (Delson 1980; Fooden and Lanyon 1989; Hoelzer and Melnick 1996). The posi-

Table 1. Four species groups of macaques

Species	
silenus-sylvanus group	
M. sylvanus	barbary macaque
M. silenus	lion-tailed macaque
M. nemestrina	pig-tailed macaque
M. nigra*	crested macaque
M. nigrescens*	Gorontaro macaque
M. hecki*	Heck's macaque
M. tonkeana*	tonkean macaque
M. ochreata*	booted macaque
M. brunnescens*	Muna-Butung macaque
M. maurus*	moor macaque
sinica group	
M. sinica	toque macaque
M. radiata	bonnet macaque
M. assamensis	Assamese macaque
M. thibetana	Tibetan macaque
arctoides group	
M. arctoides	stumptail macaque
fascicularis group	
M. fascicularis	long-tailed macaque, crab-eating macaque
M. mulatta	rhesus macaque
M. cyclopis	Taiwan macaque, Formosan macaque
M. fuscata	Japanese macaque

*Sulawesi macaques.

tions of barbary macaques and stumptail macaques are still debatable. Delson (1980) considered barbary macaques as an independent species group and assigned stumptail macaques to the *sinica* group. Recent molecular studies seemed to support Delson's (1980) view (Hayasaka et al. 1996; Moracle and Melnick 1998).

Geographic ranges of macaque species overlap extensively with one another. Within species-groups, however, ranges of species are essentially allopatric (Fooden 1976). The most disjunct distribution of extant species of the *silenus-sylvanus* group suggests that the dispersal of this group occurred earliest (Fooden 1976; Eudey 1980). It was followed by the dispersal of the *sinica* group and the *arctoides* group. Continuous and widespread distribution of the *fascicularis* group suggests that this group dispersed relatively recently. Pleistocene glacial events, as well as competition between species, seem to have had a great impact on the distribution of each species (Eudey 1980).

Fooden (1982) proposed a hypothesis that explains the speciation and present distribution of macaque species from the viewpoint of their habitat preference and utilization patterns of the habitats. If macaques were divided into those living in broadleaf evergreen forests and those living in non-broadleaf evergreen forests, the ranges of species within each habitat type would not overlap with each other. This implies strong competition between species during their dispersion. Habitat differentiation between pig-tailed macaques and long-tailed macaques has been studied well (Crockett and Wilson 1980; Rodman 1991). Ranges of the two species overlap with each other widely in Southeast Asia. The habitat of the former species is essentially limited to tropical rain forests, whereas the latter is found mainly in riverine forests and secondary forests.

2.2 General Features of Ecology and Social Organization

Macaques are essentially frugivorous (Clutton-Brock and Harvey 1977; Lindburg 1977; Berkovitch and Huffman 1999). However, some species depend on food items other than fruits. Seeds and leaves together represented an annual average of 75% and 59% of time spent feeding in two populations of Barbary macaques (Menard and Vallet 1996). Tibetan macaques consume mainly bamboo shoots and fruits in autumn, while they are supposed to eat mainly mature leaves and bark in other seasons (Zhao 1996). Japanese macaques living in the northern part of Japan heavily depend on bark and twigs in winter (Suzuki 1965; Nakagawa et al. 1996).

Macaques live in groups ranging in size from a few individuals to more than one hundred individuals in natural habitats. Most groups are in the vicinity of twenty to fifty individuals (Caldecott 1986; Berkovitch and Huffman 1999). The number of males within a group shows a considerable variation between species (Oi 1996). Males leave their natal groups while females remain in their natal groups.

3 Dominance and Kinship in Macaques

3.1 Dominance and Kinship as Organizing Principles

It has long been believed that macaque societies are characterized by strong dominance and kinship. Although macaques exhibit a great diversity in habitats, they are believed to live in similar types of social groups. Pioneer studies on Japanese macaques that started soon after World War II might have played an important role in the formation of this belief (Reynolds 1992). Japanese primatologists carried out detailed observations on social behavior at artificial provisioning sites on the basis of individual identification, and continued observations for many years (Imanishi 1960).

Earlier studies argued that the dominance rank order is one of the most important organizing principles of Japanese macaque groups (Itani 1954; Mizuhara 1957; Imanishi 1960). The dominance rank order was considered to have a function of maintaining the structure and integration of groups. Dominance relationships among adult females appeared less clear than that among adult males (Itani 1954). Most Japanese primatologists tended to devote much effort to studies on the dominance relationship among adult males, "leadership" behavior by dominant males, or "social structures" expressed in spatial distribution of group members (Itani 1954; Imanishi 1960; Kawai 1964).

Nevertheless, several important findings about female social relationships were made during the pioneering stage. Kawai (1958) found two kinds of dominance rank in the Koshima group: the "basic" rank and the "dependent" rank. The basic rank reflects differences in fighting ability, experience, or physical conditions between two individuals. The dependent rank is dominance between two individuals under the influence of third parties. Kawai (1958) reported many cases where dominance between two individuals varied according the presence of the other individuals. For example, smaller juveniles can be dominant over larger adult females when their mothers were staying in proximity to them. Lower-ranking females may use potential support from their intimate adult males to dominate higher-ranking females. Dominance relationships in a certain dyad may change by repeated winning of the "basically" subordinate individuals against dominants under the influence of third parties. In other words, the dependent rank may turn to the basic rank.

Kawamura (1958) studied the Minoo-B group and found patterns in the dominance order among mothers and their daughters. The dominance rank of daughters was just below that of their mothers. Among daughters, the dominance order was inversely correlated to their birth order. As mothers usually support their younger daughters when conflicts among sisters occur, the dependent rank of younger sisters would be higher than that of older sisters. As the younger become mature, the dependent rank will turn to the basic rank. These patterns were termed "Kawamura's rules" or "youngest ascendancy" (Datta 1988).

Kawamura's rules suggest the importance of nepotistic supports in Japanese macaques (Watanabe 1979; Mori et al. 1989). It has also been known that kinship has a great influence on affiliative interactions within a group. Nepotistic tendencies were found in grooming, co-feeding, and playmate relationships between individuals (Furuya 1957; Yamada 1963; Oki and Maeda 1973 Koyama 1985, 1991). A series of studies at Arashiyama clarified that social groups of Japanese macaques were essentially combinations of matrilineal kin groups (Koyama 1967, 1970; Takahata 1991). A dominance relation between matrilineal kin groups was easily recognized. When group fission occurred, group members were divided into two groups according to the kin groups (Koyama 1970).

In parallel with studies by Japanese primatologists, active studies were carried out on free-ranging groups of rhesus macaques in Cayo Santiago Island, Puerto Rico (Altmann 1962; Sade 1965; Rawlins and Kesseler 1986). It has been demonstrated that dominance and kinship also play an important role in rhesus macaques (Koford 1963; Sade 1965, 1967; Missakian 1972) and other primates (reviewed in Gouzoules and Gouzoules 1987; Walters and Seyfarth 1987). Kawamura's rules generally hold in rhesus macaques (Sade 1967; Missakian 1972) and other primates such as baboons and vervet monkeys (Hausfater et al. 1982; Horrcks and Hunte 1983; Walter and Seyfarth 1987; see also Hill and Okayasu 1996). Processes and proximate mechanisms of the acquisition and maintenance of dominance rank have been studied by many Western primatologists both in free-ranging groups (e.g., Cheney 1977; Datta 1988) and in captivity (e.g., de Waal and Luttrell 1985; Chapais 1992), as well as from theoretical viewpoints (e.g., Chapais and Schulman 1980; Datta and Beauchamp 1991).

3.2 Dominance and Kinship from an Evolutionary Viewpoint

A turning point in studies on dominance and kinship in primates came in the late 1970s, corresponding to the emergence of sociobiology theory (Wilson 1975). For a long time, it had been believed that animals behave for the advantages of species or groups (Lorenz 1966). However, theoretical studies made it clear that the evolution of behavior to the advantage of species or groups was difficult (Maynard-Smith 1964; Williams 1966). Behavior that improves individual survival and increases the number of offspring was expected to evolve.

From the standpoint of individual selection, altruistic behavior, which increases the recipients' fitness at the cost of the actors', appears to be difficult to evolve. The kin-selection theory (Hamilton 1964) gave an answer to the question of why altruistic behavior is widely seen in many animals. If altruistic behavior is directed to relatives and improves their fitness, the frequency of the gene for this behavior would increase. This is because kin-related pairs share identical genes with a higher probability than unrelated pairs. In other words, the gene spreads throughout the population through relatives.

Sociobiology altered views on dominance and kinship in primates. Dominance order came to be considered an outcome of competition among individuals rather than an organizing principle of social groups. Higher-ranking individuals were regarded as winners of competition. They were expected to attain a higher reproductive success than lower-ranking individuals. Thus, positive correlation between dominance and reproductive success was expected. Kinship became an evolutionarily significant factor rather than an ontogenetic, proximate factor that concerned social relationships. The simplest prediction from kin-selection theory is that altruistic behavior would be directed toward relatives more frequently than toward non-relatives.

Many researchers made attempts to test such predictions by studying primates. Long-term observations since the 1950s provided demographic and reproductive records for years, and information about matrilineal kin relations. Much effort was devoted to finding a correlation between dominance rank and reproductive success (reviewed by Fedigan 1983; Harcourt 1987; Silk 1987; Ellis 1995), or to examining the relationship between the degree of genetic relatedness and social interactions (e.g., Kurland 1977; Silk et al. 1981, reviewed by Gouzoules and Gouzoules 1987; Silk 1987).

3.3 Dominance and Kinship in Natural Habitats

The relationship between the social features of each species and its habitat has been one of the major topics since the early stages of primatology research (e.g., Crook and Gartlan 1966). Advances in evolutionary theories on cooperation and ecological studies of primates in natural habitats produced important socioecological models in the late 1970s and 1980s. In such models, female primates were expected to form groups for the benefit of effective acquisition of food resources as well as predator avoidance (Wrangham 1980; van Schaik 1983). Each species was expected to exhibit different types of social groups according to their major food items, which might differ in their distribution and abundance.

Macaques are essentially frugivorous. Fruits are usually found in patches, and fruit availability fluctuates considerably. Macaques are expected to range larger areas than folivorous or insectivorous primates and to defend food patches cooperatively. They should prefer relatives as allies. This type of social group was termed a "female-bonded" group (Wrangham 1980) and was characterized by female philopatry, cooperation among matrilineal relatives, and nepotistic dominance hierarchies. Macaques were regarded as a typical female-bonded species. Knowledge from studies on provisioned groups of Japanese and rhesus macaques seemed to agree with the model. A series of studies on a wild population of long-tailed macaques (e.g., van Schaik et al. 1983; van Schaik and van Noordwijk 1988) played an important role in the further development of socioecological models (van Schaik 1989; Sterck et al. 1997).

Independently of socioecology, Japanese primatologists started studying wild Japanese macaque groups without provisioning (Sugiyama 1992; Yamagiwa and Hill 1998). In the 1950s and 60s, most studies were carried out at a few provisioned groups such as those at Takasakiyama, Koshima, and Arashiyama. Some researchers pointed out that provisioning modified social behavior as well as group size and group composition (Izawa 1982; Maruhashi et al. 1986). In the 1970s, several researchers tried to habituate wild Japanese macaques in Yakushima Island. Some researchers started continuous observations of wild Japanese macaques on Kinkazan Island. The ecology and social behavior of Japanese macaques in natural habitats have gradually been clarified.

Dominance seemed to play different roles in their natural habitats. In wild groups of Japanese macaques, agonistic interactions during feeding occurred infrequently (Mori 1977; Furuichi 1983; Saito 1996). Agonistic interactions occurred only when the inter-individual distance was below a certain value, which was called the "Tolerance/Intolerance" (T/I) distance. In the case of Japanese macaques, T/I distance was 1 m (Furuichi 1983; Ihobe 1989). Although the frequency of agonistic interactions was low, linear dominance orders were found in wild groups of Japanese macaques as well as in provisioned groups (Furuichi 1983; Takahata et al. 1994; Hill and Okayasu 1995). However, Kawamura's rule might not hold in wild groups of Japanese macaques: elder sisters tend to dominate younger sisters (Furuichi 1983; Hill and Okayasu 1995; Hill 1999). This may result from the fact that agonistic support by mothers to their offspring occurred infrequently (Hill and Okayasu 1995).

In contrast, effects of kinship on social interactions in the wild did not appear to differ from those in provisioned groups. Social grooming was seen much more frequently between related individuals than between unrelated individuals (Furuichi 1984; Mitani 1986; Takahashi and Furuichi 1998). Tolerance during feeding was high between relatives (Furuichi 1983; Iwamoto 1987). Spatial distribution of group members on the basis of matrilineal kin groups was found during moving and feeding (Furuichi 1984; Maruhashi 1986). Group fission occurred between kin groups (Oi 1987; Maruhashi 1992). These finding were consistent with the concept of female-bonded groups.

4 Difference in Dominance Styles Between Macaque Species

4.1 Covariation of Social Traits: Despotic vs. Egalitarian Macaques

Under the influence of earlier studies on Japanese and rhesus macaques, effects of dominance and kinship on social interactions within groups of macaques might have been overestimated. The emergence of sociobiology promoted this trend. On the other hand, several authors reported a difference in patterns of social interactions between macaque species. For example, Defler (1978) reported that distribution of grooming in a group of bonnet macaques differed from that of pig-tailed macaques. Adult female bonnet macaques groomed non-relative adult females more frequently than adult female pig-tailed macaques did. Reports from wild bonnet macaques (Sugiyama 1971; Koyama 1973) showed a similar tendency.

Thierry (1985a, 1986) conducted a comprehensive comparison among three species of macaques. He compared characteristics related to agonistic interactions among captive rhesus macaques, long-tailed macaques, and tonkean macaques. In tonkean macaques, aggression was not so severe, counter-attacks were common, and reconciliation occurred more frequently than in rhesus macaques. Another comparison at an early stage of infant development (Thierry 1985b) showed that tonkean mothers were less restrictive than rhesus mothers were, and that tonkean infants exhibited more diversity in contact partners than rhesus infants did. All behavioral traits appeared to covary along a continuum ranging from "despotism" to "egalitarianism" (Thierry 1990; c.f. Vehrencamp 1983).

De Waal and his colleagues (de Waal and Ren 1988; de Waal and Luttrell 1989) compared social interactions of rhesus macaques and stumptail macaques. They found that the social characteristics of stumptail macaques were similar to those of tonkean macaques. These differences were summarized as a contrast in dominance style (de Waal 1989; de Waal and Luttrell 1989). Rhesus macaques and Japanese macaques are said to show despotic, strict dominance styles, while tonkean macaques and stumptail macaques are said to show egalitarian, relaxed, nice, or tolerant dominance styles (Table 2: de Waal and Luttrell 1989; van Schaik 1989; Thierry 1990, 2000; Moore 1992; Butobvskaya 1993; Thierry et al. 1994; Preuschoft 1995; Sterck et al. 1997).

One of the best examples that show the contrast between despotic and egalitarian macaques is the conciliatory tendency (Thierry 2000). In many primate species, non-agonistic contact between former opponents is observed immediately

Table 2. Female dominance styles of each macaque species

	Despotic	Egalitarian
silenus-sylvanus group	pig-tailed	barbary silenus crested tonkean moor (Gorontaro) (Heck's) (booted) (Muna-Butung)
sinica group		bonnet (toque) (Assamese) (Tibetan)
arctoides group		stumptail
fascicularis group	long-tailed rhesus Japanese (Taiwan)	

Parentheses indicate less known species (see Thierry 2000).

after the end of agonistic interactions. Such reunion is called "reconciliation" (de Waal and Roosmalen 1979). Reconciliation has been studied most intensively among macaques (barbary: Aureli et al. 1994; lion-tailed: Abegg et al. 1996; pig-tailed: Judge 1991; Castles et al. 1996; crested: Petit and Thierry 1994; tonkean: Thierry 1986; moor: Matsumura 1996; stumptail: de Waal and Ren 1988; rhesus: de Waal and Yoshihara 1983; Japanese: Aureli et al. 1993; Chaffin et al. 1995; long-tailed: Cords 1988; Aureli et al. 1989). Reconciliation occurred more frequently in egalitarian macaques than in despotic macaques. Almost half of agonistic interactions were reconciled in stumptail, lion-tailed, and Sulawesi macaques (Thierry 1986; de Waal and Ren 1988; Petit and Thierry 1994; Abegg et al. 1996; Matsumura 1996). By contrast, the rate of reconciliation was apparently lower in rhesus and Japanese macaques (de Waal and Yoshihara 1983; Aureli et al. 1994; Chaffin et al. 1995).

Another difference between egalitarian and despotic macaques was found in kinship effect on reconciliation (Aureli et al. 1997). The rate of reconciliation did not differ between non-kin dyads and kin dyads in barbary macaques. By contrast, Japanese macaques reconciled with relatives more frequently than non-relatives.

Egalitarian species are reported to exhibit (1) less severe (but sometimes more frequent) aggression, (2) greater symmetry in contests (more counter-attacks), (3) a higher rate of affiliation such as grooming, (4) a higher rate of peaceful post-conflict contacts or "reconciliation", and (5) less effect of kinship on social interactions than despotic species (de Waal and Luttrell 1989; Thierry et al. 1994). Differ-

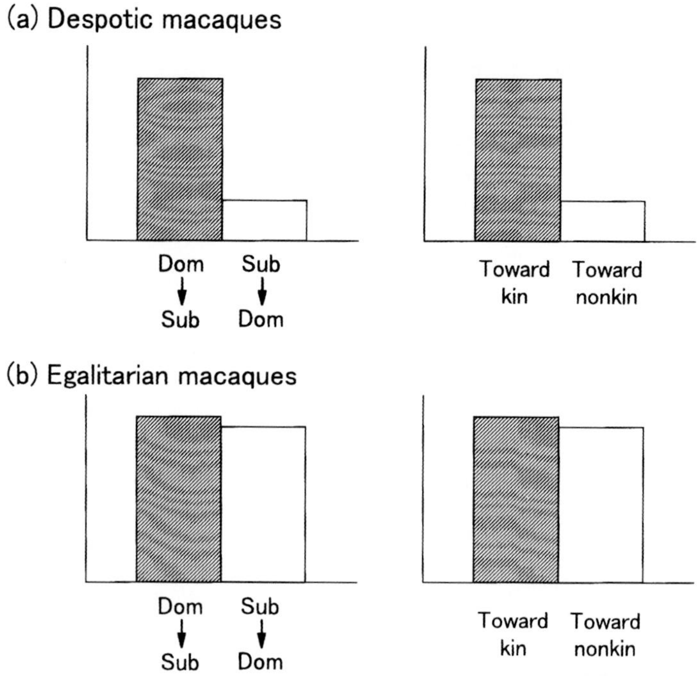

(a) Despotic macaques

(b) Egalitarian macaques

Fig. 1. Difference in dominance styles of despotic and egalitarian macaques. Despotic macaques (**a**) direct most non-agonistic approaches down the hierarchy and toward kin. By contrast, egalitarian macaques (**b**) direct approaches almost evenly toward dominants *(dom)* and subordinates *(sub)*, and toward kin and non-kin

ences between despotic and egalitarian macaques might be summarized as a difference in the effect of dominance and kinship on their social interactions. The effect of dominance and kinship on social interactions is smaller in egalitarian species than in despotic species. Power differentiation between high-ranking and low-ranking individuals is relatively small, and affiliative and cooperative interactions are commonly seen among non-kin dyads (Fig. 1). In the case of macaques, "egalitarian" does not mean a lack of formal dominance hierarchy. Some authors seemed to avoid using the term "egalitarian" and used the word "tolerant" instead (Sterck et al. 1997).

4.2 Egalitarian Macaques in Natural Habitats

Captive studies under controlled conditions have definitely contributed to advances in our knowledge of comparative behavior among macaque species. However, field studies are indispensable to ascertain the expression of the difference in dominance style in natural habitats and to examine the ecological and evolutionary significance of this difference. In contrast to captive studies, few field studies have been conducted on the ecology and social behavior of egalitarian macaque species.

Moor macaques live in the southwestern part of Sulawesi Island and have been studied in the Karaenta Nature Reserve since 1981 (Watanabe and Brotoisworo 1982; Watanabe and Matsumura 1996). Intensive observation of their ecology and social behavior started in 1988 on the basis of individual identification. Moor macaques are categorized into the egalitarian dominance style group. Forty-three percent of agonistic interactions were followed by non-agonistic contacts between the former opponents (Matsumura 1996). Intense aggression and nepotistic support were rarely observed even at the feeding site (Matsumura 1998). Only six cases out of 367 agonistic interactions involved biting. Although immature monkeys were attacked 92 times, nepotistic support was seen only one case.

Studies on moor macaques suggested that tolerance during feeding among non-related individuals characterizes their egalitarian dominance style in the natural habitat. Encounters within 1 m among adult females during feeding were frequent and symmetrical (Matsumura 1998). The frequency and degree of symmetry of such encounters differs from that of Japanese macaques (Fig. 2). In a group of Japanese macaques, an individual comes within 1 m of another only 0.7 times per hour, and subordinate individuals rarely approach non-kin dominants (Furuichi 1983). By contrast, one individual comes within 1 m of another 3.7 times per hour in the group of moor macaques. Approaches by the subordinate accounted for 42 percent of all the approaches.

Difference in tolerance during feeding results in difference in spacing of group members during feeding. The frequency distribution of spatial proximity between adult females during feeding is shown in Fig. 3. In moor macaques, kinship had little effect on spatial proximity among group members during feeding (Matsumura and Okamoto 1997). Co-feeding of unrelated individuals in a food patch was commonly observed. This presents a sharp contrast to spatial distribution on the basis of matrilineal kin groups in Japanese macaques (Furuichi 1984; Maruhashi 1986).

A high level of tolerance was also found in the social contexts. Figure 4 shows the distribution of grooming among adult females. Grooming was distributed rather evenly among adult females as compared with Japanese macaques (Furuichi 1984; Mitani 1986). No concentration of grooming was found on either related individuals or individuals of adjacent dominance rank. Moor macaques had various grooming partners from the beginning of their life. Although mothers groom their own infants and juveniles frequently, they also commonly groom others (Matsumura unpublished data).

At a proximate level, attractiveness of mothers holding their infants to other females (Matsumura 1997) makes the kinship effect weaker. Almost all females in the group frequently approached mothers with a particular behavior or vocalization. Thus, spatial proximity among females was strongly influenced by the presence of newborn infants even during feeding and moving as well as resting (Matsumura and Okamoto 1997). As mothers received more grooming when they were holding infants than when they were not (Matsumura 1997), a nepotistic tendency in grooming was less clearly recognized. A difference in attractiveness of mothers holding infants seems to exist between egalitarian and despotic macaques.

(a)

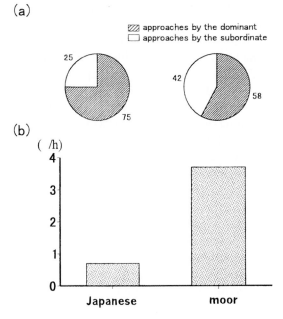

☒ approaches by the dominant
☐ approaches by the subordinate

25

75

42

58

(b)

(/h)

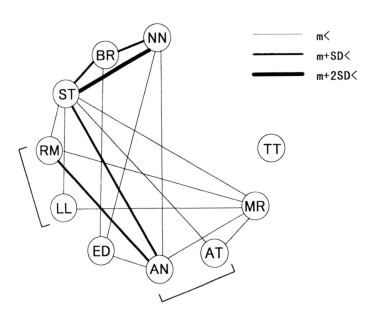

Fig. 2. Difference between Japanese (Furuichi 1983) and moor macaques (Matsumura 1998) in encounters within 1 m among adult females during feeding. **a** The proportions of approaches conducted by the dominant and the subordinate. **b** The frequency of encounters

Fig. 3. Proximity during feeding for adult females of Group B. Data were obtained by the scan sampling method (Altmann 1974) in 1993. For details of the sampling method, see Matsumura (1998). Adult females are arranged in descending rank order, anti-clockwise from NN. LL and AT were presumed to be daughters of RM and AN, respectively

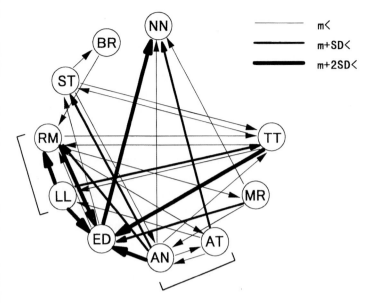

Fig. 4. Grooming among adult females of Group B. (See Fig. 3)

Mothers receive affiliative behaviors from adult females as frequently as from juvenile females in egalitarian macaques (Sugiyama 1971; Estrada and Estrada 1984; Thierry 1985b; Small 1990; Bauers 1993; Matsumura 1997; Silk 1999), but mostly from juvenile females in despotic species (Thierry 1985b; Hiraiwa 1981).

4.3 Why Do Dominance Styles Differ Between Macaque Species?

Several hypotheses or models have been proposed to explain the difference in dominance styles among macaque species both from evolutionary perspectives and from ontogenetic or epigenetic explanations.

4.3.1 Socioecological Explanation

The most influential explanations came from general socioecological models on primate female relationships (van Schaik 1989; Sterck et al. 1997). They argued that strong between-group contest competition would make female relationships more egalitarian. They assumed that communal defense against other groups may be the main selective force in the evolution of the egalitarian dominance style. Dominant females need the cooperation of subordinate females. Subordinate females can force dominant females not to exert their full power to suppress the subordinates food intake.

However, no direct support can be found for this hypothesis from the available literature (reviewed by Matsumura 1999). This may be partly because field data for egalitarian macaques are still insufficient.

4.3.2 Epigenetic Model: Nepotistic vs. Non-Nepotistic

Thierry (1990) proposed a model that claims the difference in dominance styles emerges from a difference in the degree of nepotism. In macaques, dominance is greatly influenced by third-party intervention on dyadic agonistic interactions. His model showed that if nepotism in such agonistic supports is weak, that is, supports for relatives are relatively rare or supports for non-relatives are relatively common, dominance would be less skewed. In other words, the power difference between the dominant individuals and the subordinate individuals is small. He also stressed the feedback loop between dominance and kinship. Less skewed dominance would make for a tendency for non-kin supports to be greater.

This model was criticized from mathematical and evolutionary viewpoints (Das and van Hooff 1993) because it does not explain the origin of weak nepotism. However, the idea of a positive feedback loop between dominance and kinship should not be ignored.

4.3.3 Individual-Oriented Computer Simulation: Fierce vs. Mild

Recently, Hemerlijk (1999) reported stimulating results of individual-oriented simulations in a "virtual" world. The basic concepts of her studies resembled Thierry's (1990) model. She introduced a difference in the intensity of aggression ("fierce" vs. "mild" types) instead of a difference in nepotism. Computer simulations were carried out where virtual individuals interacted with each other in a virtual world. After each interaction, individuals modified their behavior according to a learning rule. The result of her simulations showed that differentiation in individual power was smaller when the intensity of aggression was weak. The pattern of interactions between mild types resembled egalitarian macaques in the real world. This implies that various differences in social characteristics may result from a difference in the intensity of aggression.

4.3.4 Game Theoretical Model: Tolerant vs. Intolerant

On the basis of a game theory model for dominance (Matsumura and Kobayashi 1998), Matsumura (1999) proposed the hypothesis that a difference in dominance style reflects two different strategies for contest over resources: the despotic strategy and the egalitarian strategy. The despotic strategy is characterized by avoidance of escalated fights according to an existing asymmetry. When two despotic strategists meet over a resource, one chooses avoidance according to an existing asymmetry and the other obtains the resource. Animals should choose avoidance if the expected benefits of escalation are outweighed by the risk of injury. The egalitarian strategy is characterized by sharing of resources. When two egalitarian strategists meet over a resource, they share the resource with each other. If one escalates, the opponent also escalates. Coexistence around the resource can be achieved by high risk of injury.

This model agrees with reported differences between despotic and egalitarian macaques. In wild Japanese macaques, subordinate adult females tend to avoid unrelated dominants during feeding (Furuichi 1983; Saito 1996). A captive study showed that the proximate mechanism for reducing social disruption may rely on the avoidance of aggression (Chaffiln et al. 1995). In egalitarian species, a high level of tolerance during feeding and a high possibility of counter-attack during agonistic interactions was reported (Thierry 1985a; de Waal and Luttrell 1989; Petit et al. 1992; Matsumura 1998). Although the origin of the different strategies was not explained sufficiently by this model, the ecological and social features of each species can be compared with the conditions under which the strategies are evolutionarily stable (Matsumura and Kobayashi 1998). These conditions may correspond to ecological and social conditions among macaques.

4.3.5 Phylogenetic Factors

When comparing certain characteristics between species, their phylogenetic relationships must be taken into consideration (Harvey and Pagel 1991). Differences in dominance style agree considerably with the phylogenetic relationships among macaque species (Thierry et al. 1994; Matsumura 1999). With the exception of pig-tailed macaques, despotic species appear to be concentrated in the *fascicularis* group (Table 2).

5 Future of Comparative Studies on Dominance Styles in Macaques

Macaques have some potential advantages as subjects of comparative studies from the evolutionary viewpoint. In morphology, for example, a clear relationship between their habitats and physical features such as body size or tail length was demonstrated among macaque species (Fa 1989). In contrast, relationships between their habitats and social/behavioral characteristics appear to be very complex. Many interesting topics, however, must exist within this complexity. Recent studies have gradually revealed the significance of comparative social/behavioral studies of macaques.

The view of dominance and kinship has changed gradually through the development of the theories of sociobiology or behavioral ecology. Simple studies to seek correlations between dominance and feeding, mating, or reproductive success, or correlation between kinship and cooperative behavior have become outdated. Different effects of dominance and kinship are expected if ecological and social conditions differ. We should elucidate what kind of ecological and social factors give rise to such differences. Despotic and egalitarian macaques may be one of the best subjects to study in this respect.

At present, we need much more information. Socioecological studies on egalitarian macaques are indispensable to test the validity of current socioecological models. Less-known species should be studied both in natural habitats and in captivity, and more theoretical studies are needed to explain the difference in dominance styles. Present socioecological models seem to require further refinement. They have their theoretical origin in the model for reproductive skew (Vehrencamp 1983). Reproductive skew theory has made great advances in the last few years (Reeve 1998; Clutton-Brock 1998). Other advances in evolutionary theories should also be incorporated into new models.

Existing explanations for differences in dominance styles appear compatible with one another. As Tinbergen (1963) summarized, there are four ways of answering the question "why?" in biology: in terms of function, causation, development, and evolutionary history. Macaques provide opportunities to study the interplay between evolutionary and ontogenetic or epigenetic processes. Ontogeny of dominance and nepotistic behavior has been studied intensively in macaques. Many studies suggest the significance of nepotistic support in acquisition and maintenance of dominance rank (Chapais 1992). Association during early infancy is known to influence social preference and to work as an indirect mechanism of kin recognition among primates (Gouzoules 1984; Walters 1987). As many authors have pointed out, ontogenetic processes might play an important role in existing variations in dominance styles. Rhesus juveniles co-housed with stumptails showed a dramatic increase in reconciliation tendency (de Waal and Johanowicz 1993). A multivariate analysis of social interactions in captive rhesus macaques suggested that affiliative relations among mothers were an excellent predictor of affiliative relations among daughters (de Waal 1996). Egalitarian relationships among adult females may be transmitted from mothers to daughters.

In summary, further studies on the difference in dominance styles among macaques can tell us how evolutionary and ontogenetic processes work together under certain ecological conditions. Cooperation among researchers in different areas, such as those of feeding ecology, behavioral ecology, theoretical biology, developmental psychology and cognitive sciences, will bring us fruitful results.

Acknowledgments

The author is very grateful to T. Kano, Y. Sugiyama, K. Watanabe, A. Mori, H. Ohsawa, Y. Muroyama, T. Kobayashi, K. Okamoto, and other members of the seminar for ecology and social behavior, Primate Research Institute, Kyoto University, for their helpful comments, and to T Matsuzawa for giving me an opportunity to write this manuscript. This study was supported in part by a grant-in-aid from the Japanese Ministry for Education, Science, Sport, and Culture.

References

Altmann J (1974) Observational study of behaviour: sampling methods. Behaviour 49:227–267
Altmann SA (1962) A field study of the sociobiology of rhesus monkeys (*Macaca mulatta*). Ann New York Acad Sci 102:338–345
Abegg C, Thierry B, Kaumanns W (1996) Reconciliation in two groups of lion-tailed macaques (*Macaca silenus*). Int J Primatol 17:803–816
Aureli F, van Schaik CP, van Hooff JARAM (1989) Functional aspects of reconciliation among captive long-tailed macaques (*Macaca fascicularis*). Am J Primatol 19:39–51
Aureli F, Veenema HC, van Panthaleon van Eck CJ, van Hooff JARAM (1993) Reconciliation, consolation, and redirection in Japanese macaques (*Macaca fuscata*). Behaviour 124:1–21
Aureli F, Das M, Verleur D, van Hooff JARAM (1994) Postconflict social interactions among barbary macaques (*Macaca sylvanus*). Int J Primatol 15:471–485

Aureli F, Das M, Veenema HC (1997) Differential kinship effect on reconciliation in three species of macaques (*Macaca fascicularis, M. fuscata*, and *M. sylvanus*). J Comp Psychol 111:91–99

Bauers KA (1993) A functional analysis of staccato grunt vocalizations in the stumptailed macaque (*Macaca arctoides*). Ethology 94:147–161

Bercovitch FB, Huffman MA (1999) The macaques. In: Dolhinow P, Fuentes A (eds) The nonhuman primates. Mayfield, Mountain View, CA, pp 77–85

Butobvskaya M (1993) Kinship and different dominance styles in groups of three species of the genus *Macaca* (*M. arctoides, M. mulatta, M. fascicularis*). Folia Primatol 60:210–224

Caldecott JO (1986) Mating patterns, societies and the ecogeography of macaques. Anim Behav 34:208–220

Castles DL, Aureli F, de Waal FBM (1996) Variation in conciliatory tendency and relationship quality across group of pigtail macaques. Anim Behav 52:389–403

Chaffin CL, Friedlen K, de Waal FBM (1995) Dominance style of Japanese macaques compared with rhesus and stumptail macaques. Am J Primatol 35:103–116

Chapais B (1992) The role of alliances in the social inheritance among female primates. In: Harcourt AH, de Waal FBM (eds) Coalitions and alliances in humans and other animals. Oxford University Press, Oxford, pp 29–60

Chapais B, Schulman S (1980) An evolutionary model of female dominance relations in primates. J Theor Biol 82:47–89

Cheney DL (1977) The acquisition of rank and development of reciprocal alliances among free-ranging immature baboons. Behav Ecol Sociobiol 2:303–318

Clutton-Brock TH (1998) Reproductive skew, concessions and limited control. TREE 13:288–292

Clutton-Brock TH, Harvey PH (1977) Species differences in feeding and ranging behaviour in primates. In: Clutton-Brock TH (ed) Primate ecology. Academic, London, pp 557–584

Cords M (1988) Resolution of aggressive conflicts by immature long-tailed macaques *Macaca fascicularis*. Anim Behav 36:1124–1135

Crockett CM, Wilson WL (1980) The ecological separation of *Macaca nemestrina* and *M. fascicularis* in Sumatra. In: Lindburg DG (ed) The macaques: studies in ecology, behavior and evolution. Van Nostrand Reinhold, New York, pp 148–181

Crook JH, Gartlan JS (1966) Evolution of primate societies. Nature 210:1200–1203

Das M, van Hooff JARAM (1993) A non-adaptive feedback loop account for the differences between the macaques' social systems—is it convincing? J Theor Biol 160:399–402

Datta SB (1988) The acquisition of dominance among free-ranging rhesus monkey siblings. Anim Behav 36:754–772

Datta SB, Beauchamp G (1991) Effects of group demography on dominance relationships among female primates: I. Mother-daughter and sister-sister relations. Amer Nat 138:201–226

Defler TR (1978) Allgrooming in two species of macaque (*Macaca nemestrina* and *Macaca radiata*). Primates 19:153–167

Delson E (1980) Fossil macaques, phyletic relationships and a scenario of development. In: Lindburg DG (ed) The macaques: studies in ecology, behavior and evolution. Van Nostrand Reinhold, New York, pp 10–30

de Waal FBM (1989) Dominance 'style' and primate social organization. In: Standen V, Foley R (eds) Comparative socioecology: the behavioural ecology of humans and other mammals. Blackwell Scientific, Oxford, pp 243–263

de Waal FBM (1996) Macaque social culture: development and perpetuation of affiliative networks. J Comp Psychol 110:147–154

de Waal FBM, Johanowicz DL (1993) Modification of reconciliation behavior through social experience: an experiment with two macaque species. Child Dev 64:898–908

de Waal FBM, Luttrell LM (1985) The formal hierarchy of rhesus macaques: an investigation of the bared-teeth display. Amer J Primatol 9:73–85

de Waal FBM, Luttrell LM (1989) Toward a comparative socioecology of the genus *Macaca*: different dominance styles in rhesus and stumptail monkeys. Am J Primatol 19:83–109

de Waal FBM, Ren RM (1988) Comparison of the reconciliation behavior of stumptail and rhesus macaques. Ethology 78:129–142

458   S. Matsumura

de Waal FBM, van Roosmalen A (1979) Reconciliation and consolation among chimpanzees. Behav Ecol Sociobiol 5:55–66
de Waal FBM, Yoshihara D (1983) Reconciliation and redirected affection in rhesus monkeys. Behaviour 85:224–241
Ellis L (1995) Dominance and reproductive success among nonhuman animals: A cross-species comparison. Ethol Sociobiol 16:257–333
Estrada A, Estrada R (1984) Female-infant interactions among free-ranging stumptail macaques (*Macaca arctoides*). Primates 25:48–61
Eudey AA (1980) Pleistocene glacial phenomena and the evolution of Asian macaques. In: Lindburg DG (ed) The macaques: studies in ecology, behavior and evolution. Van Nostrand Reinhold, New York, pp 52–83
Fa, JE (1989) The genus *Macaca*: A review of taxonomy and evolution. Mammal Rev 19:45–81
Fedigan LM (1983) Dominance and reproductive success in primates. Yearbook Phys Anthrop 26:85–123
Fooden J (1976) Provisional classification and key to the living species of macaques (Primates: *Macaca*). Folia Primatol 25:225–236
Fooden J (1982) Ecogeographic segregation of macaque species. Primates 23:574–579
Fooden J, Lanyon SM (1989) Blood-protein allele frequencies and phylogenetic relationships in *Macaca*: a review. Am J Primatol 17:209–241
Furuichi T (1983) Interindividual distance and influence of dominance on feeding in a natural Japanese macaque troop. Primates 24:445–455
Furuichi T (1984) Symmetrical patterns in non-agonistic social interactions found in unprovisioned Japanese macaques. J Ethol 2:109–119
Furuya Y (1957) Grooming behavior in the wild Japanese monkeys. Primates 1:47–68 (in Japanese)
Gouzoules S (1984) Primate mating systems, kin associations, and cooperative behavior: evidence for kin recognition? Yearbook Phys Anthropol 27:99–134
Gouzoules S, Gouzoules H (1987) Kinship. In: Smuts BB, Cheney DL, Seyfarth RM, Wrangham RW, Struhsaker TT (eds) Primate societies. University of Chicago Press, Chicago, pp 299–305
Hamilton WJ (1964) The genetical evolution of social behavior. J Theor Biol 7:1–51
Harcourt AH (1987) Dominance and fertility among female primates. J Zool 213:471–487
Harvey PH, Pagel MD (1991) The comparative method in evolutionary biology. Oxford University Press, Oxford
Hausfater G, Altmann J, Altmann SA (1982) Long-term consistency of dominance relations among female baboons (*Papio cynocephalus*). Science 217:752–755
Hayasaka K, Fujii K, Horai S (1996) Molecular phylogeney of macaques: implication of nucleotide sequences from an 896-base pair region of mitochondorial DNA. Mol Biol Evol 13:1044–1053
Hemerlijk CK (1999) An individual-oriented model of the emergence of despotic and egalitarian societies. Proc R Soc Lond B 266:361–369
Hill DA (1999) Effects of provisioning on the social behavior of Japanese and rhesus macaques: implications for socioecology. Primates 40:187–198
Hill DA, Okayasu N (1995) Absence of 'youngest ascendancy' in the dominance relations of sisters in wild Japanese macaques (*Macaca fuscata yakui*). Behaviour 132:267–279
Hill DA, Okayasu N (1996) Determinants of dominance among female macaques: nepotism, demography and danger. In: Fa JE, Lindburg DG (eds) Evolution and ecology of macaque societies. Cambridge University Press, Cambridge, pp 459–472
Hiraiwa M (1981) Maternal and alloparental care in a troop of free-ranging Japanese monkeys. Primates 22:309–329
Hoelzer GA, Melnick DJ (1996) Evolutionary relationships of the macaques. In: Fa JE, Lindburg DG (eds) Evolution and ecology of macaque societies. Cambridge University Press, Cambridge, pp 3–19
Horrocks JA, Hunte W (1983) Maternal rank and offspring rank in vervet monkeys: an appraisal of the mechanisms of rank acquisition. Anim Behav 31:772–782
Ihobe H (1989) How social relationships influence a monkey's choice of feeding site in the troop of Japanese macaques (*Macaca fuscata fuscata*) on Koshima Islet. Primates 30:17–25

Imanishi K (1960) Social organization of subhuman primates in their natural habitat. Curr Anthropol 1:393–407

Itani J (1954) The monkeys of Takasakiyama. Kobunsha, Tokyo (in Japanese)

Iwamoto T (1987) Feeding strategies of primates in relation to social status. In: Ito Y, Brown JL, Kikkawa J (eds) Animal societies. Japan Scientific, Tokyo, pp 243–252

Izawa K (1982) Ecology of Japanese monkeys. Dobutsusha, Tokyo (in Japanese)

Judge PJ (1991) Dyadic and triadic reconciliation in pigtail macaques (*Macaca nemestrina*). Am J Primatol 23:225–237

Kawai M (1958) On the system of social ranks in a natural group of Japanese monkeys. Primates 1:111–148 (in Japanese with English summary)

Kawai M (1964) Ecology of Japanese monkeys. Kawadeshobo, Tokyo (in Japanese)

Kawamura S (1958) Matriarchal social order in the Minoo-B group: a study of the rank system of Japanese macaques. Primates 1:149–156 (in Japanese with English summary)

Koford (1963) Rank of mothers and sons in bands of rhesus monkeys. Science 141:356–357

Koyama N (1967) On the dominance rank and kinship of a wild Japanese monkey troop in Arashiyama. Primates 8:189–216

Koyama N (1970) Changes in dominance rank and division of a wild Japanese monkey troop in Arashiyama. Primates 11:335–390

Koyama N (1973) Dominance, grooming, and clasped-sleeping relationships among bonnet monkeys in India. Primates 14:225–244

Koyama N (1985) Playmate relationships among individuals of the Japanese monkey troop in Arashiyama. Primates 26:390–406

Koyama N (1991) Grooming relationships in the Arashiyama group of Japanese monkeys. In: Fedigan LM, Asquith PJ (eds) The Monkeys of Arashiyama. State University of New York Press, Albany, pp 211–226

Kurland JA (1977) Kin selection in the Japanese macaque. Contributions to Primatology, vol 12 S, Karger, Basel

Lindburg DG (1977) Feeding behaviour and diet of rhesus monkeys (*Macaca mulatta*) in a Siwalik forest in North India. In: Clutton-Brock TH (ed) Primate ecology. Academic, London, pp 223–249

Lorenz K (1966) On aggression. Methuen, London

Maruhashi T (1986) Feeding ecology of Japanese monkeys in Yakushima Island. In: Maruhashi T, Yamagiwa J, Furuichi T (eds) The wild monkeys on Yakushima Island. Tokai Daigaku Shuppankai, Tokyo, pp 13–59 (in Japanese)

Maruhashi T (1992) Fission, takeover, and extinction of a troop of Japanese monkeys (*Macaca fuscata yakui*) on Yakushima Island, Japan. In: Itoigawa N, Sugiyama Y, Sackett GP, Thompson RKR (eds) Topics in Primatology, Vol.2. University of Tokyo Press, Tokyo, pp 47–56

Maruhashi T, Yamagiwa J, Furuichi T (1986) The wild monkeys on Yakushima Island. Tokai Daigaku Shuppankai, Tokyo (in Japanese)

Matsumura S (1996) Post-conflict affiliative contacts between former opponents among wild moor macaques (*Macaca maurus*). Am J Primatol 38:211–219

Matsumura S (1997) Mothers in a wild group of moor macaques (*Macaca maurus*) are more attractive to other group members when holding their infants. Folia Primatol 68:77–85

Matsumura S (1998) Relaxed dominance relations among female moor macaques (*Macaca maurus*) in their natural habitat, South Sulawesi, Indonesia. Folia Primatol 69:346–356

Matsumura S (1999) The evolution of "egalitarian" and "despotic" social systems among macaques. Primates 40:23–31

Matsumura S, Kobayashi T (1998) A game model for dominance relations among group-living animals. Behav Ecol Sociobiol 42:77–84

Matsumura S, Okamoto K (1997) Factors affecting proximity among members of a wild group of moor macaques during feeding, moving, and resting. Int J Primatol 18:929–940

Maynard-Smith J (1964) Group selection and kin selection. Nature 201:1145–1146

Menard D, Vallet D (1996) Demography and ecology of barbary macaques (*Macaca sylvanus*) in two different habitats. In: Fa JE, Lindburg DG (eds) Evolution and ecology of macaque societies. Cambridge University Press, Cambridge, pp 106–131

Missakian EA (1972) Genealogical and cross-genealogical dominance relations in a group of free-ranging rhesus monkey (*Macaca mulatta*) on Cayo Santiago. Primates 13:169–180

Mitani M (1986) Voiceprint identification and its application to sociological studies of wild Japanese monkeys (*Macaca fuscata yakui*). Primates 27:397–412

Mizuhara H (1957) Japanese monkeys. San'ichishobo, Kyoto (in Japanese)

Morales JC, Melnick DJ (1998) Phylegenetic relationships of the macaques (Cercopithecidae: *Macaca*), as revealed by high resolution restriction site mapping of mitochondorial ribosomal genes. J Hum Evol 34:1–23

Mori A (1977) Intra-troop spacing mechanism of the wild Japanese monkeys of the Koshima troop. Primates 18:331–357

Mori A, Watanabe K, Yamaguchi N (1989) Longitudinal changes of dominance rank among the females of the Koshima group of Japanese monkeys. Primates 30:147–173

Moore J (1992) Dispersal, nepotism, and primate social behavior. Int J Primatol 13:361–378

Nakagawa N, Iwamoto T, Yokota N, Soumah AG (1996) Inter-regional and inter-seasonal variations of food quality in Japanese macaques: constraints of digestive volume and feeding time. In: Fa JE, Lindburg DG (eds) Evolution and ecology of macaque societies. Cambridge University Press, Cambridge, pp 207–234

Oi T (1987) Sociological study on the troop fission of wild Japanese monkeys (*Macaca fuscata yakui*) on Yakushima Island. Primates 29:1–19

Oi T (1996) Sexual behavior and mating system of the wild pig-tailed macaques in West Sumatra. In: Fa JE, Lindburg DG (eds) Evolution and ecology of macaque societies. Cambridge University Press, Cambridge, pp 342–368

Oki J, Maeda Y (1973) Grooming as a regulator of behavior in Japanese macaques. In: Carpenter CR (ed) Behavioral regulators of behavior in primates. Bucknel University Press, Lewisburg, pp 149–163

Petit O, Thierry B (1994) Reconciliation in a group of black macaques (*Macaca nigra*). Dodo J Wildl Pres Trusts 30:89–95

Petit O, Desportes C, Thierry B (1992) Differential probability of "coproduction" in two species of macaques (*Macaca tonkeana, M. mulatta*). Ethology 90:107–120

Preuschoft S (1995) "Laughter" and "smiling" in macaques: an evolutionary perspective. Faculteit Biologie-Universiteit Utrecht, Utrecht

Rawlins RG, Kessler MJ (1986) The Cayo Santiago macaques: History, behavior and biology. SUNY Press, New York

Reeve HK (1998) Game theory, reproductive skew, and nepotism. In: Dugatkin LA, Reeve, HK (eds) Game theory and animal behavior. Oxford University Press, Oxford, pp 118–145

Reynolds V (1992) The discovery of primate kinship systems by Japanese anthropologists. In: Itoigawa N, Sugiyama Y, Sackett GP, Thompson RKR (eds) Topics in Primatology, Vol 2 University of Tokyo Press, Tokyo, pp 79–87

Rodman PS (1991) Structural differentiation of microhabitats of sympatric *Macaca fascicularis* and *M. nemestrina* in East Kalimantan, Indonesia. Int J Primatol 12:357–375

Sade DS (1965) Some aspects of parent-offspring and sibling relations in a group of rhesus monkey, with a discussion of grooming. Am J Phys Anthropol 23:1–18

Sade DS (1967) Determinants of dominance in a group of free-ranging rhesus monkeys. In: Altmann SA (ed) Social communication in primates. University of Chicago Press, Chicago, pp 99–114

Saito C (1996) Dominance and feeding success in female Japanese macaques, *Macaca fuscata*: effects of food patch size and inter-patch distance. Anim Behav 51:967–980

Silk JB (1987) Social behavior in evolutionary perspective. In: Smuts BB, Cheney DL, Seyfarth RM, Wrangham RW, Struhsaker TT (eds) Primate societies. University of Chicago Press, Chicago, pp 318–329

Silk JB (1999) Why are infants so attractive to others? The form and function of infant handling in bonnet macaques. Anim Behav 57:1021–1032

Silk JB, Samuels A, Rodman P (1981) The influence of kinship, rank, and sex of affiliation and aggression between adult female and immature bonnet macaques (*Macaca radiata*). Behaviour 78:111–137

Small MF (1990) Alloparental behaviour in barbary macaques, *Macaca sylvanus*. Anim Behav 39:297–306

Sterck EHM, Watts DP, van Schaik CP (1997) The evolution of female social relationships in nonhuman primates. Behav Ecol Sociobiol 41:291–309

Sugiyama Y (1971) Characteristics of the social life of bonnet macaques (*Macaca radiata*). Primates 12:247–266

Sugiyama Y (1992) Behavioral studies of Japanese monkeys in artificial feeding ground and natural environment. In: Itoigawa N, Sugiyama Y, Sackett GP, Thompson RKR (eds) Topics in Primatology, Vol 2, University of Tokyo Press, Tokyo, pp 3–9

Suzuki A (1965) An ecological study of wild Japanese monkeys in snowy areas, focused on their food habits. Primates 6:31–72

Takahashi H, Furuichi T (1998) Comparative study of grooming relationships among wild Japanese macaques in Kinkazan A troop and Yakushima M troop. Primates 39:365–374

Takahata Y (1991) Diachronic changes in the dominance relations of adult female Japanese monkeys of the Arashiyama B group. In: Fedigan LM, Asquith PJ (eds) The monkeys of Arashiyama. State University of New York Press, Albany, pp 123–139

Takahata Y, Sprague DS, Suzuki S, Okayasu N (1994) Female competition, co-existence, and the mating structure of wild Japanese macaques on Yakushima Island, Japan. In: Jarman PJ, Rossiter A (eds) Animal societies: individuals, interactions and organization. Kyoto University Press, Kyoto, pp 163–179

Thierry B (1985a) Patterns of agonistic interactions in three species of macaques (*Macaca mulatta, M. fascicularis, M. tonkeana*). Aggr Behav 11:223–233

Thierry B (1985b) Social development in three species of macaque (*Macaca mulatta, M. fascicularis, M. tonkeana*): a preliminary report on the first ten weeks of life. Behav Proc 11:89–95

Thierry B (1986) A comparative study of aggression and response to aggression in three species of macaque. In: Else J, Lee PC (eds) Primate ontogeny, cognition and social behavior. Cambridge University Press, Cambridge, pp 307–313

Thierry B (1990) Feedback loop between kinship and dominance: the macaque model. J Theor Biol 145:511–521

Thierry B (2000) Covariation of conflict management patterns in macaque societies. In: Aureli F, de Waal FBM (eds) Natural conflict resolution. University of California Press, Berkeley, pp 106–128

Thierry B, Anderson JR, Demaria C, Desportes C, Petit O (1994) Tonkean macaque behaviour from the perspective of the evolution of Sulawesi macaques. In: Roeder JJ, Thierry B, Anderson JR, Herrenschmidt N (eds) Current Primatology, Vol. 2, Louis Pasteur University Press, Strasbourg, pp 103–117

Tinbergen N (1963) On aims and methods of ethology. Z Tierpsychol 20:410–433

van Schaik CP (1983) Why are diurnal primates living in groups? Behaviour 87:120–144

van Schaik CP (1989) The ecology of social relationships amongst female primates. In: Standen V, Foley R (eds) Comparative socioecology: the behavioural ecology of humans and other mammals. Blackwell Scientific Publications, Oxford, pp 195–218

van Schaik CP, van Noordwijk MA (1988) Scramble and contest in feeding competition among female long-tailed macaques (*Macaca fascicularis*). Behaviour 105:77–98

van Schaik CP, van Noordwijk MA, de Boer RJ, den Tonkeaar I (1983) The effect of group size on time budgets and social behaviour in wild long-tailed macaques (*Macaca fascicularis*). Behav Ecol Sociobiol 13:173–181

Vehrencamp SL (1983) A model for the evolution of despotic versus egalitarian societies. Anim Behav 31:667–682

Walters JR (1987) Kin recognition in nonhuman primates. In: Fletcher DF, Michener CD (eds) Kin recognition in animals. John Wiley, New York, pp 359–393

Walters JR, Seyfarth RM (1987) Conflict and cooperation. In: Smuts BB, Cheney DL, Seyfarth RM, Wrangham RW, Struhsaker TT (eds) Primate societies. University of Chicago Press, Chicago, pp 306–317

Watanabe K (1979) Alliance formation in a free-ranging troop of Japanese macaques. Primates 20:459–474

Watanabe K, Brotoisworo E (1982) Field observation of Sulawesi macaques. Kyoto University Overseas Res Report of Studies on Asian Non-Human Primates 2:3-9

Watanabe K, Matsumura S (1996) Social organization of moor macaques, *Macaca maurus*, in the Karaenta Nature Reserve, South Sulawesi, Indonesia. In: Shotake T, Wada K (eds) Variations in the Asian macaques. Tokai University Press, Tokyo, pp 147-162

Williams GC (1966) Adaptation and natural selection. Princeton University Press, Princeton, NJ

Wilson EO (1975) Sociobiology: the new synthesis. Harvard University Press, Cambridge

Wrangham RW (1980) An ecological model of female-bonded primate groups. Behaviour 75:262-300

Yamada M (1963) A study of blood-relationship in the natural society of the Japanese macaque: an analysis of co-feeding, grooming, and playmate relationships in Minoo-B-troop. Primates 4:43-65

Yamagiwa J, Hill DA (1998) Intraspecific variation in the social organization of Japanese macaques: past and present scope of field studies in natural habitats. Primates 39:257-273

Zhao QK (1996) Etho-ecology of Tibetan macaques at Mount Emei, China. In: Fa JE, Lindburg DG (eds) Evolution and ecology of macaque societies. Cambridge University Press, Cambridge, pp 263-289

23
Decision Making in Social Interactions by Monkeys

Yasuyuki Muroyama

1 Introduction

Nonhuman group-living primates are living in a socially complicated environment characterized by various relationships such as kin and dominance. In most of the Old World monkeys, females generally remain in their natal groups throughout their lives, while males emigrate from their natal group after sexual maturity. Females have strong social bonds with related group members, and thus they frequently interacted with them and form aggressive alliances against members of other matrilines. Dominance also affects the frequency and direction of social interactions among related and unrelated group members. What kind of social relationships an individual has with other group members may greatly affect its survival and fecundity (Dunbar 1988).

Social relationships are based on, and shaped by, everyday social interactions between group members, while these relationships in turn affect the type, quality, and frequency of interactions (Hinde 1976). Such feedback between social interactions and relationships makes the nature of interindividual relationships among group members dynamic rather than static. The death, or decline in strength, of one individual, for example, may change the content and frequency of social interactions, which may evoke a drastic change of relationships among group members (e.g., de Waal 1982).

Under such circumstance, group-living monkeys always face problems of what kind of behavior they should choose during social interactions with others. Monkeys must obtain information about their social environment where they are, and select useful information to make appropriate behavioral decisions. Memorized information, such as past interactions with others, may also modify such decision making. Previous studies have revealed that monkeys in some primate species are able to identify each group member and to recognize kin and/or affiliative relationships between others and themselves as well as among third parties (e.g., Bachmann and Kummer 1980; Dasser 1988a, b). To remember how often, and what kind of, social interactions are had and with whom, and to recall and make use of

Field Research Center, Primate Research Institute, Kyoto University, 41 Kanrin, Inuyama, Aichi 484-8506, Japan

the memory of these past interactions in another situation, might promote "good" relationships with others.

How do monkeys obtain and retain the necessary information from various interactions with others and make good use of it? Many studies about the memories or cognitive abilities of nonhuman primates have been conducted in an experimental situation where they are alone, whereas few such studies have been carried out during social interactions with animals of the same species (reviewed in Seyfarth and Cheney 1994). Numerous questions remain unanswered: e.g., how monkeys recognize and memorize their own behavior and that of others, how exactly they time or count their partners' social behavior as well as their own, how they process the information obtained during interactions and make behavioral decisions accordingly, and so on. In this chapter, I review a series of studies which examined how monkeys recognize social grooming (Muroyama 1991, 1994, 1996), focusing on the cognitive aspects of decision making in the turn-taking of grooming between two monkeys, and that during grooming sequences by one individual.

2 Exchange of Social Behaviors

2.1 Social Grooming as a "Currency"

Social grooming is the most common form of affiliative behavior in nonhuman primates (Goosen 1987). Various intensive studies have been conducted on its functions (e.g., Terry 1970; Hutchins and Barash 1976; McKenna 1978; Boccia 1983; Barton 1985; Schino et al. 1988; Boccia et al. 1989; Tanaka and Takefushi 1993). In contrast, few empirical studies have addressed the temporal organization of this behavior. Recent studies have reported that there is a sequential dependency between the behavior of participants during grooming interactions (Boccia et al. 1982; Tsukahara 1990). Furthermore, these studies revealed that such dependency was variable, depending on the social attributes of the performer and the recipient (Tsukahara 1990).

In most nonhuman primates, social grooming is conducted by two animals: one of them takes the part of the groomer, and the other takes the part of the groomee. Grooming another monkey appears to be costly in terms of time and effort (e.g., Kurland 1977; see also Seyfarth and Cheney 1988). Being groomed, in contrast, may be beneficial in terms of cleaning (McKenna 1978; Boccia 1983; Tanaka and Takefushi 1993), or reducing tension in the recipient (Schino et al 1988; Boccia et al. 1989). Thus grooming could be an altruistic behavior to be exchanged, as well as a "currency" with which the animal "pays" for the receipt of benefits other than being groomed itself.

Most studies on nonhuman primates have demonstrated that beneficial acts such as grooming and support in agonistic interactions are indeed exchanged reciprocally over long periods of time in long-lasting relationships (e.g., Packer 1977; de Waal 1978; de Waal and Luttrell 1988; Hemelrijk 1990), as well as over the short term (Seyfarth and Cheney 1984; de Waal 1989). It is obvious that various behavioral "currencies" may be involved in such exchanges, although it remains unclear how behavioral currencies relate to the ultimate biological currency, i.e., fitness (Seyfarth and Cheney 1988). Do monkeys really recognize grooming as a "currency" to be exchanged with other behavior? If this is the case, with whom do they exchange such a currency, and how? The answers to these questions will also provide some insights into how monkeys classify their grooming partner in terms of reciprocal relationships, or "reciprocity."

2.2 Turn-Taking in Grooming in Japanese Macaques

The first study of decision making during grooming interactions was conducted with female Japanese macaques (*Macaca fuscata fuscata*) (Muroyama 1991). To investigate whether the participants in grooming interactions recognize which is the recipient of a benefit (i.e., grooming), and then make a behavioral decision accordingly, resumptions of grooming interactions were observed. If the groomer expects her action to be reciprocated, she will resume the interaction by soliciting grooming from her partner. In contrast, if the groomee is aware of the principle of reciprocity and recognizes that her partner has the next turn at being groomed, she is likely to resume the interaction with grooming.

The subject group was the main troop of Japanese macaques that inhabit Koshima, Miyazaki, Japan. They were observed over a total of 4 months in 1985–1986. Kin relationships among the members of the troop, who were well habituated and identified, were known. When the study started, the troop consisted of eight adult males (>6 years old), 28 adult females, seven adolescent males (4–6 years old), four adolescent females, nine juvenile males (1–3 years old), six juvenile females, and four infants (<1 year old). From these individuals, five adult females, one adolescent male, and one adolescent female were selected as target animals. For each target animal, a total of 60 h of observation was undertaken by the focal animal sampling method (Altmann 1974). All dyadic grooming pairs observed were classified into four categories, according to relationships between the members of the pairs, as follows: affiliated, frequently interacting nonrelated ($r < 1/4$) pairs; unaffiliated, rarely interacting nonrelated ($r < 1/4$) pairs; mother and her adult female offspring pairs; mother and her immature offspring pairs. Focal animal sessions lasted all day when possible. Whenever a session was interrupted, it was resumed at approximately the same time on another day. The observation hours were almost evenly distributed between 0700 and 1700 hours.

In this study, both grooming and "soliciting" of grooming were recorded. "Grooming" was defined as a continuous act of picking and/or brushing by one animal through the fur of another with no interruption by pauses of longer than 1 s. "Soliciting" was defined as presenting the neck, face, rump, dorsum, ventrum, or flank, or lying down in front of the individual from which the monkey was soliciting grooming (Oki and Maeda 1973). Grooming interactions were defined as having been resumed when either of the participants in the interaction again performed grooming or soliciting within 1 min after the end of the previous grooming. Soliciting without actual grooming was also recorded as one resumption of a grooming interaction.

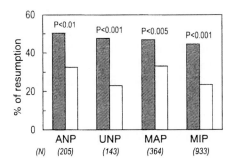

Fig. 1. Probabilities with which a grooming interaction was resumed by a previous groomer (*hatched bars*) or groomee (*open bars*). *ANP,* Affiliated and nonrelated pairs; *UNP,* unaffiliated, nonrelated pairs; *MAP,* mother and adult female offspring pairs; *MIP,* mother and immature offspring pairs. (From Muroyama 1991, with permission)

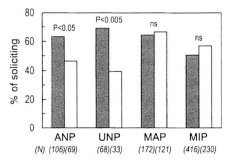

Fig. 2. Probabilities with which a grooming interaction was resumed by soliciting behavior of a previous groomer (*hatched bars*) or groomee (*open bars*). *ANP,* Affiliated and nonrelated pairs; *UNP,* unaffiliated, nonrelated pairs; *MAP,* mother and adult female offspring pairs; *MIP,* mother and immature offspring pairs. (From Muroyama 1991, with permission)

Figure 1 shows the probabilities with which, when an interaction was resumed, the previous groomer, or a groomee, resumed the interaction. The previous groomer was more likely to resume the interaction than vice versa in all grooming pairs. Thus providing or receiving a grooming affected the subsequent behavioral decision of whether to resume the interaction. This indicates that female Japanese macaques clearly distinguish between providing and receiving grooming, and make their decisions accordingly. The fact that the previous groomee was less likely to resume than the groomer suggests that the receipt of a benefit (grooming) may be one of the main goals in grooming interactions.

The subsequent decision of whether to groom or to solicit at the resumption of the grooming interaction, however, differed between related and unrelated pairs. In both of the unrelated pairs, the groomer tended to resume the interaction by soliciting grooming from the partner, while the groomee resumed by grooming the partner (Fig. 2). In related pairs, however, the groomer resumed the interaction by soliciting as frequently as by grooming (Fig. 2). Consequently, the reciprocated receipt of a benefit just given, in the form of reciprocation of grooming, is immediate in unrelated pairs, but not in related pairs. Unrelated pairs actually appear to be aware of the principle of reciprocity and behave most efficiently. By exhibiting a soliciting posture, the groomer might notify the groomee that her "turn" at being groomed has ended, or may try to ensure that the next "turn" at being groomed is hers. In any case, such behavior would certainly serve as a constraint against possible cheats, which would not like to reciprocate grooming.

This study suggests that female Japanese macaques recognize grooming as a beneficial act to be exchanged with unrelated members of the group. In other words, they appear to categorize other females into at least two groups: females with which they are expected to exchange grooming reciprocally and those with which they are not. Such categorization may be applied when females exchange various kind of social behavior, both beneficial and harmful, with other group members.

2.3 Exchange of Grooming for Allomothering in Patas Monkeys

If monkeys really recognize grooming as a "currency" to be exchanged for other behaviors, what kind of behaviors do they exchange with such a currency? A variety of studies have suggested possible currencies, but few attempts have been made to elucidate the relations between currencies (Seyfarth and Cheney 1988). Muroyama (1994) investigated the exchange of the same behavior (grooming) as well as that of different behaviors (grooming and allomothering) between female patas monkeys (*Erythrocebus patas*), in a similar way to the previous study of Japanese macaques (Muroyama 1991), and demonstrated that such exchanges do actually occur.

The study group was a troop of patas monkeys that inhabit the Kala Maloué National Park, Cameroon, West Africa. The study periods were October and November 1988 (a nonbirth season), and January and February 1989 (a birth season). Kin relationships among the members of the troop, who had been well habituated and identified, were known. When the study started, the troop consisted of 1 adult male, 6 adult females (>5 years old), 1 juvenile female (1.5 years old), and 5 infants (0.5 years old). One infant disappeared on November 27, 1988. Three adult females gave birth during the birth season. All six adult females were selected as target animals. All dyadic grooming pairs observed were classified into three categories according to the relationships between the members of the pairs, as follows: affiliated, frequently interacting nonrelated ($r < 1/4$) pairs; unaffiliated, rarely interacting nonrelated ($r < 1/4$) pairs; mother and her adult female offspring pairs. During the birth season, each of the grooming pairs was placed in one of three subcategories, namely, mother/mother dyads, mother/nonmother dyads, and nonmother/nonmother dyads. During the nonbirth season, each target animal was observed over a total of about 20 h by the focal animal sampling method (Altmann 1974). During the birth season, a total of about 30 h observation was undertaken for each mother of a newborn infant, and a total of about 20 h for each nonmother. Focal animal sessions lasted all day when possible. Whenever a session was interrupted, it was resumed at approximately the same time on another day. Observation hours were almost evenly distributed between 0630 and 1730 hours.

The definitions of grooming and soliciting were the same as those in the study of Japanese macaques, except for grooming resumption, which was defined as when

Table 1. Percentage of grooming interactions that were resumed within 2 min by a previous groomer or groomee (Muroyama 1994, with permission)

Grooming pair	Nonbirth season				Birth season				Season difference P^a
	N	Groomer	Groomee	Terminated	N	Groomer	Groomee	Terminated	
Unaffiliated nonrelated pairs									
Mother–mother	–	–	–	–	43	44.2	11.6	44.2	< 0.01
Mother–nonmother	–	–	–	–	51	33.3	2.0	64.7	< 0.001
Nonmother–nonmother	62	43.5	35.5	21.0	26	11.5	50.0	38.5	< 0.02
Affiliated nonrelated pairs									
Mother–mother	–	–	–	–	93	23.6	9.7	66.7	< 0.001
Mother–nonmother	–	–	–	–	106	33.0	9.4	57.6	< 0.001
Nonmother–nonmother	162	47.5	41.4	11.1	–	–	–	–	–
Mother and adult female offspring pairs									
Mother–mother	–	–	–	–	47	23.4	8.5	68.1	< 0.001
Nonmother–nonmother	111	43.3	40.5	16.2	23	39.1	39.1	21.8	n.s.

^{a} Comparison between the nonbirth season and each subcategory of the birth season (χ^2 test, df = 2).

either of the participants in the interaction again performed grooming or soliciting within 2 min after the end of the previous grooming. Allomothering was defined as nuzzling, sniffing, touching, pulling, grooming, holding, or kidnapping another female's infant (Zucker and Kaplan 1981).

Table 1 shows the probabilities with which, when an interaction was resumed, a previous groomer or a groomee resumed the interaction. During the nonbirth season, a groomer and groomee resumed a grooming interaction about equally often. In contrast, during the birth season, and when the interaction was with the mother of a newborn infant, the previous groomer resumed significantly more often than did the groomee in all cases.

In mother and nonmother dyads, the nonmother resumed more frequently than did the mother (rarely interacting nonkin pairs: 16/51 vs 2/51, $\chi^2 = 13.222$, df = 1, $P < 0.001$; frequently interacting nonkin pairs: 36/106 vs 9/106, $\chi^2 = 20.565$, df = 1, $P < 0.001$), and the mother received nearly all grooming before resumptions (rarely interacting nonkin pairs, 46/51; frequently interacting nonkin pairs, 99/106). These findings suggest that mothers of a newborn infant were less likely to reciprocate grooming after being groomed. This was the case for both mother and nonmother pairs. In the dyads of nonmothers, in contrast, this was not the case (Table 1). This indicates that the effects given above were due to the partner being a mother and not to the season.

If an individual who has groomed resumes the interaction by soliciting, a request for reciprocation of grooming is indicated. During the nonbirth season, in nonkin, reciprocation by soliciting was more common (Table 2). In kin, the two types of resumption occurred equally frequently. During the birth season, in contrast, nonkin pairs resumed the interaction with grooming again more often if one or two mothers were involved: so they performed grooming more often than they asked for reciprocation (Table 2). This trend is even found for mother/mother pairs in related pairs.

During the nonbirth season, and only in rarely interacting unrelated pairs, the groomee resumed the interaction by grooming more frequently than by soliciting. During the birth season, more grooming by the previous groomee was only found for N–N pairs in rarely interacting pairs. However, if one or two mothers were involved, no differences were found between resumption by grooming and that by soliciting.

If monkeys groom in order to be allowed to allomother, one expects extra allomothering after grooming. This prediction was tested by examining whether a groomer (giver of a benefit to the mother) subsequently allmothered more often than a groomee (recipient of a benefit from the mother). Table 3 shows that this was generally the case, although the difference was only significant for nonkin mother/mother dyads. In mother/nonmother pairs, the association between giving grooming and allomothering was greater in rarely interacting pairs (66.7%) than in frequently interacting pairs (46.7%) ($\chi^2 = 5.352$, df = 1, $P < 0.05$). This means that the less affiliated female pairs were, the more they invested by means of preceding grooming. When two mothers interacted, the percentages of giving grooming followed by allomothering were significantly lower than when a nonmother inter-

Table 2. Percentage of grooming resumptions that started with either grooming or soliciting by the same individual (Muroyama 1994, with permission)

Grooming pair	After grooming				After being groomed			
	N	Grooming	Soliciting	P^a	N	Grooming	Soliciting	P^a
Nonbirth season								
Unaffiliated nonrelated pairs	27	37.0	63.0	n.s.	22	68.2	31.8	n.s.
Affiliated nonrelated pairs	77	29.9	70.1	< 0.001	67	49.3	50.7	n.s.
Mother and adult female offspring pairs	48	54.2	45.8	n.s.	45	51.1	48.9	n.s.
Birth season								
Unaffiliated nonrelated pairs								
Mother–mother	19	68.4	31.6	n.s.	5	40.0	60.0	n.s.
Mother–nonmother	17	94.1	5.9	< 0.001	1	100.0	0.0	n.s.
Nonmother–nonmother	3	33.3	66.7	n.s.	13	84.6	15.4	< 0.05
Affiliated nonrelated pairs								
Mother–mother	22	90.9	9.1	< 0.001	9	66.7	33.3	n.s.
Mother–nonmother	35	94.3	5.7	< 0.001	10	30.0	70.0	n.s.
Mother and adult female offspring pairs								
Mother–mother	11	90.9	9.1	< 0.05	4	50.0	50.0	n.s.
Nonmother–nonmother	9	55.6	44.4	n.s.	9	66.7	33.3	n.s.

[a] Binomial test.

Table 3. Percentage of groomings that were followed by allomothering behavior by the same individual (Muroyama 1994, with permission)

Grooming pair	After grooming		After being groomed		P^a
	N	Allomothering	N	Allomothering	
Mother–mother					
Unaffiliated nonrelated pairs	37	27.0	37	8.1	< 0.05
Affiliated nonrelated pairs	81	32.1	81	16.0	< 0.05
Mother and adult female offspring pairs	39	20.5	39	12.8	n.s.
Mother–nonmother					
Unaffiliated nonrelated pairs	45	66.7	4	25.0	n.s.
Affiliated nonrelated pairs	92	46.7	1	0.0	n.s.

[a] Comparison between after grooming and after being groomed (χ^2 test, df = 1, or Fisher's exact probability test).

Table 4. Percentage of groomings that were preceded by allomothering behavior by the same individual (Muroyama 1994, with permission)

Grooming pair	N	Allomothering
Mother–mother		
Unaffiliated nonrelated pairs	21	38.1
Affiliated nonrelated pairs	52	48.1
Mother and adult female offspring pairs	24	33.3
Mother–nonmother		
Unaffiliated nonrelated pairs	33	90.9
Affiliated nonrelated pairs	63	58.7

acted with a mother (rarely interacting pairs: 27.0 vs 66.7 %, $\chi^2 = 12.770$, df = 1, $P < 0.001$; frequently interacting pairs: 32.1 vs 46.7 %, $\chi^2 = 3.851$, df = 1, $P < 0.05$). This may be understandable, because in mother/mother dyads both interactants can offer allomothering in return.

It is possible that a mother who allows allomothering is subsequently rewarded by the allomother in the form of grooming. If this is the case, many grooming interactions will be preceded by allomothering. The need for this may be greater for nonmothers than for mothers, because nonmothers cannot offer allomothering in exchange. Table 4 shows that nonmothers had a very high percentage (90.9) of groomings that were preceded by allomothering if they were unaffiliated with the mother whom they interacted with. If the nonmother was affiliated with the mother, this percentage was significantly lower ($\chi^2 = 10.636$, df = 1, $P<0.002$). The percentages in mother/mother dyads appeared to be lower than in mother/nonmother dyads (rarely interacting pairs: $\chi^2 = 17.168$, df = 1, $P < 0.001$; frequently interacting pairs: $\chi^2 = 1.301$, df = 1, ns). This suggests that the mothers may offer allomothering in return for allomothering, as was found in the previous analysis.

In patas monkeys, the nature of the benefits of allomothering is not obvious. Practicing maternal behavior (Hrdy 1976; Fairbanks 1990) is not a sufficient explanation in the present case, because the allomothering was performed by parous (experienced) females rather than by juveniles; it was even shown by mothers who had a young infant themselves (Zucker and Kaplan 1981; Muroyama 1994). Zucker and Kaplan (1981) suggested that allomothering could be one of various affiliative behaviors such as grooming.

However, it is obvious that female patas monkeys do exchange allomothering for grooming, which suggests that they may recognize allomothering as a beneficial act which can be exchanged for another kind of benefit, such as grooming. They also appear to classify other females into at least two groups, as do Japanese macaques, and such classification was reflected to the exchange of different acts, namely grooming and allomothering, as well as that of the same act, grooming.

2.4 Reciprocal Exchanges of Social Behavior

The analyses suggest that Japanese macaques and patas monkeys may recognize that some beneficial acts such as grooming are "currencies" with which they "pay" for the receipt of benefits other than grooming itself, for example, allomothering. Furthermore, monkeys appear to classify their partners in terms of reciprocal relationships, or "reciprocity," and then make appropriate behavioral decisions accordingly. Although it is not known whether monkeys have the concept of "reciprocity" in their mind, they at least appear to know "how" to behave towards nonkin with reciprocity.

The kin selection hypothesis predicts that individuals provide benefits to their relatives without any expectation of reciprocity (Hamilton 1964; Wrangham 1982). In contrast, the reciprocal altruism hypothesis predicts that individuals expect the partner to reciprocate benefits just given (Trivers 1971). The results of the analyses are quite consistent with the predictions based these hypotheses from the viewpoint of behavioral ecology. The classification between kin and nonkin, and decision making based on such classification in terms of reciprocity, appears to be "adaptive" in many cases. This is why monkeys are expected to apply the principle of "reciprocity" broadly to various kinds of social behavior with nonkin, although what kind of mechanism underlies such a principle is still unknown. The exchange of beneficial and harmful behaviors, or that of harmful ones (i.e., retaliation), may also be expected to occur in some cases based on the principle of reciprocity, because negative reciprocity can also work in theory (Clutton-Brock and Parker 1995).

3 Decision Making in Grooming by Japanese Macaques

3.1 Sequential Analysis of Grooming

In the previous section, temporal rules of exchange of social grooming were discussed. Social grooming, however, is not a single event but a series of states involving bouts and their durations. Estimates of time spent grooming may be important when grooming is performed as an investment or to be exchanged with the partner, and may be more likely to be precise when reciprocation is expected. However, it is still not known whether, and to what extent, monkeys recognize time spent grooming by themselves, and decide upon their own subsequent behavior accordingly (Muroyama 1996).

To answer this question, the temporal dependency in grooming sequences by one individual Japanese macaque (*Macaca fuscata*) was examined in terms of a Markov chain model (Goosen and Metz 1980; Haccou and Meelis 1994). In this model, it is assumed that transitions between acts are independent of both the time a bout has already lasted (the constant termination rate in the state) and previous behavior (the independence of the order and duration of preceding acts). If monkeys decide upon their own behavior according to the preceding one, then this model is less likely to fit. In this case, the termination rate of grooming bouts may vary with the bout length, and/or the occurrence or duration of their preceding bouts may affect that of their subsequent one (i.e., sequential dependency between bouts). The extent of such temporal dependency is expressed as that of the departure from the model.

In this study, patterns of grooming sequences were analyzed for mothers and offspring of different age/sex classes and for nonkin females. We examined (1) the relation between the bout length and the termination rate of grooming bouts, and (2) the effects of preceding bouts on subsequent bouts for each individual subject. To examine the influences of kinship and the age and sex of offspring, comparisons were also made between these subjects. Observations were made in the same troop of Japanese macaques as in the previous section. To compare kin and nonkin, the data of two frequently interacting nonkin pairs were used. In this analysis, when one animal that had been groomed by another started to groom within 1 min after the end of a grooming bout, it was decided that another grooming session had been initiated. Thus, by the current definition, one animal groomed in each grooming session.

Data were analyzed separately for each member of dyads involving one of the target animals. Thus, 14 data sets of grooming sessions by mothers, 13 data sets of grooming sessions by offspring (one male infant did not groom his mother during

Table 5. Number of first, second, and third bouts in grooming sessions of all subjects (From Muroyama 1996, with permission)

Subject	Age and sex of offspring (year:sex)[a]	Mother as groomer First	Second	Third	Offspring as groomer First	Second	Third
Mother–offspring pair							
URI–CHA	> 8:F	67	30	7	45	11	3
TUT–TAK	> 8:F	49	24	11	35	11	5
UME–MBR	6:M	28	7	3	17	3	0
URI–CHN	6:M	20	6	3	11	2	0
TUT–TAR	6:F	3	1	1	4	0	0
GAK–BUN	4:M	53	11	4	49	10	3
KRM–MMJ	4:F	11	3	1	9	3	1
KED–YUR	4:F	62	11	3	78	33	13
URI–CHS	3:F	68	24	10	46	13	6
TUT–TAD	2:F	30	10	2	22	6	2
KED–YUD	1:M	72	23	11	3	1	1
URI–CHG	1:F	39	13	10	6	0	0
GAK–BNA	1:F	54	20	9	8	4	3
TUT–TAI	0:M	52	26	14	0	0	0
Nonkin female							
ZKR	–	43	19	6			
GAK	–	27	1	0			
NEM	–	36	18	4			
BOK	–	21	2	1			

[a] M, Male; F, female.

the study period), and 4 data sets of grooming sessions by nonkin females were analyzed (Table 5). In addition, all grooming sessions observed in mother/offspring pairs and those in nonkin female pairs were classified into the following categories and analyzed separately in order to see the overall trends in each category: adult (>6 years old, male = 0, female = 2); adolescence (4–6 years old, male = 3, female = 3); immature (< 3 years old, male = 2, female = 4); nonkin females (N = 4).

3.2 Termination Rates of Grooming Bouts

Figure 3 shows the log-survivor plots of the first, second, and third bouts of all individual subjects. When the termination rate of grooming during a bout is independent of the time that has already been spent grooming, such a plot should be a straight line (i.e., the exponential distribution). A concave-up plot (i.e., a negative Z value in Fig. 3) indicates a decreasing termination rate, whereas a concave-down plot (i.e., a positive Z value in Fig. 3) indicates an increasing termination rate. Therefore, if an animal is more likely to stop grooming as the time spent on it becomes longer, the plot shows a concave-down shape; if the animal is less likely to stop grooming, the plot shows a concave-up shape.

In all cases of the first bouts of mothers' grooming, there were no statistically significant deviations from the exponential distribution (Fig. 3a), although the plots of female offspring more than 1 year old showed concave-down shapes. In contrast, most of the plots of grooming received by male or 1-year-old female offspring showed concave-up curves. With regard to the second and third bouts, most of the plots of mothers' grooming showed concave-down curves, irrespective of the ages and sexes of offspring. When the data were pooled within each age/sex class, significant deviations from the exponential distribution were found for the first bouts of grooming received by adult female offspring ($N = 116, Z = 2.079, P < 0.05$) and for the second bouts received by adolescent female offspring ($N = 15, Z = 2.026, P < 0.05$).

In contrast to mothers, almost all plots of the first bouts of grooming given by offspring represented concave-down shapes (Fig. 3b). The plots of grooming given by two adolescent males and one adolescent female showed statistically significant deviations from the exponential distribution. Concave-down curves of the plots of offspring's grooming were also found in most of the second and third bouts, irrespective of the ages and sexes of individuals, although this finding was not significant. Pooled data indicated that there were significant deviations from the exponential distribution for the first bouts given by adolescent males ($N = 77, Z = 4.313, P < 0.01$) and by juvenile females ($N = 82, Z = 2.128, P < 0.05$).

In the case of nonkin females, three individuals showed conspicuous concave-down curves of the plots of first bouts, whereas the other showed a weak concave-down curve (Fig. 3c). With regard to the second and third bouts, there was no significant deviation from the exponential distribution. Pooled data of the four females showed a significant deviation from the exponential distribution for the first bouts ($N = 127, Z = 4.944, P < 0.01$).

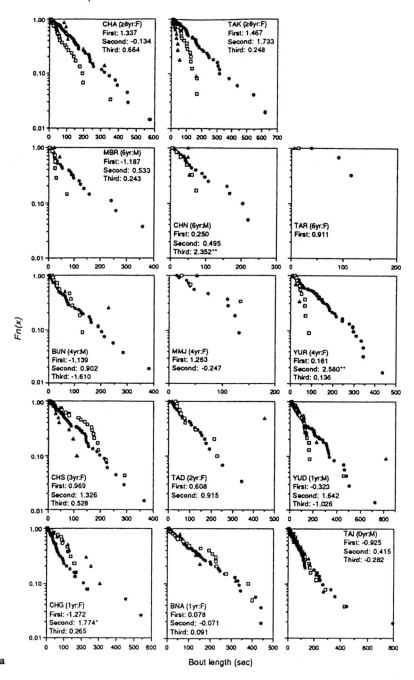

Fig. 3. a–c. Log-survivor plot of the duration of the first (*solid circles*), second (*open squares*), and third (*open triangles*) bouts in grooming sessions by the same individual. Figures indicate the Z values of Barlow's tests for the exponential distribution (Haccou and Meelis 1994). Positive (negative) Z values indicate increasing (decreasing) termination rates. *P < 0.05; **P < 0.01. **a** Mothers groomed their offspring. **b** Offspring groomed their mothers. **c** Females groomed nonrelatives. (From Muroyama 1996, with permission)

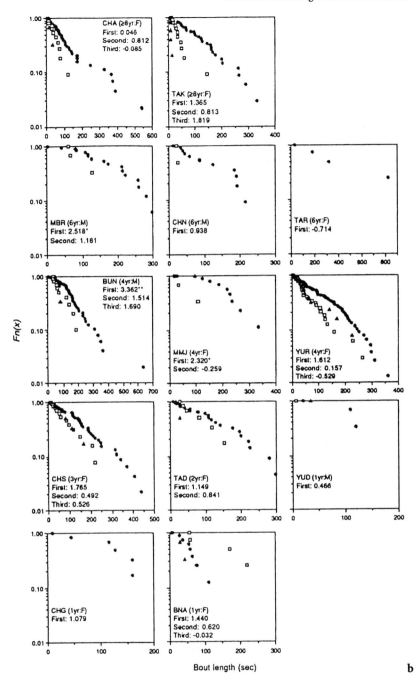

Bout length (sec)

b

Fig. 3. *Continued*

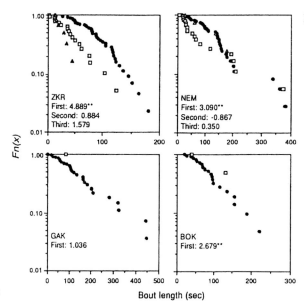

Fig. 3. *Continued*

c

Bout length (sec)

Thus grooming bouts were more likely to be terminated as the bout length became longer, when females groomed nonkin or adult female offspring, or when adolescent offspring or juvenile female offspring groomed their mothers. These findings imply that an animal may estimate the duration of grooming in some way, and is likely to avoid the continuous expenditure of effort in grooming during a brief period. This is consistent with the facts that most grooming sessions involved only one bout (Table 5).

3.3 Effects of Preceding Bouts on Subsequent Bouts

If each bout in a grooming session occurs independently, the mean duration of each bout should be equal irrespective of whether the bout is preceded by another (i.e., the second and third bouts) or not (i.e., the first bout). As shown in Table 6, however, the mean duration of the first, second, and third bouts differed significantly in most cases. There were no significant differences between the second and third bouts in all subjects.

In the cases when mothers groomed their offspring, the differences between the durations of first bouts and those of the other bouts appeared to vary with the age and sex of the offspring (Table 6). When mothers groomed their adult female offspring, subsequent bouts were likely to be short. This tendency was also seen in cases where mothers groomed their adolescent female offspring, when the data were pooled (Kruskal–Wallis one-way ANOVAs, $N = 96$, $H = 11.31$, $P < 0.01$). In contrast, when mothers groomed their male or juvenile female offspring, the duration of first bouts did not differ significantly from those of the others (pooled data:

Table 6. Mean duration of first, second, and third bouts in grooming sessions of all subjects (From Muroyama 1996, with permission)

Subject	Mother as groomer				Offspring as groomer			
	First	Second	Third	H^a	First	Second	Third	H^a
Mother–offspring pair								
URI–CHA	141.8	77.5	129.7	9.238**	113.8	47.7	19.3	8.862*
TUT–TAK	160.1	74.1	40.6	18.614***	111.6	42.4	8.0	14.142***
UME–MBR	77.0	26.1	50.3	2.108	161.7	85.0	–	(12)
URI–CHN	73.9	37.8	50.0	1.108	107.6	23.0	–	(2)⁺
TUT–TAR	80.7	13.0	5.0	3.200	333.8	–	–	–
GAK–BUN	72.4	59.6	77.3	0.236	155.3	92.2	71.0	5.824⁺
KRM–MMJ	66.5	66.0	52.0	0.085	196.6	53.7	97.0	5.407⁺
KED–YUR	124.5	49.4	24.0	8.986*	110.3	64.2	67.2	8.835*
URI–CHS	86.4	107.3	46.6	4.765⁺	144.6	85.6	77.8	4.069
TUT–TAD	95.5	94.4	233.5	0.086	119.4	77.3	21.5	3.436
KED–YUD	147.1	107.8	160.9	0.118	82.3	6.0	34.0	2.133
URI–CHG	105.4	105.4	127.7	4.763⁺	102.5	–	–	–
GAK–BNA	128.4	138.2	104.2	0.221	54.6	126.3	25.7	5.273⁺
TUT–TAI	109.8	103.3	90.1	0.370	–	–	–	–
Nonkin female								
ZKR	83.6	43.3	25.7	19.603***				
GAK	149.7	86.0	–	–				
NEM	132.8	87.0	100.1	5.701⁺				
BOK	92.2	85.5	14.0	2.805				

[a] Kruskal-Wallis one-way ANOVAs and Mann-Whitney U tests (in parentheses). ⁺$P < 0.1$; *$P < 0.05$; **$P < 0.01$; ***$P < 0.001$.

adolescent males, $N = 135$, $H = 1.042$, n.s.; juvenile males, $N = 198$, $H = 1.320$, n.s.; juvenile females, $N = 289$, $H = 2.153$, n.s.).

Both male and female offspring's grooming, except for those of 1-year-old animals, followed the same pattern of decline in the duration of successive bouts as that of mothers, although this was not statistically significant in all cases (Table 6). When the data were pooled within each age/sex class, all classes except that of juvenile males showed statistical significance (adult females, $N = 117$, $H = 29.806$, $P < 0.001$; adolescent males, $N = 96$, $H = 9.578$, $P < 0.01$; adolescent females, $N = 152$, $H = 22.38$, $P < 0.001$; juvenile males, $N = 5$, $H = 2.133$, n.s.; juvenile females, $N = 121$, $H = 10.45$, $P < 0.01$).

In two cases of nonkin dyads, the duration of first bouts was significantly longer than that of subsequent bouts (Table 6). In the two other cases, small numbers of subsequent bouts did not allow statistical tests (see Table 5); in these cases, first bouts were unlikely to be followed by another bout (see Table 6). Pooled data for the four females also indicated statistical significance ($N = 178$, $H = 23.55$, $P < 0.001$).

To examine the effects of the duration of preceding bouts on whether to resume grooming or not, a comparison was made of the mean duration of the first bout followed by the second bout with that followed by termination of the session.

There were no significant differences for any individual, except for the case of a mother who was more likely to resume grooming of her male juvenile offspring after longer first bouts (211.7 s vs. 118.6 s, Mann–Whitney U test, $N_1 = 24, N_2 = 44$, $z = -2.360, P < 0.02$). In contrast, pooled data suggested that nonkin adult females were likely to resume grooming after shorter first bouts (93.3 s vs. 122.3 s, $N_1 = 40$, $N_2 = 87, z = 1.950, P < 0.06$).

The correlation was also examined between the durations of the first bout and those of the second bout by the same individual. Only one significant result was found, which was for a mother who groomed her adult female offspring (TUT-TAK: Kendall's rank correlation test, $N = 24$, t $= 0.288, z = 1.970, P < 0.05$). No pooled data of any age/sex classes were statistically significant.

If an animal monitors precisely the amount of grooming given in a session, the duration of preceding grooming and/or the number of preceding bouts would be expected to affect subsequent grooming. However, the bout length of prior grooming hardly affected the occurrence or duration of subsequent grooming bouts in all subjects, with a few exceptions. The number of preceding bouts also had no effect.

3.4 Estimation of Grooming Amounts

The results suggest that the model of a Markov chain is less likely to fit the patterns of grooming sequences by one individual. This means that decision making during grooming interactions is affected by the preceding behavior. In other words, monkeys may make some kind of estimates about "investment" in the preceding behavior, and make behavioral decisions accordingly.

The precision of such estimations seems to be rather low, however. The length of time spent in a grooming bout affects the decision of whether or not monkeys continue that grooming further, which indeed implies that monkeys estimate the length in some way. When a bout of grooming is terminated and resumed after a short interval, however, monkeys do not appear to retain information about grooming length, but only about the fact that they were grooming before the interruption. In other words, they do not appear to recognize a bout of grooming as an event that has a duration. Similarly, the results suggest that monkeys may not count the number of bouts during grooming interactions if they repeat several bouts of grooming with short intervals in between. This might be due either to the lack of mental capacity in monkeys for precisely counting or timing preceding bouts, or to the fact that precise recognition is not useful, or to both, although it is also likely that the sample sizes in this study were too small to detect subtle differences (see Table 5).

The degree of departure from the Markov model appears to have varied with the relationships between participants in grooming. Between nonrelatives, a preceding grooming bout was most likely to affect subsequent bouts. The effects of preceding grooming were less in grooming by offspring and least in that by mothers, although the degree of these effects varied with the age and sex of the offspring. The difference found between kin and nonkin appears to be consistent with the results in the previous sections and those of studies on reciprocity over the short term (Seyfarth and Cheney 1984).

Monkeys appear to vary the way they use their estimation of their "investment" during social interactions according to very different information or knowledge, such as their social relationships with their partners. In other words, monkeys may use different decision-making algorithms to process information during social interactions, based on their classification of their partners. Again, such algorithms appear to relate to the principle of reciprocity: the degree of precision of the estimation appears to be affected by the expected degree of reciprocity between the participants.

4 Decision Making in Social Behaviors

Social relationships between one individual and others are based on, and shaped by, various social interactions of that individual with the others (Hinde 1976). In other words, social relationships are some kind of database which reflects the history of the individual's past social interactions with others, and can be referred to only by the individual itself. Such databases may store various kinds of knowledge or information, i.e., kin relationships, dominance, and affiliation with others, etc., which are abstracted and arranged in particular ways, based on the cognitive architecture of the species. As presented in this chapter, such social relationships relate to what kind of behaviors will be chosen, and how they will be chosen, i.e., the mechanism of decision making under different circumstances.

Animals obtain and process necessary information from various complicated environments around them, and then make decisions about their subsequent behavior. Internal mechanisms, such as a nervous system related to decision making, have evolved through natural selection, like physical characteristics or behaviors. Studies on internal mechanisms relating to decision making have generally been neglected in the paradigms of behavioral ecology. Recently, however, some studies have been carried out relating to information processing or cognitive ability concerning decision making (Real 1994). Indeed, experimental studies about feeding behaviors have developed greatly in recent years. In contrast, most of the studies on decision making during social interactions, many of which are not experimental but observational, are in the preliminary stages, as presented here. Furthermore, it is still not known, for example, whether or not monkeys can estimate the effort or time spent on social behaviors by other monkeys.

Although the studies presented here are preliminary and their scientific validity is limited, it is time to take their implications seriously and begin observations and experiments to construct theories and examine hypotheses based on those theories. To investigate how nonhuman primates make decisions, and what kind of information they use, during social interactions may be crucial in any consideration of the evolution of cognitive capacity, or "mind," from mammals to human beings. Animals may know what they do and how they do it much better than we expect.

References

Altmann J (1974) Observational study of behavior: sampling methods. Behavior 49:227–265

Bachmann C, Kummer H (1980) Male assessment of female choice in hamadryas baboons. Behav Ecol Sociobiol 6:315–321

Barton RA (1985) Grooming site preferences in primates and their functional implications. Int J Primatol 6:519–532

Boccia ML (1983) A functional analysis of social grooming patterns through direct comparison with self-grooming in rhesus monkeys. Int J Primatol 4:399–418

Boccia ML, Rockwood B, Novak MA (1982) The influence of behavioral context and social characteristics on the physical aspects of social grooming in rhesus monkeys. Int J Primatol 3:91–108

Boccia ML, Reite M, Laudenslager M (1989) On the physiology of grooming in a pigtail macaque. Physiol Behav 45:667–670

Clutton-Brock TH, Parker GA (1995) Punishment in animal societies. Nature 373:209–216

Dasser V (1988a) Mapping social concepts in monkeys. In: Byrne RW, Whiten A (eds) Machiavellian intelligence: social expertise and the evolution of intellect in monkeys, apes, and humans. Oxford University Press, Oxford, pp 85–93

Dasser V (1988b) A social concept in Java monkeys. Anim Behav 36:225–230

Dunbar RIM (1988) Primate social systems. Croom Helm, London, Sydney

Fairbanks LA (1990) Reciprocal benefits of allomothering for female vervet monkeys. Anim Behav 40:553–562

Goosen C (1987) Social grooming in primates. In: Mitchell G, Erwin J (eds) Comparative primate biology, vol 2B. Behavior, cognition, and motivation. Alan R Liss, New York, pp 107–132

Goosen C, Metz JAJ (1980) Dissecting behavior: relations between autoaggression, grooming and walking in a macaque. Behavior 75:97–132

Haccou P, Meelis E (1994) Statistical analysis of behavioral data. An approach based on time-structured models. Oxford University Press, Oxford

Hamilton WD (1964) The genetical evolution of social behavior. I. J Theor Biol 7:1–16

Hemelrijk CK (1990) Models of, and tests for, reciprocity, unidirectionality and other social interaction patterns at group level. Anim Behav 39:1013–1029

Hinde RA (1976) Interactions, relationships, and social structure. Man 11:1–17

Hrdy SB (1976) Care and exploitation of non-human primate infants by conspecifics other than the mother. Adv Stud Behav 6:101–158

Hutchins M, Barash DP (1976) Grooming in primates: implications for its utilitarian functions. Primates 17:145–150

Kurland JA (1977) Kin selection in the Japanese monkeys. Contributions to primatology. vol 12. S Karger, Basel

McKenna JJ (1978) Biosocial functions of grooming behavior among the common Indian langur monkeys (Presbytis entellus). Am J Phys Anthropol 48:503–510

Muroyama Y (1991) Mutual reciprocity of grooming in female Japanese macaques. Behavior 119:161–170

Muroyama Y (1994) Exchange of grooming for allomothering in female patas monkeys. Behavior 128:103–119

Muroyama Y (1996) Decision making in grooming by Japanese macaques (Macaca fuscata). Int J Primatol 17:817–830

Oki J, Maeda Y (1973) Grooming as a regulator of behavior among the common Indian langur monkeys (Presbytis entellus). Am J Phys Anthropol 48:503–510

Packer C (1977) Reciprocal altruism in Papio anubis. Nature 265:441–443

Real LA (1994) Behavioral mechanisms in evolutionary ecology. University of Chicago Press, Chicago, London

Schino G, Scucchi S, Maestripieri D, Turillazzi PG (1988) Allogrooming as a tension reduction mechanism: a behavioral approach. Am J Primatol 16:43–50

Seyfarth RM, Cheney DL (1984) Grooming, alliances and reciprocal altruism in vervet monkeys. Nature 308:541–543

Seyfarth RM, Cheney DL (1988) Empirical tests of reciprocity theory: problems in assessment. Ethol Sociobiol 9:181–187

Seyfarth RM, Cheney DL (1994) The evolution of social cognition in primates. In: Real LA (ed) Behavioral mechanisms in evolutionary ecology. University of Chicago Press, Chicago, London, pp 371–389

Tanaka I, Takefushi H (1993) Elimination of external parasites (lice) is the primary function of grooming in free-ranging Japanese macaques. Anthropol Soc 101:187–193

Terry RL (1970) Primate grooming as a tension reduction mechanism. J Psychol 76:129–136

Trivers RL (1971) The evolution of reciprocal altruism. Q Rev Biol 46:35–57

Tsukahara T (1990) Initiation and solicitation in male–female grooming in a wild Japanese macaque troop on Yakushima island. Primates 31:147–156

de Waal FBM (1978) Exploitative and familiarity dependent support strategies in a colony of semi-free living chimpanzees. Behavior 66:268–312

de Waal FBM (1982) Chimpanzee politics. Jonathan Cape, London

de Waal FBM (1989) Food sharing and reciprocal obligations among chimpanzees. J Hum Evol 18:433–459

de Waal FBM, Luttrell LM (1988) Mechanisms of social reciprocity in three primate species: symmetrical relationship characteristics or cognition? Ethol Sociobiol 9:101–118

Wrangham RW (1982) Mutualism, kinship and social evolution. In: King's College Sociobiology Group (ed) Current problems in sociobiology. Cambridge University Press, Cambridge, pp 269–289

Zucker EL, Kaplan JR (1981) Allomaternal behavior in a group of free-ranging patas monkeys. Am J Primatol 1:57–64

Part 7
Culture

24
"Sweet-Potato Washing" Revisited

Satoshi Hirata[1], Kunio Watanabe[1], and Masao Kawai[2]

1 Introduction

Japanese monkeys (*Macaca fuscata*) on Koshima Island wash sweet potatoes. This fact is quite well known, and is often discussed as an aspect of "culture" in nonhuman animals. However, the full picture is not always entirely understood. The finding of sweet-potato washing among Koshima monkeys followed theoretical considerations of culture in nonhuman animals presented by Imanishi, and from the beginning research on Japanese monkeys has developed from anthropological concerns about human evolution. In this chapter, we would like to illustrate the history of sweet-potato washing and other characteristic behaviors shown by Koshima monkeys (for reviews, see also Itani and Nishimura 1973; Nishida 1987).

2 Prehistory of Sweet-Potato Washing

In 1948, when Kinji Imanishi, Shunzo Kawamura, and Junichiro Itani were conducting research on semiwild horses at Toi Peninsula in Miyazaki Prefecture, Japan, they happened to find a group of wild Japanese monkeys. This was the start of many studies of wild Japanese monkeys. Two points characterized these studies. The first is that they are a long-term research project which still continues after more than a half a century. The second is the individual identification of the monkeys. Each monkey was given a nickname, and not labeled like a rat or mouse in a laboratory. The researchers began to understand the cognition, behavior, and society of Japanese monkeys in the wild through long-term research with individual identification.

In November 1948, soon after their encounter with the wild Japanese monkeys at Toi Peninsula, Imanishi, Itani, and Kawamura visited Koshima Island in the same prefecture for the first time (Fig. 1). Koshima was the first, and that time the only, monkey habitat designated a natural monument in Japan. Intensive study of the Koshima troop of Japanese monkeys started since then with the success of

[1] Primate Research Institute, Kyoto University, 41 Kanrin, Inuyama, Aichi 484-8506, Japan
[2] Museum of Nature and Human Activities, 6 Yayoigaoka, Sanda, Hyogo 669-1546, Japan

Fig. 1. Koshima Island (center)

provisioning in 1952 (see Chapter 20 by Watanabe, this volume). These workers and their colleagues also began to focus their research on monkeys at several other sites in Japan. These included Takasakiyama in 1950 by Kawamura and Itani, Minoo in 1950 by Kawamura, Arashiyama in 1952 by Kawamura and Itani, Yakushima in 1952 by Kawamura and Itani, Shimokita in 1952 by Itani, Kinkazan in 1959 by Yoshiba, and Jigokudani in 1960 by Azuma, Sugiyama, and Wada.

In the meantime, in 1952, Imanishi published a landmark paper entitled "The evolution of human nature," written in Japanese. At the very beginning of the research on Japanese monkeys, or even before that, Imanishi's mind was oriented toward anthropology. The paper is rather an imaginary theoretical one, as only a little was known about nonhuman primates at that time. The paper takes the form of fantasy, with some imaginary characters debating on several subjects. It goes as follows:

Evolutionist "First of all, I think it's good to start with a commonplace topic. Well, people say, animals live only by instinct, while humans have culture."
Layman "To be a human means to have a culture."
Evolutionist "... Instinct is inherited through a genetic channel, while culture is transmitted through a nongenetic channel. Culture is acquired through learning and teaching, so that the model and the pedagogy are necessary. Therefore, a group life is inevitably required for the establishment of a culture. Thus, to maintain a culture as culture, the group life must be a perpetual one."
Layman "If the condition for the establishment of a culture is just like you described, then culture may be seen not only in humans but also in other animals that live in a perpetual social group. How is it in monkeys? Do you have culture, Monkey?"
Monkey " We live in a perpetual social group... but it is not made clear yet how much of our behavior is determined by instinct and how much is determined by culture..."

In the subsequent discussion, Imanishi let Monkey say "a change in food habits can be considered as a change in culture," and Evolutionist say "when an experience of an individual is transmitted to the next generation, it is called a culture."

The above hypothesis by Imanishi was later corroborated by colleagues in the field. The first discussion about the culture of Japanese monkeys that had a basis in actual field observations was proposed by Kawamura in "On a new type of feeding habit which developed in a group of wild Japanese monkeys" (1954), a paper written in Japanese. Kawamura later published three related papers: "Prehuman culture" (1956), "The process of subculture propagation among Japanese macaques" (1959), and "Subculture in Japanese monkeys" (1965); the first and third were written in Japanese, and the second in English. In these papers, Kawamura gave examples of possible cultural behavior in Japanese monkeys, such as the repertoire of foods (e.g., monkeys in some troops eat bird eggs, while monkeys in other troops do not), the nomadism of the troop (i.e., the maintenance of a home range by a group, and also the moving pattern within the home range, which is inherited over generations), social behavior (e.g., many adult males of one troop show paternal care towards infants, while males in other troops do not), and social structure (e.g., males are tolerant of each other in some troops, while males in other troops become much more aggressive when approached by other members). Kawamura attempted to probe the cultural behavior of monkeys by checking on the variability of the behavior across troops. He said that: "The life of higher animals closer to humans are in the mist at present. We should not dispose of the question by considering whether or not they have language or productions, but we must keep our outlook broad and explore untouched areas to shed light on the buried evolutionary path" (Kawamura 1965); "Our purpose is not to go into an endless argument about the definition of the term, but to conduct concrete research about in what way a simple behavioral mechanism has developed into a higher complex one in animals. ..., isn't it possible to cast a completely different light on human culture by carefully tracing each step and by arriving at the evolutionary perspective?" (Kawamura 1956).

The first of the four above-mentioned papers by Kawamura was the first report on the sweet-potato washing behavior of Koshima monkeys. He briefly described the invention of sweet-potato washing by a 1.5-year-old female and its propagation to three other members by 1954. Later, Kawamura, Kawai, and colleagues conducted a follow-up study of this behavior, and Kawai (1965) wrote about the process of propagation in detail, along with reports of three other newly acquired behaviors. This paper is an extremely detailed record of these well-known behaviors based on systematic data collection. We invite readers to revisit Kawai's original 1965 paper in the following sections.

3 Revisit to Kawai (1965): "Newly Acquired Precultural Behavior of the Natural Troop of Japanese Monkeys on Koshima Islet"

Before going to the paper itself, one may wonder why Kawai used a coined word "preculture." Let us first give an explanation (Kawai 1964). Imagine the following example: Japanese people eat sea cucumbers, but people in Western countries do not. Such a phenomenon can be labeled the "culture" of food. Culture can be loosely defined as a mode of life that is invented in a group, shared by group members, and transmitted to subsequent generations through social media. However, when we say "culture," its substances and levels are, of course, different in monkeys and humans. The word "culture" often reminds us of literature, art, music, pictures, academe, religion, etc. It cannot be denied that there is a big gap between human culture and monkey culture. In the case of monkeys, they do not have the ability to teach something. The way infant monkeys acquire a behavior is that they somehow become familiar with a certain behavior while being with their mothers, and acquire it in some way. In the case of humans, those who do not follow the culture of their own community will be treated as unorthodox and will be socially punished; human culture has the power to restrict or constrain the behavior of a person who belongs to that society. Thus, human culture of any kind is systematized in some way. On the other hand, a monkey who does not follow a culture will not be blamed, and no social restriction works against a violation. We must not overestimate the situation and say that "monkeys have culture," and then confuse it with human culture. Cultural behavior in monkeys must always be discussed in the light of evolution. This is the reason why the term "preculture," which takes such differences into consideration, was used in Kawai's paper. An abridged edition of the paper follows.

3.1 Sweet-Potato Washing (SPW) Behavior

3.1.1 Acquisition of SPW Behavior

Sweet-potato washing (SPW) is a behavior in which monkeys take a sweet potato to the edge of the water and wash the sand off the potato with water (Fig. 2). This behavior was begun in September 1953 by a female named Imo, who was one and a half years old at that time.

This behavior gradually spread to other monkeys. Table 1 shows the process of propagation during the period from 1953 to March 1958. In 1958, the acquisition rate in adults was 18.1%: i.e., 2 out of 11 animals (6 males and 5 females). The rate in monkeys aged between 2 and 7 years was 78.9%: i.e., 15 out of 19 (10 males and 9 females). After that, most newborns began to show this behavior. In August 1962, 36 out of 49 monkeys over 2 years old showed SPW behavior (73.4%). There were 13 monkeys who did not show SPW behavior. Out of 11 monkeys over 12 years old, i.e., those born before 1950, only two females showed SPW behavior (Eba and Nami). On the other hand, among the monkeys born after 1951, only 4 individuals did not perform this behavior. Interestingly, they were all Nami's children.

3.1.2 Process of Propagation

The acquisition of SPW behavior could be divided into two periods: before and after 1958. The author calls them the first and second period, respectively.

3.1.2.1 The First Period (The Period of Individual Propagation)

This is the period when monkeys born before 1956 acquired SPW behavior. The times and processes of acquisition were diverse. Adult monkeys who did not acquire the behavior during this period could not acquire it later.

Figure 3 and Table 1 show us the importance of age, sex, and kinship. Most

Fig. 2. Sweet-potato washing (SPW) behavior

Table 1. The year and age when the monkeys acquired SPW behavior

| Year | Age | | | | | |
	1–1.5	2–2.5	3	5	6	Adult
1953	Imo ♀	Semushi ♂				Eba ♀
1954	Uni ♂					
1955	Ei ♂	Nomi ♂	Kon ♂			
1956		Jugo ♂		Sango ♀, Aome ♀		
1957	Hama ♀, Enoki ♀				Harajiro ♀	Nami ♀
1958		Zabon ♀, Nogi ♀	Sasa ♀			

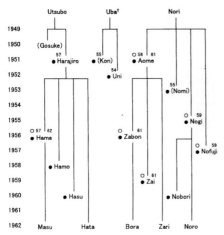

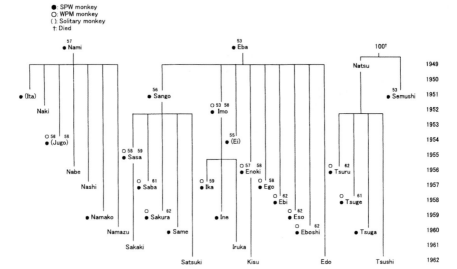

Fig. 3. The lineages of the Koshima troop and the acquisition of SPW and WPM behavior as of 1962. Individuals who died after 1949 are not included. Numerals to the left above the names indicate the year when the monkeys acquired SPW behavior. Numerals to the right above the names indicate the year when the monkeys acquired WPM behavior (snatching behavior is not included)

competent in acquiring SPW behavior were juveniles between 1 and 2.5 years old. Males older than 4 years had great difficulty in acquiring this behavior. Females, on the other hand, could acquire SPW behavior even if they were older than 4 years, but judging by the fact that only 2 females out of 11 born before 1950 did acquire the behavior, old age was obviously a great obstacle in both sexes for the acquisition of this behavior.

What causes the difference between males and females in the acquisition of this behavior? In order to acquire SPW behavior, a close social interaction with monkeys engaging in it (SPW-monkey) at feeding time seemed to play an important role. In this lies a difference between males and females which is due to their social status. When a male becomes 4 years old, he generally begins to move from the center to the periphery of the troop. Therefore, the social interactions of adolescent and adult males with females and juveniles in the center of the group become very limited, and they seldom feed with females and juveniles. On the other hand, all females live closely with each other in the center of the troop. In particular, mother and child often move together. Kawamura (1954, 1956) suggested the importance of mother–child relationships in acquiring this behavior by describing a typical example of the behavior of Imo and her mother Eba. Eba, the mother of the originator of SPW behavior, acquired it early on although she was an adult female. The same can be said for the four females who acquired the behavior when they were older than 4 years.

The order of acquisition of the behavior was the young before the old within the same family: i.e., the child rather than the mother, and the younger brother or sister rather than the older ones. Kawamura (1954, 1956) pointed out that propagation of the behavior had two courses: through playmate relations and through kinship. The propagation in this period was done mostly from child to mother, and from younger sibling to older one. As propagation was done through the relationship of individuals in the first period, this period was labeled the period of individual propagation.

3.1.2.2 The Second Period (The Period of Precultural Propagation)

After 1959, some aspects of propagation were different from those of the first period. SPW behavior was no longer a new behavior to the troop; it had been fixed in the troop during 1958–1959. Monkeys born during this period accepted SPW behavior as a normal feeding behavior and learned it without any resistance.

Sweet potato skins fall to the bottom of the water while SPW-monkeys are eating. This means that babies have the experience of eating their potatoes in water at the beginning of the development of their feeding behavior. The babies are always with their mothers, and they stare at their mother while she is doing SPW behavior. In this manner infants acquire SPW behavior. Therefore, the process of propagation in this period was always from mother to child, which was different from that of the first period. The process of acquiring this behavior by infants and juveniles in this period is described below.

(a) Strengthening affinity to water. Infants are taken to the edge of the water during the period when they are dependent solely on their mother's milk for nourishment. While the mothers are engaging in SPW behavior, the babies strengthen their affinity to water by being dipped in water, or by splashing water by hand.

(b) Eating potato in the water. Infants eat fragments of potatoes that their mothers drop in the water. This begins at about 6 months old.

(c) Acquisition of SPW behavior. Infants acquire SPW behavior when they are 1–2.5 years old.

Thus, the acquisition of SPW behavior begins in infants in the second period. In this period, acquisition or propagation of SPW behavior occurred independently of sex. It can be said that precultural pressure is working toward acquiring this behavior. Therefore, the author calls the second period the period of precultural propagation.

3.1.3 Variations of SPW Behavior

3.1.3.1 From Fresh Water to Salt Water

During 1953–1954, SPW behavior was done on the edge of a brook running into the sea. Monkeys never washed potatoes with salt water, but used the fresh water of the brook. During the surveys of 1957 and 1958, many monkeys began to wash potatoes in salt water. During the survey of December 1961, all the SPW-monkeys washed potatoes in both salt and fresh water. However, when they used fresh water there were particular reasons for doing so: for example, when they were given potatoes in the close vicinity of fresh water, or when subordinate monkeys avoided coming near the seashore for fear of dominant individuals. In other words, SPW monkeys preferred salt water to fresh water. For one thing, the quantity of fresh water was limited. In dry periods, the brook that ran into the sea dried up. Another reason was that if the monkeys became familiar with salt water, it would make the potatoes taste good. The author presumes that these are the two main reasons why the monkeys preferred salt water.

3.1.3.2 Seasoning Behavior

SPW behavior is, as described above, to dip a piece of potato into water by holding it in one hand and brushing off any sand with the other hand. Imo, the originator of SPW behavior, showed this typical behavior, but monkeys do not always brush the pieces of potato. Often they let a piece of potato fall into shallow water and wash the sand off by rolling it with one hand on the bottom. Among the monkeys of the first period, monkeys such as Eba, Sango, and Sasa rolled the potato more frequently than they brushed it. However, during the second period another type of behavior appeared. This consisted of dipping the potato into the water, gnawing it once or twice, and then repeating this behavior. Monkeys collected potatoes and

Table 2. SPW monkeys classified by three types of SPW behavior

Birth year	Age	B-type	BS-type	S-type
Before 1949	Over 13	Eba, Nami		
1951	11	Semushi, Sango, Aome, Harajiro		
1952	10	Imo		
1953	9			
1954	8			
1955	7	Nogi	Sasa	
1956	6	Hama, Zabon	Enoki	Tsuru
1957	5		Ika, Ego, Nofuji	Saba
1958	4		Ebi, Hamo	Tsuge
1959	3	Zai	Namako, Eso, Sakura	Ine
1960	2		Same, Eboshi	Nobori, Hasu, Tsuga

B-type, Brushing; BS-type, brushing–seasoning; S-type, seasoning.

took them to the seashore to carry out this type of behavior. The author considers the function of this behavior to be seasoning the potatoes. Table 2 classifies individual monkeys according to these three types of SPW behavior (B-type, brushing; BS-type, brushing–seasoning; S-type, seasoning). B-type monkeys who retained the old style were those who acquired SPW behavior in the first period, while younger monkeys who acquired SPW behavior in the second period belonged to the BS-type or the S-type. These monkeys were acquainted from their infancy with eating potatoes in water, so they had first learned to eat potatoes with salt seasoning or wet with water, and then acquired brushing behavior to remove sand.

3.2 Wheat Placer Mining (WPM) Behavior

3.2.1 Wheat Placer Mining Behavior

3.2.1.1 Acquisition of Wheat Placer Mining Behavior

The Koshima troop also has another precultural behavior: wheat placer mining behavior (Fig. 4). When grains of wheat were scattered about on the beach, the monkeys ate them by painstakingly picking up one grain after another. However, if a monkey gathers up the grains of wheat together with some sand and then throws them into the water, it succeeds in separating the grains of wheat from the sand more easily. The grains float to the surface of the water, whereas the sand sinks. This is called wheat placer mining behavior, owing to its resemblance to the mechanism of gold mining.

This behavior, which was begun by Imo, was first observed by Kawamura in 1956. Imo was then 4 years old and was well acquainted with SPW behavior. Table

Fig. 4. Wheat placer mining (WPM) behavior

In Kawai's (1965) original paper, this behavior is called "wheat washing" behavior. However, it is actually a bit different from "washing" the wheat, and therefore we use the term "wheat placer mining" to avoid confusion.

3 shows the propagation of WPM behavior from 1956 to 1962. By August 1962, a total of 19 monkeys had acquired this behavior (38.7%).

3.2.1.2 Process of Propagation

In the 1957 survey, the only WPM-monkey was Imo, who was also the originator of this behavior. However, during that year, the researchers used wheat in a test to analyze dominance rank. Because of this, the monkeys grew familiar with wheat, and this seemed to contribute to the rapid spread of WPM behavior. Yoshiba and Azuma noticed Imo's WPM behavior and gave her the conditions which would strengthen this behavior by burying wheat grains in sand and stamping them down. This resulted in an increase in the number of monkeys who learned WPM behavior. The propagation of WPM behavior can be considered to be individualistic up to 1962. It is of interest to examine the course by which WPM behavior was propagated from its initiation by Imo in 1956 until August 1962, and also the conditions in which the acquisition of this behavior took place.

(a) Lineage and playmate relationships. The propagation process for WPM behavior was similar to that for SPW behavior. Especially noticeable was the effect of lineage. Of 15 monkeys (not including 1-year-old infants) of Eba's lineage, to which Imo belonged, 13 used either WPM behavior or snatching behavior (see Sect. 3.2.2). Nori's family also showed a high percentage of WPM behavior. As with SPW behavior, in Nami's lineage, only Jugo used WPM behavior.

(b) Age and sex. According to the data shown in Table 3, WPM behavior was mostly acquired by monkeys aged 2, 3, or 4 years. Monkeys aged 1 year or older than 6 years were not as good as others at acquiring this behavior. In particular, none of the

Table 3. Age and year when monkeys acquired wheat placer mining (WPM) behavior and snatching behavior

Year	1.5	2.0–2.5	3.0–3.5	4.0–4.5	5.0–5.5	6.0	Adult
				Age			
1956				Imo ♀			
1957			(Jugo ♂)				
1958	Ego ♀	Enoki ♀		Jugo ♂			
1959		Nofuji ♀, Ika ♂, Ego^(s) ♀, Saba^(s) ♂	(Zabon ♀), Enoki^(s) ♀	Nogi ♀, Sasa ♀			Eba^s ♀
1961–62		Zai ♂ (Eso ♂)	Tsuge ♀, (Ebi ♂)	Saba ♂	Zabon ♀		Aome ♀
1962 (Aug.)		Eboshi ♂, Same^s ♂	Sakura ♀, Eso ♂	Ebi ♂		Hama ♀, Tsuru ♂	Sango^s ♀

Names in parentheses indicate incomplete acquisition.
^s snatching behavior only.
^(s) WPM behavior + snatching behavior.

monkeys over 12 years of age (born before 1950) were seen to exhibit WPM behavior. In adolescents and adults, sex differences became important, as they did with SPW behavior, because of the differences in social status. Such sex differences were not important in juveniles.

3.2.2 Snatching Behavior

3.2.2.1 Snatching Behavior

Some monkeys watched the WPM-monkeys and snatched the grains of wheat when they were thrown into the water. The author calls this snatching behavior. In July 1959, an adult female, Eba, and 2-year-old Saba showed this behavior. Enoki and Ego showed WPM behavior, but they often showed snatching behavior also (Table 3). During the period from December 1961 to January 1962, Enoki and Ego did not snatch any wheat from other monkeys, but only showed WPM behavior. In August 1962, Sango and Same began snatching behavior in addition to Eba. Eba and Sango were ranked as No. 1 and No. 2 among the females. When they approached other monkeys aggressively, the other monkeys ran away. Therefore Eba and Sango could eat the wheat in the water without any effort. Same, on the other hand, who was given a high degree of tolerance during cofeeding because he was still only 2 years old, came close to monkeys engaging in WPM behavior and ate wheat with them, or collected the grains which floated toward him, or consumed leftovers after WPM-monkeys had gone.

3.2.2.2 Two Types of Snatching Behavior

As suggested above, there were two types of snatching behavior. One was collecting the leftovers, as shown by Same. This was peculiar to juveniles, and developed into WPM behavior later. The other type was plundering, as seen in the two adults Eba and Sango. They did not perform WPM behavior themselves, but let WPM-monkeys throw the wheat and sand into the water for them. Their behavior was far more effective than WPM behavior itself because they could monopolize the fruit of other animals' labor by plundering.

3.3 Bathing Behavior

3.3.1 Acquisition of Bathing Behavior

The monkeys of Koshima, although they had been living on a small islet surrounded by the sea, were never seen to go into the sea before 1959. Even after they were accustomed to salt water by SPW behavior, all that they did was just to dip their hands and feet in water. None of them bathed in the water.

However, in the summer of 1959, Mrs. Mito attracted monkeys into the water of Otomari Bay by throwing peanuts into the sea. Since then, some monkeys have gone into the sea to get peanuts. The first monkey who went into the sea was, according to Mrs. Mito, 2-year-old Ego. The author calls this behavior bathing behavior (B-behavior, Fig. 5).

In the summer of 1960, several monkeys born after 1954 were observed to bathe.

Fig. 5. Bathing behavior

A thorough survey made in January and in the summer of 1962 produced the following results: (1) of the 49 monkeys available for the investigation, 31 were observed doing B-behavior (63.2%); (2) all the monkeys born after 1955, except for two 1-year-old monkeys and Namazu, bathed in the water; (3) of the monkeys born before 1955, only 6 individuals performed B-behavior (namely 2 solitary males (Gosuke and Jugo) out of 14 males, and 4 out of 9 females). Gosuke probably drowned while he was trying to swim to the other shore across the sea; Jugo swam to the other shore in 1960 and swam back to the islet in the fall of 1964). No other adults ever bathed in the sea, and they also hated even to dip their feet in.

3.3.2 Process of Propagation

3.3.2.1 Acquisition and Propagation

B-behavior quickly propagated to other members of the troop. Within only 3 years after the first appearance of bathing monkeys, almost all juveniles and adolescents began to show this behavior. The level of acquisition in those who were born after 1955, excluding monkeys of less than 2 years old, was 96.1%, but the level of acquisition in adults was very low at 26.0% (6 out of 23 adults).

There are two probable reasons for the speed of propagation. One is that the event which gave rise to this behavior was humans throwing their favorite food into the sea. The other is that, unlike SPW and WPM behavior, bathing is a matter of adaptation to a new habitat and of changing conservatism. This gives us a strong interest in monkeys' conservatism, which is especially marked in adult males. The rate of adaptability to a new habitat seems to be high in juveniles and adolescents, medium in adult females, but very low in adult males.

3.3.2.2 Precultural Propagation

Infants are offered many opportunities to acquire B-behavior because the mothers go into the sea with their infants clinging to their fur. When the mothers go into the sea, they do not pay any particular attention to whether or not their infants are dipped into the water. Often infants are completely submerged in the water, and sometimes come close to being drowned. In this way, not long after their birth, infants become adapted to bathing in water. As a result, infants accept the sea as a part of their habitat, as they do with the mountains, and feel no reluctance in bathing. Therefore, the acquisition of B-behavior is a precultural acquisition for infants.

3.3.3 Variation of Bathing Behavior

In B-behavior, almost all monkeys used quadrupedal locomotion in the ankle-deep shallows. In deeper places, monkeys used bipedal locomotion. Often they bathed, dipped themselves in water up to their shoulders, or used quadrupedal locomotion with only their head held above the water. With bipedal locomotion, the degree of dipping themselves differed among monkeys: for some the water only came up to their knees, others went in up to their waist, and others went in up to breast height. There were 10 monkeys who swam: eight 2 to 5-year-old monkeys and two solitary adult males. Some juveniles began to take a strong interest in bathing itself. They dived from rocks and enjoyed swimming. It can be said that these juveniles have developed the original bathing behavior into new practices of avoiding the heat and just playing in the hot summer. They also dived under water skillfully, and sometimes took seaweed from the bottom at a depth of 1 or 1.5 m.

3.3.4 Comparison with Other Troops

It is not unusual for Japanese monkeys to go into the river or the sea. Monkeys of several other troops bathe in the river or in a pool. An interesting case is the Jigokudani troop, where some monkeys go into a hot spring in the cold of winter (Suzuki 1965). It is considered to be a general habit of Japanese monkeys to bathe or swim, but, as seen in the Koshima troop, this habit should be recognized as a characteristic of the troop rather than of the individuals. Another point to be noted is the adaptability and tradition of the troop in Japanese monkeys. It is surprising that until 1959, the strong traditions of the Koshima troop meant that they had never gone into the water. However, once that strong tradition began to break down for one reason or another, it could easily be removed.

3.4 "Give-Me-Some" Behavior

3.4.1 Give-Me-Some Behavior

When a human observer put his hand into his pocket to take out some peanuts, the monkeys waited, sitting in front of him, taking up a posture of let-me-have-some-please. This behavior closely resembles that of a human child when he is given sweets or cookies. The author calls this give-me-some behavior (GM behavior). GM behavior could not always be seen when monkeys were given food. When they were not psychologically calm, that is, in a situation where they would easily be disturbed by others or by nearby dominant monkeys, they did not show GM behavior.

3.4.2 Acquisition and Propagation

The initiator, the date of first occurrence, and the process of propagation of GM behavior are unknown. In 1960, Azuma noticed that Kaminari, one of the leader males, showed GM behavior. Many other monkeys must have performed this behavior at a much earlier date. Out of 47 monkeys available for the test, 37 performed GM behavior (two 1-year-old monkeys were not tested). This was 78.8%, which was higher than that for SPW behavior. Among 24 males, 19 (79.1%) showed GM behavior, and among 23 females, 18 (78.2%) did so. No difference due to sex could be observed. The high percentage of acquisition by adults was characteristic of this behavior. There were 10 monkeys who did not perform GM behavior (5 males and 5 females). Included in these were three children of Nami (Nasi, Namako, and Namazu).

3.4.3 Meaning of GM Behavior

Why did GM behavior begin, and how did it propagate? Judging from the intellectual faculties of monkeys, the acquisition of this behavior cannot simply be ascribed to imitation. The remarkable differences between this behavior and SPW and WPM behaviors is that it was performed by all the adult males, including the solitary males. This behavior pattern is believed to be quite general among Japanese monkeys, because GM behavior can be witnessed in other troops, although it remains individualistic and is not a general behavior pattern in these troops.

Compared with other monkeys in Japanese monkey parks, the monkeys at Koshima are distinguished by the fact that they have never bitten or attacked anyone, and have seldom threatened humans since the start of provisioning. In other monkey parks, relationships between monkeys and humans are not always peaceful. At Koshima, there are few sightseers, and it is mainly local monkey lovers or researchers who make contact with monkeys. Thus, the monkeys do not need to snatch their food by threat or attack, and they have learned to wait. In short, what is characteristic of this troop is their gentle, friendly attitude towards humans. Thus, GM behavior is one manifestation of an attitude toward humans born out of friendship and composure on the part of monkeys. The friendly attitude of all the monkeys towards humans can be taken as a kind of "implicit" preculture (Kawamura 1956).

3.5 The Four Behaviors Compared

In each section, the author has suggested that three factors, i.e., age, sex, and kinship, were important in the acquisition and propagation of the four newly acquired behaviors. The author would like to compare the four behaviors with each other. Figure 6 shows the change in the ratio of monkeys who performed each behavior as of 1962 as a function of age. The curves of SPW, WPM, and B-behaviors are similar, while that of GM behavior is of a different form. The ratio of monkeys who showed the first three behaviors began to increase in juveniles, reached a peak in adolescents, and decreased in adults in 1962. However, the highest ratio of GM behavior (100%) was seen in monkeys of 6–11 years old (adolescent and adult),

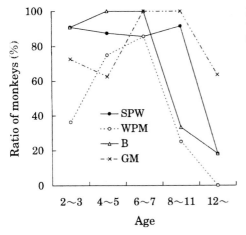

Fig. 6. The change in the ratio of monkeys who performed each behavior as of 1962 as a function of age

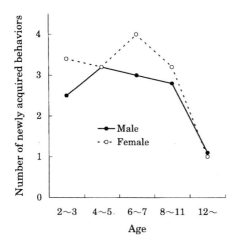

Fig. 7. Average number of newly acquired behaviors by males and females in each age class

and even adults more than 12 years old showed a high percentage. The similarity between SPW behavior and WPM behavior is shown in Fig. 6. The only difference is that the age of acquisition of SPW behavior was lower than that of WPM behavior in the period of individual propagation. This difference is due to the relative difficulty of acquisition. SPW behavior has sometimes been observed in other troops such as Takasakiyama, Ohirayama, Arashiyama, and Gagyusan, although it remains individualistic and has not been propagated to other monkeys in these troops. It seems to be possible to acquire SPW behavior incidentally. In WPM behavior, monkeys have to go through the procedure of collecting wheat by hand and taking it, together with any sand they have picked up, to the water before they can select the wheat they want by putting the mixture into the water. WPM behavior is a more complex activity. Figure 7 shows the differences in behavior acquisition in relation to the sex and age of the animals. In juveniles and adolescents, sex differences had no effect. In adolescents and 8 to 11-year-old adults, the males showed a lower

Table 4. Mean number of newly acquired behaviors (MNNB) by monkeys of each lineage

Mother	Sango	Eba	Imo	Aome	Harajiro	Natsu	Nori	Utsubo	Nami
No. of children	4	8	2	2	3	3	4	2	7
MNNB	4.0	3.6	3.3	3.3	3.0	2.6	2.6	1.6	1.6

percentage of acquisition than the females. This is due, as has been explained, to the differences in status of males and females.

In each section, the author has referred to the differences in acquisition caused by lineage. For example, the children of Sango and Eba acquired these new behaviors very readily, while Nami's children were slow (Table 4). In a society of Japanese monkeys, mother–child relations are known, but it is difficult to determine which animal is the father. Therefore, it is difficult to pursue this genetic aspect with scientific rigor. Nevertheless, it is interesting that a difference in acquisition ability according to lineage has been suggested by the data.

3.6 Environmental Basis of Precultural Phenomena

The precultural behaviors seen above all developed from the monkeys' foraging behavior and their relation to humans. That is, they were all derived from provisioning. Provisioning made great changes in the natural life of the troop. Another important factor is the natural environment. In Koshima, three different environments—precipitous mountains with thick woodland, the sandy beach, and the sea—are beautifully integrated. Before the monkeys were provisioned, the mountains were their only habitat. After provisioning, they came to know contrasting environments such as the sandy beach and the sea. That is, a new, different environment, or niche, was introduced into their natural life. The major precultural behaviors of the Koshima troop are connected with the sands and the sea. It is doubtful whether such inventive behaviors as SPW and WPM would have been developed in their previous niche. Therefore the monkeys have invented adaptive behavior in response to changes in the environmental conditions. Behavioral adaptability or plasticity in response to changes in the environment are important when we think about the evolution of behavior in animals.

4 Return from Revisit: Koshima Monkeys Afterwards

Here we return from Kawai's (1965) paper. What happened to the Koshima monkeys afterwards? A total of more than 450 monkeys were recorded from 1952 until 1999. None of the monkeys who experienced the emergence of these precultural behaviors is alive now, but their descendants are still dipping sweet potatoes into the sea, throwing grains of wheat into the water, and bathing in the sea. These behaviors have been transmitted over the generations.

Table 5. Examples of the variation in SPW behavior within individuals

Name (sex, age)	Se	Ri	Rw	Br	Rh	Pl	Ga	Rr	Others
				Type of SPW behavior					
Uri (♀ , 15)	18	12	8			1	1	1	1
Cha (♀ , 7)	30	14	1				1		
Chinu (♂ , 4)	2	3			2		1		
Chisha (♀ , 1)	1	2					8		
Ume (♀ , 12)	33	25	2		8	1		1	
Mebaru (♂ , 4)	39	13			3		1	1	
Megi (♀ , 1)	1						10		
Utsugi (♀ , 6)	39	31	3		11		2		
Udo (♀ , 3)	7	4	1				3		
Kuri (♀ , 16)	35	21	1						
Kemushi (♂ , 5)							1		
Keshi (♀ , 2)	1								
Kurumi (♀ , 13)	40	14	1		4			1	2
Mono (♀ , 5)	2	3		3	2		2		
Momiji (♀ , 2)	8	5			1		4	1	
Kuma (♂ , 9)		1			1		4	1	

Se, Seasoning; Ri, rinsing; Rw, rubbing in water; Br, brushing; Rh, rubbing between hands; Pl, plundering; Ga, gathering; Rr, rubbing on rock.
The data in the body of the table show the number of times each behavior type was observed. Uri, Ume, Utsugi, and Udo are siblings. Kuri, Kurumi, and Kemushi are also siblings. An indented name indicates a mother–child relationship. The individual whose name is indented is a child of the previous individual whose name is not indented.

Kawai and colleagues conducted an intensive follow-up study on these behaviors (Kawai et al. 1992; Watanabe 1994). Five new behavioral patterns have been added to the repertoire of sweet-potato washing, and six to wheat placer mining. An interesting case is "pool-making," which is efficient for wheat placer mining. When grains of wheat are scattered on the beach while it is still wet at low tide, some monkeys dig out some sand and make small pools from the water that oozes up. They then dip a piece of sweet potato or sweep nearby grains of wheat into the pool before they eat them.

By reanalyzing the systematic data taken during this follow-up study, we can infer the process of monkeys' acquiring SPW and WPM behaviors. Table 5 shows examples of the variation in SPW behavior within and between closely related individuals. Different behavioral patterns of SPW were distributed throughout almost all members of the group, and no monkey engaged persistently in any one specialized behavior (for details of the methods of study and descriptions of each behavior type, see Kawai et al. 1992; Watanabe 1994). Although it is difficult to interpret these data exactly in terms of the mechanisms underlying the process of acquisition, we have the impression that these behavior types are established by trial and error by each individual rather than by copying every method from other

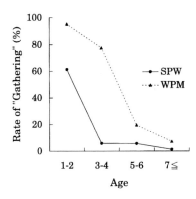

Fig. 8. The ratio of gathering behavior as a SPW- and WPM-related behavior by age class. The data were collected from 69 individuals. A total of 1662 samples of SPW behavior were divided into nine categories (see Table 5 and Kawai et al. 1992). A total of 2805 samples of WPM behavior were divided into eight categories: throwing, sweeping, pool-making, plundering, removing, screening, screening while walking, and gathering (see Kawai et al. 1992)

monkeys. The motor patterns used in these SPW behaviors seem to be general in Japanese monkeys, and can appear spontaneously in each individual. Figure 8 supports Kawai's (1965) hypothesis about the development of SPW and WPM behavior. In infants and juveniles, the gathering of other monkeys' leftovers was the predominant SPW- and WPM-related behavior. This type of behavior decreased in adults, and they engaged in their own methods of processing sweet potatoes or grains of wheat. Infants first learn to eat seasoned or washed pieces of sweet potato, or floating grains of wheat, by gathering others' leftovers. They then begin to engage in SPW or WPM behavior by themselves.

Now, SPW behavior can be seen everywhere on the beach whenever sweet potatoes are given. However, wheat placer mining has become less common. The main reason for this is that the amount of provisioning has been reduced because the population of monkeys on Koshima Island became too large (Watanabe et al. 1992). Potatoes are now given only a few times per year. Sweet potato washing and wheat placer mining began under provisioning, so it is quite natural that these behaviors should have changed as the degree of provisioning has changed.

There is another reason why wheat placer mining behavior has become less common. Imagine the snatching behavior described above. Now that the amount of provisioning is reduced, the competition for wheat is much more severe. Monkeys lose wheat if they throw grains into the water because dominant individuals will snatch them away; it is better to pick them up and eat them directly from the ground. Wheat placer mining can be seen when the monkeys have almost finished eating and there are still a few grains left on the ground mixed with sand. In the case of sweet potato washing, however, the animals can run away carrying the sweet potato in their hands if a dominant monkey approaches. The fact that one or two pieces of sweet potato per monkey are enough is also relevant. For these reasons, wheat placer mining has become less common but sweet potato washing can be seen even under greatly reduced provisioning. In addition, seasoning is now the predominant behavior type in sweet potato washing, and real washing has become very rare. In the 1960s, all sweet potatoes were covered with soil, while now they have already been washed in the market, and there is no need for the monkeys to wash off the soil.

Table 6. Precultural behaviors and the sand-digging test

Result of sand-digging test	No. of newly acquired behaviors				
	0	1	2	3	4
A, B			2	7	13
C, D		6	2	3	
E	3	3	1		1[a]

[a] Uncertain result because of a failure to do an appropriate test.
The results of the tests were classified from A (best) to E (worst).
The numbers of individuals are shown in each category.

Another change for the Koshima monkeys is that they have begun to eat fish (Watanabe 1989). Fish-eating was first reported to happen sporadically among adult males, and initially the habit did not spread to other troop members. In 1982, about 3 years after the first observation, four adult females were observed to eat raw fish. This was the turning point for this habit to spread among the troop. Most members then began to eat raw fish, and this continued until 1986. The reason may be that the population had increased to about 100 at that time, which caused a scarcity of food.

Besides the intensive observation of the behavior of the monkeys, some field experiments were conducted on Koshima. In 1961, Kawai developed a simple test called the "sand-digging test." A human experimenter buried a peanut on the beach in front of a monkey. Then all traces of disturbed sand were carefully smoothed away. When Ego was tested, she immediately dug out the peanut. Later, Kawai and colleagues conducted a systematic study of this sand-digging test on the Koshima monkeys (Tsumori et al. 1965). Eighteen monkeys easily found the peanut. Five other monkeys found it in the second trial. The results for each monkey were classified into five grades from A (best) to E (worst). A correlation was found between the results of the sand-digging test and the number of precultural behaviors they acquired (Table 6). Another experimental study was conducted by Higuchi (1992). He introduced an open operant box to the island, and examined the acquisition and propagation of panel-pressing behavior. Out of 74 monkeys in the troop, 34 acquired the panel-pressing behavior. He concluded that most of the transmission of this behavior occurred through local enhancement, along with individual trial and error.

Imo, the originator of SPW and WPM behavior, died on May 21, 1972, but she is still alive in current literature (e.g., de Waal 1999; Vogel 1999). Ironically, discussions about cultural phenomena in animals have become heated since her death. Galef (1990, 1992) picked up the example of sweet-potato washing to argue about imitation and culture in animals, Visalberghi and Fragaszy (1990) conducted experimental studies on the emergence and diffusion of food-washing behavior in captive capuchin monkeys and crab-eating macaques, and Tomasello et al. (1993) discussed cultural learning in human and nonhuman primates.

Longitudinal observations of free-ranging Japanese monkeys have added two other examples of social transmission of behavior. The first is "stone-handling," which was seen in the Arashiyama troop (Huffman 1984, 1996; Huffman and Quiatt 1986). A 3-year-old female began to handle stones in a peculiar manner in 1979, and this stone-handling behavior propagated to other troop members. This behavior initially spread only among individuals of the same age-class as the initiator, but subsequently passed down from older to younger individuals in successive generations. It is now a commonly seen behavior among most individuals of this troop. The second example is the technique of grooming (Tanaka 1995, 1998). Tanaka investigated the techniques of grooming in the Shiga-A troop, and identified four patterns of grooming behavior. An analysis of the distribution of each technique in three maternal lineages showed the possibility of social transmission of the grooming technique from adults to infants.

By looking at animals other than Japanese monkeys, we can find very conspicuous examples of "chimpanzee cultures." Longitudinal investigations of wild chimpanzees at several different sites in Africa have been clarifying the behavioral diversity between different study sites, which cannot be explained by ecological or genetic theories (Goodall 1973; McGrew 1992; Wrangham et al. 1994; Sugiyama 1997; Boesch 1996; Matsuzawa 1998; Whiten et al. 1999).

5 Back to the Question of Culture

Half a century has passed since the start of the research on Koshima monkeys. The study of nonhuman primates has accumulated data from both the field and the laboratory, and by both naturalistic observation and experimental manipulation. Have we succeeded in casting a different light on human culture by keeping our outlook broader? The examples from Koshima Island clearly show that the social transmission of behavior occurs in Japanese monkey society, but it seems to be inappropriate to answer yes or no to the question of whether nonhuman primates have culture. A valuable way forward would be, as Kawamura (1956, 1965) remarked, to trace each step and consider the evolutionary pathway (e.g., van Shaik et al. 1999). In the first place, intensive and longitudinal observations have revealed highly organized permanent social structures in many species of monkeys and apes. Such permanent group living is, as Imanishi (1952) suggested, one condition for the establishment of a culture. In the second place, studies of social learning in animals have revealed more precise mechanisms of social transmission of a behavior. The process of social learning is now divided into several categories, such as social facilitation, stimulus enhancement, emulation, program-level imitation, action-level imitation, etc. (Tomasello 1990; Byrne and Russon 1998). In these terms, the growing body of knowledge has been making it clear that imitation is more difficult for nonhuman primates than was expected in the 1950s: monkeys do not imitate, but are influenced by stimulus enhancement (Visalberghi and Fragaszy 1990); chimpanzees do emulate, and much less frequently they imitate (Whiten et al. 1996). In addition, monkeys do not teach others to do anything, and chimpanzees only do so extremely rarely (Boesch 1991). Therefore, it can be said that imitation and teaching, the two most important elements in human material culture, cannot be found in monkeys, and is only seen in chimpanzees to a very limited extent (Myowa-Yamakoshi and Matsuzawa 1999). However, the transmission of material culture in humans is not always accompanied by teaching and imitation either. The fact that human children imitate does not mean that they always imitate (Whiten et al. 1996). It is quite natural to assume that stimulus enhancement and emulation are involved in some cases of the transmission of cultural phenomena in humans as well as imitation and teaching. Although recent experimental studies have been trying to distinguish and separate every social learning process, it is also possible to look at the problem in reverse, that is, we could also try to find a link between stimulus enhancement, emulation, and imitation in terms of cognitive mechanisms.

Drawing a line between culture and nonculture is beyond the scope of this paper. Instead, we would like to conclude by saying that humans and monkeys might share some traits in the social mechanism and cognitive capacity which comprise human culture, and this is more true for apes and humans. The invention and propagation of precultural behaviors shown by Koshima monkeys are valuable examples when thinking of such phenomena. In Koshima, the sixth generation descendants of the monkeys in the initial study periods are still engaging in sweet-potato washing.

References

Boesch C (1991) Teaching among wild chimpanzees. Anim Behav 41:530–532
Boesch C (1996) The emergence of cultures among wild chimpanzees. In: Runciman WG, Maynard Smith J, Dunbar RIM (eds) Evolution of social behaviour patterns in primates and man. Oxford University Press, Oxford, pp 251–268
Byrne RW, Russon AE (1998) Learning by imitation: a hierarchical approach. Behav Brain Sci 21:667–721
de Waal FBM (1999) Cultural primatology comes of age. Nature 399:635–636
Galef BG (1990) Tradition in animals: field observation and laboratory analyses. In: Beckoff M, Jamieson D (eds) Interpretation and explanation in the study of animal behavior. Westview Press, Boulder, pp 74–95
Galef BG (1992) The question of animal culture. Hum Nat 3:157–178
Goodall J (1973) Cultural elements in a chimpanzee community. In: Menzel EW (ed) Precultural primate behavior. Karger, Basel, pp 144–184
Higuchi Y (1992) Cultural behavior of Japanese monkeys (in Japanese). Kawashima-shoten, Tokyo
Huffman MA (1884) Stone-play of Macaca fuscata in Arasiyama B troop: transmission of a non-adaptive behavior. J Hum Evol 13:725–735
Huffman MA (1996) Acquisition of innovative behaviors in nonhuman primates: a case study of stone handling, a socially transmitted behavior in Japanese macaques. In: Heyes CM, Galef BG (eds) Social learning in animals: the roots of culture. Academic Press, San Diego, CA, pp 267–289
Huffman MA, Quiatt D (1986) Stone handling by Japanese macaques (Macaca fuscata): implications of tool use of stone. Primates 27:427–437
Imanishi K (1952) The evolution of human nature (in Japanese). In: Imanishi K (ed) Ningen. Mainichi-shinbunsha, Tokyo, pp 36–94
Itani J, Nishimura A (1973) The study of infrahuman culture in Japan: a review. In: Menzel EW (ed) Precultural primate behavior. Karger, Basel, pp 26–50
Kawai M (1964) Ecology of Japanese monkeys (in Japanese). Kawade-shobo, Tokyo
Kawai M (1965) Newly acquired pre-cultural behavior of the natural troop of Japanese monkeys on Koshima Islet. Primates 6:1–30

Kawai M, Watanabe K, Mori A (1992) Pre-cultural behaviors observed in free-ranging Japanese monkeys on Koshima islet over the past 25 years. Primate Rep 32:143–153

Kawamura S (1954) On a new type of feeding habit which developed in a group of wild Japanese monkeys (in Japanese). Seibutsu-shinka 2:11–13

Kawamura S (1956) Prehuman culture (in Japanese). Shizen 11:28–34

Kawamura S (1959) The process of sub-culture propagation among Japanese macaques. Primates 2:43–60

Kawamura S (1965) Sub-culture in Japanese monkeys. In: Kawamura S, Itani J (eds) Monkeys and apes: sociological studies (in Japanese). Chuokoron-sha, Tokyo, pp 237–289

Matsuzawa T (1998) Chimpanzee behavior: a comparative cognitive perspective. In: Greenberg G, Haraway MM (eds) Comparative psychology: a handbook. Garland, New York, London, pp 360–375

McGrew WC (1992) Chimpanzee material culture: implications for human evolution. Cambridge University Press, Cambridge

Myowa-Yamakoshi M, Matsuzawa T (1999) Factors influencing imitation of manipulatory actions in chimpanzees (Pan troglodytes). J Comp Psychol 113:128–136

Nishida T (1987) Local traditions and cultural transmission. In: Smuts BB, Cheney DL, Seyfarth RM, Wrangham RW, Struhsaker TT (eds) Primate society. University of Chicago Press, Chicago, pp 462–474

Sugiyama Y (1997) Social tradition and the use of tool-composites by wild chimpanzees. Evol Anthropol 6:23–27

Suzuki A (1965) An ecological study of wild Japanese monkeys in snowy areas: focused on their food habits. Primates 6:31–72

Tanaka I (1995) Matrilineal distribution of louse egg-handling techniques during grooming in free-ranging Japanese macaques. Am J Phys Anthropol 98:197–201

Tanaka I (1998) Social diffusion of modified louse egg handling techniques in free-ranging Japanese macaques. Anim Behav 56:1229–1236

Tomasello M (1990) Cultural transmission in the tool use and communicatory signaling of chimpanzees? In: Parker S T, Gibson K R (eds) "Language" and intelligence in monkeys and apes: comparative developmental perspectives. Cambridge University Press, New York, pp 274–311

Tomasello M, Kruger AK, Ratner HH (1993) Cultural learning. Behav Brain Sci 16:495–552

Tsumori A, Kawai M, Motoyoshi R (1965). Delayed response of wild Japanese monkeys by the sand-digging test. I. Case of the Koshima troop. Primates 6:195–212

van Shaik CP, Deaner RO, Merrill MY (1999) The conditions for tool use in primates: implications for the evolution of material culture. J Hum Evol 36:719–741

Visalberghi E, Fragaszy DM (1990) Food-washing behavior in tufted capuchin monkeys, Cebus apella, and crab-eating macaques, Macaca fascicularis. Anim Behav 40:829–836

Vogel G (1999) Chimps in the wild show stirrings of culture. Science 284:2070–2073

Watanabe K (1989) Fish: a new addition to the diet of Koshima monkeys. Folia Primatol 52:124–131

Watanabe K (1994) Precultural behavior of Japanese macaques: longitudinal studies of the Koshima troops. In: Gardner RA, Gardner B, Chiarelli B, Plooij FX (eds) The ethological roots of culture. Kluwer, Dordrecht, pp 81–94

Watanabe K, Mori A, Kawai M (1992) Characteristic features of the reproduction in Koshima monkeys, Macaca fuscata fuscata: a 34-year summary. Primates 33:1–32

Whiten A, Custance D, Gomez JC, Teixidor P, Bard KA (1996) Imitative learning of artificial fruit processing in children (Homo sapiens) and chimpanzees (Pan troglodytes). J Comp Psychol 110:3–14

Whiten A, Goodall J, McGrew WC, Nishida T, Reynolds V, Sugiyama Y, Tutin CEG, Wrangham RW, Boesch C (1999) Cultures in chimpanzees. Nature 399:682–685

Wrangham RW, McGrew WC, de Waal FBM, Heltne PG (1994) Chimpanzee cultures. Harvard University Press, Cambridge

25
Tube Test in Free-Ranging Japanese Macaques: Use of Sticks and Stones to Obtain Fruit from a Transparent Pipe

Ichirou Tanaka[1], Eishi Tokida[2], Haruo Takefushi[2], and Toshio Hagiwara[2]

1 Introduction

Apes excel in tool using (Yerkes and Yerkes 1929). Captive chimpanzees (*Pan troglodytes*) learned to use bamboo poles as rakes and, when one was too short, to fit two together to form a longer rake (Köhler 1927). They also climbed up onto boxes to reach a banana hanging from a ceiling, and if one box was too small, would stack one box on another to gain a higher platform. This deployment of previous experience in other contexts such as using poles as rakes and boxes as a ladder is known as perceptual reorganization or insight (Köhler 1927; Manning 1972; Mackintosh 1983).

We report here a case of perceptual reorganization in toolusing in Japanese macaques (*Macaca fuscata*). A free-ranging Japanese macaque female named Tokei invented the technique of throwing a small stone underhand into a pipe to gain an apple, after a series of instrumental conditioning trials using sticks as tools. We investigated whether free-ranging Japanese macaques threw stones at random or selected stones beforehand, as wild chimpanzees do when selecting stones for nut-cracking (Boesch and Boesch 1983), and how this stonethrowing influenced the other naive macaques in their troop.

2 Methods

2.1 Study Area and Subjects

Free-ranging Japanese macaques were studied at Jigokudani Monkey Park, Japan from December 1983 to April 1992. The study focused on the Shiga A-1 troop, which has been provisioned since 1962. The troop was initially composed of 108 individuals, increasing to 252 by the end of the study. The monkeys are not fed by sightseers. The ecology of the study area has been described by Wada and Ichiki (1980). During winter (from December to April), the study area is covered with snow more than 1 m deep.

---

[1] Section of Language and Intelligence, Section of Ecology, Primate Research Institute, Kyoto University, 41 Kanrin, Inuyama, Aichi 484-8506, Japan
[2] Jigokudani Monkey Park, Yamanouchi-machi, Nagano 381-0401, Japan

2.1 Conditioning Stick Use

We fixed an acrylic transparent tube (102 cm long and 12 cm in internal diameter) horizontally onto a large log in the middle of the provisioning site and then placed as much as half an apple in the middle of the pipe where the macaques could see but not reach it. A hooked stick, 60 cm long, was inserted into the tube so that the hook touched the apple.

On December 29, 1983, we started instrumental conditioning. First, after some macaques had learned to obtain an apple by pulling the stick, we removed the stick from the apple within the pipe. Second, when the macaques learned to push and attach the stick to the apple before pulling it, we moved the stick away from the pipe. Third, after they learned to insert the stick, we moved it more than 5 m away from the pipe. Fourth, after the macaques learned to fetch the hooked stick, we gave them other sticks including some too short to reach the fruit and some too long to handle. Finally, we removed all the sticks within 10 m of the pipe so the macaques had only natural sticks available.

2.2 Analyzing Choice of Stones

We collected 100 stones at random from our experimental site (a randomly selected 1-m^2 square by the river) and randomly sampled them on March 6, 1992. We weighed each stone and numbered it with ink from 1 to 100. The following day we fixed the acrylic transparent tube with the log onto a concrete foundation to install the pipe above snow level. To record the detailed position of apples and thrown stones, we fastened a ruler (accurate to 1 mm) onto the tube. We placed a steel plate 180 cm long and 90 cm wide on the snow in front of the tube and put the stones on the plate (see Fig. 2). Monkeys took stones only from the steel plate because snow covered the other stones. After each trial, we brought the used stones back to the plate and jumbled them up to randomize them.

From March 9 to 18, 1992, to investigate whether macaques threw stones at random or selected them beforehand, we recorded stone throwing by the three females (Tokei, Towano, and Kerimi) who had learned stone throwing by then. Other monkeys were kept away with a supply of food. The mean weight ($\pm$SEM) of the apple in the tube was 25.4 $\pm$ 0.5 g ($n = 130$). We recorded the number on the stone thrown by the macaques and which hand the monkeys used. We recorded throwing behavior on video cameras (SONY CCD-TR75 and CCD-TR705), one camera (CCD-TR75) looking over the experimental site and recording the interaction between the thrower and the other macaques, and the other camera (CCD-TR705 with 1/2000 shutter speed) looking at the transparent pipe with the scale of the ruler and recording the movement of the thrown stones (see Fig. 2). From the latter video record taken at intervals of 0.03 s, we calculated the speed of the thrown stones between 0.06 and 0.03 s before the collision with the apple and their momentum (mass multiplied by speed). We excluded from the analysis of momentum those observations in which thrown stones rotated and decelerated abruptly.

After almost 20 trials, the subject had eaten enough apples (more than two) and was uninterested in stone throwing for almost 2 h. We then recorded the other subjects. We recorded the thrown stone weight for each macaque thrower 40 times. Because monkeys sometimes retrieved and threw the same stones several times and threw different stones at the same apple in one trial until they obtained the fruit, the total number of throws differed from that we used for weighing the stones and that of trials.

3 Results

3.1 Conditioning of Stick Use

Six macaques (Tokei, Torani, Togura, Towano, Kerimi, and Tachimi) learned to pull the stick only when it touched an apple. Tokei, Torani, Towano, and Kerimi also learned to insert a stick into the pipe to remove an apple. By September 1984, Torani and Tokei had learned to fetch the hooked stick and in July 1985 they began to select an appropriate stick (about 60 cm long) to insert into the pipe, discarding any that were too short or too long. In August 1985, Torani and Tokei began to search for a natural appropriate stick and bring it back to the pipe. If we gave them only an inadequate stick, the two macaques tried to compensate by stretching their arms if the sticks were short and by pushing out the apple if the sticks were too long. Torani also threw short sticks at the apple but failed to push the apple out in this way.

On August 11, 1986 when there were no sticks near the pipe and only long sticks at the provisioning site, Tokei brought a long stick back to the pipe and bit it to shorten it. We observed her doing this five times. Once she pulled up a shrub from the ground, plucked the leaves, and bit off the root to make a stick. After August 11, 1986, she always used stones (see following) if there were no adequate sticks near the pipe, instead of shortening sticks.

3.2 Invention of Stone Throwing

On August 4, 1986, Tokei, then 7 years old, began to throw a small stone underhand into the pipe, push out an apple, go to the opposite end of the pipe, and take the apple (see Fig. 2), after she had learned to bring back and insert natural sticks into the tube. Unlike Torani's short sticks, thrown stones had enough momentum to collide with the fruit and push it out of the tube. After Tokei's innovation, in June 1987 6-year-old Towano and (in August 1987) 4-year-old Kerimi also came to throw stones. On March 12, 1992, for the first time, 10-year-old Torani threw a stone, which Tokei had already thrown and left in the pipe, although he failed to obtain an apple. In April 1992 he removed fruits from the pipe only with stones.

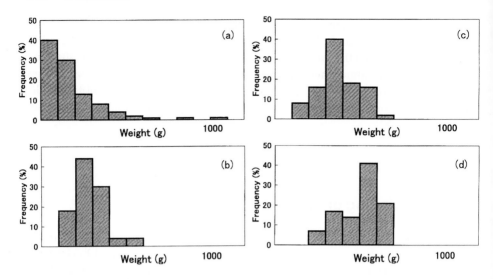

Fig. 1. Distribution of the weight of stones available and those selected for throwing. **a** Random sampling (X ± SEM, 252.0 ± 4.6 g, n = 100). **b** Stones thrown by Tokei (X ± SEM, 285.4 ± 10.1 g, n = 72). **c** Stones thrown by Towano (X ± SEM, 379.6 ± 12.3 g, n = 83). **d** Stones thrown by Kerimi (X ± SEM, 491.6 ± 15.2 g, n = 43)

3.3 Selection of Stones

For each stone-thrower, the weight distribution of thrown stones differed from that of random sampling (Kolmogorov-Smirnov two-sample test, $P < 0.01$; Fig. 1). Tokei threw lighter stones than Towano (Kolmogorov-Smirnov two-sample test, $P < 0.01$) and Kerimi (Kolmogorov-Smirnov two-sample test, $P < 0.01$); Towano also threw lighter stones than Kerimi (Kolmogorov-Smirnov two-sample test, $P < 0.01$). These females sometimes picked up a stone, held it, and discarded it, choosing a different one to bring to the pipe. Tokei discarded 7.7% of the stones (6 of 78); Towano discarded 6.7% of the stones (6 of 89), and Kerimi discarded 20.4% of the stones (11 of 54). Towano carried two stones simultaneously on 19.2% of the occasions (15 of 78 cases). If she failed to obtain an apple with the first stone, she threw the second at once (5 of 15 cases); if she succeeded with the first stone, she abandoned the second (10 of 15 cases).

Tokei threw stones mainly with her left hand (88 of 89 cases); Towano threw stones only with her left hand (184 cases); and Kerimi threw stones only with her right hand (132 cases). Tokei used her right hand once but with small momentum [0.13 kg(m/s); see Fig. 3]. Towano once tried to throw a stone with her right hand, but she changed hands and threw it with her left hand.

3.4 Influence of Stone Throwing on the Other Naive Macaques

Although the troop members other than the conditioned four macaques could observe stone throwing, they did not throw a stone into the tube. Some macaques, however, obtained the fruit by a different way from the four throwers; they came to get the reward by waiting for the conditioned macaque to throw a stone at the opposite end of the pipe. As the reward was pushed out from the end opposite to the thrower, the fruit just after throwing was free. These monkeys could obtain the apple if they reached it before the thrower went to the opposite end of the pipe. On the contrary, the throwers did not always obtain the reward after their successful tasks. To the conditioned four macaques, this failure spontaneously (beyond our instrumental conditioning) became the sixth problem in addition to the five steps of our instrumental conditioning.

3.5 Variation in the Force Used to Throw Stones

Other macaques sometimes went to the end of the pipe and stole the apple before the stone-thrower could retrieve it. Tokei threw stones at an apple with less momentum when other macaques were at the end (Fig. 2a) than she did when there were no macaques waiting (Mann-Whitney U-test, $U \geq 0$, $n = 38$, $P = 0.004$, Fig. 3a,b). Owing to the lesser momentum, the apple moved only a little and remained in the pipe. Thus, the thieves could not get at before Tokei (who was the second-ranking female) could chase off the subordinates and get the apple herself (Fig. 2b). Towano, however, did not throw stones with less momentum when there were macaques at the end of the pipe than she did when there were no macaques there (Mann-Whitney U-test, $U \geq 134.5$, $n = 40$, $P = 0.112$; Fig. 3c,d). Thus, her fruits were often stolen. Kerimi never threw stones when there were macaques at the end of the pipe; she always waited until they left.

3.6 Use of Infants to Obtain Apples

Until they were 6 months old, infants were small enough to enter the pipe and grasp the apple. On July 30, 1986, Tokei pushed her 3-month-old infant into the pipe when it was nearby. When the baby bit the apple, she pulled it out and took the fruit. Togura and Tomato also used their babies to get the apple when the babies had entered by themselves. Only Tokei, however, brought her infants to the pipe and actively pushed them in. She did this with all four of her infants. If she failed to make her infants go into the pipe, she used a stick or stone instead.

Fig. 2. Stone throwing by a female Japanese macaque, Tokei, to remove an apple from a pipe. **a** Tokei (the macaque on the *right*) looked at a macaque at the left end of the pipe and pitched the stone into the tube toward an apple in the middle with little momentum. **b** She ran to the opposite end of the pipe where the apple was pushed out and chased off the subordinate. The other macaque (*front*) watched the interaction. (These photographs were taken from video replays on an external color LCD 3.5-in. monitor (SONY DCR-TRV9) by I. Tanaka)

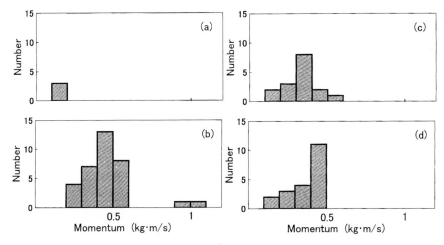

Fig. 3. Distribution of the momentum of thrown stones. **a** Stones thrown by Tokei when other macaques were at the opposite end of the pipe (X ± SEM, 0.157 ± 0.028 kg (m/s), $n = 3$). **b** Stones thrown by Tokei when no macaques were present (X ± SEM, 0.453 ± 0.027 kg (m/s), $n = 35$). **c** Stones thrown by Towano when other macaques were at the opposite end of the pipe (X ± SEM, 0.327 ± 0.025 kg (m/s), $n = 16$). **d** Stones thrown by Towano when no macaques were present (X ± SEM, 0.381 ± 0.024 kg (m/s), $n = 24$). The occasion when Tokei used her right hand was excluded (see text)

4 Discussion

Captive tufted capuchin monkeys (*Cebus apella*) learned by trial and error within 2 h to use sticks to remove food from a transparent pipe, although species of *Cebus* also appear to have a strong propensity for spontaneous tool using (Visalberghi and Trinca 1989). Japanese macaques, however, solved the transparent tube problem much more slowly than the capuchins.

After August 1985, when Tokei and Torani had learned to fetch natural appropriate sticks, if we gave them short sticks the two macaques tried to use them and only after several unsuccessful trials did they go off to find more appropriate tools. Tokei then used stones while Torani searched for another stick. If they were given long sticks, however, the macaques used these to push out the apple. This behavior may be connected with throwing stones and pushing out apples with the stones. This process is similar to the acquisition of aimed throwing in pigtailed macaques (*Macaca nemestrina*), which learned to throw a stick (given to them by humans) toward a food item after many unsuccessful attempts (Beck 1976).

On August 11, 1986, Tokei made an appropriate stick from a live shrub. Beck (1980) defined four modes of tool manufacture by animals: detach, subtract, add plus combine, and reshape. Tokei's tool manufacture corresponds to the "subtract" mode rather than the simple "detach" mode. However, it was easier for Tokei to search for stones than for sticks because there were more of the former at our experimental site.

Macaques and baboons are known to drop or throw down objects (review in

Beck 1980). Hamilton et al. (1975) discovered that wild chacma baboons (*Papio ursinus*) threw stones underhand to strange humans at the Kuiseb Canyon in South West Africa, where there were free and loose rocks in the cliffs. Hamilton et al. (1975) suggested that if they were in stone-rich environment monkeys would learn to throw stones. This is true of Tokei, who combined stone-handling (see Weinberg and Candland 1981; Huffman and Quiatt 1986) with pushing sticks into the pipe spontaneously as new successful tool using with different materials. This behavior is perceptual reorganization, the deployment of previous experience in other contexts, or insight (Köhler 1927; Manning 1972; Mackintosh 1983; McFarland 1985), like Köhler's chimpanzees, who perceived objects in a new light, poles as rakes and boxes as the elements of a crude ladder. In a similar manner, a famous Japanese macaque female named Imo applied washing of sweet potatoes to washing of wheat (Kawamura 1959). All the stick-users (Tokei, Towano, Kerimi, and Torani) combined their experience in pushing sticks with their ability to handle stones whereas the non-stick-users did not learn to throw stones. This observation supports the idea that previous experience is important in insight learning (McFarland 1985).

The macaque selected stones before throwing. Similarly, chacma baboons selected relatively large stones to throw at intruders (Hamilton et al. 1975). This selectivity indicates that the macaques were aware of the appropriate weight for stone throwing. They appear to have been able to judge different weights (Essock-Vitale and Seyfarth 1987) and to learn the principle of the preservation of momentum in collision. This is like the learning sets (Harlow 1949) of rhesus macaques (*Macaca mulatta*), who learned the principle of two-choice discrimination problems with different objects. In stick using, two macaque siblings (Torani and Tokei) selected an appropriate stick for the pipe; this is similar to selecting an appropriate weight in stone throwing. Moreover, the macaques sometimes picked up a stone, held it, and dropped it; they then chose a different stone and brought it to the pipe. Thus, macaques appeared to weigh stones by hand. This behavior supports the idea that they can judge and categorize stimuli.

If the momentum is constant, a low velocity results in low kinetic energy (0.5 × momentum × velocity). A decrease in velocity reduces kinetic energy more than a decrease in mass does. In other words, if the momentum is fixed, a large mass means less kinetic energy. Thus, by selecting heavier stones Towano used less energy when throwing than did Tokei. Kerimi, by using the heaviest stones, appears to have used the least energy. Although the final momentums of the stones were similar (see Fig. 3b,d), the three stone-throwers chose different weights. Based on Whiten and Ham's (1992) classification of social learning, these variable preceding elements (criteria of choice) with the same final act (throwing) correspond to "goal emulation." Thus, if stone throwing is socially transmitted, the mechanism is not "imitation" at least. Moreover, the variation suggests that Towano and Kerimi learned most stone-throwing individually. Because our field experiments could not control the opportunities of witnessing, the probability of social transmission of behavior in the diffusion of stone-throwing remains unresolved.

Naive monkeys other than the conditioned four throwers did not spontaneously imitate the behavior of stone-throwing either. Some macaques came to get the reward by waiting for the conditioned macaque to throw a stone at the opposite end of the pipe. Tokei could control the momentum of thrown stones depending on whether other macaques, who might steal her fruit, were present. Thus, she appears to infer which tactics will be successful from the conservation of momentum, the arm reach of the other individuals, and her social status as the beta (β) female who could supplant most members of her troop. This behavior is perceptual reorganization or insight (Köhler 1927; Manning 1972; Mackintosh 1983; McFarland 1985). In a provisioned large troop, only high-ranking monkeys can reach the experimental apparatus containing valuable food such as apples. Low-ranking Kerimi's inactivity in the presence of dominant macaques shows that social status can hide the performance of behavior even though the animals have learned the task (see Visalberghi and Trinca 1989). This choice is the converse of "social support" based on Whiten and Ham's (1992) classification of social influence.

Kerimi's inactivity also indicates that most of the troop members adopted the alternate solution. Thus, this alternate solution appeared to be diffused rapidly, although there were no systematic records. Thus, the partial information of stone throwing might be transmitted socially. These macaques appeared to comprehend an environmental change before and after their throwing. This behavior corresponds to the "emulation learning" mechanism of the social transmission (Whiten and Ham 1992) because the information conveyed in emulation learning is an environmental change before and after the behavior (Tomasello 1996). This alternate solution is easier than stone throwing, which needed the complex instrumental conditioning. Thus, the former appeared to diffuse rapidly and might hinder the spontaneous acquisition of stone throwing.

Tokei extended perceptual reorganization to animate and reliable objects (her infants) beyond inanimate objects (sticks and stones). Although the other two mothers, Togura and Tomato, passively pulled their infants out after they had entered the pipe, Tokei actively pushed her infants into the pipe as if she were inserting sticks. This observation also agrees with the previous experience being important in insight learning (McFarland 1985) and is also a countermeasure against the waiters at the opposite end. Thus, Japanese macaques can deploy previous experience in other contexts, form concepts, change tactics, and handle other individuals.

Acknowledgments

This report is modified from Tokida E, Tanaka I, Takefushi H, Hagiwara T (1994) Tool using in Japanese macaques: use of stones to obtain fruit from a pipe. Animal Behaviour 47:1023-1030. We thank Professor Tetsuro Matsuzawa, Dr. Elisabetta Visalberghi, Professor Kenichi Aoki, Dr. Tomo'o Enomoto, Ms. Masako Yamaguchi, Dr. Frans de Waal, Mr. Mitsuo Shida, and the anonymous referees of *Animal Behaviour* for helpful advice; Mr. Sougo Hara for permission to study at Jigokudani Monkey Park; and Mr. Shigenori Nishizawa and the other staff of Jigokudani Monkey Park for their assistance. This study was supported in part by a grant-in-aid from the Japanese Ministry of Education, Science and Culture, Japan.

References

Beck BB (1976) Tool use by captive pigtailed macaques. Primates 17:301–310

Beck BB (1980) Animal tool behavior: the use and manufacture of tools by animals. Garland, New York

Boesch C, Boesch H (1983) Optimisation of nut-cracking with natural hammers by wild chimpanzees. Behaviour 83:265–286

Essock-Vitale S, Seyfarth RM (1987) Intelligence and social cognition. In: Smuts BB, Cheney DL, Seyfarth RM, Wrangham RW, Struhsaker TT (eds) Primate societies. University of Chicago Press, Chicago, pp 452–461

Hamilton WJ III, Buskirk RE, Buskirk WH (1975) Defensive stoning by baboons. Nature (Lond) 256:488–489

Harlow HF (1949) The formation of learning sets. Psychol Rev 56:51–65

Huffman MA, Quiatt D (1986) Stone handling by Japanese macaques (*Macaca fuscata*): implications for tool use of stone. Primates 27:413–423

Kawamura S (1959) The process of sub-culture propagation among Japanese macaques. Primates ii:43–60

Köhler W (1927) The mentality of apes, 2nd edn. Kegan Paul, Trench and Trubner, London

Mackintosh NJ (1983) General principles of learning. In: Halliday TR, Slater PJB (eds) Animal behaviour, Vol 3: Genes, development and learning. Blackwell, Oxford, pp 149–177

Manning A (1972) An introduction of animal behaviour, 2nd edn. Arnold, London

McFarland D (1985) Animal behaviour: psychobiology, ethology and evolution. Longman, Harlow

Tomasello M (1996) Do apes ape? In: Heyes CM, Galef BG (eds) Social learning in animals: the roots of culture. Academic Press, San Diego, pp 319–346

Visalberghi E, Trinca L (1989) Tool use in capuchin monkeys: distinguishing between performing and understanding. Primates 30:511–521

Wada K, Ichiki Y (1980) Seasonal home range use by Japanese monkeys in the snowy Shiga Heights. Primates 21:468–483

Weinberg SM, Candland DK (1981) "Stone-grooming" in *Macaca fuscata*. Am J Primatol i:465–468

Whiten A, Ham R (1992) On the nature and evolution of imitation in the animal kingdom: reappraisal of a century of research. In: Slater P, Rosenblatt J (eds) Advances in the study of behavior, Vol 21. Academic Press, New York, pp 239–283

Yerkes RM, Yerkes AW (1929) The great apes. Yale University Press, New Haven

26
Tool Use by Chimpanzees (*Pan troglodytes*) of the Arnhem Zoo Community

HIDEKO TAKESHITA[1,2] and JAN A.R.A.M. VAN HOOFF[2]

1 Introduction

Tool use has been reported for a number of nonhuman primates. However, chimpanzees are the only consistent and habitual tool users and tool makers (McGrew 1992). Certain objects easily obtained in their natural habitats are used, and some tools are even manufactured (Beck 1980; Boesch and Boesch 1990; Goodall 1986; McGrew 1992; Nishida 1973; Sugiyama and Koman 1979; Wrangham et al. 1994). Moreover, the use of tool composites or "meta-tools" has been reported (Brewer and McGrew 1990; Matsuzawa 1994, 1996; Suzuki et al. 1995).

Reports from various study sites show that there are regional differences in tool use by wild chimpanzees (Whiten et al. 1999). These differences cannot be related directly and solely to ecological variables, such as habitat, which suggests that cultural factors play a major role in the establishment and propagation of tool-use behaviors (McGrew 1992; Sugiyama 1993). Chimpanzee tool use is nearest to that of humans in its diversity and complexity. Understanding the nature of tool use in chimpanzees may illuminate our understanding of the evolution of human material culture.

To establish what factors explain the variation in tool use, it is important to collect more data from various groups and individuals, in various habitats and living conditions. This chapter describes the repertoire of tool-use behaviors of the captive chimpanzee group at Arnhem Zoo, in The Netherlands. It consists of between 25 and 30 individuals and has a composition comparable to that of communities in the wild. The group lives in spacious and varied surroundings.

The chimpanzees in this community have not been taught certain tool uses explicitly. However, their environment provides numerous incentives, some of which differ from those experienced in the wild. For example, they have been able to watch complex human instrumental behaviors. Thus, they are familiar with the concept of container because they see their food brought and presented to them in containers. Given the capacity of chimpanzees for imitation (Custance et al. 1995;

[1] School of Human Cultures, The University of Shiga Prefecture, 2500 Hassakacho, Hikone, Shiga 522-8533, Japan
[2] Ethology and Socio-Ecology Group, Utrecht University, Centrumgebouw Noord, Padualaan 14, Pb. 80.086, 3508TB, Utrecht, The Netherlands

Hayes and Hayes 1952; Gardner et al. 1989; Myowa-Yamakoshi and Matsuzawa 1999; Nagell et al. 1993; Tomasello et al. 1987, 1993; Whiten 1998; Whiten et al. 1996), these incentives may affect the development of spontaneous tool use in this community.

Our aim is to show to what extent the chimpanzees in this enriched captive setting perform tool use spontaneously, what tool-use behaviors they perform, and which classes of chimpanzees are involved.

The captive group was established in Burgers' Zoo, Arnhem, in 1971. It has a system of social relationships established for almost 30 years (Adang et al. 1987; van Hooff 1973; Kats 1994; de Waal 1982, 1989). Although the description of tool use by these chimpanzees has been anecdotal (van Hooff 1973; de Waal 1982), there is interesting cinematographic documentation by Haanstra et al. (1984).

2 Methods

2.1 Subjects

At the beginning of the observation period, June 1993, there were 26 chimpanzees in the group: 5 infants (1 male, 4 females, 3 months to 4 years old), 5 juveniles (1 male, 4 females, 5 to 7 years old), 7 adolescents (2 males, 5 females, 9 to 14 years old), and 9 adults (2 males, 7 females, over 16 years old). Seven of the adults still present in the community came from other zoos between 1971 and 1973, just after the colony had been established, and 2 of the 7 were born wild. Exact data on the origin and history of the other individuals were for the most part unobtainable (van Hooff 1973). All other individuals were born in the colony. Three neonates (1 male and 2 females) were born during the 6-month observation period in 1993. Each infant was reared by its own mother.

2.2 Housing

During the day, the chimpanzees were usually kept outdoors, on an extensive grassy and sloped terrain (approximately 7000 m^2). The enclosure is surrounded by ditches 8 m wide, which gently slope down to a maximum depth of 2 m, and by an adjoining building that contains the night cages, winter halls, an observation room, and a visitor's corridor. There are many trees in the outside enclosure, which are protected by electric wire from being damaged by the chimpanzees. A tall dead tree in the center of the field has no electric wire and can be climbed. The enclosure is also provided with some large scaffold constructions with dead tree trunks, ropes, and truck tires. Various objects such as pieces of wood, rope, and truck tires were also lying loose in the enclosure. When it rained heavily the chimpanzees were kept in an inside hall (380 m^2). It has a cement floor with a variety of climbing facilities made up of metal pipes, wooden platforms, and ropes. It is also provided with truck tires and fresh straw.

At night the chimpanzees were kept in seven night cages. The females and their offspring were grouped into families and stayed in five cages. The adolescent and adult males remained apart from them and stayed in two separate cages. Some nights all the females and their offspring stayed together in the inside hall.

Every morning, the chimpanzees received food (milk, fruit, and vegetables) in their night cages. After eating their food at about 930, they went outside or into the inside hall. They received additional vegetables there. At 1300 and at 1500, they received food again. They remained outside or in the inside hall until 1730, when they went into the night cages and were fed once more. The night cages and the inside hall were cleaned every day.

2.3 Materials for Tool Use

The chimpanzees had free access to leaves, sand, branches, ropes, and water in the ditch in the outside enclosure. They also had free access to straw, paper sacks, tires, and ropes in the inside hall. In addition to those familiar objects, several other sorts of novel objects were added during the period of this study to facilitate tool-use behaviors. Wooden spoons (30 cm long), metal bowls (16 cm diameter, 7 cm depth), plastic boxes (43 × 35 × 23 cm), cotton towels (55 × 35 cm), wooden sticks (120 cm), volleyballs, and egg cups were introduced in the outside enclosure. Plastic boxes (43 × 35 × 23 cm) were also introduced in the inside hall. Ten of each object type, one object type per day, were provided, as a rule, during the observation period. Stuffed toys, which were goal objects for tool use rather than the materials for tools, were hung on an outside wall of the adjoining building and on a wall of the inside hall.

2.4 Procedures

The first author recorded all the tool-use behaviors of the subjects in a written protocol by the ad libitum sampling method. The observations were made in three study periods: 6 months in 1993, 2 months in 1994, and 2 months in 1995. Data were collected for at least 2 hours a day, 4 days a week, in principle. The present report is mainly based on the data of the 29 individuals in the first study period of 6 months, June to November 1993. We define tool use as the behaviors in which an individual uses a single or multiple detached environmental object(s) as an intermediary to efficiently change the environment in obtaining a goal.

3 Results

The introduced novel objects, as well as many of the familiar objects, were used as tools. We identified various goals in the tool-use behaviors performed by the chimpanzees, that is, drinking water, getting food, maintaining comfort, threatening or teasing, attracting the attention of other individuals, reaching a higher place, moving an object, self-decoration, and playing blindman's buff. The chimpanzees used various kinds of detached objects as an intermediary to obtain the goals. We distinguished three components in tool-use behaviors: the goal, the object(s) used as tool(s), and the actions involved. Based on these, we categorized 13 types of tool use observed in the community. Table 1 shows the distribution of the different types of tool use among the individual members of the community. Bullets in the table indicate at least one instance of performance of the indicated tool use during the observation period in 1993.

3.1 The Repertoire of Tool Use

This section summarizes each type of tool use observed in the Arnhem Zoo community.

3.1.1 Ladle to Scoop Up Water

Most of the chimpanzees (20 of 23 chimpanzees 3 years old or older) drank water from the ditch around the outside enclosure using an introduced object as a ladle (Fig. 1). The objects used for scooping up water were of different sizes and shapes and included a spoon, a bowl, or an egg cup. Two chimpanzees used a volleyball that had been transformed into the shape of a bowl.

3.1.2 Container to Carry Water

Seventeen chimpanzees used various objects to get water and to carry it to other places. They might walk many meters before drinking or playing with the water. In the inside hall, chimpanzees drank water by putting their mouths directly to a tap.

Fig. 1. A 16-year-old female scoops up water with a transformed volleyball

Fig. 2. A 17-year-old male rakes inward food floating on the water of the ditch

This tap extended from the wall, and water poured from it continually. In the case of two adolescent females, an individual collected water in a tire that she had hung on the tap or in a box which she had put under the tap. Then, they carried the container to another place before drinking.

3.1.3 Sponge to Absorb Water

In addition to the ladles described above, six chimpanzees used sponges for drinking water. They tore up a piece of paper, a towel, or a stuffed toy and used them as a sponge to absorb water.

3.1.4 Rake to Obtain an Out-of-Reach Incentive

When something edible was out of reach, ten chimpanzees were seen to use a stick, for instance, to rake inward food floating on the water of the ditch, which had been thrown to them from the opposite side of the ditch but had not reached them. Sticklike material was usually available in the outdoor enclosure (Fig. 2).

3.1.5 Missile for Aimed Throwing

Eleven chimpanzees were seen to throw a stick up into the foliage of a tree to break loose leaves. They threw the sticks while standing bipedally and then ate the leaves.

3.1.6 Stool for Resting

Six chimpanzees were seen to use an object as a stool to avoid the wetness or coldness of the floor of the inside hall. They placed an object such as a box or tire on the floor and sat down on it.

3.1.7 Weapon

Most of the young chimpanzees from 3 to 9 years old (10 of 11 chimpanzees) were seen to use objects as weapons (Adang 1984, 1985). They threw sticks, stones, sand, or leaves at other individuals, or they used a branch or stick to flail or club others. Thus, they threatened or teased group members, keepers, or observers. The alpha male of the community also used to throw a box about for bluffing in a charging display.

3.1.8 Signal for Invitation to Mating

Two adult males and four youngsters including females threw sand or leaves at other individuals to get their attention. This behavior was observed as a precursor to mating, mounting, or presenting.

3.1.9 Step or Ladder to Reach a Target

Fourteen chimpanzees were seen to use object(s) as a ladder or step (Fig. 3). Young chimpanzees often used a stick as a ladder or a box as a step to reach a windowsill, which was about 2 m above the ground. The window allowed a look from the outside enclosure into the inside, where they could watch, for instance, the keepers working. They seemed highly motivated to do so. There might have been another motivation: they could be alone on the narrow sill in front of the window. In addition to a stick or a box, they could also use another individual's body as a stepping-stone to reach the window. In these cases, one chimpanzee made another one crouch, and put him or her by the wall under the window, and climbed up to the window using his or her back as a step. The supporting individuals adjusted to this role reluctantly in most cases, and in a few cases they ran away. When a stuffed toy was hung on the walls of the adjoining building or the inside hall, chimpanzees used a box, a tire, or a stack of straw as a step to reach it.

3.1.10 Lever to Move a Trunk

Five young chimpanzees, from 5 to 7 years old, put a stick into a crack between two fallen tree trunks that were in contact with each other. Then, they pushed the stick up and down as a lever to move the upper tree, evidently with the intention of getting access to the space between the logs.

3.1.11 Ornament

Five chimpanzees were seen to drape their backs or shoulders with a towel, ropes, leaves, or a stick and to walk around with this ornamentation. This they did often, especially when provided with towels.

3.1.12 Playing Blindman's Buff

Seven of 11 young chimpanzees, from 3 to 9 years old, were seen to put a towel on their face and then walk on the ground or in a tree, apparently enjoying blindman's buff (Fig. 4).

Fig. 3. A 5-year-old female uses a stick as a ladder

3.1.13 Peek-a-Boo-Like Game

Putting a towel on the face was also observed in a different context of social interaction. There were three chimpanzees, more than 9 years old, who covered their faces with a towel and then held out their hands toward another individual. This behavior initiated social play between the two chimpanzees. In some cases, they put a towel on another's face. Other chimpanzees around the individuals showed a play face (Fig. 5). This unique behavior has been reported only in the Arnhem Zoo community.

Fig. 4. A 9-year-old female puts a towel on her face and then walks on a tree, apparently enjoying blindman's buff

3.2 Developmental Change of the Repertoire of Tool Use

Figure 6 shows the number of tool use types shown by chimpanzees in each age class. No examples of tool use were observed in any subjects 1 year old or younger during the study period in 1993. On the other hand, all individuals 3 years old or older exhibited at least one type of tool use, with only one exception (a 34-year-old female). The additional observations in 1994 and 1995 revealed tool use by individuals younger than 3 years. The earliest spontaneous use of an object as a tool was scooping water from the ditch by an infant aged 1 year and 10 months (Geisha). Most of the infants mastered this skill by the age of 3 years. In general, young chimpanzees from 5 to 9 years old showed a greater repertoire of tool use than infants under 4 years old and adults of 10 years and older.

Fig. 5. A 33-year-old female covers her face with a towel; the other chimpanzees react with play faces

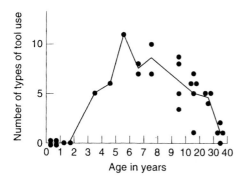

Fig. 6. The number of tool use types shown by the individual chimpanzees according to their age

Some types of tool use were popular in the community while the others were not. The most popular was scooping water with an object. Twenty chimpanzees of the community showed this type of tool use. Although the sample is too small for definitive conclusions, there seems to be a developmental change in the repertoire of tool use. Thus, there were two types of tool use, "lever" (repertoire #10) and "playing blindman's buff" (#12), which were performed by only individuals 9 years old and younger. By contrast, one type of tool use was performed exclusively by chimpanzees 7 years old and older; this was "missile" (#5), throwing an object into a tree to break loose leaves. Six of the nine adults performed this type of tool use. "Peek-a-boo-like game" (#13) was performed only by chimpanzees 9 years old and older.

3.3 Multivariate Analysis of Tool Use

Multivariate analysis was applied to examine the relations among the individual chimpanzees and among the types of tool use. The 29 (individuals) × 13 (types of tool use) one-zero matrix shown in Table 1 was subjected to Hayashi's quantification model III analysis (Hayashi 1952). This analysis is a modification of the principal-factor analysis for qualitative nonparametric data. It is a useful tool for rearranging similar items nearer to one another and different items farther away from one another in a multidimensional space (Torigoe 1985).

Two axes were extracted. The eigenvalues of each axis were 0.4051 and 0.2879 (the correlation coefficients were 0.6365 and 0.5365, respectively). Figure 7 shows the distribution of types of tool use in the two-dimensional space. Two types of tool use, "ornament" (#11) and "peek-a-boo-like game" (#13), occupied a unique position in the similarity space. To show the relationships among the other types of tool use, we adopted another procedure, called a hierarchical cluster analysis, to form clusters of similar items. In the cluster analysis, the similarity matrix data were analyzed to give a dendrogram, that is, a treelike structure unconstrained by the dimensions shown in Fig. 7.

Figure 8 is the result of the cluster analysis following Ward's method. The branch length in the tree structure indicates the degree of similarity for any given cluster. Figure 8 clearly shows that the two types of tool use, "ladle" (#1) and "container" (#2), are the most similar among the 13 types. The next similar pair was "rake" (#4) and "step or ladder" (#9), and the pair was similar to "weapon" (#7). The two types of tool use performed by youngsters, "lever" (#10) and "playing blindman's buff" (#12), were also classified in a cluster. The 13 types of tool use were thus classified into four groups. The largest group (group I) consisted of the types of tools used in a practical or substantial context. The remaining three groups (groups II, III, and IV) reflect unique types of tools mainly used in a nonpractical or play context.

Each individual was also plotted in this two-dimensional space. The cluster analysis grouped the individuals in the dendrogram (Figs. 9, 10). The 29 individuals were classified into five groups. The first three groups (groups I, II, and III) from left to right in the dendrogram (Fig. 10) represent three age groups: young (infant and juvenile), adolescent (subjects #14 and #17), and adult individuals. The remaining two groups (groups IV and V) consist of only 1 member each, #18 and #26, respectively. These two individuals showed a unique profile of tool use.

These results from multivariate analysis, which provided the quantitative data profiling the nature of various types of tool use and also the variety of individual differences, further confirmed the findings about developmental changes in the use of tools. The tool-use repertoire of each individual was more similar among the individuals in the same age class than those in the same kin relationship.

Table 1. Tool-use types found in individual chimpanzees of the Arnhem Zoo community in 1993

Subject No.	Name	Sex	Age	Mother	(1)	(2)	(3)	(4)	(5)	(6)	(7)	(8)	(9)	(10)	(11)	(12)	(13)	Total
1	Goya	f	0	Gaby	–	–	–	–	–	–	–	–	–	–	–	–	–	0
2	Geisha	f	0	Gorilla	–	–	–	–	–	–	–	–	–	–	–	–	–	0
3	Zouly	m	0	Zaira	–	–	–	–	–	–	–	–	–	–	–	–	–	0
4	Zombi	f	0	Zola	–	–	–	–	–	–	–	–	–	–	–	–	–	0
5	Tushi	f	1	Tepel	–	–	–	–	–	–	–	–	–	–	–	–	–	0
6	Avanti	f	1	Amber	–	–	–	–	–	–	–	–	–	–	–	–	–	0
7	Roani	f	3	Roos	•	•	•	–	–	–	•	–	–	–	–	•	–	5
8	Giambo	m	4	Gorilla	•	•	•	–	–	–	•	–	•	–	–	•	–	6
9	Zizwa	f	5	Zwart	•	•	•	•	•	•	•	–	•	•	•	•	–	11
10	Morami	f	6	Moniek	•	•	–	•	–	–	•	•	•	•	–	•	–	8
11	Jelle	m	6	Jimmie	•	•	–	–	–	–	•	•	•	•	–	•	–	7
12	Tesua	f	7	Tepel	•	•	–	•	–	•	•	–	•	•	–	–	–	7
13	Sabra	f	7	Jimmie[1]	•	•	•	•	•	–	•	–	•	•	•	•	–	10
14	Marka	f	9	Mama	•	•	–	•	–	–	•	•	•	–	•	•	•	9
15	Ayo	m	9	Amber	•	•	–	•	–	•	–	–	•	–	–	–	–	5
16	Gaby	f	9	Gorilla	•	•	–	–	•	–	–	–	–	–	–	–	–	3
17	Zaira	f	9	Zwart	•	•	–	•	•	•	–	•	•	–	–	–	•	8
18	Jing	m	12	Jimmie	–	–	–	–	–	–	–	–	–	–	•	–	–	1
19	Zola	f	13	Zwart	•	•	–	•	•	•	–	•	•	–	–	–	–	7
20	Roos	f	14	Gorilla[2]	•	•	•	–	•	–	–	–	•	–	–	–	–	5
21	Moniek	f	16	Mama	•	•	•	•	–	–	•	–	•	–	–	–	–	6
22	Fons	m	17	Franje[3]	•	•	–	•	•	–	•	•	–	–	–	–	–	6
23	Zwart	f	23	wild born	•	•	–	–	•	•	–	–	•	–	–	–	–	5
24	Amber	f	24	wild born	•	•	–	–	•	–	•	–	–	–	–	–	–	4
25	Dandy	m	27	unknown	•	–	–	–	•	–	•	–	•	–	•	–	–	5
26	Jimmie	f	33	unknown	–	–	–	–	–	–	–	–	–	–	–	–	•	1
27	Tepel	f	34	unknown	–	–	–	–	–	–	–	–	–	–	–	–	–	0
28	Gorilla	f	36	unknown	•	–	–	–	•	–	–	–	–	–	–	–	–	2
29	Mama	f	37	unknown	•	–	–	–	–	–	–	–	–	–	–	–	–	1
Total					20	17	6	10	11	6	12	6	14	5	5	7	3	122

Distribution of tool use types:

(1) Ladle: scooping up water with an object.
(2) Container: carrying water in an object.
(3) Sponge: absorbing water into a sponge produced from an object.
(4) Rake: poking or raking a target with an object to get it.
(5) Missile: throwing an object at a tree to get leaves.
(6) Stool: utilizing something as a stool after putting it on a spot of the floor to avoid wetness or coldness.
(7) Weapon: flailing, clubbing or throwing an object against another individual.
(8) Signal for invitation to mating: throwing sand or leaves against another individual.
(9) Step or ladder: putting an object as a step or ladder to reach a target on a spot of the substrate.
(10) Lever: utilizing an object as a lever to move a tree trunk.
(11) Ornament: walking while one's body is covered by an object.
(12) Playing blindman's buff: walking with one's face covered by an object.
(13) "Peek-a-boo-like" game: holding out one's hand to another individual while one's face is covered with a towel.

[1,2] These are foster mothers of the individuals whose biological mothers died.
[3] Died.

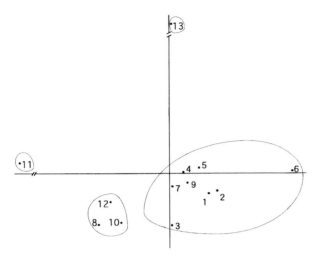

Fig. 7. Distribution of types of tool use in the two-dimensional space extracted by Hayashi's quantification model III analysis

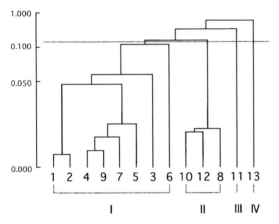

Fig. 8. Hierarchical cluster analysis of types of tool use. The branch length in the tree structure indicates the degree of similarity for any given cluster

3.4 Combination of Tool Use

The most complex type of tool use ever found in wild chimpanzees is the use of a meta-tool, namely, a set of three stones for nut-cracking (Matsuzawa 1994). The chimpanzees at Bossou, West Africa, use a pair of stones as a hammer and anvil to crack open oil-palm nuts. Three chimpanzees were observed to use a third stone as a wedge to make the anvil stone horizontal and stable. Such a complex type of tool use was observed neither in the Arnhem Zoo community nor in other wild populations.

Although there were no examples of using a set of objects that had different and complementary functions, there were a few examples of tools consisting of multiple objects of the same kind. There were two juveniles who stacked more than two boxes. A 5-year-old female was highly motivated to reach a window in a high place, and she successfully stacked up to four boxes. The other was a 7-year-old female who was motivated to get an object hung out of reach. She also stacked

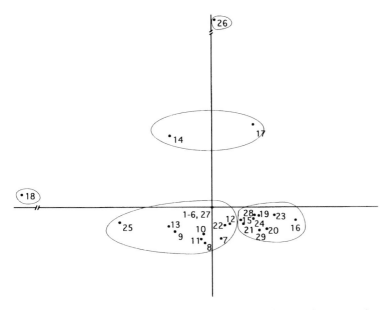

Fig. 9. Distribution of each individual chimpanzee in the two-dimensional space extracted by Hayashi's quantification model III analysis

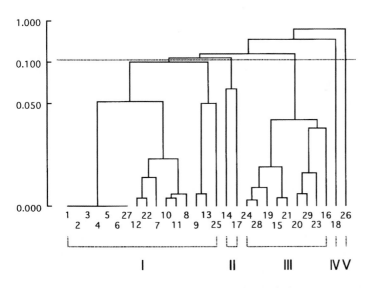

Fig. 10. Hierarchical cluster analysis of each individual chimpanzee, in terms of the similarity of the tool use

boxes, one on top of the other. Other chimpanzees used only one box in the same situation.

There were also examples of combining different types of tool use in a sequence. Some chimpanzees used a stick to rake a target object after they had climbed on top of a box that had been placed under the target by themselves in some cases or by other individuals.

In general, however, the tool use by the chimpanzees in the Arnhem Zoo community was congruent with that reported in the previous studies. The long list of tool use by the chimpanzees is characterized by the general rule of "a single tool for a specific goal".

4 Discussion

Chimpanzees are the only nonhuman species for which there are reports of a long list of using different tools for solving different problems (Goodall 1986; McGrew 1992). The present study shows the variety of tool use by chimpanzees in the Arnhem Zoo community. It reveals some fundamental features similar to those found in communities in the wild.

In wild chimpanzees, most of the habitual tool-use patterns are subsistence activities for acquiring or processing several kinds of food or water (McGrew 1994). The chimpanzees in the Arnhem Zoo community also engaged in tool use concerning food and water: all the types of tool use that more than 50% of adults performed involved food and water. Despite the full supply of food and water, their desire for something edible seems to be the primary cause of tool use in captive chimpanzees, as well as those in the wild.

The following is an interesting example of the use of tools to drink water, from the point of view of embodied cognition. There is a popular theory that a tool is like an attachment to the body or like a substitute for a body part. But to which body part does a "ladle (#1)" or "container (#2)" correspond? An aluminum bowl and plastic box were used as a ladle or container. The chimpanzees drank water from the ditch surrounding the outside enclosure. When the chimpanzees drank water without using a tool, they put their mouth to the surface of the water or used their hands. The chimpanzee's mouth seemed to suck up the water, and their hands were used to scoop up the water. The mouth corresponded to a sponge and the hands corresponded to a ladle. However, although the chimpanzee's hands worked as if scooping up water, the hands were not round-shaped like a ladle. Also, the chimpanzee's hands, which conveyed water from the ditch to the mouth, could not store water for a long time like a ladle. In fact, chimpanzees only lapped up the water from their fingers, and their hands worked like a rake. In a similar case, humans always scoop up and carry water with their palms placed together and round-shaped, but none of the chimpanzees of this group have learned to handle water in this manner yet. Therefore, the ladle and container are not substitutes for the hands in chimpanzees. Are the functions of tools used by chimpanzees related to the functions of body parts?

The actions of delivering some drops from a gutter into their own wide-open mouth has been often observed among the chimpanzees in Arnhem (de Waal 1982). The elder individuals can efficiently store water in the pocket between the lower lip and teeth in their mouth. This finding proves that, although humans and chimpanzees may use the same material as a ladle for the same purpose, the used material corresponds to the hand of a human or the mouth of a chimpanzee.

It was observed that the chimpanzees used a volleyball as a ladle or sponge. Although the transformed volleyball was actually used in this way, the volleyball was not transformed intentionally by the chimpanzees; rather, the alpha male transformed the volleyball into an object like a container while he was displaying.

It has already been described that chimpanzees do not use their hands as a container to drink water. That is, the chimpanzees of this group used only existing container-shaped materials. It has been reported that a captive bonobo ate the flesh of a pineapple and then used the empty pineapple as a container to collect water from a water tap. The authors stated the possibility that the pineapple may have been processed to produce a container (Van Elsacker and Walraven 1994).

Nakagawa (1996) reported that one captive chimpanzee subject produced a clay model having a concave part while playing with clay in her study. Some chimpanzee subjects then produced a clay model from a given block of clay in various, unique manners. Several chimpanzees processed the clay into a rope and a ball in succession. One of them produced a model having a concave section, then took a piece from the floor and repeated the action of putting the piece in and out of the concave section. More than 75.1% of high school students or older people judged the concave model to be a container. This rate of 75.1% almost equals the rate at which an object shaped like a container produced by a child aged 5 years old is actually judged as a container. Chimpanzees cannot move their hands to use them as a container. However, once a chimpanzee finds an appropriate material, it can fashion the material into a container through interaction with the material's attributes.

When did humans start using their hands as a vessel in ancient times? Was it before they invented containers that were defined as tools or materials? Or was it later? Sasaki (1990) suggested that there are many trends in which after a person finds a tool that has been produced artificially, he or she may start using their hands as a tool, with regard to the history of civilization.

Humans may not discover the method of using their hands as a container by handling such natural materials as water or sand that can be stored in a container. That is, humans may instead discover the method by handling a container that is actually used as a tool or material.

It is clear that some types of tool use are specific for this community. We mentioned "playing blindman's buff" (#12) and the "peek-a-boo-like game" (#13). Their development obviously depends on the availability of suitable material. Other remarkable patterns are those involving throwing things. The adoption of throwing in a social context, "signal for invitation to mating (#8)", would not seem to be hampered by availability of suitable materials in the wild. Its frequent use in the Arnhem Zoo community may, therefore, be a cultural idiosyncrasy.

A unique pattern is also "missile" (#5). To throw a heavy object upward with sufficient force to break loose leaves requires much motor coodination (see Haanstra et al. 1984). This pattern is clearly an adaptive development in response to the specific ecological condition, namely, attractive food being outside climbing range. Unlike "ladle" (#1), which could have been facilitated by the use of containers by the human caretakers, "missile" (#5) cannot be other than a colony-specific invention.

The repertoire of tool-use behaviors varied as a function of age. Juveniles and adolescents performed a greater variety of tool-use behaviors than adult chimpanzees. By the age of 10, all types of tool use had appeared, although there was enormous individual variation. Thus the chimpanzees acquire the community-common repertoire of tool use before adulthood.

Increases in the variation and qualitative differences are also suggested. In general, many playful actions using objects are observed in young individuals, and actions that are associated with subsistence are observed mainly in older individuals. Regarding the playful actions, the action of playing peek-a-boo was observed among individuals who were 9 years or older. Regarding the actions that were associated with subsistence, the action of throwing a "missile" was observed only among individuals who were 7 years or older. These actions seem to be related to a higher cognitive ability.

The peek-a-boo play includes the element of relationship to others, in addition to the blindman's buff's element of relationship to an invisible world. The chimpanzee's peek-a-boo play can be described as the equivalent in humans of a person wearing a towel on his or her head when trying to play with a mate. The mate's face then shows a playful face or grimaces; thus, the mate is thought to be smiling or to be puzzled. An individual that wears a towel on its head does not try to take off the towel, unlike when a human is playing peek-a-boo. However, the chimpanzee's peek-a-boo is thought to be an action in preparation for the mate's reaction toward the player's own face covered with a towel. Although the action of throwing sand or leaves at the mate is also done to attract the mate's attention, the peek-a-boo action seems to be based on a higher ability of cognition.

The action of throwing something like a ball at a target was observed mainly among individuals who were 7 years or older. This may be understood as detailed elaborate processing of the cerebral cortex is necessary for performing an action such as a throwing motion (Calvin 1996). Nakamichi (1998; 1999) recently observed how three individuals (each 8 years old) from a group of captive gorillas threw a "missile."

It was reported that the minimum age for showing the "meta-tool" skill, cracking oil-palm nuts shell using a stone hammer, stone anvil, and wedge-shaped stone support, was at least 6.5 years (Matsuzawa 1994). Chimpanzees develop cognitive ability from infancy through juvenile age to acquire techniques of using useful materials in various manners in their environment.

A question of major interest left unanswered is when and how the youngsters acquire each type of tool use typical of, or unique to, the individuals in the community. What is the learning mechanism of the acquisition? Are there specific acquisition processes depending on the types of tool use? What kind of imitative and educational processes are involved in the acquisition?

The learning mechanisms of acquiring the tool use skills are likely to be affected by environmental and social factors as well as by cognitive development (McGrew 1994; Tomasello et al. 1993; Matsuzawa 1996, 1999). These factors cannot be easily manipulated or identified in wild chimpanzees. Chimpanzees kept in a social group in a seminatural environment can provide a unique opportunity to clarify the acquisition mechanisms. Further research in this community should be done through longitudinal observation and experimental manipulation of tool-use behaviors.

Acknowledgments

We are very grateful to the staff of Burgers' Zoo, especially to the chimpanzee keepers Jacky Hommes, Inge Stevens-Beerma, Rene Klein-Nulant, Bianca Klein, and Maarten Houtriet for their kind help during the observations. The first author has been supported by grants from the Overseas Research Program of Shiga Prefectural Junior College and The University of Shiga Prefecture, and Scientific Research Programs of the Ministry of Education, Science, and Culture, Japan. promoted by Shozo Kojima, Shigeru Kiritani, and Tetsuro Matsuzawa. We thank Tetsuro Matsuzawa, Linda Van Elsacker, Hilde Vervaecke, Vera Walraven, and Christel Muller for their helpful suggestions during the study and concerning the manuscript. We also thank Astrid Kappers, Mirjam van Gool, and Ruud Derix for their kind advice for introducing the first author to the chimpanzees in Burgers' Zoo, and Beatriz Kats and Loes Schmeink for their support in various forms throughout her study in The Netherlands.

References

Adang OMJ (1984) Teasing in young chimpanzees. Behaviour 88:98–122
Adang OMJ (1985) Exploratory aggression in chimpanzees. Behaviour 95:138–163
Adang OMJ, Wensing JAB, van Hooff JARAM (1987) The Arnhem Zoo colony of chimpanzees: development and management techniques. Int Zoo Yearbook 26:236–248
Beck B (1980) Animal tool behavior. Garland, New York
Boesch C, Boesch H (1990) Tool use and making in wild chimpanzees. Folia Primatol 54:86–99
Brewer SM, McGrew WC (1990) Chimpanzee use of a tool-set to get honey. Folia Primatol 54:100–104
Calvin WH (1996) How brains think. Basic Books, Harper Collins, New York
Custance D, Whiten A, Bard K (1995) Can young chimpanzees (*Pan troglodytes*) imitate arbitrary actions? Hayes and Hayes (1952) revisited. Behaviour 132:839–858
de Waal F (1982) Chimpanzee politics. Jonathan Cape, London
de Waal F (1989) Peacemaking among primates. Harvard University Press, Cambridge
Gardner BT, Gardner RA, Nichols SG (1989) The shapes and uses of signs in a cross-fostering laboratory. In: Gardner, RA, Gardner BT, Van Cantfort TE (eds) Teaching sign language to chimpanzees. SUNY Press, Albany, pp 55–180
Goodall J (1986) The chimpanzees of Gombe. Harvard University Press, Cambridge
Haanstra B, Adang OMJ, van Hooff JARAM (1984) The family of chimps (film). Haanstra Productions, Laren (N-H), The Netherlands
Hayashi C (1952) On the prediction of phenomena from quantitative data and the quantification of qualitative data from the mathematico-statistical point of view. Ann Inst Stat Math 3:69–98

Hayes KJ, Hayes C (1952) Imitation in a home-raised chimpanzee. J Comp Physiol Psychol 45:450–459

Kats B (1994) Post conflict behavior among chimpanzees: interactions with the aggressor. Master's thesis, Utrecht University, The Netherlands

Matsuzawa T (1994) Field experiments on use of stone tools by wild chimpanzees. In: Wrangham RW, McGrew WC, de Waal FBM, Heltne PG (eds) Chimpanzee cultures. Harvard University Press, Cambridge, pp 351–370

Matsuzawa T (1996) Chimpanzee intelligence in nature and in captivity: isomorphism of symbol use and tool use. In: McGrew WC, Marchant LF, Nishida T (eds) Great ape societies. Cambridge University Press, Cambridge, pp 196–209

Matsuzawa T (1999) Communication and tool use in chimpanzees: cultural and social contexts. In: Hauser M, Konishi M (eds) The design of animal communication. Cambridge University Press, Cambridge, pp 645–671

McGrew WC (1992) Chimpanzee material culture. Cambridge University Press, Cambridge

McGrew WC (1994) Tools compared. In: Wrangham RW, McGrew WC, de Waal FBM, Heltne PG (eds) Chimpanzee cultures. Harvard University Press, Cambridge, pp 25–35

Myowa-Yamakoshi M, Matsuzawa T (1999) Factors influencing imitation of manipulatory actions in chimpanzees (Pan troglodytes). J Comp Psychol 113:128–136

Nagell K, Olguin R, Tomasello M (1993) Processes of social learning in the imitative leaning of chimpanzees and human children. J Comp Psychol 107:174–186

Nakagawa O (1996) A comparative study on the plastic art of clay in chimpanzees and human children. Doctoral dissertation, Japan Women's University, Tokyo

Nakamichi M (1998) Stick throwing by gorillas (Gorilla gorilla gorilla) at the San Diego Wild Animal Park. Folia Primatol 69:291–295

Nakamichi M (1999) Spontaneous use of sticks as tools by captive gorillas (Gorilla gorilla gorilla). Primates 40:487–498

Nishida T (1973) The ant gathering behavior by the use of tools among chimpanzees of Mahale Mountains. J Hum Evol 2:357–370

Sasaki M (1990) When postures change. In: Saeki Y, Sasaki M (eds) Active mind. University of Tokyo Press, Tokyo, pp 87–109

Sugiyama Y, Koman J (1979) Tool-using and -making behavior in wild chimpanzees at Bossou, Guinea. Primates 20:513–524

Sugiyama Y (1993) Local variation of tools and tool use among wild chimpanzee populations. In: Berthelet A, Chavaillon J (eds) The use of tools by human and nonhuman primates. Clarendon Press, Oxford, pp 175–187

Suzuki S, Kuroda S, Nishihara T (1995) Tool-set for termite-fishing chimpanzees in the Ndoki Forest, Congo. Behaviour 132:219–235

Tomasello M, Davis-Dasilva M, Camak L, Bard K (1987) Observational learning of tool-use by young chimpanzees. Hum Evol 2:175–183

Tomasello M, Savage-Rumbaugh S, Kruger A (1993) Imitative learning of actions on objects by chimpanzees, enculturated chimpanzees, and human children. Child Dev 64:1688–1705

Torigoe T (1985) Comparison of object manipulation among 74 species of nonhuman primates. Primates 26:182–194

Van Elsacker L, Walraven V (1994) The spontaneous use of a pineapple as a recipient by a captive bonobo. Mammalia 58:159–162

van Hooff JARAM (1973) The Arnhem Zoo chimpanzee consortium: an attempt to create an ecologically and socially acceptable habitat. Int Zoo Yearbook 13:195–205

Whiten A (1998) Imitation of sequential structure of action by chimpanzees (Pan troglodytes). J. Comp Psychol 112:270–281

Whiten A, Custance DM, Gomez JC, Teixidor P, Bard KA (1996) Imitative leaning of artificial fruit processing in children (Homo sapiens) and chimpanzees (Pan troglodytes). J Comp Psychol 110:3–14

Whiten A, Goodall J, McGrew WC, Nishida T, Reynold V, Sugiyama Y, Tutin CEG, Wrangham RW, Boesch C (1999) Cultures in chimpanzees. Nature 399:682–685

Wrangham RW, McGrew WC, de Waal FBM, Heltne PG (eds) (1994) Chimpanzee cultures. Harvard University Press, Cambridge

27
Ecology of Tool Use in Wild Chimpanzees: Toward Reconstruction of Early Hominid Evolution

GEN YAMAKOSHI

1 Introduction

The emergence and evolution of human technology is a central issue in evolutionary anthropology. To date, most evidence has come from stone artifacts found together with hominid fossils (e.g., Leakey 1971). The oldest known stone tools manufactured by hominids date back to about 2.5 million years ago (Semaw et al. 1997). According to recent molecular studies, however, the human lineage diverged from the chimpanzee–bonobo lineage about 6 million years ago (e.g., Sibley and Ahlquist 1984). It follows that we have no evidence of tool use by early hominids from the period 6–2.5 million years ago, more than half of hominid history. Should we assume that they did not use tools? The answer is probably no. Because organic matter such as sticks and leaves is the material most likely to have been used by early hominids for tools, any artifacts would quickly have decomposed and are unlikely to have been fossilized. Many textbooks end their discussion simply by saying that early hominids must have used tools to the same extent as modern chimpanzees. However, chimpanzees' tool use has been regarded as opportunistic, and is thought to have only a trivial effect on their subsistence (e.g., Mann 1972).

There have been many recent studies of the feeding ecology of wild chimpanzees. These studies have elucidated complex interactions between environmental conditions and chimpanzee behavior (Tutin et al. 1991; Wrangham et al. 1992; Kuroda et al. 1996; Yamagiwa et al. 1996; Yamakoshi 1998). With regard to tool use, interesting relationships between tool-using behavior and environmental changes have been discovered, some of which were not anticipated. In this chapter, I review these new findings, and discuss their implications for reconstructing the ecological niche of early hominids.

2 Chimpanzee Cultures

Ever since Jane Goodall (1963, 1964) confirmed earlier anecdotes of wild chimpanzees' tool use (Savage and Wyman 1843–44; Beatty 1951; Merfield and Miller

Center for African Area Studies, Kyoto University, 46 Yoshida Shimoadachi-cho, Sakyo-ku, Kyoto 606-8501, Japan

1956, pp 43–44) with her detailed descriptions of a variety of tool-using behaviors by wild chimpanzees at Gombe, Tanzania, tool use has been the focus of intensive investigations at almost every wild chimpanzee study site in tropical Africa. At the time of McGrew's exhaustive review (McGrew 1992), more than 30 behavioral patterns had been reported from approximately 20 research sites, and the number has increased with recent observations (e.g., Nishida and Nakamura 1993; Alp 1997). Chimpanzees use tools for various behaviors, including feeding, aggression, communication, and hygiene.

The most notable aspect of tool use in wild chimpanzees is perhaps the regional variation in the behavioral repertoire. Local differences are found between subspecies (only the West African subspecies, *Pan troglodytes verus*, has a nut-cracking tradition; Sugiyama 1993; Boesch et al. 1994), within subspecies (Mahale chimps have never been observed to dip commonly found driver ants, although Gombe chimps, 170 km to the north, regularly do so; McGrew 1974; Nishida 1987), and even between neighboring communities (between Bossou and Nimba, Matsuzawa and Yamakoshi 1996; between Mahale B and K, see below).

It has been claimed that environmental differences can explain some of this observed variation. For instance, researchers have found many termite-fishing tools in the range of the Mahale-B community, while this behavior has only been seen once in the well-habituated Mahale-K community, which has been observed for many years. The reason is simple. In the range of the K community, there are no *Macrotermes* termites, the genus that wild chimpanzees most commonly fish for (McGrew et al. 1979), owing to the difference in annual rainfall (Nishida and Uehara 1980; Uehara 1982).

However, there are also many behavioral differences that ecological conditions fail to explain. For example, a close ecological examination revealed that the environment at Lopé, Gabon, provides chimpanzees with all the conditions necessary for nut cracking (nuts, stones, and so forth), but the chimpanzees there have never been observed to crack nuts with tools (McGrew et al. 1997). The only factor thought to be absent is an adequate tradition of the accumulated knowledge found in other communities. In addition, a recent exhaustive comparison of seven long-term wild chimpanzee studies identified regional variations in 39 behavioral patterns (of which 23 involved tool use) that cannot be explained by differences in local environmental conditions (Whiten et al. 1999).

3 Ecological Importance of Tool Use in Wild Chimpanzees

Although the context of chimpanzee tool use varies, the majority of instances relate to feeding (e.g., McGrew 1992). However, there have been few attempts to estimate the adaptive significance of wild chimpanzees' tool-using behaviors which focus on nutritional benefits. After an extensive survey of the literature, I listed 54 cases of reported tool use from 14 study sites (Table 1), using the following criteria. The behaviors had to be in a feeding context, directly observed in detail by the authors, observed in wild populations, and the species involved and the location of the

Table 1. Tool use by wild chimpanzees in a feeding context

No.	Site	Description	Freq.	Target	References
1	Bossou, Guinea	Cracking hard nuts with stones	H	n	Sugiyama and Koman 1979b, Yamakoshi 1998
2	Bossou, Guinea	Extracting water from a tree hole with a leaf	H	w	Sugiyama and Koman 1979b, Yamakoshi 1998
3	Bossou, Guinea	Pounding termites in a tree hole with a stick	A	s	Sugiyama and Koman 1979b
4	Bossou, Guinea	Pounding resin in a tree hole with a stick	A	p	Sugiyama and Koman 1979b
5	Bossou, Guinea	Hooking a fruiting branch with a twig	A	p	Sugiyama and Koman 1979b
6	Bossou, Guinea	Dipping for ants on or under the ground with a stick	H	s	Sugiyama et al. 1988, Yamakoshi 1998
7	Bossou, Guinea	Pushing a "leaf sponge" deeper into a tree hole with a twig	A	w	Matsuzawa 1991, pp. 256–257
8	Bossou, Guinea	Pounding the apical meristem of an oil palm using a palm frond as a pestle	H	p	Sugiyama 1994a, Yamakoshi 1998, Yamakoshi and Sugiyama 1995
9	Bossou, Guinea	Dipping for arboreal ants with a stick	A	s	Yamakoshi pers. obs. 1995
10	Bossou, Guinea	Scooping algae from water with a stick	H	p	Matsuzawa et al. 1996, Yamakoshi 1998
11	Bossou, Guinea	Fishing for termites with a stick	A	s	Humle 1999
12	Liberia	Cracking hard nuts with stones	A	n	Beatty 1951
13	Taï, Ivory Coast	Cracking hard nuts with stones or branches	H	n	Boesch and Boesch 1981, 1983, Günther and Boesch 1993
14	Taï, Ivory Coast	Dipping bone marrow with a stick	H	v	Boesch and Boesch 1989
15	Taï, Ivory Coast	Dipping ants from a subterranean nest with a stick	H	s	Boesch and Boesch 1990
16	Taï, Ivory Coast	Pulling out wood-boring bees with a stick	A	s	Boesch and Boesch 1990
17	Taï, Ivory Coast	Dipping honey from a bee hive with a stick	H	h	Boesch and Boesch 1990
18	Taï, Ivory Coast	Dipping the remains from a colobus skull with a stick	A	v	Boesch and Boesch 1990
19	Taï, Ivory Coast	Removing the remains from cracked hard nuts with a stick	H	n	Boesch and Boesch 1990
20	Taï, Ivory Coast	Extracting insect larvae from a nest tunnel with a stick	A	a	Boesch 1995
21	Taï, Ivory Coast	Extracting mushrooms from a termite nest with a stick	A	p	Boesch 1995

Table 1. *Continued*

No.	Site	Description	Freq.	Target	References
22	Cape Palmas, Ivory Coast	Pounding nuts with stones	A	n	Savage and Wyman 1843–1844
23	Yaounde Forest, Cameroon	Dipping honey from a subterranean bee hive with a long twig	A	h	Merfield and Miller 1956, pp. 43–44
24	Bai Hokou, Central Africa	Pounding an arboreal bee hive with a dead piece of branch	A	h	Fay and Carroll 1994
25	Ndakan, Central Africa	Pounding an arboreal bee hive with a dead piece of branch	A	h	Fay and Carroll 1994
26	Ndoki, Congo	Fishing for termites with a stick	H	s	Suzuki et al. 1995
27	Lossi, Congo	Digging a hole with a stick for later termite fishing	A	s	Bermejo and Illera 1999
28	Lossi, Congo	Fishing for termites with a stick	A	s	Bermejo and Illera 1999
29	Lossi, Congo	Pounding an arboreal bee hive with a dead piece of branch	A	h	Bermejo and Illera 1999
30	Lossi, Congo	Puncturing an arboreal bee hive wall with a sharp-pointed end of a dead piece of branch	A	h	Bermejo and Illera 1999
31	Lossi, Congo	Dipping honey from an arboreal bee hive with a stick	A	h	Bermejo and Illera 1999
32	Lopé, Gabon	Extracting water from a tree hole with a leaf sponge	A	w	Tutin et al. 1995
33	Lopé, Gabon	Dipping honey from an arboreal bee hive with a stick	H	h	Tutin et al. 1995
34	Tongo, D.R. Congo	Extracting water from a tree hole with a moss sponge	H	w	Lanjouw, in Wrangham and Peterson 1996, p. 58
35	Gombe, Tanzania	Fishing for termites with a stick	H	s	Goodall 1963, 1986, p. 253
36	Gombe, Tanzania	Dipping for ants on or under the ground with a stick	H	s	Goodall 1963, McGrew 1974, Goodall 1986, p. 252
37	Gombe, Tanzania	Extracting water from a tree hole or a stream with a leaf sponge	H	w	Goodall 1964, 1986, p. 542
38	Gombe, Tanzania	Digging or prying open subterranean bee hives with sticks	A	s	Goodall 1970, 1986, p. 540

39	Gombe, Tanzania	Sponging out the inside of a skull cavity with a wad of leaves	A	v	Teleki 1973a
40	Gombe, Tanzania	Sponging out the inside of *Strychnos* fruit with a wad of leaves	A	p	Wrangham 1977
41	Gombe, Tanzania	Throwing a rock at a bushpig or baboon during a hunt	A	v	Plooij 1978
42	Gombe, Tanzania	Prying open a tree hole to extract a fledgling	A	v	Goodall 1986, p. 540
43	Gombe, Tanzania	Inserting a stick forcefully into a tree cavity to expel ants or termites	A	s	Goodall 1986, pp. 541–542
44	Gombe, Tanzania	Wiping bees away with a handful of leaves to access honey	A	h	Goodall 1986, p. 542
45	Kasakati, Tanzania	Dipping honey from an arboreal bee nest with a stick	A	h	Izawa and Itani 1966
46	Kasakati, Tanzania	Fishing for termites with a stick	A	s	Suzuki 1966
47	Mahale-K, Tanzania	Wiping ants from a tree trunk with a clump of leafy boughs	A	s	Nishida 1973
48	Mahale-K, Tanzania	Fishing for arboreal ants with a stick	H	s	Nishida 1973, Nishida and Hiraiwa 1982
49	Mahale-K, Tanzania	Fishing for termites with a stick	A	s	Nishida and Uehara 1980, Uehara 1982
50	Mahale-K, Tanzania	Dipping a stick into a hole to extract honey	A	h	Nishida and Hiraiwa 1982
51	Mahale-K, Tanzania	Dipping for termites with a liana shoot	A	s	Uehara 1982
52	Mahale-B, Tanzania	Fishing for termites with a stick	H	s	McGrew and Collins 1985
53	Mahale-M, Tanzania	Fishing for arboreal ants with a stick	H	s	Kawanaka 1990, Nishida and Turner 1996
54	Mahale-M, Tanzania	Inserting a stick forcefully into a tree-hole to expel a squirrel	A	v	Huffman and Kalunde 1993

A, Anecdotal or idiosyncratic; H, habitual or customary. The definitions of these terms are from McGrew and Marchant (1997). v, Vertebrates; s, social insects; h, honey; a, other animal matter; n, nuts; p, other plant matter; w, water; o, others.

study site specified. With reference to this table, I will examine (1) what kind of food and (2) how much food is obtained through tool-using behaviors, and (3) the relationship of these behaviors to seasonal changes in the environment.

3.1 Food Categories Exploited by Tool Use

Chimpanzees are omnivorous; their diet consists of a variety of foods including fruits, leaves, the pith of herbs, bark, resin, insects, animal meat, etc. (Teleki 1973a; Nishida and Uehara 1983; Sugiyama and Koman 1987). Based on time spent feeding, the largest proportion of the diet consists of fruit (the pulp of ripe fruits, in particular), followed by leaves and the pith of herbs (Fig. 1). In general, these foods are not physically protected and can be categorized as "easy-to-get" foods.

What kinds of food are obtained using tools? By definition, tools are unlikely to be used for "easy-to-get" foods, since they are accessible without tools. In fact, tools were used to obtain "easy-to-get" food categories (i.e., fruits) in only 2 of the 54 cases listed (Nos. 5 and 40 in Table 1), although in these two cases the edible parts were somewhat "hard-to-get." Typically, chimpanzees use tools to obtain "hard-to-get" foods (Table 2).

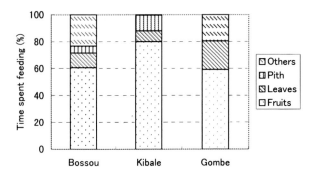

Fig. 1. The time wild chimpanzees spent feeding on different food types. "Easy-to-get" foods made up the majority of the chimpanzee diet. Only data collected over at least an entire year are included. The sources are: Bossou, Yamakoshi (1998); Kibale, Chapman et al. (1994); Gombe, Goodall (1986)

Table 2. Types of chimpanzee foods obtained by the use of tools

Food categories	Number of cases
Vertebrates	6
Social insects	20
Honey	11
Other animal matter	1
Nuts	5
Other plant matter	6
Water	5
Total	54

Table 3. Time spent feeding, by tool use

Site	Food categories	TSF (%)	TSF-tool (%)	C-tool (%)	LS (months)	Tool use pattern
Bossou[a]	Seeds	6.7	6.1	91.3	12	Nut-cracking
	Pith[f]	10.1	4.3	43.2		Pestle-pounding
	Insects	1.8	0.8	44.3		Ant-dipping
	Algae	0.7	0.7	100		Algae-scooping
Taï[b]	Seeds	–	15.0[h]	–	?	Nut-cracking
Gombe[c]	Insects	4.3	3.5	87.4[i]	24	Termite-fishing
			0.2			Ant-dipping
Mahale[d]	Insects	10.8	7.0	64.8	36.5	Ant-fishing
Mahale[e]	Insects[g]	4.1	0.7	18.1	5	Ant-fishing

TSF, Time spent feeding on each food category; TSF-tool, time spent feeding on each food category by tool use; C-tool, contribution of tool use (TSF-tool /TSF); LS, length of study.
[a] From Yamakoshi (1998).
[b] Calculated from Günther and Boesch (1993).
[c] Calculated from Figs. 10.1 and 10.5 of Goodall (1986).
[d] Calculated from Nishida and Turner (1996); data only for females who had infants.
[e] Calculated from Kawanaka (1990); data only for males.
[f] Pith of woody plants.
[g] Only ants.
[h] See text for calculation method.
[i] Termites and ants combined.

Obviously, social insects (ants, termites, and bees) are the most common targets of tool use. Since they usually stay in a well-protected nest or mound, it is difficult for chimpanzees to obtain them with their bare hands. Chimpanzees use flexible plant parts such as leaf stalks, vines, or grass stems to penetrate the colony through a small entrance, or stout sticks or branches to break or pry open the hard surface of the colony. The situation is essentially the same for honey, which is protected deep inside beehives. Certain other foods, such as nuts, water in tree holes, and bone marrow are also difficult to access without using tools.

In terms of time spent feeding, the use of tools contributes to the utilization of these "hard-to-get" foods (Table 3). In Gombe, chimpanzees spend 3.7% of their total feeding time termite fishing and ant dipping. This represents 87.4% of the total time spent feeding on insects at Gombe. In other words, without tool use the consumption of insects at Gombe would be one-eighth the level observed. At Bossou, four types of tool use enable the chimpanzees to obtain almost half of the insects and woody pith, and nearly all of the nuts and algae consumed. Interestingly, the chimpanzees at Kibale do not have any ecologically important tool behaviors, and 99.8% of their total feeding time is spent feeding on "easy-to-get" foods (Fig. 1).

These results suggest that by using tools wild chimpanzees are able to obtain foods that are not typical for them. In the majority of cases, tools are used to consume social insects and honey. This suggests that tool use, as a behavioral adaptation, allows chimpanzees to overcome limitations of physical adaptation and to open a new feeding niche, from which a new set of physical adaptations may evolve.

3.2 Nutritional Intake by Tool Use

At Gombe, termites of the genus *Macrotermes* usually stay deep within firmly constructed mounds in the dry season. In November, at the beginning of the rainy season, they come closer to the surface of the mound to prepare for the departure of reproductives. Termite fishing by Gombe chimpanzees occurs exclusively in this season. In November, the chimpanzees spend 17% of their total feeding time on termite fishing, although termite fishing amounts to only 3.5% of the annual total (Table 3). Although no quantitative estimation has yet been made, there is no doubt that Gombe chimpanzees obtain a considerable amount of protein through this tool behavior.

The chimpanzees at Gombe and Taï use quite different methods for ant dipping. Gombe chimpanzees use longer sticks, averaging 66 cm, which they dip into a nest entrance; they then sweep their other hand along the stick to remove a mass of swarming ants, which they quickly eat (McGrew 1974). However, Taï chimpanzees use a shorter stick, averaging 24 cm, which they dip into a nest entrance and then use their lips to remove the ants directly from the stick (Boesch and Boesch 1990). At Gombe, 760 ants (1.17 g) per minute were seen to be consumed, whereas at Taï, only 180 ants per minute were consumed (McGrew 1974; Boesch and Boesch 1990). Even at Gombe, however, only 0.2% of the total annual feeding time is allotted to this behavior (Table 3). Although no nutritional analysis has been made of driver ants, making intake estimation difficult, it is unlikely that the chimpanzees obtain a significant amount of energy or protein by ant dipping.

Mahale chimpanzees seem to eat ants most frequently. They do not eat driver ants, but instead use tools to fish for arboreal *Camponotus* ants (Nishida 1973). This behavior can be observed throughout the year. Mothers with infants in Mahale spend 7.0% of their total feeding time feeding on *Camponotus* ants (Table 3). Since females are more frequent ant-fishers than males (T. Nishida, personal communication, 1999), the percentage for the whole community would be lower. For instance, in Kawanaka's 5-month study, 10 males of various ages spent only 0.7% of their feeding time ant fishing (Kawanaka 1990). One hour of ant fishing was observed to yield only about 600 ants, which is equivalent to no more than 1 g protein (Nishida and Hiraiwa 1982). From this calculation, the nutritional benefit of ant-fishing behavior seems trivial, and we should therefore consider alternate hypotheses to explain it. It may occur in order to acquire essential amino acids (Hladik 1977), or perhaps chimpanzees eat ants for nonnutritional reasons; the ants are simply a tasty treat, like spice or a snack (Nishida and Hiraiwa 1982). At Mahale, however, the frequency of feeding on *Camponotus* ants decreases during the termite season (Nishida and Hiraiwa 1982), and there is an obvious sex difference in feeding on animal matter: females spend more time ant fishing, whereas males tend to hunt more (Uehara 1997). These facts reveal a complementary relationship between the consumption of ants and other protein sources, and suggest that some aspects of protein intake affect fishing frequency.

Taï chimpanzees crack open five species of nut with wooden or stone hammers. *Coula edulis* nuts are consumed most often, and these are available for the 4 months from November to March (Boesch and Boesch 1983). During this season, the chimpanzees spend an average of 2 h 17 min per day cracking nuts (Günther and Boesch 1993). Taï chimpanzees spend approximately 43% of the day feeding (Doran 1997), so it is estimated that they spend 44% of their feeding time cracking nuts during the *Coula* season, or about 15% of the annual total (Table 3). A biomechanical study revealed that a Taï chimpanzee obtains 3762 kcal per day from 2.3 h nut cracking, which is nine times greater than the energy expended in nut cracking (Günther and Boesch 1993). The San people of the Kalahari Desert, who have the honor of working less time each day to survive than people in modern societies, needed 4 h 39 min (including searching time), on average, to obtain about 2000 kcal food per day per person (Tanaka 1980). No doubt, the Taï chimpanzees' nut cracking is a highly efficient way to gain energy.

At Bossou, Guinea, five types of tool use (nut cracking, pestle pounding, algae scooping, ant dipping, and leaf sponging) are regularly observed. A year-round study revealed that the chimpanzees spent 12% of their total annual feeding time on these tool-using behaviors; 32% during the peak month (Yamakoshi 1998). Although no biomechanical estimate of nutritional intake has been made, the efficiency of nut cracking at Bossou must be similar to that at Taï, since the nutritional value of oil-palm nuts, the only nuts cracked at Bossou, is almost identical to that of *Coula* nuts (Wu Leung 1968).

In summary, the nutrition obtained through tool use can be substantial. When consuming insects such as termites or ants, the likely nutritive gain is protein. In the case of termite fishing, protein intake appears to be important, but the benefit of consuming ants is less clear. Chimpanzees obtain a huge amount of energy from nut cracking, which occurs at two study sites in West Africa.

3.3 Seasonality and Tool Use

Even in tropical forests, there is a marked seasonal fluctuation in flowering and fruiting, caused by both biotic and abiotic factors (van Schaik et al. 1993). Under such conditions, the importance of certain food-gathering behaviors may differ seasonally. For example, capuchin monkeys in Cocha Cashu, Peru, eat ripe fruit when fruit is abundant, but depend on hard palm nuts when fruit is scarce (Terborgh 1983). Capuchin monkeys normally bite the nuts open, but they sometimes bang the nuts open against branches (Thorington 1967; Izawa and Mizuno 1977). The monkeys' manual dexterity clearly plays an important role in facilitating nut consumption, and such behavior must become more important in the season when fruit is scarce.

Likewise, tool use is ecologically important for chimpanzees at Bossou. Not only is the overall frequency of tool use high, as stated above, but their tool use has a special function: to facilitate survival during a critical period of the year when ripe

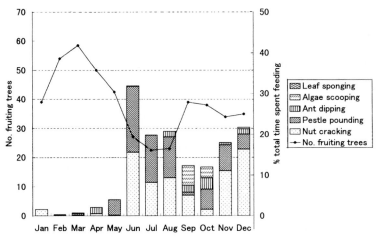

Fig. 2. The relationship between fruit availability and the frequency of tool use at Bossou, Guinea. The scale on the left Y-axis refers to the line, while that on the right Y-axis refers to the bars. The data are from Yamakoshi (1998)

fruit becomes quite scarce. In the home range of Bossou chimpanzees, there are plenty of oil palm trees, which provide a variety of foods. The chimpanzees eat the pulp of the oil-palm nuts without difficulty. They also pick up fallen nuts from the ground, and crack them open with stones (Sugiyama and Koman 1979b). In addition, they pull out palm fronds to eat the base, and pound the apical meristem of palm trees ("palm heart") with palm fronds to extract the juicy pith (pestle pounding; Sugiyama 1994a; Yamakoshi and Sugiyama 1995). Oil palm nuts for cracking and pith for pestle pounding are both available in all seasons, yet they are rarely consumed during the fruit-rich seasons. It is when the availability of fruit in the chimpanzee's home range decreases that the frequency of these two types of tool use increases (Fig. 2). These parts of the palm tree are well protected, and very difficult to access without tools. Thus, the existence of a tool-using tradition at Bossou actually mitigates the effects of fruit scarcity by allowing the utilization of otherwise inaccessible food items, such as oil palm nuts and pith.

What are the consequences of this? If a critical food scarcity in a particular period is reduced, there should be effects on reproductive or social parameters. Interestingly, the interbirth interval of Bossou females is shorter than in any other chimpanzee population (Sugiyama 1994b). As far as social organization is concerned, it has been claimed that food abundance and female gregariousness are correlated in chimpanzee society (e.g., Wrangham et al. 1992). Indeed at Bossou, females frequently stay together (Sugiyama and Koman 1979a) and groom each other regularly (Sugiyama 1988), unlike females in other populations. These correlations remain largely speculative, but continued ecological investigations will certainly provide further details.

The central subspecies of chimpanzees provides additional evidence of the active use of a tool tradition to cope with environmental uncertainty. As stated above,

Gombe chimpanzees fish for termites almost exclusively in the swarming season. In contrast, at Okorobikó, Equatorial Guinea, fishing tools are used relatively constantly throughout the year (Jones and Sabater Pi 1969; Sabater Pi 1974). The reason for this difference in behavior was once sought, but no clear answer was found (McGrew et al. 1979). Indirect evidence implies that the central subspecies might use two different types of stick to consume termites: one for perforating nests and the other for fishing (Sabater Pi 1974; McGrew and Rogers 1983; Sugiyama 1985; Fay and Carroll 1994). A recent ecological study at Ndoki, Republic of Congo, also confirmed the use of these two types of tool, with the first substantial direct observations of the use of fishing tools (Suzuki et al. 1995). (Recently, Bermejo and Illera (1999) also directly observed termite-obtaining behavior on one occasion at Lossi, Republic of Congo.) Moreover, fecal analysis revealed termite remains in the feces throughout the year, with no relationship to rainfall or termite activity. It has been suggested that Ndoki chimpanzees are able to access termites remaining deeper within a mound by using a stick to perforate the mound before fishing, even in the season when chimpanzees with only a fishing tradition, like the Gombe chimpanzees, cannot obtain termites. This addition of a different tool (perforating stick plus fishing stick) forms a so-called "tool-set" (Brewer and McGrew 1990), and clearly increases the annual termite intake. Incidentally, it is interesting to note that the use of perforating sticks has only been reported from a limited geographical area (Sugiyama 1993), which may form another "cultural area" in addition to the "nut-cracking area" in West Africa described above.

These two cases clearly demonstrate that chimpanzees' tool use has a dynamic interaction with the seasonal availability of food resources. In the Bossou case, two types of tool use allow the chimpanzees access to protected, but seasonally stable, food resources during the fruit-scarce season. In the Ndoki case, tool use appears to make a highly seasonal food resource available all year round. These instances suggest that tool use can be a strong weapon to change and overcome the food availability of a given environment.

4 Implications for Early Hominid Evolution

The systematic examination of the tool-using behavior of wild chimpanzees in a feeding context revealed that (1) chimpanzees use tools almost exclusively to access "hard-to-get" foods that are not included in their typical food repertoire, (2) tool use increases the nutritional intake of chimpanzees much more than was previously assumed, particularly in the case of the West African subspecies with a nut-cracking tradition, and (3) chimpanzee tool use functions to mitigate or manipulate the given availability of food resources, thereby increasing their ability to cope with a seasonal environment.

These results clearly demonstrate that chimpanzee tool use has a dynamic function to expand their food repertoire, especially during times of food scarcity, and thus to create a new feeding niche without any change in morphology. This is rather surprising, because such adaptive significance for tool use has been regarded as a uniquely human trait, which could have been acquired during the adaptation process as humans extended their habitat into savanna (Bartholomew and Birdsell 1953; Dart and Craig 1959; Oakley 1963; Kortlandt 1972; but see Boesch-Achermann and Boesch 1994). This study strongly suggests the contrary: tool-using ability that could facilitate niche expansion must have originated in the common ancestor of the human–chimpanzee clade. I discuss the possible implications of this study for early hominid evolution in the following sections.

4.1 Early Hominid Dietary Niche

Every evolutionary model (hypothesis, scenario, or whatever; Moore 1996) of early hominid evolution depends more or less on the ecology and behavior of the extant chimpanzee and bonobo, our closest living relatives (Stanford and Allen 1991; McGrew 1992). Therefore, the recent findings of chimpanzee tool use described above must have some impact on existing models for the evolution of the hominid food niche, and for the role of tool use in particular.

The common ancestor of chimpanzees and humans is thought to have lived around 6 million years ago, although no fossils have been found. The creature is assumed to have been a forest fruit-eater, with chimpanzee-like morphology (Pilbeam 1996), and to have formed a closed, patrilineal society (Wrangham 1987).

The earliest hominid, *Ardipithecus*, lived in a forested habitat (WoldeGabriel et al. 1994), and its diet must have consisted mainly of fruits. The habitat of *Australopithecus* species was likely more arid woodland (Reed 1997). Their diet is also assumed to have been fruit-centered, but some evidence suggests the significant inclusion of meat (Sponheimer and Lee-Thorp 1999; de Heinzelin et al. 1999) or the underground storage organs (USO) of plants (Peters and O'Brien 1981; Conklin-Brittain et al. 1998).

The ecological niche of the first *Homo* species is quite controversial. Paleo-environmental studies suggest that it was not until the emergence of *Homo* (about 2–2.5 million years ago) that any hominids could survive in a habitat that was totally savanna (Vrba 1985; Reed 1997). In such a habitat, where the density and seasonal availability of fruits is reduced, the *Homo* species must have depended on other foods. Their enlarged brain volume also implies a significant change in their feeding niche, since this organ requires enormous amounts of energy and thus cannot be maintained without utilizing high-quality foods (Aiello and Wheeler 1995). As was the case with *Australopithecus*, it is now disputed whether meat or USO was the primary staple food for *Homo erectus* (O'Connell et al. 1999; Wrangham et al. 1999).

4.2 Ecological Role of Early Hominid Tool Use

The early hominids appear to have adapted their fruit-centered diet to woodland or savanna-type resources, such as meat or USO, as their habitat became drier. In other words, early hominids could have established a totally new food habit; one that was formerly atypical. How did they achieve these changes? Interestingly, as we have already seen, extant wild chimpanzees cope with similar environmental conditions by consuming protected, but constantly available, foods, which they could not normally consume without the aid of tools (Fig. 2, Table 3). The "referential model" (Tooby and DeVore 1987) based on modern chimpanzee ecology, which was reviewed in this chapter, suggests that early hominids also relied heavily on tools, in just such situations as described above.

It seems difficult for extant chimpanzees to acquire and process meat and USO; presumably, the same was true for early hominids. Chimpanzees hunt for many species of animals, but only in exceptional cases is any tool use observed during the hunt (only one out of six instances of tool use for vertebrates is "habitual or customary"; see Tables 1 and 2). Moreover, the relative time spent feeding and the estimated nutritional intake of meat remain trivial (Teleki 1973b; Stanford et al. 1994). Scavenging on large carcasses is a likely substitute for direct hunting, but chimpanzees seem reluctant to scavenge (Kortlandt 1967; Muller et al. 1995). On the other hand, USOs are not included in the food repertoire of chimpanzees (Peters and O'Brien 1981), although a few exceptions have been reported (Lanjouw, cited in Wrangham and Peterson 1996, pp 57–59; McGrew et al. 1988; Sugiyama 1978). Evidence suggests that chimpanzee-like early hominids would have needed certain types of tools to make substantial use of these food resources.

Therefore, the invention of a "hunting tool" (e.g., Dart and Craig 1959) or a "digging tool" (e.g., Nishida 1974, 1981) must have had a strong impact on the feeding ecology of early hominids, and it has been proposed that these were key events that differentiated hominids from their chimpanzee-like ancestors. That is, the hominids that first accessed new resources with the aid of a tool could have been the first to invade a new niche.

4.3 Habitat, Feeding Niche, and Tool Use

It is disputed whether there is a causal relationship between niche invasion and tool use. The direction of causation can be reversed; invasion into a new niche can augment the necessity for tool use, or establishment of a tool tradition can allow niche invasion. Some authors have suggested the former possibility: tool use may evolve when a species without special morphological adaptation invades a new niche (Washburn 1960; Alcock 1972). Although this seems possibly true for the case of early hominids, Beck (1980, pp. 186–187) pointed out that the idea lacks generality, because chimpanzees are skillful tool users, but they have not invaded a new niche.

This study has shown that although chimpanzees may not have invaded a new geographical niche, they have invaded and occupied new feeding niches by using tools (i.e., invasion of the anteater and nutcracker niches). It is therefore more plausible to assume the latter possibility; tool use made niche expansion possible through the development of new food habits, for both early hominids and chimpanzees, although the hominids geographically expanded their habitat into woodland and savanna, while chimpanzees stayed in the forest.

Instead of the "niche invasion hypothesis," Parker and Gibson (1979) extended

Hamilton's idea (Hamilton 1973) and proposed the "extractive foraging hypothesis." In this scenario, the factor that enhanced the evolution of tool use was the feeding niche for "embedded foods." They went on to argue that "intelligent" tool users, such as hominids and apes, should depend on seasonal embedded resources, while "context-specific" tool users, such as sea otters, should utilize aseasonal resources, because the former are presumed to be more complex and to require more intelligence.

Since this study dealt with tool-using behaviors only in a feeding context, it is difficult to evaluate the validity of the extractive foraging hypothesis per se. As for seasonality, however, the most important types of tool use by chimpanzees clearly serve to mitigate seasonal food scarcity, rather than to utilize seasonal resources more efficiently. That is, extractive foraging skills may be ecologically less important if they are applied only to very seasonal resources, regardless of the degree of intelligence needed for such an omnivorous foraging style.

4.4 Tool Use and Social Variables

If a technical invention had a significant effect on the hominid feeding niche, then changes in morphology, reproductive systems, and social structures could be expected as a consequence (e.g., enlarged brain volume, Aiello and Wheeler 1995; extended life span, O'Connell et al. 1999; emergence of pair-bonding, Wrangham et al. 1999). As we have seen, chimpanzee tool use plays a similar ecological role in extending their feeding niche. The most prominent case seems to be that of nut cracking by West African chimpanzees, which enables them to utilize otherwise inaccessible, but seasonally stable and enormously energy-rich, food resources. As was suggested for early hominids, this technology may explain some of the West African subspecies' unique reproductive and social characteristics (e.g., shorter interbirth intervals, Sugiyama 1994b, Boesch 1997; extensive between-female sociality, Sugiyama 1988, Boesch 1991).

Recently, van Schaik et al. (1999) suggested a positive correlation between tool use frequency and gregariousness among wild orangutans and wild chimpanzees. They went on to argue that gregarious and tolerant social relationships could facilitate social learning of tool use. This study, however, suggests the opposite; that increased food intake brought about by tool use may reduce feeding competition among individuals and thus enhance sociality and gregariousness. Considering the huge amount of energy that West African chimpanzees obtain by nut cracking, the correlation that van Schaik et al. found with chimpanzee populations would more reasonably be explained by the latter argument. In the context of hominid evolution, however, these seemingly opposite phenomena could have caused positive feedback, leading to profound and irreversible changes in the subsistence of early hominids (van Schaik et al. 1999).

In summary, tool use can be an effective ecological tactic, allowing both modern chimpanzees and early hominids to overcome seasonal instability in the food supply. From this perspective, it is quite interesting that tool use for feeding is absent in wild bonobos, which are excellent tool users in captivity (e.g. Toth et al.

1993). Bonobo habitat appears to provide a more stable food supply than that of chimpanzees, either spatially (Badrian and Badrian 1984), temporally (Malenky 1990), or due to the abundance of terrestrial herbs (Wrangham 1986). This is also true for gorillas; they are good tool users in captivity (e.g. Nakamichi 1999), but have a more or less specialized herbivorous diet (Fossey and Harcourt 1977) that frees the animals from seasonal food shortage in the wild. These facts also demonstrate that the presence or absence of ecological pressure strongly affects the frequency of tool use among hominoids.

It is to be hoped that future studies on extant chimpanzees (in savanna habitat, in particular) and other great apes will provide more evidence of an ecological relationship between feeding technique and subtle environmental changes, so that we can infer more about the evolutionary origin of our technology.

Acknowledgments

This study was financed by a 1996 research grant from the Nakayama Foundation for Human Science, and a grant under Research Fellowships of the Japan Society for the Promotion of Science for Young Scientists to the author from 1997 to 1999 (No. 2670). Thanks are due to T. Matsuzawa, J. Yamagiwa, M. Hayashi, and T. Humle for their helpful comments on an earlier draft of this paper. I thank Y. Sugiyama, T. Nishida, H. Ihobe, M.A. Huffman, S. Suzuki, K. Adachi, and M. Myowa-Yamakoshi for their support and advice for the study.

References

Aiello LC, Wheeler P (1995) The expensive-tissue hypothesis: the brain and the digestive system in human and primate evolution. Curr Anthropol 36:199–221

Alcock J (1972) The evolution of the use of tools by feeding animals. Evolution 26:464–473

Alp R (1997) "Stepping-sticks" and "seat-sticks": new types of tools used by wild chimpanzees (*Pan troglodytes*) in Sierra Leone. Am J Primatol 41:45–52

Badrian AJ, Badrian NL (1984) Group composition and social structure of *Pan paniscus* in the Lomako Forest. In: Susman RL (ed) The pygmy chimpanzee: evolutionary biology and behavior. Plenum Press, New York, London

Bartholomew GA Jr, Birdsell JB (1953) Ecology and the protohominids. Am Anthropol 55:481–498

Beatty H (1951) A note on the behavior of the chimpanzee. J Mammal 32:118

Beck BB (1980) Animal tool behavior. Garland STPM Press, New York

Bermejo M, Illera G (1999) Tool-set for termite-fishing and honey extraction by wild chimpanzees in the Lossi Forest, Congo. Primates 40:619–627

Boesch C (1991) The effects of leopard predation on grouping patterns in forest chimpanzees. Behaviour 117:220–242

Boesch C (1995) Innovation in wild chimpanzees (*Pan troglodytes*). Int J Primatol 16:1–16

Boesch C (1997) Evidence for dominant wild female chimpanzees investing in more sons. Anim Behav 54:811–815

Boesch C, Boesch H (1981) Sex differences in the use of natural hammers by wild chimpanzees: a preliminary report. J Hum Evol 10:585–593

Boesch C, Boesch H (1983) Optimization of nut-cracking with natural hammers by wild chimpanzees. Behaviour 83:265–286

Boesch C, Boesch H (1989) Hunting behavior of wild chimpanzees in the Taï National Park. Am J Phys Anthropol 78:547–573

Boesch C, Boesch H (1990) Tool use and tool making in wild chimpanzees. Folia Primatol 54:86–99

Boesch C, Marchesi P, Marchesi N, Fruth B, Joulian F (1994) Is nut cracking in wild chimpanzees a cultural behavior? J Hum Evol 26:325–338

Boesch-Achermann H, Boesch C (1994) Hominization in the rainforest: the chimpanzee's piece of the puzzle. Evol Anthropol 3:9–16

Brewer SM, McGrew WC (1990) Chimpanzee use of a tool-set to get honey. Folia Primatol 54:100–104

Chapman CA, White FJ, Wrangham RW (1994) Party size in chimpanzees and bonobos: a reevaluation of theory based on two similarly forested. In: Wrangham RW, McGrew WC, de Waal FBM, Heltone PG (eds) Chimpanzee cultures. Harvard University Press, Cambridge, pp 169–180

Conklin-Brittain NL, Wrangham WR, Smith CC (1998) Relating chimpanzee diets to potential *Australopithecus* diets. Online. http://www.cast.uark.edu/local/icaes/conferences/wburg/posters/nconklin/conklin.html Accessed August 4, 1999

Dart RA, Craig D (1959) Adventures with the missing link. Harper & Row, New York

de Heinzelin J, Clark JD, White T, Hart W, Renne P, WoldeGabriel G, Beyene Y, Vrba E (1999) Environment and behavior of 2.5-million-year-old Bouri hominids. Science 284:625–629

Doran D (1997) Influence of seasonality on activity patterns, feeding behavior, ranging and grooming patterns in Taï chimpanzees. Int J Primatol 18:183–206

Fay JM, Carroll RW (1994) Chimpanzee tool use for honey and termite extraction in Central Africa. Am J Primatol 34:309–317

Fossey D, Harcourt AH (1977) Feeding ecology of free-ranging mountain gorilla (*Gorilla gorilla beringei*). In: Clutton-Brock TH (ed) Primate ecology: studies of feeding and ranging behaviour in lemurs, monkeys and apes. Academic Press, London, pp 415–447

Goodall J (1963) Feeding behaviour of wild chimpanzees: a preliminary report. Symp Zool Soc London 10:39–48

Goodall J (1964) Tool-using and aimed throwing in a community of free-living chimpanzees. Nature 201:1264–1266

Goodall J (1986) The chimpanzees of Gombe: patterns of behavior. Harvard University Press, Cambridge

Goodall J van Lawick (1970) Tool-using in primates and other vertebrates. In: Lehrman DS, Hinde RA, Shaw E (eds) Advances in the study of behavior, Vol 3. Academic Press, London, pp 195–249

Günther MM, Boesch C (1993) Energetic cost of nut-cracking behaviour in wild chimpanzees. In: Preuschoft H, Chivers DJ (eds) Hands of primates. Springer-Verlag, Vienna, pp 109–129

Hamilton WJ III (1973) Life's color code. McGraw-Hill, New York

Hladik CM (1977) Chimpanzees of Gabon and chimpanzees of Gombe: some comparative data on the diet. In: Clutton-Brock TH (ed) Primate ecology: studies of feeding and ranging behaviour in lemurs, monkeys and apes. Academic Press, London, pp 481–501

Huffman MA, Kalunde MS (1993) Tool-assisted predation on a squirrel by a female chimpanzee in the Mahale Mountains, Tanzania. Primates 34:93–98

Humle T (1999) New record of fishing for termites (*Macrotermes*) by the chimpanzees of Bossou (*Pan troglodytes verus*), Guinea. Pan Africa News 6:3–4

Izawa K, Itani J (1966) Chimpanzees in Kasakati Basin, Tanganyika (1): ecological study in the rainy season 1963–1964. Kyoto University African Studies 1:73–156

Izawa K, Mizuno A (1977) Palm-fruit cracking behavior of wild black-capped capuchin (*Cebus apella*). Primates 18:773–792

Jones C, Sabater Pi J (1969) Sticks used by chimpanzees in Rio Muni, West Africa. Nature 223:100–101

Kawanaka K (1990) Age differences in ant-eating by adult and adolescent males. In: Nishida T (ed) The chimpanzees of the Mahale Mountains: sexual and life history strategies. University of Tokyo Press, Tokyo, pp 207–222

Kortlandt A (1967) Experimentation with chimpanzees in the wild. In: Starck D, Schneider R, Kuhn HJ (eds) Neue Ergebnisse der Primatologie—Progress in Primatology. Fischer, Stuttgart, pp 208–224

Kortlandt A (1972) New perspectives on ape and human evolution. University of Amsterdam, Amsterdam

Kuroda S, Nishihara T, Suzuki S, Oko RA (1996) Sympatric chimpanzees and gorillas in the Ndoki Forest, Congo. In: McGrew WC, Marchant LF, Nishida T (eds) Great ape societies. Cambridge University Press, Cambridge, pp 71–81

Leakey MD (1971) Olduvai gorge, Vol 3: excavations in beds I and II, 1960–1963. Cambridge University Press, Cambridge

Malenky RK (1990) Ecological factors affecting food choice and social organization in *Pan paniscus*. PhD dissertation, State University of New York, Stony Brook

Mann A (1972) Hominid and cultural origins. Man 7:379–386

Matsuzawa T (1991) Chimpanzee mind (in Japanese). Iwanami-Shoten, Tokyo

Matsuzawa T, Yamakoshi G (1996) Comparison of chimpanzee material culture between Bossou and Nimba, West Africa. In: Russon AE, Bard KA, Parker S (eds) Reaching into thought: the mind of the great apes. Cambridge University Press, Cambridge, pp 211–232

Matsuzawa T, Yamakoshi G, Humle T (1996) A newly found tool-use by wild chimpanzees: algae scooping (abstract). Primate Res 12:283

McGrew WC (1974) Tool use by wild chimpanzees in feeding upon driver ants. J Hum Evol 3:501–508

McGrew WC (1992) Chimpanzee material culture: implications for human evolution. Cambridge University Press, Cambridge

McGrew WC, Collins DA (1985) Tool use by wild chimpanzees (*Pan troglodytes*) to obtain termites (*Macrotermes herus*) in the Mahale Mountains, Tanzania. Am J Primatol 9:47–62

McGrew WC, Marchant LF (1997) Using the tools at hand: manual laterality and elementary technology in *Cebus* spp. and *Pan* spp. Int J Primatol 18:787–810

McGrew WC, Rogers ME (1983) Chimpanzees, tools, and termites: new record from Gabon. Am J Primatol 5:171–174

McGrew WC, Tutin CEG, Baldwin PJ (1979) Chimpanzees, tools, and termites: cross-cultural comparisons of Senegal, Tanzania, and Rio Muni. Man 14:185–214

McGrew WC, Baldwin PJ, Tutin CEG (1988) Diet of wild chimpanzees (*Pan troglodytes verus*) at Mt. Assirik, Senegal: I. composition. Am J Primatol 16:213–226

McGrew WC, Ham RM, White LJT, Tutin CEG, Fernandez M (1997) Why don't chimpanzees in Gabon crack nuts? Int J Primatol 18:353–374

Merfield FG, Miller H (1956) Gorilla hunter: the African adventures of a hunter extraordinary, Farrar, Straus and Cudahy, New York

Moore J (1996) Savanna chimpanzees, referential models and the last common ancestor. In: McGrew WC, Marchant LF, Nishida T (eds) Great ape societies. Cambridge University Press, Cambridge

Muller MN, Mpongo E, Stanford CB, Boehm C (1995) A note on scavenging by wild chimpanzees. Folia Primatol 65:43–47

Nakamichi M (1999) Spontaneous use of sticks as tools by captive gorillas. Primates 40:487–498

Nishida T (1973) The ant-gathering behaviour by the use of tools among wild chimpanzees of the Mahali Mountains. J Hum Evol 2:357–370

Nishida T (1974) Origin of tools (in Japanese). Languages (in Japanese) 3:1084–1092

Nishida T (1981) Comment on Peters and O'Brien (1981). Curr Anthropol 22:137

Nishida T (1987) Local traditions and cultural transmission. In: Smuts BB, Cheney DL, Seyfarth RM, Wrangham RW, Struhsaker TT (eds) Primate societies. University of Chicago Press, Chicago, pp 462–474

Nishida T, Hiraiwa M (1982) Natural history of a tool-using behavior by wild chimpanzees in feeding upon wood-boring ants. J Hum Evol 11:73–99

Nishida T, Nakamura M (1993) Chimpanzee tool use to clear a blocked nasal passage. Folia Primatol 61:218–220

Nishida T, Turner LA (1996) Food transfer between mother and infant chimpanzees of the Mahale Mountains National Park, Tanzania. Int J Primatol 17:947–968

Nishida T, Uehara S (1980) Chimpanzees, tools, and termites: another example from Tanzania. Curr Anthropol 21:671–672

Nishida T, Uehara S (1983) Natural diet of chimpanzees (*Pan troglodytes schweinfurthii*): long-term record from the Mahale Mountains, Tanzania. Afr Study Monogr 3:109–130

Oakley KP (1963) Man the tool-maker, 5th edn. British Museum, London

O'Connell JF, Hawkes K, Blurton Jones NG (1999) Grandmothering and the evolution of *Homo erectus*. J Hum Evol 36:461–485

Parker ST, Gibson KR (1979) A developmental model for the evolution of language and intelligence in early hominids. Behav Brain Sci 2:367–408

Peters CR, O'Brien EM (1981) The early hominid plant-food niche: insights from an analysis of plant exploitation by *Homo*, *Pan*, and *Papio* in Eastern and Southern Africa. Curr Anthropol 22:127–140

Pilbeam D (1996) Genetic and morphological records of the Hominoidea and hominid origins: a synthesis. Mol Phyl Evol 5:155–168

Plooij FX (1978) Tool-use during chimpanzees' bushpig hunt. Carnivore 1:103–106

Reed KE (1997) Early hominid evolution and ecological change through the African Plio-Pleostocene. J Hum Evol 32:289–322

Sabater Pi J (1974) An elementary industry of the chimpanzees in the Okorobikó Mountains, Rio Muni (Republic of Equatorial Guinea), West Africa. Primates 15:351–364

Savage TS, Wyman J (1843–44) Observation on the external characters and habits of the *Troglodytes niger* Geoff. Boston J Nat Hist 4:362–386

Semaw S, Renne P, Harris JWK, Feibel CS, Bernor RL, Fesseha N, Mowbray K (1997) 2.5-million-year-old stone tools from Gona, Ethiopia. Nature 385:333–336

Sibley CG, Ahlquist JE (1984) The phylogeny of the hominid primates, as indicated by DNA–DNA hybridization. J Mol Evol 20:2–15

Sponheimer M, Lee-Thorp JA (1999) Isotopic evidence for the diet of an early hominid, *Australopithecus africanus*. Science 283:368–370

Stanford CB, Allen JS (1991) On strategic storytelling: current models of human behavioral evolution. Curr Anthropol 32:58–61

Stanford CB, Wallis J, Matata H, Goodall J (1994) Patterns of predation by chimpanzees on red colobus monkeys in Gombe National Park, 1982–1991. Am J Phys Anthropol 94:213–228

Sugiyama Y (1978) The people and the chimpanzees of Bossou village (in Japanese). Kinokuniya-Shoten, Tokyo

Sugiyama Y (1985) The brush-stick of chimpanzees found in south-west Cameroon and their cultural characteristics. Primates 26:361–374

Sugiyama Y (1988) Grooming interactions among adult chimpanzees at Bossou, Guinea, with special reference to social structure. Int J Primatol 9:393–407

Sugiyama Y (1993) Local variation of tools and tool use among wild chimpanzee populations. In: Berthelet A, Chavaillon J (eds) The use of tools by human and non-human primates. Clarendon Press, Oxford, pp 175–187

Sugiyama Y (1994a) Tool use by wild chimpanzees. Nature 367:327

Sugiyama Y (1994b) Age-specific birth rate and lifetime reproductive success of chimpanzees at Bossou, Guinea. Am J Primatol 32:311–318

Sugiyama Y, Koman J (1979a) Social structure and dynamics of wild chimpanzees at Bossou, Guinea. Primates 20:323–339

Sugiyama Y, Koman J (1979b) Tool-using and making behavior in wild chimpanzees at Bossou, Guinea. Primates 20:513–524

Sugiyama Y, Koman J (1987) A preliminary list of chimpanzees' alimentation at Bossou, Guinea. Primates 28:133–147

Sugiyama Y, Koman J, Sow MB (1988) Ant-catching wands of wild chimpanzees at Bossou, Guinea. Folia Primatol 51:56–60

Suzuki A (1966) On the insect-eating habits among wild chimpanzees living in the savanna woodland of Western Tanzania. Primates 7:481–487

Suzuki S, Kuroda S, Nishihara T (1995) Tool-set for termite-fishing by chimpanzees in the Ndoki Forest, Congo. Behaviour 132:219–235

Tanaka J (1980) The San hunter–gatherers of the Kalahari: a study in ecological anthropology. University of Tokyo Press, Tokyo

Teleki G (1973a) The omnivorous chimpanzee. Sci Am 228:32–42

Teleki G (1973b) The predatory behavior of wild chimpanzees. Bucknell University Press, Lewisburg

Terborgh J (1983) Five new world primates: a study in comparative ecology. Princeton University Press, Princeton

Thorington RW Jr (1967) Feeding and activity of *Cebus* and *Saimiri* in a Colombian forest. In: Starck D, Schneider R, Kuhn, HJ (eds) Progress in primatology. Gustav Fischer, Stuttgart, pp 180–184

Tooby J, DeVore I (1987) The reconstruction of hominid behavioral evolution through strategic modeling. In: Kinzey WG (ed) The evolution of human behavior: primate models. State University of New York Press, Albany, pp 183–237

Toth N, Schick KD, Savage-Rumbaugh ES, Sevcik RA, Rumbaugh DM (1993) *Pan* the toolmaker: investigations into the stone tool-making and tool-using capabilities of a bonobo (*Pan paniscus*). J Archaeol Sci 20:81–91

Tutin CEG, Fernandez M, Rogers ME, Williamson EA, McGrew WC (1991) Foraging profiles of sympatric lowland gorillas and chimpanzees in the Lopé Reserve, Gabon. Philos Trans R Soc London, Ser B 334:179–186

Tutin CEG, Ham R, Wrogemann D (1995) Tool-use by chimpanzees (*Pan t. troglodytes*) in the Lopé Reserve, Gabon. Primates 42:1–24

Uehara S (1982) Seasonal changes in the techniques employed by wild chimpanzees in the Mahale Mountains, Tanzania, to feed on termites (*Pseudacanthotermes spiniger*). Folia Primatol 37:44–76

Uehara S (1997) Predation on mammals by the chimpanzee (*Pan troglodytes*). Primates 38:193–214

van Schaik CP (1999) The evolution of material culture: insights from orangutans. Paper presented at COE International Symposium, Evolution of the Apes and the Origin of the Human Beings, November 18–20, Inuyama, Japan

van Schaik CP, Terborgh JW, Wright SJ (1993) The phenology of tropical forests: adaptive significance and consequences for primary consumers. Annu Rev Ecol Systematics 24:353–377

van Schaik CP, Deaner RO, Merrill MY (1999) The conditions for tool use in primates: implications for the evolution of material culture. J Hum Evol 36:719–741

Vrba ES (1985) Ecological and adaptive changes associated with early hominid evolution. In: Delson E (ed) Ancestors: the hard evidence. Alan R. Liss, New York, pp 63–71

Washburn SL (1960) Tools and human evolution. Sci Am 203:63–75

Whiten A, Goodall J, McGrew WC, Nishida T, Reynolds V, Sugiyama Y, Tutin CEG, Wrangham RW, Boesch C (1999) Cultures in chimpanzees. Nature 399:682–685

WoldeGabriel G, White TD, Suwa G, Renne P, de Heinzelin J, Hart WK, Heiken G (1994) Ecological and temporal placement of early Pliocene hominids at Aramis, Ethiopia. Nature 371:330–333

Wrangham RW (1977) Feeding behaviour of chimpanzees in Gombe National Park, Tanzania. In: Clutton-Brock TH (ed) Primate ecology: studies of feeding and ranging behaviour in lemurs, monkeys and apes. Academic Press, London, pp 504–538

Wrangham RW (1986) Ecology and social relationships in two species of chimpanzees. In: Rubenstein DI, Wrangham RW (eds) Ecological aspects of social evolution: birds and mammals. Princeton University Press, Princeton, pp 352–378

Wrangham RW (1987) The significance of African apes for reconstructing human social evolution. In: Kinzey WG (ed) The evolution of human behavior: primate models. State University of New York Press, Albany, pp 51–71

Wrangham RW, Peterson D (1996) Demonic males: apes and the origins of human violence. Houghton Mifflin, New York

Wrangham RW, Clark AP, Isabirye-Basuta G (1992) Female social relationships and social organization of Kibale Forest chimpanzees. In: Nishida T, McGrew WC, Marler P, Pickford M, de Waal FBM (eds) Topics in primatology, Vol 1: human origins. University of Tokyo Press, Tokyo, pp 81–98

Wrangham RW, Holland Jones J, Laden G, Pilbeam D, Conklin-Brittain NL (1999) The raw and the stolen: cooking and the ecology of human origins. Curr Anthropol 40:567–594

Wu Leung W (1968) Food consumption table for use in Africa. Public Health Service, US Department of Health, Education, and Welfare, Nutrition Division, Food and Agriculture Organization of the United Nations

Yamagiwa J, Maruhashi T, Yumoto T, Mwanza N (1996) Dietary and ranging overlap in sympatric gorillas and chimpanzees in Kahuzi-Biega National Park, Zaïre. In: McGrew WC, Marchant LF, Nishida T (eds) Great ape societies. Cambridge University Press, Cambridge, pp 82–98

Yamakoshi G (1998) Dietary responses to fruit scarcity of wild chimpanzees at Bossou, Guinea: possible implications for ecological importance of tool use. Am J Phys Anthropol 106:283–295

Yamakoshi G, Sugiyama Y (1995) Pestle-pounding behavior of wild chimpanzees at Bossou, Guinea: a newly observed tool-using behavior. Primates 36:489–500

28
Emergence of Culture in Wild Chimpanzees: Education by Master-Apprenticeship

Tetsuro Matsuzawa[1], Dora Biro[2], Tatyana Humle[3], Noriko Inoue-Nakamura[1], Rikako Tonooka[1], and Gen Yamakoshi[4]

1 Introduction

This chapter describes a series of field experiments aimed at investigating aspects of emergence of cultural traditions in wild chimpanzee communities. Long-term research at a number of sites in Africa has revealed that each community of chimpanzees has developed its unique set of cultural traditions (Boesch and Boesch-Achermann 2000; Goodall 1986; McGrew 1992; Nishida 1990; Whiten et al. 1999). The evidence poses an intriguing question: How did these unique cultures come into existence?

In this chapter, we define "culture" according to the definition proposed by Matsuzawa (1999). Culture is "a set of knowledge, techniques, and values that are shared by members of a community and are transmitted from one generation to the next through non-genetic channels." Hence, the key points we wish to advocate are that culture is (1) community based, (2) cross-generational, and (3) reliant on postnatal learning.

In November 1948, more than 50 years ago, Japanese primatologists led by the late Kinji Imanishi and Jun'ichiro Itani began a long-term field observation project on Koshima Island, focusing on social behavior among wild Japanese monkeys (Chapter 20 by Watanabe, this volume). One of the most widely known findings to have emerged from this work was the appearance of a novel behavior described as "sweet-potato washing." Events that took place on the island were initially reported under the title "Newly-acquired pre-cultural behavior of the natural troop of Japanese monkeys on Koshima Islet" (Kawai 1965; Chapter 24 by Hirata et al., this volume), although whether the behavior is indeed "cultural" or "precultural" is still a controversial issue. Importantly, however, this represents a very rare case documenting the invention process and social propagation of skills unique to a community of wild nonhuman primates. Chimpanzees are highly adept at a variety of tool-using techniques such as termite fishing, ant-dipping, nut-cracking,

[1] Primate Research Institute, Kyoto University, 41 Kanrin, Inuyama, Aichi 484-8506, Japan
[2] Department of Zoology, University of Oxford, UK
[3] Department of Psychology, University of Stirling, UK
[4] Center for African Area Studies, Kyoto University, 46 Yoshida Shimoadachi-cho, Sakyo-ku, Kyoto 606-8501, Japan

and so forth. However, in these cases it is nearly impossible to directly observe the emergence of such cultural behaviors because they tend to be well established within communities by the time researchers' attention is drawn to them.

The target behavior of the study reported here is nut-cracking by wild chimpanzees at Bossou, Guinea, West Africa. Members of this community use a pair of stones as hammer and anvil to crack open oil-palm nuts (see color plates, this volume). To examine aspects of such highly complex tool-using skills, we applied two distinct approaches. First, we carried out an extensive survey of neighboring communities and found that each community utilizes different species of nuts as the targets of cracking. Then, we introduced unfamiliar nuts (Coula and Panda, neither of which is naturally available at Bossou) to the Bossou chimpanzees and recorded their behavior toward these novel potential targets. We video-recorded all behaviors occurring during the repeated presentations of novel nuts, including individuals' initial responses and incidents of observing behavior, which helped to ascertain how the cracking of new nuts could spread among members of the community. Through such a field experiment, we aimed to reveal processes of acquisition as well as their associated educational aspects in chimpanzees – in other words, the mechanisms involved in the transmission of knowledge and skills from one generation to the next.

2 Extensive Survey

Previous studies have revealed notable cultural differences among different research sites. However, consider for instance Gombe in East Africa and Bossou in West Africa: these are sites located more than 4000 km apart at two ends of a continent. There can clearly be no interaction of any kind between the two communities that may lead to cultural transmission. To examine the emergence of culture within a community, it is essential that we possess a knowledge of cultural variation in neighboring communities with the potential for direct interaction. We carried out an extensive survey of areas around Bossou to assess similarities and differences in cultural traditions among adjacent communities. We focused on tool use as an important marker of cultural traditions, for two main reasons. First, tool use is well documented in studies of cultural variation in chimpanzee communities. Second, chimpanzees leave traces of tool-using activity, such as stone hammers and anvils, which allows for relatively easy evaluation of cultural variation even in nonhabituated chimpanzee communities.

2.1 Research Sites

A small group of chimpanzees (*Pan troglodytes verus*) lives at Bossou, in the southeastern corner of the Republic of Guinea, close to the country's borders with Côte d'Ivoire and Liberia. Bossou is located at 7°39' N and 8°30' W, at an altitude of approximately 550-700 m. The forest area at Bossou has been isolated from the nearby forest of the Nimba mountains (a "world heritage" site with a maximum peak of 1740 m) by a 3- to 4-km stretch of savanna vegetation. The habitat of the Bossou community is a mosaic of primary, secondary, riverine, and scrub forest, as well as farmland. The research site was established in 1976 by Sugiyama and his colleagues (Sugiyama and Koman 1979), and the community has been the subject of continuous study ever since. The number of chimpanzees within the group has remained relatively stable for more than two decades at an average of 20 individuals (minimum, 16; maximum, 23).

Matsuzawa and colleagues began a survey of neighboring communities in 1993 (Humle and Matsuzawa, 2001; Matsuzawa and Yamakoshi 1996).So far, we have succeeded in confirming the presence of three adjacent communities, at Seringbara, Yealé, and Diéké. The Seringbara community is located to the east of Bossou, about 5 km away, and inhabits the mountain forests of Mt. Nimba (Humle, personal communication; Shimada 2000). The Yealé community is located to the southeast of Bossou, about 12 km away, and resides in the mountain forest covering the opposite side of Mt. Nimba (Matsuzawa and Yamakoshi 1996). The Diéké community lives about 50 km to the west of Bossou (Matsuzawa et al. 1999), where its habitat, "Forêt classée de Diéké," is protected land covering an area of more than 700 km^2, stretching over approximately 35 km in the north-south as well as east-west directions.

2.2 Results of the Extensive Survey

Table 1 summarizes cultural differences in the nut-cracking behavior of neighboring communities. We found three species of nuts in total, oil-palm nuts (*Elaeis guineensis*), Coula nuts (*Coula edulis*), and Panda nuts (*Panda oleosa*), in the area (Fig 1). However, the four communities (Bossou, Seringbara, Yealé, and Diéké) show different patterns of utilization for the three species of nuts as target items of cracking. Of course, ecological factors such as the availability of nuts are important and critical determinants. However, such ecological factors cannot fully explain some striking observations. For instance, why do Seringbara chimpanzees not crack open

Table 1. Cultural differences in nut-cracking behavior between Bossou and the neighboring communities; for comparison, data from the Tai Forest, Côte d'Ivoire (Boesch and Boesch-Achermann 2000) are also included

Community	Location[a]	Genus of nut		
		Elaeis	*Coula*	*Panda*
Bossou	0	Yes	No	No
Seringbara/Nimba	5 km to the east	No[b]	No	No
Yealé/Nimba	12 km to the southeast	Yes	Yes	No[b]
Diéké	50 km to the west	No	Yes	Yes
Taï	240 km to the southeast	No[b]	Yes	Yes

[a] Location, Location of the community described with reference to Bossou (distance and direction).
[b] No, Nuts are available but no evidence of nut-cracking has been found.

Fig. 1. Three species of nuts used as targets for nut-cracking in our field experiment. From *left* to *right*: oil-palm nuts (*Elaeis guineensis*), Coula nuts (*Coula edulis*), and Panda nuts (*Panda oleosa*). The *upper row* shows the relatively fresh fruits in which the nuts are encased; the *lower row* shows the nuts. In the natural habitat, the fruits fall to the ground and the outer soft layer naturally decomposes to reveal the hard shell of the nut

oil-palm nuts even though the nuts are available in abundance? Similarly, why do Yealé chimpanzees neglect Panda nuts even though these are present in the forest? These facts clearly suggest that cultural factors are also at work.

3 Intensive Survey at an Outdoor Laboratory at Bossou

Bossou chimpanzees use a pair of stones to crack open a specific target food, oil-palm nuts. They do not crack Coula nuts and Panda nuts simply because these nuts are not available at Bossou. How do Bossou chimpanzees react to these novel, unfamiliar nuts that are habitually cracked and consumed by the chimpanzees living in some of the neighboring communities? We carried out a field experiment in which we presented the unfamiliar nuts to the chimpanzees of Bossou.

3.1 Field Experiment

Since 1988, Matsuzawa and his colleagues have been closely observing and video-recording the nut-cracking behavior of Bossou chimpanzees, in the framework of a novel paradigm, "field experiments on tool use" (Matsuzawa 1994; Sakura and Matsuzawa 1991). We established an outdoor laboratory in the core part of the chimpanzees' ranging area (Figs. 2, 3) where we provided a collection of stones and nuts. As the outdoor laboratory was located at the junction of paths used regularly by the chimpanzees, they soon discovered the ready supply of food and necessary

Fig. 2. Field experiment on tool use in the outdoor laboratory. In this picture, two kinds of tool use can be seen simultaneously: use of leaves for drinking water (*left*), and use of stone hammers and anvils to crack open oil-palm nuts (*right*). A third kind of tool use, "ant dipping," has also been observed here

Fig. 3. One example of experimental setup at the outdoor laboratory. There are three piles of different species of nuts in front (Coula nuts, Panda nuts, and oil-palm nuts from left to right), while the stones provided for cracking are visible in the background. Fresh oil-palm nuts in the center attracts the safari ants. All materials were provided by the experimenters to increase the chances of observing the tool-using behavior

tools, and returned often to feed. This simple experimental manipulation thus dramatically increased our opportunities for observing the nut-cracking behavior. We applied the same idea to other tools, including ant-dipping and the use of leaves for drinking water. For the former purpose, we placed on the ground of the outdoor laboratory fresh oil-palm fruits as well as dead insects such as grasshoppers and caterpillars, which attracted safari ants. The ants were in turn discovered by chimpanzees, who on occasion proceeded to catch them by using a dipping stick manufactured especially for the purpose. To observe the use of leaves for drinking, a hole was artificially drilled in a large tree trunk (damage to the tree was minimal) and subsequently filled with fresh water. Chimpanzees found the artificial tree-hole and were soon drinking the water contained inside with the aid of folded leaves. Thus, field experiments in the outdoor laboratory provided a unique opportunity to video-record and carefully analyze individual episodes of tool-using behavior. In the following section, we focus on the behavior toward unfamiliar nuts and nutlike objects by the chimpanzees of Bossou.

3.2 Data Collection

We have collected data for oil-palm nut-cracking once every year in the dry season (usually December to February) since 1988. The behavior of chimpanzees at the outdoor laboratory was always recorded simultaneously by at least two video cameras, positioned at different angles. About 20 to 30 hours of recording was accumulated in each year. Party size of chimpanzees visiting the outdoor laboratory averaged about seven individuals, with the mean duration of continuous stay at the laboratory being approximately 30 min depending on the number of oil-palm nuts provided by the experimenters. Such duration is comparable to chimpanzees' average feeding bouts in fruiting trees in this season.

Our experiment of introducing unfamiliar items as potential targets of nut-cracking entailed the presentation of three different types of items: (1) Coula nuts, (2) wooden balls, and (3) Panda nuts. The successive stages of data collection are summarized in Table 2.

Table 2. Summary of data collection for different target items presented

Target nuts	Year	Researchers and years of participation
Oil-palm nuts	1988–2000	TM-11, GY-1, RT-1, NI-3, MM-2, DB-2
Coula nuts	1993, 1996, 2000	TM-3, GY-1, RT-1, NI-2, DB-1
Wooden balls	1994	RT-1
Panda nuts	2000	TM-1, DB-1

"Year" shows the year in which the field experiments were carried out.
Abbreviations stand for the initials of the authors (except MM, Masako Myowa-Yamakoshi that collected the data in 1998); numerals after initials represent the number of years in which the researcher participated in the field experiments.

3.3 Oil-Palm Nut-Cracking: Developmental Changes in Tool Use

Numerous reports document nut-cracking at Bossou: these have focused on aspects such as hand preference (Sugiyama et al. 1993), developmental changes and critical periods in learning (Inoue-Nakamura and Matsuzawa 1997; Matsuzawa 1994), ecological determinants of tool use (Yamakoshi 1998), and so forth (see review by Matsuzawa 1999).

Table 3 presents the long-term record of stone-tool use for cracking oil-palm nuts at Bossou. Data for a total of 30 chimpanzees who are or at one time have been members of the community have been collected once a year for 13 consecutive years. As we have found perfect laterality in the hand used for hammering during nut-cracking, the identity of this hammering hand is designated in the table. Based on this record, Fig. 4 illustrates the developmental process underlying the acquisition of the skill of using a pair of stones as hammer and anvil. The data clearly illustrate that the period between 3 and 5 years of age is critical for learning the skill (Matsuzawa 1994).

To summarize the long-term record, we classified individuals into three major categories: "adult" (more than 9 years old), "young" or "juvenile" (5-8 years), and "infant" (0-4 years). The categories are based on our observations of chimpanzee life histories at Bossou: we have noted two examples of first parturition occurring at the age of 9, while interbirth intervals are just under 5 years on average.

Behaviors displayed by individuals toward nuts were also divided into three major categories: "Crack," "Explore," and "Neglect." "Crack" describes the use of a pair of stones as anvil and hammer; whether the nuts are eventually cracked is inessential, but hitting action using a stone in one hand is necessary. "Explore" incorporates investigative behaviors such as sniffing, mouthing, touching, rolling, throwing, and so forth. "Neglect" describes a lack of behaviors being directed toward target nuts even though these are located within the visual field. Here, this category also covers the behavior of simply viewing, but not touching, nuts.

Table 4 summarizes behaviors exhibited toward oil-palm nuts for individuals in each age class. The major observations in this baseline performance were threefold. (1) No chimpanzee, including all infants, neglected the nuts. (2) Infants were unable to crack nuts. They "explored", manipulating nuts and stones, and often obtained kernel fragments from their mothers. (3) No essential difference between adults and juveniles was evident in terms of behavioral categories applied to nut-cracking; however, there were contrasts in refined techniques and time-energy costs. There have been four noninfant chimpanzees who did not perform oil-palm nut-cracking. All were female: two adults (Nina and Pama) and two adolescents (Yunro and Julu). The year-by-year longitudinal record of these two adolescents revealed that they failed to learn the skill in the critical period (Matsuzawa 1999).

Table 3. Summary of stone-tool use by chimpanzees in the wild at Bossou, Guinea, during the period from 1988 to 2000

Name	Sex	Birth	Disap.	Age	Mother	88	90	91	92	93	94	95	96	97	98	99	00
																Year observed	
Tua	m	N.A.	P	adult	N.A.	?	L	L	L	L	L	L	L	L	L	L	L
Kai	f	N.A.	P	adult	N.A.	?	R	R	R	R	R	R	R	R	R	R	R
Nina	f	N.A.	P	adult	N.A.	?	X	X	X	X	X	X	X	X	X	X	X
Fana	f	N.A.	P	adult	N.A.	?	L	L	L	L	L	L	R	R	R	R	R
Jire	f	N.A.	P	adult	N.A.	?	L	L	L	L	L	L	L	L	L	L	L
Velu	f	N.A.	P	adult	N.A.	?	R	R	R	R	R	R	R	R	R	R	R
Yo	f	N.A.	P	adult	N.A.	?	L	L	L	L	L	L	L	L	L	L	L
Pama	f	N.A.	P	adult	N.A.	?	X	X	X	X	X	X	X	X	X	X	X
Kie*	f	75	Mar-91	24	Kai	?	R	R	-	-	-	-	-	-	-	-	-
Foaf	m	80L	P	19	Fana	?	R	R	R	R	R	R	R	R	R	R	R
Puru*	m	80L	Nov-91	19	Pama	R	R	R	-	-	-	-	-	-	-	-	-
Vube*	f	82	Mar-90	17	Velu	?	L	-	-	-	-	-	-	-	-	-	-
Ja*	f	83L	Feb-93	16	Jire	?	R	R	R	R	-	-	-	-	-	-	-
Yunro*	f	84L	Feb-93	15	Yo	?	X	X	X	X	-	-	-	-	-	-	-
Na*	m	85	Mar-96	14	Nina	?	R	R	R	R	R	R	-	-	-	-	-
Kakuru*	f	86L	Mar-91	13	Kie	?	A	R	-	-	-	-	-	-	-	-	-
Vui*	m	86L	Jul-99	13	Velu	?	X	X	L	L	L	L	L	L	L	L	-

Name	Sex	Birth	Status	Age	Mother													
Pili	f	87E	P	12.5	Pama	–	X	R	R	R	R	R	R	R	R	R	R	R
Jokro*	f	89E	P	11	Jire	–	X	X	–	–	–	–	–	–	–	–	–	–
Yela*	f	89M	P	10.5	Yo	–	X	–	–	–	–	–	–	–	–	–	–	–
Fotayu	f	91M	P	8.5	Fana	–	–	X	X	X	X	X	AR	R	R	R	R	R
Vuavua	f	91M	P	8.5	Velu	–	–	X	X	X	X	X	AL	L	L	L	L	L
Yoro	m	91L	P	8	Yo	–	–	X	X	X	–	X	X	L	L	L	L	L
Poni	m	Feb-93	P	7	Pama	–	–	X	–	–	–	X	X	R	R	R	R	R
Nto	f	93E	P	6.5	Nina	–	–	–	–	–	–	X	X	–	R	R	R	R
Julu	f	Nov-93	P	6	Jire	–	–	X	X	X	X	X	X	X	X	X	X	X
Pokru	m	Aug-96	P	3.5	Pili	–	–	–	–	X	X	X	–	X	X	–	X	X
Fanle	f	Oct-97	P	2.5	Fana	–	–	–	–	–	–	–	–	–	–	X	X	X
Jeje	m	Dec-97	P	2	Jire	–	–	–	–	–	–	–	–	–	–	X	X	X
Peley	f	Apr-98	P	1.5	Pama	–	–	–	–	–	–	–	–	–	–	X	–	X

The hand used to hold the hammer stone during nut-cracking has been recorded since the project began in 1988.

In general, the data were collected from January to March each year, i.e., in the dry season.

*, The individual had disappeared or died by February 2000; L, always used left hand for hammer; R, always used right hand for hammer; A, ambidextrous use of hammer; X, no successful hammer use but eating nuts cracked by others; ?, data unavailable because no observation at the cracking site; –, data unavailable because the subject had not yet been born, had disappeared, or died during the research period; P, present; N.A., not available; E, early in the year; M, midway through the year; L, late in the year. Age represents the estimate as of January 2000.

In the case of Fana, her left arm was paralysed probably by an accident in 1995. Then, Fana switched the hand for hammering from left to right.

Source: Data collection was done by Matsuzawa et al. in 1988, 1991 to 1997, 1999 and 2000; Fushimi et al. in 1990; Myowa-Yamakoshi in 1998.

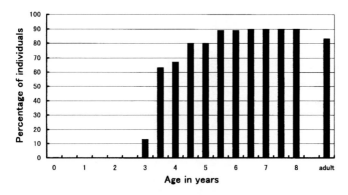

Fig. 4. Development of the skill of using a pair of stones as a hammer and anvil. Percentage of individuals who succeeded to use the tools is plotted as a function of age

Table 4. Behavior of Bossou chimpanzees toward oil-palm nuts (percentage of individuals in the three age classes)

Age class	Age (year)	n	Crack	Explore	Neglect
Adult	9 and above	10	80	20	0
Juvenile	5 to 8	6	84	16	0
Infant	0 to 4	4	0	100	0
Total		20			

Data collected in 2000 by DB and TM.
No fundamental differences in age-class distribution of oil-palm nut-cracking were noted over the years 1988 to 2000.

3.4 Presentation of Coula Nuts: Immigrant Female from Nimba?

In the next phase of the experiment, we provided Bossou chimpanzees with unfamiliar Coula nuts. Coula nuts are encased within a thick fruit. The outer appearance does not suggest the presence of a hard-shelled nut hidden inside. How did chimpanzees react to this "puzzle"? Table 5 summarizes the behavior of each individual toward Coula nuts on their initial encounters with the unfamiliar nuts in 1993.Then, the data were compiled by assigning individuals to one of the three age classes (Table 6).

The findings may be summed up as follows. (1) In the majority of cases, chimpanzees encountering Coula nuts for the first time neglected the unfamiliar objects. (2) On a few occasions, an individual would examine the outer fruit layer and attempt to bite into it, but cracking efforts were not made even by skilled oil-palm nut-crackers. (3) There was, however, a single exception. An adult female named Yo immediately placed the Coula nut on her anvil stone, proceeded to hit it without any exploration or hesitation, and finally ate the kernel without first sniffing the

Table 5. Behavior of individual chimpanzees toward newly introduced Coula nuts on the first four encounters

| Sex/Age class | Name | Age | Behavior shown to Coula nuts | | | |
			1st	2nd	3rd	4th
Adult male	Tua	N.A.	Explore	Neglect	Neglect	Neglect
	Foaf	12	Neglect	Neglect	Neglect	Neglect
Adult female	Kai	N.A.	Explore	Explore	Neglect	Neglect
	Fana	N.A.	Neglect	Explore	Neglect	Neglect
	Nina	N.A.	Neglect	Neglect	Neglect	Neglect
	Jire	N.A.	Neglect	Explore	Neglect	Neglect
	Yo	N.A.	Crack	Crack	Crack	Crack
	Velu	N.A.	Neglect	Neglect	Neglect	Neglect
	Pama	N.A.	Neglect	Neglect	Neglect	Neglect
Juvenile male	Na	7	Neglect	Explore	Explore	Neglect
	Vui	6.5	Neglect	Neglect	Neglect	Crack
Juvenile female	Yunro	8	Neglect	Neglect	Explore	Explore
	Pili	6	Neglect	Neglect	Neglect	Crack
Infant male	Yoro	1	Neglect	Neglect	Neglect	Neglect
	Poni	0	Neglect	Neglect	Neglect	Neglect
Infant female	Vuavua	1.5	Neglect	Neglect	Neglect	Neglect
	Fotayu	1.5	Neglect	Neglect	Neglect	Neglect

Data collected in 1993 by TM, NI, RT, and GY.

Table 6. Behavior of Bossou chimpanzees toward Coula nuts (percentage of individuals in the three age classes)

Age class	Age (year)	n	Crack	Explore	Neglect
Adult	9 and above	9	11	44	44
Juvenile	5 to 8	4	50	50	0
Infant	0 to 4	4	0	0	100
Total		17			

Data collected in 1993 by TM, NI, RT, and GY.

contents of the shell. A group of juveniles gathered around and peered at Yo's bout of activity; however, none of them tried to even touch the unfamiliar nuts in this first encounter. (4) In general, adults were more conservative and more neophobic than youngsters. By the fourth session of Coula presentation, two juveniles had begun to crack: they each cracked a nut open, sniffed its kernel, chewed it, then spat it out. Although we provided Coula nuts continuously for another 2 weeks, these two juveniles were the only members who, apart from Yo, attempted to crack the nuts. We repeated the Coula nut experiment in two subsequent years, 1996 and 2000. Our results show that members of the Bossou community have gradually come to accept Coula nuts as targets for nut-cracking (Biro and Matsuzawa, in preparation).

Table 7. Behavior of Bossou chimpanzees toward wooden balls (percentage of individuals in the three age classes)

Age class	Age (year)	n	Crack	Explore	Neglect
Adult	9 and above	9	0	33	66
Juvenile	5 to 8	3	100	0	0
Infant	0 to 4	6	0	33	66
Total		18			

Data collected in 1993 by RT.

3.5 Presentation of Wooden Balls: Curiosity Shown by Juveniles

We provided Bossou chimpanzees with wooden balls that resembled in their appearance the unfamiliar Coula nuts. The two were identical in size and shape, although they differed in color, texture, and smell. Hence, the presentation of wooden balls could serve as a control condition for the Coula provisions.

Table 7 lists the behavior of each chimpanzee toward the Coula-like wooden balls. Our main findings were as follows. (1) All juvenile chimpanzees attempted to crack open the wooden balls. (2) No adult and no infant ever tried to crack the wooden balls. This observation included Yo, which suggests that she clearly recognized the difference between Coula nuts and wooden balls from the start. In addition, we saw a distinct tendency for adults to be more conservative or neophobic toward unfamiliar items than youngsters.

3.6 Presentation of Panda Nuts: No Prior Knowledge

Next, we provided Bossou chimpanzees with Panda nuts. These are larger than oil-palm and Coula nuts, have the thickest shell among the three species, and are also encased within a thick fruit. This experiment was a variation on the presentation of Coula nuts. We aimed to confirm the results of the latter by introducing another species of nut that is not available at Bossou but is found at the neighboring sites.

Our results show that every member of the Bossou community neglected or showed only exploratory behavior toward Panda nuts on their first encounter. None of the chimpanzees reacted with the signs of familiarity displayed by Yo toward Coula nuts. Following initial exploratory behavior, two adults (Jire and Yo, and three juveniles (Yoro, Vuavua, and Fotayu) proceeded to attempt the cracking of Panda nuts. These nuts, however, are remarkably hard to crack. On several occasions we observed hammer stones being broken as a result of lengthy pounding. Three of the five chimpanzees (Yo, Yoro, and Vuavua) succeeded to eventually crack open the nut, tasted it, and then never attempted to crack one again. Subsequently, they gave up all cracking efforts and completely neglected Panda nuts. The remaining two of the five (Jire and Fotayu) failed to crack open Panda nuts

Table 8. Behavior of Bossou chimpanzees toward Panda nuts (percentage of individuals in the three age classes)

Age class	Age (year)	n	Crack	Explore	Neglect
Adult	9 and above	10	20	0	80
Juvenile	5 to 8	6	50	16	33
Infant	0 to 4	4	0	25	75
Total		20			

Data collected in 2000 by DB and TM.

despite their efforts. Gradually, all chimpanzees lost motivation for the cracking of Panda nuts.

Table 8 summarizes the behavior displayed by each chimpanzee toward Panda nuts. The main findings were as follows. (1) No chimpanzee had prior knowledge of Panda nuts. None showed behaviors toward Panda nuts that resembled Yo's reaction to Coula (immediately setting out to crack; no initial exploration). (2) Just as in our previous experiments with Coula nuts and wooden balls, adults appeared more conservative and more neophobic toward the unfamiliar items than did youngsters.

3.7 Transmission of Culture Between Communities and Across Generations

We introduced unfamiliar Coula and Panda nuts to the chimpanzees at Bossou, a site where oil-palm nuts are the only naturally occurring targets of nut-cracking. The results clearly showed that none of the chimpanzees, except for a single adult female named Yo, had any knowledge of the novel nuts, and that all showed neophobic responses toward the latter. However, Yo appeared to be familiar with Coula nuts and proceeded to crack them immediately without the slightest sign of hesitation. Her behavior was carefully observed by other group members, particularly those of the younger generation (Fig. 5). In turn these youngsters, who were on the whole less conservative in their reactions to the new items, began to crack open the Coula nuts by themselves. This behavior tended not to be the case for adults, who remained conservative and neophobic to the unfamiliar items.

Our interpretation of these results is as follows. In chimpanzee societies, while males remain in their natal group, adolescent females emigrate from the natal community to join one of the neighboring groups. It is highly possible that Yo was such an immigrant female. Born in another community such as nearby Yealé, 12 km from Bossou, where the group's cultural repertoire includes the cracking of Coula nuts (but not of Panda nuts even though both are available), she grew up learning to crack Coula nuts before migrating to Bossou. Our experimental manipulation reintroduced Yo to Coula nuts; as a result, she functioned as an innovator by introducing a new kind of nut-cracking to the Bossou community. A previously unseen behavior was thus transmitted from an immigrant female to mem-

Fig. 5. An adult female chimpanzee named Yo is cracking Coula nuts while two juveniles carefully observe her behavior. A pile of Coula nuts is visible to the right. The three Panda nuts in front are neglected by all those present

bers of the community that received her, demonstrating moreover the passing of knowledge from one generation to the next. Although all the Bossou chimpanzees had access to Coula nuts, only younger individuals learned to crack them in the initial stages of presentation, and they did so only after observing the informant, an adult female.

In contrast, Bossou is located 50 km from Diéké where the nearest community known to crack Panda nuts resides. The fact that no chimpanzees at Bossou were familiar with Panda nuts suggests that the distance separating Bossou and Diéké is too great for any direct interaction between the communities. In fact, the two locations have been separated by farmland, roads, and villages for decades.

In sum, we propose that neighboring chimpanzee groups may develop similar cultural traditions resulting in a kind of "cultural zone" composed of several communities. Bossou and its vicinity may be a cultural zone characterized, among other things, by the use of stone tools for the cracking of various kinds of hard-shelled nuts. However, which species of nuts are cracked in a particular community is determined by the ecological environment as well as the established cultural traditions.

4 Education

4.1 Observing Behavior: Juveniles Learn from Adults, Not the Reverse

Examining issues related to the emergence of culture leads to the following questions: When and how do young individuals acquire the tool-using techniques unique to the community? What learning mechanisms are involved in the acquisition? What kinds of imitative or educational processes are involved in the acquisition? In this section, we consider these questions in some detail.

According to our analysis of observing behavior (Biro and Matsuzawa, in preparation), adults were not only the most likely to be the targets of the observation by other members of the group but also the least likely to be observers themselves. On the other hand, juveniles were rarely targets but often observers, both of adults and of juveniles. In sum, chimpanzees showed a strong tendency to pay attention to the stone-tool use of conspecifics in their own age group or older, but not younger. This observation means that cultural innovations are more likely to spread horizontally or vertically/orthogonally downward, but not upward as is sometimes observed in the case of humans.

4.2 Spontaneity and High Levels of Tolerance: Not Active Teaching

Chimpanzee infants move around the forest together with their mothers at all times up the age of 4 to 5 years, until the birth of a younger sibling. During this period, therefore, mothers are the most important models for observation by infants. What other roles might mothers play? Boesch (1991) described examples of active teaching in wild chimpanzees, however, it must be noted that he reported only two possible episodes over the course of his long-term field observation. We never encountered episodes of active teaching in our 13 years of research; this means that active teaching in wild chimpanzees is either nonexistent or occurs only in very few and exceptional cases. We never observed chimpanzee mothers perform molding (grasping the hands of infants for guidance), or giving appropriate stone tools or good-quality nuts to be cracked to their infants. Certainly, verbal instructions or written manuals are unique to humans.

In the case of nut-cracking at least, chimpanzee education is to be characterized by spontaneity in the youngster (learner) and high levels of tolerance from the mother (teacher). We have often seen infants stealing kernels from their mothers just after she finished cracking; such episodes are in fact a 1-year-old infant's first step toward nut-cracking. Although mothers never actively give kernels to their offspring, they may often stop cracking and hold a kernel in their hand as the young attempt to take it. We have seen an infant continue to try to pick up a kernel seven times after the mother completed the cracking. Infants less than 3 years old have, in a sense, free access to kernels, nuts, and stone tools that are in use by their mothers or even by other members of the community. The mother's behavior toward her offspring is thus typified by high levels of tolerance.

4.3 Imitation

It must be noted at this point that simple accounts of conditioning with food reward cannot explain the acquisition process of stone-tool use. Infants may obtain kernels from their mothers at any time they chose, yet at the age of 1 year they begin to manipulate nuts and stones on their own. At the age of 2 they start to combine two items together: placing a stone on another stone, hitting a stone with a nut, and so forth. It takes at least 3 or even 5 years for young chimpanzees to combine the three objects (a nut, a hammer stone, and an anvil stone) in the appropriate spatial and temporal arrangement for successful nut-cracking. During this extended learning process, their manipulatory behavior is never once reinforced by an edible item. This fact clearly indicates that the infants' learning process is not supported by feeding motivation but by the motivation to copy the mothers' behavior. They spontaneously try to imitate the mothers' actions despite the fact that initially they are bound to fail in the majority of cases.

Chimpanzees in captivity seldom succeed to imitate arbitrary actions performed by a human model in a face-to-face situation. In such tasks, a human model demonstrates an action either with or without objects to an observer chimpanzee and then gives a vocal/gestural instruction meaning "Do this!" Results from such tests indicate that subjects succeed in immediately imitating the demonstrated actions only in approximately 7% of trials (Myowa-Yamakoshi and Matsuzawa 1999; Chapter 17 by Myowa-Yamakoshi, this volume). Monkeys never ape, but apes ape in a few cases, even though their imitative abilities are a long way from those of humans.

4.4 Master-Apprentice Relationship with Affection

The weaning of infant chimpanzees takes place at approximately 3 to 3.5 years of age. Chimpanzee mothers continue to breast-feed much longer than human mothers in modern societies, and this is likely to help establish a strong bond of affection between mothers and infants. Infant and juvenile chimpanzees learn a great deal from their mothers and other adult members of the community, especially in their first 4 to 5 years of life. They attempt to copy their mother's behavior with a spontaneous motivation for imitation, supported by the high levels of tolerance displayed by the mothers. In some sense, this is a form of education in wild chimpanzees.

The word "educate" originates from "educe" (Latin "educere") meaning "to extract." Education is thus the drawing forth of one's potential abilities. Many societies regard "active teaching" as the most advanced form of education. However, active teaching may be neither the best nor the only way of educating; in some cases, it may instead have an effect of "active interference." Active teaching often focuses on what should be learned and easily neglects the motivation of the learner. The so-called "master-apprentice" relationship is common in many traditional human societies, where knowledge and skills can be transmitted from one person to another without verbal explanation, molding, or written instructions. The master does not provide any form of active teaching, but the apprentice is expected to possess a strong motivation to imitate.

Through our long-term observation of nut-cracking behavior in the wild, we have encountered many interesting episodes. In one, a 4-year-old infant interrupted the mother's hammering, stole the nut to be cracked from the anvil, and then hit it on her own anvil. In another, a 3-year-old infant picked up a nut from the ground after having watched the mother's nut-cracking for a long time. The infant moved toward the mother and placed the nut on her anvil stone; the mother briefly stopped in midmotion and then proceeded to crack the nut. Finally, the infant removed the kernel from the anvil and ate it.

Behaviors of this kind are often seen and have their counterparts in human social interactions. The apprentice observes the master's behavior for extended periods of time. Such long-term repetitive exposure to a problem situation helps the apprentice learn the knowledge and skills necessary for the solving of particular problems. This interaction is reminiscent of the master-apprentice relationship in the art of Japanese sushi making. The apprentice is forbidden to even touch the utensils, rice, fish, or other ingredients for the first few years of his training. He is allowed only to carefully observe the Sushi master and wash the dishes, that is, until one day the master suddenly gives him permission to attempt making his first sushi. It is no exaggeration to say that the apprentice produces exquisite sushi from the start.

The Japanese kanji-character " 人 " is derived from the shape of two persons leaning toward each other, thus lending one another support. Altruism and mutual support may be fundamental elements of human nature. However, when observing the education process that leads to the acquisition of nut-cracking by chimpanzee infants, we are reminded of the master-apprentice relationship in human societies. The apprentice of course needs the master, but the masters need apprentices, too, if they are to transmit their knowledge and skills to the next generation. In a sense, the master-apprentice bond is an adaptation for cultural and spiritual reproduction from generation to generation in contrast to the biological. Humans and chimpanzees may thus have much in common, even in the field of education.

References

Boesch C (1991) Teaching among wild chimpanzees. Anim Behav 41:530–532

Boesch C, Boesch-Achermann H (2000) The chimpanzees of the Tai forest. Oxford University Press. Oxford

Goodall J (1986) The chimpanzees of Gombe: patterns of behavior. Harvard University Press, New York

Humle T, Matsuzawa T (2001) Cultural variants between the wild chimpanzee populations of Bossou and neighboring area, Guinea and Côte d'Ivoire, West Africa: a preliminary report. Folia Primatol 72:57–68

Inoue-Nakamura N, Matsuzawa T (1997) Development of stone tool use by wild chimpanzees (*Pan troglodytes*) J Comp Psychol 111:159–173

Kawai M (1965) Newly acquired pre-cultural behavior of the natural troop of Japanese monkeys on Koshima islet. Primates 6:1–30

Matsuzawa T (1994) Field experiment on use of stone tools by chimpanzees in the wild. In: Wrangham R, McGrew W, de Waal F, Heltne P (eds) Chimpanzee cultures. Harvard University Press, Cambridge, pp 351–370

Matsuzawa T (1999) Communication and tool use in chimpanzees: cultural and social contexts. In: Hauser M, Konishi M (eds) The design of animal communication. Cambridge University Press, New York, pp 645–671

Matsuzawa T, Yamakoshi G (1996) Comparison of chimpanzee material culture between Bossou and Nimba, West Africa. In: Russon A, Bard K, Parker S (eds) Reaching into thought. Cambridge University Press, New York, pp 211–232

Matsuzawa T, Takemoto H, Hayakawa S, Shimada M (1999) Diecke forest in Gunea. Pan Africa News 6:10–11

Myowa-Yamakoshi M, Matsuzawa T (1999) Factors influencing imitation of manipulatory actions in chimpanzees (Pan troglodytes). J Comp Psychol 113:128–136

McGrew W (1992) Chimpanzee material culture. Cambridge University Press, New York

Nishida T (1990) The chimpanzees of the Mahale Mountains. University of Tokyo Press, Tokyo

Sakura O, Matsuzawa T (1991) Flexibility of wild chimpanzee nut-cracking behavior using stone hammers and anvils: an experimental analysis. Ethology 87:237–248

Shimada M (2000) A survey of the Nimba Mountains, West Africa from three routes: confirmed new habitat and ant-catching wand use of chimpanzees. Pan Africa News 7:7–10

Sugiyama Y, Koman J (1979) Tool-using and -making behavior in wild chimpanzees at Bossou, Guinea. Primates 20:513–524

Sugiyama Y, Fushimi T, Sakura O, Matsuzawa T (1993) Hand preference and tool use in wild chimpanzees. Primates 34:151–159

Whiten A, Goodall J, McGrew W, Nishida T, Reynolds V, Sugiyama Y, Tutin C, Wrangham R, Boesch C (1999) Cultures in chimpanzees. Nature (Lond) 399:682–685

Yamakoshi G (1998) Dietary responses to fruit scarcity of wild chimpanzees at Bossou, Guinea: possible implications for ecological importance of tool use. Am J Phys Anthropol 106:283–295

Subject Index